Bacillus thuringiensis Biotechnology

Estibaliz Sansinenea
Editor

Bacillus thuringiensis Biotechnology

Editor
Dr. Estibaliz Sansinenea
Benemérita Universidad Autónoma de Puebla
Facultad de Ciencias Químicas
Blv. 14 sur y Av. San Claudio
Puebla
Mexico

ISBN 978-94-007-3020-5 e-ISBN 978-94-007-3021-2
DOI 10.1007/978-94-007-3021-2
Springer Dordrecht Heidelberg London New York

Library of Congress Control Number: 2012930388

Printed on acid-free paper

Springer is part of Springer Science+Business Media (www.springer.com)

Preface

Since its discovery about a century ago, *Bacillus thuringiensis* (Bt) has been used as a biopesticide in agriculture, forestry and mosquito control because of its advantages of specific toxicity against target insects, lack of polluting residues and safety to non-target organisms. Today Bt is the most successful commercial microbial insecticide, comprising about 90% of the biopesticide market. The insecticidal properties of this bacterium are due to the presence of insecticidal proteins, also called crystals, usually produced during sporulation. The new tools of biotechnology are changing the way scientists can address problems in the agriculture. Transgenic technology, involving a wide range of pesticidal genes from Bt, dominates the scenario of agricultural biotechnology. At the same time, Bt technology is also the most vehemently criticized area of agricultural biotechnology. Genetic improvement of Bt strains for the development of novel biopesticides entails increasing their potency against target insects, broadening the insecticidal spectra for specific crop applications, improving persistence on plants, and optimizing fermentation production.

On the other hand, Bt biotechnology comprises other aspects different from insecticidal proteins that has become an interesting topic of research such as secondary metabolites or enzymes with many applications in biotechnology which are deeply discussed in the last part of this book.

In this book, I have pretended to gather a great team of experts to bring together all recent studies regarding both fundamental and more applied research aspects related to biotechnology of Bt. The nineteen different chapters, written by the leading researchers in the field, give us a tour of the whole story of Bt biotechnology since its discovery and comprehensively update the entire subject.

- Part 1: *Bacillus thuringiensis*: an environmentally safe alternative
  The first five chapters provide a general overview of *bacillus thuringiensis* as biopesticide describing what is Bt, how the mode of action of this bacterium is, how its use today in the market is, and giving an idea about how protein engineering is used to improve the efficacy of the toxins.
  In the first chapter Sansinenea (pp. 3–18) presents a detailed history of the discovery of this bacterium and gives a brief description about Bt and its toxins.

In the second chapter George and Crickmore (pp. 19–39) continue nicely the description of the toxins of Bt and its mode of action and then explain the types of Bt products and their advantages over chemicals, disadvantages (narrow spectrum, low activity), the resistance as a threat to continue use of Bt, the use of synergists to improve activity, the use of genetic manipulation to improve activity and finally give a brief description of gene stacking.
In the third chapter Kaur (pp. 41–85) extensively and in depth describes the risk assessment of Bt transgenic crops from various aspects: the risk in non-target organisms, persistence of Bt in soil, risk analysis of Bt transgenic crops including studies in different countries to finish with conclusions and future prospect.
In the fourth chapter Abdullah (pp. 87–92) describes what are biopesticides and their market detailing the use and efficacy of *bacillus thuringiensis* as a biopesticide in integrated pest management.
To finish this first part of the book, in the fifth chapter Florez et al. (pp. 93–113) revise Cry proteins domains and analyze in an elegant way several protein modifications that have been successfully used for creating stable, functional proteins with minimal structural alterations. The understanding and proper use of protein engineering approaches may help in implementing appropriate pest management strategies by improving the efficacy of these toxins against insect pests.

- Part 2: Genetics of *Bacillus thuringiensis*
Due to the great importance that the genetic of Bt has had in biotechnological advances, this part is the longest and most extensive in the book and is composed of seven chapters. This part will provide a general view of the genetics of Bt from the beginning until today explaining all the genetic tools available for the improvement of the Bt strains.
To start this part, in the sixth chapter Økstad and Kolstø (pp. 117–129) describe with their great experience, the evolution of the *bacillus cereus* group, describing the main factors that differentiate *B. cereus*, *B. anthracis* and *B. thuringiensis.* This description provides a wide view of the characteristics of the *B. cereus* group including the sequencing of the whole genomes of Bt for the engineering of commercial *B. thuringiensis* strains with increased safety to humans.
In the seventh chapter, Sorokin (pp. 131–157) introduces in an nice way the genetics of the *B. cereus* group reviewing phage-mediated gene transduction and recombineering perspectives for the *B. cereus* group. In combination with new generation sequencing these approaches will constitute the gene identification methodologies in the post-genomics time.
In the eighth chapter, Vilas-Bôas and Santos (pp. 159–174) interestingly describe the studies focusing on conjugative transfer in *Bacillus thuringiensis*, involving the detection of *cry* genes in large conjugative plasmids, the genetic basis of the process, the main plasmids, and methodological variations of mating systems.
Continuing with a logical sequence of the book in the ninth chapter, Ochoa-Zarzosa and Lopez-Meza (pp. 175–184) describe the shuttle vectors of *B. thuringiensis* that have been constructed using essentially replicons from resident plasmids from this bacterium that replicate by the theta mechanism and also, the vectors that have been developed using plasmid replicons from other Gram-pos-

itive bacteria or RCR plasmids. The development of shuttle vectors with better characteristics and protocols with high transformation efficiency have greatly facilitated basic research and engineering of *B. thuringiensis*.

Following with the topic of the vectors, in the tenth chapter Xu et al. (pp. 185–199) take up the subject in great depth and provide a deep overview of current research and applications of *B. cereus* group plasmid vectors and prospects for further development, describing shuttle vectors, integration vectors both homologous recombination and transposons vectors, resolution vectors and expression vectors. In this way through two chapters the topic of vectors is explained at length to get started with recombination chapter.

In the eleventh chapter Abdelkefi-Mesrati and Tounsi (pp. 201–214) present the genetic recombination in *Bacillus thuringiensis*. In the first part they describe the site specific recombination, including transposition by transposons and transduction by phage, in this bacterium and its exploitation in the construction of recombinant strains of *B. thuringiensis* improving their production as bioinsecticides and their insecticidal activities and *B. thuringiensis* mutagenesis. In the second part they describe the homologous recombination and its role in the construction of improved *B. thuringiensis* strains and in gene disruption.

To finish this second part of the book, in the twelfth chapter Sanchis (pp. 215–228) describes how recombinant DNA technology has been used to improve *Bacillus thuringiensis* (Bt) products and overcome a number of the problems associated with Bt-based insect control measures giving several examples describing how biotechnology has been used to increase the production of insecticidal proteins in Bt.

- Part 3: Bt as biopesticide: Applications in biotechnology

  Once explained the second part of the book, begins the third part, which is composed of four chapters, and provides a detailed view of the applications of Bt in biotechnology until today including many details about Bt crops, such as food safety.

  Continuing with the previous chapter and to start this part of the book, in the thirteenth chapter Li and Yu (pp. 231–258), firstly, in an elegant way, realize a very interesting and detailed summary of some chapters of the previous part that are important to this topic such as vectors, recipient strains and methods for constructing genetically modified Bt strains, to enter with great vision on the topic of genetically modified Bt strains viewed from different points such as broader insecticidal spectrum, high insecticidal activity, multifunctional activity and delayed pests resistance.

  In the chapter fourteen Schnepf (pp. 259–281) retakes the recipient strains topic and describes in depth the insecticidal protein expression in gram-negative hosts, in Bt and in plants leading to biotechnological applications. Notable successes will be mentioned, however, more time will be spent on modifications that incrementally improve high production levels, unresolved issues with low-expressing proteins, or maintaining functionality of the expression host.

  In the chapter fifteen Castagnola and Jurat-Fuentes (pp. 283–304) discuss key events in the history of Bt crop development and summarize current regulations

aimed at reducing the risks associated with increased adoption of this technology. By analyzing the history of Bt transgenic crops and the current marketplace trends and issues, they examine the outlook of current and impending Bt crops as well as potential issues that may emerge during their future use.
To finish this part an interesting sixteenth chapter is presented by Hammond and Koch (pp. 305–325) who relate a history of safe consumption of Cry proteins from use of Bt microbial pesticides on vegetable food crops, and summarize the published literature addressing the safety of Cry insect control proteins found in both Bt microbial pesticides and those introduced into Bt agricultural crops. A discussion on the species-specific mode of action of Cry proteins to control target insect pests is presented. A human dietary exposure assessment for Cry proteins has also been provided. Lastly the food and feed safety benefits of Bt crops are briefly summarized including lower insecticide use and reduction in fumonisin mycotoxin contamination of grain.

- Part 4: Other *Bacillus* species in biotechnology
Finally this last part is composed of three chapters and provides a view about the all *bacillus* species that have biotechnological importance, and gives a new vision of the modern biotechnology that leads to discover new secondary metabolites of great importance in biotechnology.
To start this part of the book, in the seventeenth chapter Raddadi et al. (pp. 329–345) give an interesting survey on the most important biotechnological processes that have as effectors bacterial species in the genus *Bacillus*. Before highlighting the main biotechnological applications in which these species have been implicated, they briefly introduce the taxonomy, the ecology, the evolution and the natural variation that characterizes *Bacillus* genus.
In the eighteenth chapter Chaabouni et al. (pp. 347–366) summarize the most important secondary metabolites produced by members of genus *Bacillus*. Bacteriocins of *B. subtilis* and *B. thuringiensis* and other metabolites (as zwittermicin A, siderophore, surfactin and others) are described and the potential application of antimicrobial peptides in food, agriculture and pharmaceutical industries are discussed. This biotechnological potential will be highlighted and the safety evaluation of the metabolites and the species of the producer will be discussed.
To finish this last part of the book, in the nineteenth chapter Barboza-Corona et al. (pp. 367–384) review in depth the different kinds of chitinases that are synthesized by *B. thuringiensis,* their roles in nature, and their application in environment, agriculture and food industry. Additionally they, analyze bacteriocins of *B. thuringiensis* reported to date, how to enhance their production, and the methods for screening the bacteriocin activity. Finally, the future challenges and prospects of the antimicrobial peptides as biopreservatives, antibiotics, and nodulation factors are presented.

The biotechnological applications of Bt as biopesticide have increased in the last ten years and a new way of expanding the biotechnology of Bt has started. The challenge is great and there is still a need for research in different areas. I hope that this book will help to continue expanding the knowledge of Bt applied to biotechnology.

Finally the book is ready. From these lines, I would like to thank firstly Mr Max Haring from Springer SBM BV Publishing Editor who invited me to this interesting and exciting project and Mrs Marlies Vlot senior assistant for her very useful assistance in this project.

With a special mention, the editor gratefully acknowledges all the authors involved in this project, for their enthusiasm and their courage, for contributing chapters on the subjects of their expertise and giving so much of their time to make this book possible. I am deeply grateful for their generous and collegial spirit and their willingness to match their contributions to the view of the book and for all aspects of this adventure that, at the end, have crystallized in this book. For me, as Editor of this book, it has been really a privilege to interact with such a collection of scientists and present this useful work. I am very happy to have had the opportunity to contact and gather a great team of experts to bring together all recent studies related to biotechnology of Bt that I am completely sure will made this book an exit.

Contents

Contributors

Dr. Estibaliz Sansinenea Facultad de Ciencias Químicas, Benemérita Universidad Autónoma de Puebla, San Claudio, Puebla, México

Dr. Zenas George Department of Biochemistry, School of Life Sciences, University of Sussex, Falmer, Brighton, UK

Dr. Neil Crickmore Department of Biochemistry, School of Life Sciences, University of Sussex, Falmer, Brighton, UK

Dr. Sarvjeet Kaur National Research Centre on Plant Biotechnology, Indian Agricultural Research Institute Campus, New Delhi, India

Dr. Mohd Amir Fursan Abdullah InsectiGen Inc.ADS, Athens, GA, USA

Dr. Alvaro M. Florez Laboratorio de Biología Molecular y Biotecnología, Facultad de Medicina, Universidad de Santander, Bucaramanga, Colombia.

Dr. Cristina Osorio Systems Proteomics Center, Program in Molecular Biology and Biotechnology, School of Medicine,University of North Carolina, Chapel Hill, NC, USA

Dr. Oscar Alzate Associate Professor, Department of Cell and Developmental Biology, School of Medicine, University of North Carolina, Chapel Hill, NC, USA

Dr. Ole Andreas Økstad Laboratory for Microbial Dynamics (LaMDa), Department of Pharmaceutical Biosciences, University of Oslo, Blindern, Oslo, Norway

Dr. Anne-Brit Kolstø Laboratory for Microbial Dynamics (LaMDa), Department of Pharmaceutical Biosciences, University of Oslo, Blindern, Oslo, Norway

Dr. Alexei Sorokin Genome Analysis Team ANALGEN, UMR1319 MICALIS, Centre de Recherche de Jouy-en-Josas, Institut National de la Recherche Agronomique (INRA), France

Dr. Gislayne Fernandes Lemes Trindade Vilas-Bôas Depto. Biologia Geral - CCB Universidade Estadual de Londrina, CEP, Londrina, PR, Brazil

Dr. Clelton Aparecido Santos Depto. Biologia Geral - CCB, Universidade Estadual de Londrina, CEP, Londrina, PR, Brazil

Dr. Alejandra Ochoa-Zarzosa Centro Multidisciplinario de Estudios, en Biotecnología-FMVZ, Universidad Michoacana de San Nicolás de Hidalgo, km 9.5 Carretera Morelia-Zinapécuaro, Posta Veterinaria, Morelia, Michoacán, México

Dr. Joel Edmundo López Meza Centro Multidisciplinario de Estudios en Biotecnología-FMVZ, Universidad Michoacana de San Nicolás de Hidalgo, km 9.5 Carretera Morelia-Zinapécuaro, Posta Veterinaria, Morelia, Michoacán, México

Dr. Chengchen Xu Laboratory of Agriculture Microbiology, National Engineering Research Center of Microbial Pesticides, College of Life Science and Technology, Huazhong Agricultural University, Wuhan, Hubei Province, China

Dr. Yan Wang Laboratory of Agriculture Microbiology, National Engineering Research Center of Microbial Pesticides, College of Life Science and Technology, Huazhong Agricultural University, Wuhan, Hubei Province, China

Dr. Chan Yu Laboratory of Agriculture Microbiology, National Engineering Research Center of Microbial Pesticides, College of Life Science and Technology, Huazhong Agricultural University, Wuhan, Hubei Province, China

Dr. Lin Li Laboratory of Agricultural Microbiology, Huazhong Agricultural University, Wuhan, Hubei, China

Dr. Minshun Li Laboratory of Agriculture Microbiology, National Engineering Research Center of Microbial Pesticides, College of Life Science and Technology, Huazhong Agricultural University, Wuhan, Hubei Province, China

Dr. Jin He Laboratory of Agriculture Microbiology, National Engineering Research Center of Microbial Pesticides, College of Life Science and Technology, Huazhong Agricultural University, Wuhan, Hubei Province, China

Dr. Ming Sun Laboratory of Agriculture Microbiology, National Engineering Research Center of Microbial Pesticides, College of Life Science and Technology, Huazhong Agricultural University, Wuhan, Hubei Province, China

Dr. Ziniu Yu Laboratory of Agricultural Microbiology, National Engineering Research Center for Microbial Pesticides, Huazhong Agricultural University, Wuhan, Hubei Province, China

Dr. Lobna Abdelkefi-Mesrati Laboratory of Biopesticides, Centre of Biotechnology of Sfax, Sfax, Tunisia

Dr. Slim Tounsi Biopesticides Team (LPIP), Centre of Biotechnology of Sfax, Sfax, Tunisia

Dr. Vincent Sanchis INRA, UMR1319-MICALIS, Equipe GME, La Minière, Guyancourt Cedex, France

Dr. Harry Ernest Schnepf San Diego, CA, USA

Dr. Anais S. Castagnola Department of Entomology and Plant Pathology, University of Tennessee, TN, USA

Dr. Juan Luis Jurat-Fuentes Department of Entomology and Plant Pathology, University of Tennessee, Knoxville, TN, USA

Dr. Bruce G. Hammond Product Safety Center, Monsanto Company, St. Louis, MO, USA

Dr. Michael S. Koch Product Safety Center, Monsanto Company, St. Louis, MO, USA

Dr. Noura Raddadi Dipartimento di Ingegneria Civile, Ambientale e dei Materiali (DICAM)- ,Unità di Ricerca di Biotecnologie Ambientali e Bioraffinerie, Università di Bologna, Bologna, Italy

Dr. Elena Crotti, Dipartimento di Scienze e Tecnologie Alimentari e Microbiologiche (DISTAM) University of Milan, Milano, Italy

Dr. Eleonora Rolli Dipartimento di Scienze e Tecnologie Alimentari e Microbiologiche (DISTAM) University of Milan, Milano, Italy

Dr. Ramona Marasco Dipartimento di Scienze e Tecnologie Alimentari e Microbiologiche (DISTAM) University of Milan, Milano, Italy

Dr. Fabio Fava Dipartimento di Ingegneria Civile, Ambientale e dei Materiali (DICAM)-, Unità di Ricerca di Biotecnologie Ambientali e Bioraffinerie, Università di Bologna, Bologna, Italy

Dr. Daniele Daffonchio Professor of Microbial Ecology and Biotechnology Dipartimento di Scienze e Tecnologie Alimentari e Microbiologiche (DISTAM), University of Milan, Milano, Italy

Dr. Ines Chaabouni Laboratoire Microorganismes et Biomolécules Actives, Faculté des Sciences de Tunis, Université de Tunis El Manar, Tunis, Tunisia

Dr. Amel Guesmi Laboratoire Microorganismes et Biomolécules Actives, Faculté des Sciences de Tunis, Université de Tunis El Manar, Tunis, Tunisia

Dr. Ameur Cherif Institut Supérieur de Biotechnologie, Université de la Manouba, BioTechPole Sidi Thabet, Sidi Thabet, Ariana, Tunisie

Dr. J. Eleazar Barboza-Corona División Ciencias de la Vida, Departamento de Alimentos, University of Guanajuato,Campus Irapuato-Salamanca, Irapuato, Guanajuato, México

Dr. N. M. de la Fuente-Salcido Escuela de Ciencias Biológicas, Universidad Autónoma de Coahuila, Torreón, Coahuila, México

Dr. M. F. León-Galván División Ciencias de la Vida, Departamento de Alimentos, University of Guanajuato, Campus Irapuato-Salamanca, Irapuato, Guanajuato, México

Part I
Bacillus thuringiensis: An Environmentally Safe Alternative

Chapter 1
Discovery and Description of *Bacillus thuringiensis*

Estibaliz Sansinenea

Abstract The use of biopesticides, as a component of integrated pest management (IPM), has been gaining acceptance over the world. An entomopathogenic organism should be highly specific and effective against the target pest and should demonstrate the potential to be successfully processed by continuous production technology. *Bacillus thuringiensis* (Bt) was discovered as a soil bacterium, which fulfills all these requirements and due to it has been used as a biopesticide in agriculture, forestry and mosquito control. Studies of the basic biology of Bt have shown that the insecticidal activity of Bt is due to the presence of parasporal protein inclusion bodies, also called crystals, produced during sporulation that determines its activity for insect species belonging to different orders, which act like a stomach poison causing larval death. Environmentally safe-insect control strategies based on Bt and their insecticidal crystal proteins are going to increase in the future, especially with the wide adoption of transgenic crops. In this chapter, I have summarized the discovery and the description of Bt.

Keywords *Bacillus thuringiensis* · Soil bacterium · Discovery of Bt · Biopesticide · Cry proteins

1.1 Introduction

Since World War II, insect disease control methods have relied heavily on broad-spectrum synthetic chemical insecticides to reduce vector populations. However, synthetic chemical insecticides are being phased out in many countries due to insecticide resistance in mosquito populations. Furthermore, many governments have restricted chemical insecticide use due to their environmental effects on non-target beneficial insects and, especially, on vertebrates through contamination of food and water supplies. To counteract this contamination, attention and efforts were directed to the use of biological control agents including insect pathogens. As a result, the

E. Sansinenea (✉)
Facultad de Ciencias Químicas, Benemérita Universidad Autónoma de Puebla,
Boulevard 14 sur y Av. San Claudio, 72570 Puebla, Pue, México
e-mail: estisan@yahoo.com

E. Sansinenea (ed.), *Bacillus thuringiensis Biotechnology*,
DOI 10.1007/978-94-007-3021-2_1, © Springer Science+Business Media B.V. 2012

use of biopesticides, as a component of integrated pest management (IPM), has been gaining acceptance over the world. However, an entomopathogenic organism must fulfill several requisites before being released to the environment as a potential control agent. It should be highly specific and effective against the target pest. The organism should demonstrate the potential to be successfully processed by continuous production technology. The control agent should be available in formulations with a reasonable shelf life, should be stable, and should be harmless to human and non-target flora and fauna.

As an entomopathogenic organism, *Bacillus thuringiensis* (Bt) fulfills all these requirements. Bt has been used as a biopesticide in agriculture, forestry and mosquito control. Its advantages are specific toxicity against target insects, lack of polluting residues and safety to non-target organisms such as mammals, birds, amphibians and reptiles. Although several proteins and other compounds produced by Bt contribute to its insecticidal activity, by far the most important components are the proteins that form parasporal crystalline inclusions during sporulation. Transgenic crops based on insecticidal crystal proteins of Bt are now an international industry with revenues of several billion dollars per year. In this chapter, I briefly review the discovery of this bacterium and follow this with section on its description including the basic biology of this bacterium and a brief description of insecticidal proteins and the mode of action of them.

1.2 A Brief History and Discovery of *Bacillus thuringiensis*

Bt was first discovered in Japan by Shigetane Ishiwata (1901) as the causal agent of sotto disease in silkworms (*Bombyx mori*) larvae (Ishiwata 1901). He named it Sottokin, which means "sudden death *bacillus*," and described the pathology it causes in silkworm larvae and its cultural characteristics (Ishiwata 1905a). He also noted that many of the larvae that did not die when exposed to the *bacillus* were very weak and stunted. In a subsequent report (Ishiwata 1905b) he stated that "From these experiments the intoxication seems to be caused by some toxine, not only because of the alimentation of *bacillus*, but the death occurs before the multiplication of the *bacillus*…" This showed that from the very beginning it was realized that a toxin was involved in the pathogenicity of Bt. His identification was not complete, however, and the first morphologically valid description was made by the German bacteriologist Ernst Berliner (1915), who isolated the *bacillus* from the Mediterranean flour moth (*Anagasta kuehniella*) (Berliner 1915). He named it *Bacillus thuringiensis* (Bt), which is derived from Thuringia, the German town where the moth was found. Subsequently, Aoki and Chigasaki (1916) published a detailed description of Ishiwata's organism, which they named *Bacillus sotto*, noting that its activity was due to a toxin present in sporulated cultures, but not in young cultures of vegetative cells (Aoki and Chigasaki 1916). The toxin was not an exotoxin because it was not found in culture filtrates. It is obvious from their data on inactivation of the toxin by acids, phenol, mercuric chloride, and heat that they had a protein. In 1915, Berliner reported the presence of parasporal inclusions within Bt, but the activity of this crystal was not discovered until much later. Even though little was understood about the basic biology of Bt in this time, it was shown to be highly pathogenic for larvae of certain species of lepidopterous pests.

As synthetic chemical insecticides had not yet been developed, farmers started to use Bt as a pesticide in 1920. The first record of its application to control insects was in Hungary at the end of 1920 (Husz 1928), and in Yugoslavia at the beginning of 1930s, it was applied to control the European corn borer (Vouk and Klas 1931). During the following two decades, several field tests were conducted to evaluate its effectiveness against lepidopterans, both in Europe and in the United States, and results favored the development of formulations against on this pathogen. Subsequently, the first commercial product, spore based formulations called Sporeine, was produced in 1938 by Laboratoire Libec in France (Aronson et al. 1986). Sporeine, at the time was used primarily to kill flour moths. Unfortunately, the product was used only for a very short time, due to World War II.

After World War II, the Green Revolution provided great agricultural advantages via the use of agrochemicals, chemical fertilizers, highly productive cultivars, and mechanization. The result was a considerable decrease in a great variety of insect populations, and as a consequence, synthetic insecticidal compounds became popular due to the long residual action and the wide toxicity spectrum. However, fully synthetic chemical insecticides appeared in 1940, when organochlorinated and organophosphate insecticides were discovered. These insecticides were applied during all growing seasons to attack all the developmental stages of insect pests. The indiscriminate use of these compounds caused, by 1950, a resurgence of pests, due to the elimination of their natural enemies and the appearance of pest populations showing resistance to insecticides. Also serious environmental and health issues began to be recognized by the presence of chemical residues in food, water, and air. To counteract this contamination, attention and efforts were directed to the use of biological control agents including insect pathogens (Margalith and Ben-Dov 2000).

A resurgence of interest in Bt has been attributed to Edward Steinhaus, who obtained a culture in 1951 and attracted attention to the potential of Bt through his subsequent studies (Steinhaus 1951). Steinhaus used successfully Bt preparations against *Colias eurytheme* larvae. Edward Steinhaus was perplexed that at sporulation in Bt the spores were not centrally located, but they were rather displaced to one end. In 1953 Steinhaus sought the advice of the Canadian bacterial morphologist C. Hannay regarding this phenomenon (Steinhaus and Jerrel 1954). Coincidentally in 1951 Steinhaus published a picture of a sporulated and lysed Bt culture showing bipyramidal crystals, but did not make note of them in the text. Upon examining Bt sporulated cells, Hannay noticed a second body in the sporangium, as had Berliner and Mattes. But Hannay went one step further and speculated that the parasporal inclusion bodies had some role in the pathogenicity of the bacterium toward susceptible lepidopterous larvae (Hannay 1953). Angus quickly proved that Hannay was correct, that the parasporal crystal was responsible for the toxicity of Bt (Angus 1954). Angus showed that spores by themselves had not effect, and that dialyzed supernatant of alkali-dissolved crystals had the same toxic action as did the spore-crystal complex when fed to *B. mori* larvae. Angus also noted that toxicity varied with crystal count, and was independent of the number of spores present (Angus 1954). In 1956 Steinhaus (Steinhaus 1956) and R. A. Fisher (Fisher and Rosener 1959) met with the president of Pacific Yeast Products (later Bioferm Corporation), J. M. Sudarsky, to explore the practicality of producing a Bt-based product. The decision was made to produce Bt and by 1957 a product called Thuricide was available for testing. Thuricide was formulated as liquid concentrates, dusts, and wettable powders and now owned by Sandoz.

In 1956, researchers, Hannay and Fitz-James (1955) and Angus (1954) found that the main insecticidal activity against lepidoteran (moth) insects was due to the crystalline protein inclusions formed in the course of sporulation. With this discovery came increased interest in the crystal structure, biochemistry, and general mode of action of Bt. In the US, Bt was used commercially starting in 1958. By 1961, Bt was registered as a pesticide to the EPA (Environmental Protection Agency) and several U.S. Companies (Merck, Agritrol; Rohm & Haas, Bakthane; and Grain Producers, Parasporine) produced Bt for short periods (van der Geest and van der Laan 1971; van der Laan 1967). In this time, there was also a development of Bt production and usage in the European socialist countries. In The Soviet Union the All Union Institute for Microbial Products for Plant Protection produced the product Entobaktirin using a bt subs. galleriae (Isakova 1958). In Moscow, the government agency for the Direction of Microbial Industry started producing the product Dendrobacillin for use against larvae of the Siberian silkworm (Talalaev 1956). In the same way the product Insektin also was used for forest protection (Weiser 1986).

The presence of Beta-exotoxin in some of the products resulted in the confusing host range and safety data that exist in the literature for those subsp. *thuringiensis*-based products. In 1962, Edouard Kurstak isolated another subspecies of Bt from diseased *A. kuehniella* larvae from a flour mill at Bures sur Yvette near Paris, France (Kurstak 1962). He gave the isolates to A. M. Heimpel of the U. S. Dept. of Agriculture in 1962 and again in 1963 (Beegle and Yamamoto 1992). In 1970 Dulmage reported isolating from diseased *Pectinophora gossypiella* (Saunders) larvae, an isolate he named HD-1 (Dulmage 1970). De Barjac and Lemille (1970) examined the isolates from Kurstak, from Heimpel, and from Dulmage, and found that on the basis of flagellar serotyping they were a new subspecies of Bt, which they named *kurstaki* (de Barjac and Lemille 1970).

In 1970 Abbott Laboratories entered the market with Dipel, which was the first commercial preparation based on the new Bt subp. *kurstaki* (HD1) isolate and was used to control many lepidopterous pests, such as the cabbage looper (*Trichoplusia ni*), corn earworm (*Helicoverpa zea*), and tobacco budworm (*Heliothis virescens*) in vegetable and field crops, and major forest pests, mainly the gypsy moth (*Lymantria dispar*) and spruce budworm (*Choristoneura fumiferana*) (Dulmage 1970). During the late 1970s and early 1980s, years after its initial commercial success, the application of new molecular biological techniques to research on Bt demonstrated that HD1's broad spectrum of insecticidal activity was due to its complex parasporal body, which was shown to consist of two crystals that together contain four proteins, each with a different lepidopteran target spectrum and specific activity. It was not long before all companies producing Bt-based produced in the United States were subsp. *kurstaki*. However, the market for *kurstaki*-based products in agriculture fell to about 20% of its peak in the mid-1970s, due largely to competition from synthetic pyrethroids. In the late 1970s a considerable amount of *kurstaki*-based Bt product was used on cotton for *Heliothis* spp. In the mid-1980s Sandoz registered the product Javelin with EPA. The worldwide market for *kurstaki*-based products for forestry and agriculture was estimated at $ 20–25 million in the U.S. in 1992 (Beegle and Yamamoto 1992). Keio Aizawa of Kyushu University in Japan discovered an isolate in 1962 (Aizawa and Iida 1963), a new subspecies, naming it *aizawai*. Subsp. *aizawai* isolates were particularly effective against *Galleria mellonella* and *Spodoptera* spp. larvae. Sandoz developed their product Certan, based on an *aizawai* isolate, for control of *G. mellonella* larvae in honey comb. In 1976, Zakharyan reported the presence of a plasmid in a strain of Bt and suggested its involvement in endospore and crystal formation (Zakharyan et al. 1976).

Up until 1977, only 13 Bt strains had been described. All 13 subspecies were toxic only to certain species of lepidopteran larvae. Then in 1976, Goldberg and Margalit discovered a new subspecies, subsequently named Bt subsp. *israelensis*, in the Negev desert of Israel, highly toxic to larvae of a wide range of mosquito species (Goldberg and Margalit 1977). Goldberg obtained a U.S. patent on the organism and assigned the patent to the U.S. government (Goldberg 1979). By 1992 both U.S. and European companies were producing and marketing subsp. *israelensis*-based products for use against mosquito and blackfly larvae, such as Vectobac, Bactimos, and Teknar. This subspecies was subsequently shown to also be insecticidal for larvae of other species of flies in the dipteran suborder Nematocera, including blackflies and chironomid midges. Owing to their high efficacy and narrow target spectrum, these products replaced many broad-spectrum chemical insecticides used for mosquito and blackfly control in developed countries, and are currently under development for control of the major anopheline vectors of malaria is Africa and South America. Moreover, products based on Bt subsp. *israelensis* proved of particular importance in the Onchocerciasis Control Program in West Africa, which significantly reduced onchocerciasis, commonly known as river blindness, a debilitating human eye disease caused by *Onchocerca volvulus*, a nematode transmitted by blackflies of the *Simulium damnosum* complex (Guillet et al. 1990).

Bt subs. *tenebrionis*, which is highly toxic to larvae and adults of coleopterous insects, i.e., beetles, was discovered in Germany by Krieg and associates in diseased *Tenebrio molitor* (Krieg et al. 1983). Owing to its toxicity to certain important coleopteran pests, such as the Colorado potato beetle (*Leptinotarsa decemlineata*), this isolate was developed as a bacterial insecticide for control of beetle pests. In 1988 Mycogen brought out its product M-One based on its isolate, primarily for use against larvae of *L. decemlineata* (Say) on potatoes. Unlike Bt subsp. *kurstaki* and *israelensis*, however, the efficacy of commercial products based on the original and similar isolates of Bt toxic to beetles was not as effective as new chemical insecticides, such as the neonicotinoids, and thus the products were not a commercial success. Ecogen received registration for its product Foil in 1992, which was a combination of subsp. *kurstaki* and subsp. *tenebrionis* for use against both lepidopterous and coleopterous pests on potatoes.

The first transgenic plants to express Bt toxins were tobacco and tomato plants (Van Frankenhuyzen 1993). The first Bt plant-pesticide, Bt field corn, was registered with the United States EPA in 1995. Today, major Bt transgenic crops also include corn, cotton, potatoes, and rice. The engineering of plants to express Bt delta-endotoxins has been especially helpful against pests that attack parts of the plant that are usually not well-protected by conventional insecticide application. A prime example of this is protection against *Ostrinia nubilalis*, the European corn borer. Larvae of this lepidopteran bore into the stalk of a corn plant and destroy its structural integrity. In the stalk, the pest is relatively safe from pesticide application. With toxins engineered into the plant, *O. nubilalis* is exposed and its damage becomes easier to control (Ely 1993). Overall, because of benefits such as these, Bt has become a major presence in agriculture. In 1997, Bt cotton, corn, and potatoes covered nearly 10 million acres of land in the United States alone. These crops have also been commercialized and are in wide use in Canada, Japan, Mexico, Argentina, and Australia (Frutos et al. 1999). While using Bt in the form of transgenic crops is now very common, the more traditional spray form of Bt is still widely used (Liu and Tabashnik 1997). Environmental benefits are derived from the insecticidal specificity of these crops, which unlike synthetic chemical insecticides, kill only target species and closely related insects, as well as from reductions in chemical insecticide usage (O'Callaghan et al. 2005). The relevant works on the screening and isolation of new Bt strains have been performed and finally resulted in the production of commercial products (Table 1.1).

Bt-based biopesticide production depends on high quality and high-efficiency formulation processes. Formulations must be safe and effective products, must be easy to use, and should have a long shelf life. The active ingredient in commercial formulations is the spore-crystal complex, which is more effective to use and cheaper to obtain than the crystals alone, which are frequently used in experimental tests. The spore-crystal complex must be carried by suitable inert substance that can function to protect the spore-crystal complex or to increase palatability to insects. In this sense, many studies reveal the inclusion of some kinds of components, related to enhance toxicity, or to enhance the attraction of insects. Bt sprays are used sporadically and typically over small areas. Sprayable Bt formulations have penetrated cotton, fruit and vegetable, aquatic, and other insecticide markets. However, the use of Bt spray as an insecticide has several disadvantages; (1) Bt spray cannot be applied uniformly to all parts of the plant, (2) it cannot be delivered to pests that are inside plant tissues, and (3) Bt is susceptible to rapid degradation by UV light and removal by water runoff. Therefore, multiple applications are required to provide extended pest protection New Bt formulations have consistently made gains in a limited number of fruit and specialty vegetable markets over the last number of years (Cerón 2001). The persistence of Bt spores in the laboratory, greenhouse, and field or forest environment has been reasonably well studied (Petras and Casida 1985). Bt spores can survive for several years after spray applications, although rapid declines in population and toxicity have been noted (Addison 1993). There has been a

Table 1.1 Bt-based biopesticides available in the world

Bt subspecies	Product name	Company	Target insect
kurstaki	Dipel	Abbott Labs. (Now Valent Bioscience Co.)	Lepidopteran
	Biobit, Foray	Valent Bioscience Co. and Novo Nordisk	Lepidopteran
	Condor, Cutlass, Crymax, Lepinox	Ecogen Inc.	Lepidopteran
	Javelin, Thuricide	Sandoz Agro, Inc.	Lepidopteran
	Bactospeine, Futura	Solvay & Cie/duphar B.V.	Lepidopteran
	Bernan Bt	Bactec	Lepidopteran
	Bactis	Compagnia di Recerca chim. CRC	Lepidopteran
	Biospor	Farbwerbe-Hoechst	Lepidopteran
	Larvo-Bt	Knoll Bioproducts	Lepidopteran
	Bt	Korea Explosives	Lepidopteran
	Sporoine	LIBEC	Lepidopteran
	M-peril	Mycogen	Lepidopteran
	SOK	Nor-Am Chemical	Lepidopteran
	Plantibac	Procida	Lepidopteran
	Baturad, Nubilacid	Radonja	Lepidopteran
	Able, delfin, CoStar, Steward, Vault	Thermo Trilogy Corporation	Lepidopteran
	Bactur	Thompsoni Hayward Co	Lepidopteran
	Toaro, Toaro, Ct	Towagosei Chem	Lepidopteran
aizawai	Agree[a]	Sandoz Agro Inc.	Lepidopteran
	Florbac, Xentari	Valent Bioscience Co.	Lepidopteran
	Tobaggi	Dongbu Hannong Chemicals	Lepidopteran
	Solbichae	Gree Biotech Co.	Lepidopteran
	Selectgyn	Kyowa-Hakko Kogyo Co.	Lepidopteran
israelensis	Gnatrol	Valent Bioscience Co.	Dipteran
	Bactimos, VectoBac, Teknar	Valent Bioscience Co.	Dipteran
	Skeetal	Novo Nordisk	Dipteran
	Baktokulicid	VPO Biopreparat	Dipteran
	Moskitur	JZD Slusovice	Dipteran
morrisoni or tenebrionis	M-One, M-Trak	Mycogen	Coleopteran
	Trident, TridentII	Sandoz Agro Inc.	Coleopteran
	Di Terra	Valent Bioscience Co.	Coleopteran
	Novodor	Novo Nordisk	Coleopteran
	Foil[b]	Ecogen Inc.	Lepidopteran Coleopteran
galleriae	Entobaktirin	Glavmikro-Bioprom	Lepidopteran
	Spicturin	Tuticorin Alcali Chemicals and Fertilisers Limited	Lepidopteran
thuringiensis	Bathurin	Chamapol-Biokrna	Lepidopteran
	Muscabac	Farmos	Lepidopteran
	Insektin	Glavmikro-Bioprom	Lepidopteran
	Bacillex	Shionogi Co.	Lepidopteran
dendrolimus	DendroBacillin	Glavmikro-Bioprom	Lepidopteran

[a] Combination of subsp. *kurstaki* and *aizawai*

[b] Combination of subsp. *kurstaki* and *tenebrionis*

standardization procedure for Bt-based products. Before 1970, the standardization procedure was carried out through the use of spore counts. However, there was no relationship between the number of spores in a preparation and its insect killing power. Because of this actually standardization procedure is carried out using insect bioassay. Insect bioassay of Bt products is expensive, time consuming, fraught with problems, and takes a relatively long time (4–7 days) to furnish information on the potency of the material. For those reasons it has been replaced insect bioassays with chemical assays. There are several things to take in consideration using chemical assays. The desired information is the killing power of the preparation toward target pest insects. The killing power of a preparation is determined by both the quality and quantity of crystal toxin present. Chemical methods only measure the quantity of toxin present, but not measure the quality of toxin present. In addition, there are a number of pest insects that require the presence or spores for maximum toxin activity. Chemical methods can not measure the presence, number, or viability of Bt spores. Because of this many laboratories still standardizes their Bt preparations by insect bioassay (Beegle and Yamamoto 1992).

1.3 A Description of *B. thuringiensis*

1.3.1 Basic Biology of Bt

Bacteria belonging to the genus *Bacillus* have a long and distinguished history in the realms of biotechnology. The bacilli include many versatile bacteria and the most effective bacterial control agents for various insect pests. Bt is an aerobic, Gram-positive, spore-forming facultative bacterial pathogen that can be readily isolated on simple media such as nutrient agar from a variety of environmental sources including soil, water, plant surfaces, grain dust, dead insects, and insect feces (Federici 1999). Its life cycle is simple. When nutrients and environmental conditions are sufficient for growth, the spore germinates producing a vegetative cell that grows and reproduces by binary fission. Cells continue to multiply until one or more nutrients, such as sugars, amino acids, or oxygen, become insufficient for continued vegetative growth. Under these conditions, the bacterium sporulates producing a spore and parasporal body, the latter, as noted above, composed primarily of one or more insecticidal proteins in the form of crystalline inclusions (Fig. 1.1). These are commonly referred to in the literature as insecticidal crystal proteins (ICP), or δ-endotoxins, which are selectively toxic to different species of several invertebrate phyla. ICPs include the more prevalent Cry (crystal) proteins, as well as the Cyt (cytolytic) proteins produced by some Bt strains. Cry proteins are solubilized upon ingestion and in most cases cleaved by proteolytic enzymes to active toxins. Because of their high specificity and their safety for the environment, crystal proteins are a valuable alternative to chemical pesticides for control of insect pests in agriculture and forestry and in the home.

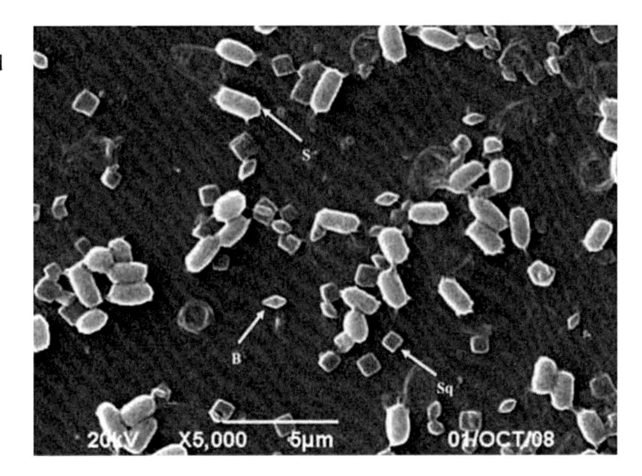

Fig. 1.1 Scanning electron micrograph of spores (*S*) and crystals of *Bacillus thuringiensis* HD-125 strain. The strain synthesizes bypiramidal (*B*) and square (*Sq*) crystals. The photograph is provided by N.M. Rosas-García. (Rosas-García 2009)

As a species, Bt is subdivided into more than 70 subspecies, which are not based on insecticidal protein complements or target spectrum, but rather on the antigenic properties of the flagellar (H) antigen (Lecadet et al. 1999). Each new isolate that bears a flagellar antigen type that differs detectably from the others in immunological assays is assigned a new H antigen serovariety number and subspecific name. there are four main subspecies: Bt subsp. *kurstaki* (H 3a3b3c) and Bt subsp. *aizawai* (H 7) used against lepidopteran pests; Bt subsp. *israelensis* (H 14) used against mosquitoes and blackfly larvae; and Bt subsp. *morrisoni* strain tenebrionis (H 8a8b), used against certain coleopteran pests, such as the Colorado potato beetle, *Leptinotarsa decemlineata*.

Over the past few decades, there has been an ongoing discussion regarding the validity of Bt as a species separate from *B. cereus* and *B. anthracis*. However, the insecticidal crystals formed by Cry and Cyt proteins are the principal characteristic that differentiates Bt from *B. cereus* as well as *B. anthracis*. As far as is known, most if not all Cry and Cyt proteins are encoded on plasmids present in Bt, i.e., not on the bacterial chromosome. Thus, if these plasmids are lost from a strain, or deliberately eliminated by plasmid curing, the resulting strain would be identified as *B. cereus* (Rasko et al. 2005). In some studies, it has been suggested that *B. cereus*, Bt, and *B. anthracis* are all members of the same species (Helgason et al. 2000). However, the two plasmids that encode the toxins of *B. anthracis* do not occur naturally in Bt or *B. cereus*, nor have parasporal bodies containing Bt Cry proteins been found naturally in *B. anthracis*. Therefore, this supports maintaining *B. anthracis* as a species different from *B. cereus* and Bt.

1.3.2 Toxicity and Mode of Action of Its Insecticidal Proteins

Individual Cry toxin has a defined spectrum of insecticidal activity, usually restricted to a few species in one particular order of Lepidoptera (butterflies and moths), Diptera (flies and mosquitoes), Coleoptera (beetles and weevils), Hymenoptera (wasps and bees), and nematodes, respectively (de Maagd et al. 2001). A few toxins have an activity spectrum that spans two or three insect orders due to the combination of toxins in a given strain. The Cry proteins comprise at least 50 subgroups with more than 200 members (Bravo et al. 2007) and the toxins are classified only on the basis of amino acid sequence homology, where each protoxin acquired a name consisting of the mnemonic Cry (or Cyt) and four hierarchical ranks consisting of numbers, depending on its place in a phylogenetic tree (Crickmore et al. 1998). The members belong to a three-domain family, and the larger group of Cry proteins is globular molecules with three structural domains connected by single linkers. The protoxins are characteristic of this family and have two different lengths. The C-terminal extension found in the long protoxins is necessary for toxicity and is believed to play a role in the formation of the crystal within the bacterium (de Maagd et al. 2001). Knowing the precise complement of insecticidal proteins produced by a specific isolate of Bt can go a long way to explaining its toxicity and lethality to a particular insect or nematode species. However, several Bt components other than endotoxins contribute to the activity of a particular isolate against a specific insect species.

Owing to the overwhelming interest in Cry and Cyt proteins (Fig. 1.2a, b), most of some other factors have received relatively little attention. Among the most important of these are the spore, *β*-exotoxin (also called thuringiensin), antibiotics such as zwittermicin, vegetative insecticidal proteins (Vip's) (Fig. 1.2c), phopholipases, chitinases, and various proteases (Federici et al. 2010). Cyt proteins (Fig. 1.2b) are hemolytic and cytolytic toxins produced by some Bt strains, particularly by those that show insecticidal activity against mosquitoes. Cyt toxins are also pore-forming proteins, and they are not related phylogenetically to Cry toxins. They have a single α–β domain comprised of two outer layers of α-helix hairpins wrapped around a β-sheet (Li et al. 1996). Cyt toxins do not bind to protein receptors and directly interact with membrane lipids inserting into the membrane to form pores (Promdonkoy and Ellar 2000) or destroy the membrane by a detergent like interaction (Butko 2003). *β*-exotoxin I is produced in the exponential growth phase, is thermostable and is not specific and thus, may have detrimental effects on nontarget organisms; it is particularly active against dipteran species, but it is also active against coleopteran, lepidopteran, and some nematode species. This toxin inhibits the synthesis of RNA by competing with ATP for binding sites, thereby affecting insect molting and pupation and causing teratogenic effects at sublethal doses (Espinasse et al. 2002, 2004). The genetic determinants responsible for *β*-exotoxin I production found on Cry-dependent plasmids are likely to be regulatory elements and large amounts of *β*-exotoxin I can be produced in the absence of such plasmids. A putative ABC transporter, encoded by *berAB*, is essential for *β*-exotoxin I production. *β*-exotoxin I production can be activated in response to particular environmental conditions (Espinasse et al. 2002, 2004). Vegetative insecticidal proteins (Vip's) (Fig. 1.2c) are produced by some strains of Bt during vegetative growth. These proteins do not form parasporal crystal proteins and are apparently secreted from the cell and show a great activity spectrum to lepidopterous and some coleopterous. Its structure and its mode of action are similar of the cry protein; however, the receptor to bind it,

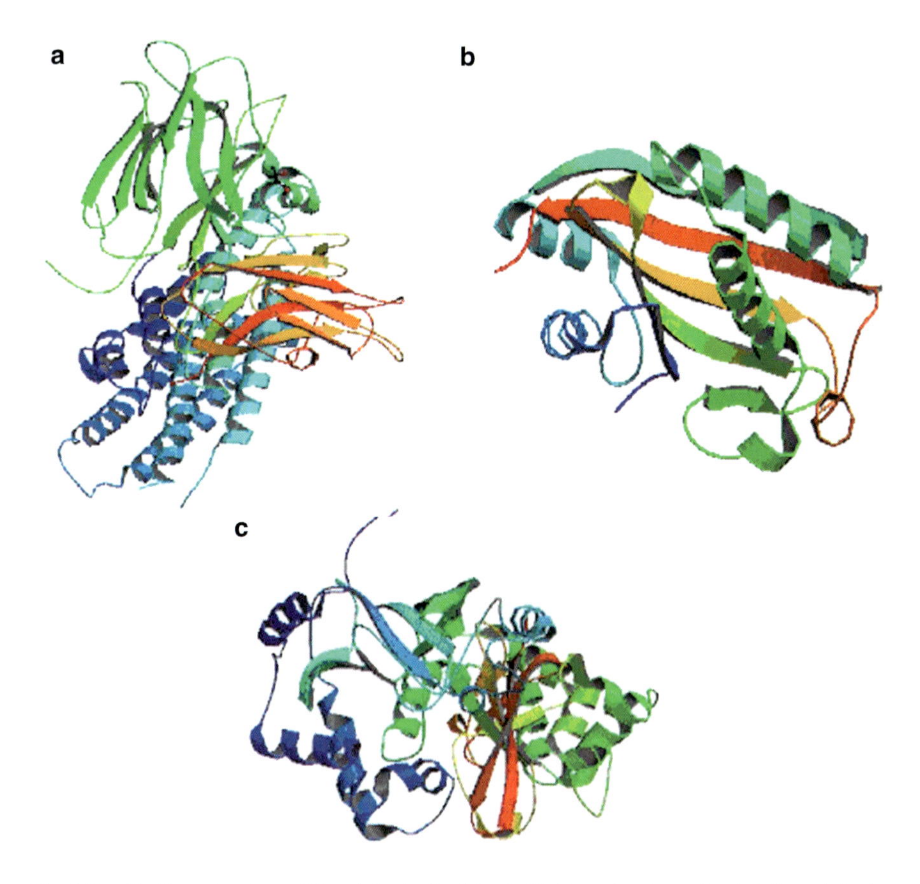

Fig. 1.2 **a** Tertiary structure of δ-endotoxin Cry4Aa. **b** Tertiary structure of cytolytic endotoxin Cyt2Ba. **c** Tertiary structure of Vip2 protein. Taken from RCSB protein data bank. (http://www.rcsb.org/pdb/home/home.do)

the union sites and the ionic channels are different (Lee et al. 2003). Zwittermicin A is a highly polar, water-soluble aminopolyol antibiotic isolated from the soil-born bacterium *B. cereus* and also identified in some strains of Bt with significant activity against phytopathogenic fungi as a secondary metabolite (Silo-Suh et al. 1994). More importantly, enhances the activity of the δ-endotoxin from Bt, the active ingredient in biocontrol agents used against crop pests and for the eradication of gypsy moth from forest trees (Broderick et al. 2000).

Analysis of *cry* gene sequences combined with the three dimensional structures of Cry3A, Cry1Aa, and Cry2A showed that the active portion of Cry toxins is a wedge shaped molecule of three domains, and typically consists of approximately 600 amino acids (de Maagd et al. 2003). Domain I of this protein is composed of amino acids 1–290, and contains a hydrophobic seven-helix amphipathic bundle, with six helices surrounding a central helix. This domain contains the first conserved amino acid block and a major portion of the second conserved block. Theoretical computer models of the helix bundle show that after insertion and rearrangement, aggregations of six of these domains likely form a pore through the microvillar membrane. Domain II extends from amino acids 291–500 and contains three antiparallel β-sheets around a hydrophobic core. This domain contains most of the hypervariable region and most of conserved blocks three and four. The crystal structure of the molecule together with recombinant DNA experiments and binding studies indicate that the three extended loop structures in the β-sheets are responsible for initial recognition and binding of the toxin to binding sites on the microvillar membrane. Domain III is comprised of amino acids 501–644 and consists of two antiparallel β-sheets, within which are found the remainder of conserved block number three along with blocks four and five. The Cry3Aa structure indicated that this domain provides structural integrity to the molecule. Site-directed mutagenesis studies of conserved amino acid block 5 in the Cry1 molecules show that this domain also plays a role in receptor binding and pore formation (de Maagd et al. 2003).

Endotoxin crystals must be ingested to have an effect. This is the reason sucking insects and other invertebrates such as spiders and mites are not sensitive to Cry proteins used in Bt insecticides or Bt crops. Their mode of action involves several events that must be completed several hours after ingestion in order to lead to insect death. Following ingestion, the crystals are solubilized by the alkaline conditions in the insect midgut and are subsequently proteolytically converted into a toxic core fragment (de Maagd et al. 2003). Under the highly acidic conditions in stomachs of many vertebrates, including humans, Cry and Cyt protein crystals may dissolve, but once in solution they are rapidly degraded to non-toxic peptides by gastric juices, typically in less than 2 min. During proteolytic activation, peptides from the N terminus and C terminus are cleaved from the full protein. Activated toxin binds to receptors (glycoprotein or glycolipid) located on the apical microvillus membranes of epithelial midgut cells (Griffiths et al. 2005). For some toxins, at least four different binding sites have been described in different lepidopteran insects: a cadherin-like protein (CADR), a glycosylphosphatidyl-inositol (GPI)-anchored aminopeptidase-N (APN), a GPI-anchored alkaline phosphatase (ALP) and a 270 kDa glycoconjugate (Lee et al. 1996; Jurat-Fuentes and Adang 2004). After binding, toxin adopts a conformation allowing its insertion into the cell membrane and form a cation-selective channel. Subsequently, oligomerization occurs, and this oligomer forms a pore or ion channel induced by an increase in cationic permeability within the functional receptors contained on the brush borders membranes (Bravo et al. 2004). Complete nature of this process still remains unknown (Federici et al. 2010); however, it is believed that toxin aggregation occurs at the membrane surface after receptor binding, or alternatively only after the toxin inserts itself into the membrane. Once a sufficient number of these channels have formed, a surplus of cations, K^+ for example, enter the cell. This causes an osmotic imbalance within the cell, and the cell compensates by taking in water. This process, referred to as colloid-osmotic induced lysis, continues until the cell ruptures and exfoliates from the midgut microvillar membrane. When a sufficient number of cells have been destroyed, the midgut epithelium loses its integrity. This allows the alkaline gut juices and bacteria to cross the midgut basement membrane, resulting in death, the latter caused by Bt bacteremia and tissue colonization in lepidopteran species. In mosquito and black fly larvae, midgut bacteria do not cross the midgut epithelium until after death, thus in these the cause of paralysis and death is apparently due only to the insecticidal Cry and Cyt proteins (Bravo et al. 2002).

Cyt proteins have received little study in comparison to Cry proteins, as they typically only occur in mosquitocidal strains of Bt. Nevertheless, these proteins are extremely important in the biology of mosquitocidal strains because they synergize mosquitocidal Cry proteins, and delay the phenotypic expression of resistance to these (Wu et al. 1994; Wirth et al. 2005). As far as is known, Cyt proteins do not require a protein receptor, but instead bind directly to the non-glycosylated lipid portion of the microvillar membrane. Once within the membrane, they appear to aggregate, forming lipid faults that cause an osmotic imbalance resulting in cell lysis (Butko 2003).

1.4 Conclusions

Since the discovery of Bt as soil bacterium with specific insecticidad activity about a century ago, Bt has developed into an important tool for pest control. Today Bt is the most successful commercial microbial insecticide, comprising about 90% of the biopesticide market. The insecticidal activity of Bt is due to the presence of parasporal protein inclusion bodies, also called crystals, produced during sporulation, which act like a stomach poison causing larval death. Their mode of action is enough complex and involves several events that must be completed several hours after ingestion in order to lead to insect death. Environmentally safe-insect control strategies based on Bt and their insecticidal crystal proteins are going to increase in the future, especially with the wide adoption of transgenic crops. Genetic improvement of Bt strains for the development of novel biopesticides is the tool of the future to increase their potency against target insects, broadening the insecticidal spectra for specific crop applications, improving persistence on plants, and optimizing fermentation production.

In conclusion, there is a great amount of scientific research on the bacterium Bt, involving aspects ranging from its molecular biology to its activity in a bioinsecticide. Although Bt has been studied for more than a century, it is amazing to know that scientists always find something new in it. Although there is an enormous possibility to control many pests, mainly in agriculture and forestry, and to obtain more organic pest-control products, the ignorance about advantages of use of Bt products limits its application. The discovery of new toxins and new ways of presenting the toxin to the target insects, which includes the development of recombinant microorganisms and proteomic technology, could be adapted to the study of Bt crystal proteins; additionally, interaction studies between Bt and target insects involving modes of action of Bt Cry proteins and resistance mechanisms should be carried out, all of which are fundamental studies that will allow for improvement of existing Bt application strategies and the ability to design alternative options.

References

Addison JA (1993) Persistence and non-target effects of *Bacillus thuringiensis* in soil: a review. Can J Forest Res 23:2329–2342

Aizawa K, Iida S (1963) Nucleic acids extracted from the virus polyhedra of the silkworm, *Bombyx mori* (Linnaeus). J Insect Path 5:344–348

Angus TA (1954) A bacterial toxin paralyzing silkworm larvae. Nature 173:545–546

Aoki K, Chigasaki Y (1916) Über die Pathogenität der sog. Sottobacillen (Ishiwata) bei Seidenraupen. Bull Imp Sericult Expt Sta 1:97–139

Aronson A, Beckman W, Dunn P (1986) *Bacillus thuringiensis* and related insect pathogens. Microbiol Rev 50:1–24

Beegle CC, Yamamoto T (1992) Invitation paper (C.P. Alexander Fund): history of *Bacillus thuringiensis* Berliner research and development. Can Entomol 124:587–604

Berliner E (1915) Ueber die schlaffsucht der *Ephestia kuhniella* und *Bac. thuringiensis* n. sp. Z Angew Entomol 2:21–56

Bravo A, Sanchez J, Kouskovra T, Crickmore N (2002) N-terminal activation is an essential early step in the mechanism of action of the *Bacillus thuringiensis* Cry1Ac insecticidal toxin. J Biol Chem 27:23985–23990

Bravo A, Gomez I, Conde J, Muñoz-Garay C, Sánchez J, Miranda R, Zhuang M, Gill SS, Soberón M (2004) Oligomerization triggers binding of a *Bacillus thuringiensis* Cry1Ab pore-forming toxin to aminopeptidase N receptor leading to insertion into membrane microdomains. Biochem Biophys Acta 1667:38–46

Bravo A, Gill SS, Soberón M (2007) Mode of action of *Bacillus thuringiensis* Cry and Cyt toxins and their potential for insect control. Toxicon 49:423–435

Broderick NA, Goodman RM, Raffa KF, Handelsman J (2000) Synergy between zwittermicin A and *Bacillus thuringiensis* subsp. *kurstaki* against gypsy moth (Lepidoptera: lymantriidae). Environ Entomol 29:101–107

Butko P (2003) Cytolytic toxin Cyt1Aa and its mechanism of membrane damage: data and hypothesis. Appl Environ Microbiol 69:2415–2422

Cerón JA (2001) Productos comerciales nativos y recombinantes a base de *Bacillus thuringiensis*. In: Caballero P, Ferré J (eds) Bioinsecticidas: fundamentos y aplicaciones de *Bacillus thuringiensis* en el control integrado de plagas. Phytoma-España, Valencia, pp 153–168

Crickmore N, Zeigler DR, Feitelson J, Schnepf E, Van Rie J, Lereclus D, Baum J, Dean DH (1998) Revision of the nomenclature for the *Bacillus thuringiensis* pesticidal crystal proteins. Microbiol Mol Biol Rev 62:807–813

de Barjac H, Lemille F (1970) Presence of flagellar antigenic subfactors in Serotype 3 of *Bacillus thuringiensis*. J Invertebr Pathol 15:139–140

de Maagd RA, Bravo A, Crickmore N (2001) How *Bacillus thuringiensis* has evolved specific toxins to colonize the insect world. Trends Genet 17:193–199

de Maagd RA, Bravo A, Berry C, Crickmore N, Schnepf HE (2003) Structure, diversity, and evolution of protein toxins from spore forming entomopathogenic bacteria. Ann Rev Genet 37:409–420

Dulmage HT (1970) Insecticidal activity of HD-1, a new isolate of *Bacillus thuringiensis* var. *alesti*. J Invertebr Pathol 15:232–239

Ely S (1993) The engineering of plants to express *Bacillus thuringiensis* delta-endotoxins. In: Entwistle PF, Cory JS, Bailey MJ, Higgs S (eds) *Bacillus thuringiensis*, an experimental biopesticide: theory and practice. Wiley, Chichester, pp 105–124

Espinasse S, Gohar M, Lereclus D, Sanchis V (2002) An ABC transporter from *Bacillus thuringiensis* is essential for β-exotoxin I production. J Bacteriol 184:5848–5854

Espinasse S, Gohar M, Lereclus D, Sanchis V (2004) An extracytoplasmic-function sigma factor is involved in a pathway controlling β-Exotoxin I production in *Bacillus thuringiensis* subsp. *thuringiensis* Strain 407–1. J Bacteriol 186:3108–3116

Federici BA (1999) *Bacillus thuringiensis*. In: Bellows TS, Gordh G, Fisher TW (eds) Handbook of biological control. Academic Press, San Diego, pp 517–548

Federici BA, Park H-W, Bideshi DK (2010) Overview of the basic biology of *Bacillus thuringiensis* with emphasis on genetic engineering of bacterial larvicides for mosquito control. Open Toxinol J 3:83–100

Fisher R, Rosener L (1959) Toxicology of the microbial insecticide, Thuricide. Agric Food Chem 7:686–688

Frutos R, Rang C, Royer M (1999) Managing insect resistance to plants producing *Bacillus thuringiensis* toxins. Crit Rev Biotechnol 19:227–276

Goldberg LJ (1979) Mosquito larvae control using a bacterial larvicide. US Patent 4(166):112

Goldberg LJ, Margalit J (1977) A bacterial spore demonstrating rapid larvicidal activity against *Anopheles sergentii*, *Uranotaenia unguiculata*, *Culex univeritattus*, *Aedes aegypti*, and *Culex pipiens*. Mosq News 37:355–358

Griffiths JS, Haslam SM, Yang T, Garczynski SF, Mulloy B, Morris H, Cremer PS, Dell A, Adang MJ, Aroian RV (2005) Glycolipids as receptors for *Bacillus thuringiensis* crystal toxin. Science 307:922–925

Guillet PD, Kurtak C, Philippon B, Meyer R (1990) Use of *Bacillus thuringiensis israelensis* for onchocerciasis control in West Africa. In: Barjac H de, Sutherland D (eds) Bacterial control of mosquitoes and blackflies; biochemistry, genetics, and applications of *Bacillus thuringiensis* and *Bacillus sphaericus*. Rutgers Univ Press, New Brunswick, pp 187–201

Hannay CL (1953) Crystalline inclusions in aerobic sporeforming bacteria. Nature 172:1004

Hannay CL, Fitz-James PC (1955) The proteins crystals of *Bacillus thuringiensis* Berliner. Can J Microbiol 1:694–710

Helgason E, Okstad OA, Caugant DA, Johansen HA, Fouet A, Mock M, Hegna I, Kolsto AB (2000) *Bacillus anthracis*, *Bacillus cereus*, and *Bacillus thuringiensis*—One species on the basis of genetic evidence. Appl Environ Microbiol 66:2627–2630

Husz B (1928) *Bacillus thuringiensis Berl*. A bacterium pathogenic to corn borer larvae. Int Corn Borer Invest Sci Rep 1:191–193

Isakova NP (1958) A new variety of bacterium of the "*cereus*" type pathogenic for insects. Dokl Akad Sci Nauk Selsk 3:26–27

Ishiwata S (1901) On a new type of severe flacherie (sotto disease) (original in Japanese). Dainihon Sansi Kaiho 114:1–5

Ishiwata S (1905a) Concerning 'Sotto-Kin' a *bacillus* of a disease of the silkworm. Rept Assoc Seric Japan 160:1–8

Ishiwata S (1905b) About "Sottokin," a *bacillus* of a disease of the silk-worm. Dainihon Sanshi Kaiho (Rept Assoc Seric Japan) 161:1–5

Jurat-Fuentes JL, Adang MJ (2004) Characterization of a Cry1Ac receptor alkaline phosphatase in susceptible and resistant *Heliothis virescens* larvae. Eur J Biochem 271:3127–3135

Krieg A, Huger A, Lagenbruch G, Schnetter W (1983) *Bacillus thuringiensis* var. *tenebrionis*: a new pathotypes effective against larvae of Coleoptera. Z Angew Entomol 96:500–508

Kurstak E (1962) Donnees sur l'epizootie bacterienne naturelle provoguee par un Bacillus du type *Bacillus thuringiensis* sur *Ephestia kuhniella* Zeller. Entomophaga Mem Hors Ser 2:245–247

Lecadet M-M, Frachon E, Dumanoir VC, Ripouteau H, Hamon S, Laurent P, Thiery I (1999) Updating the H-antigen classification of *Bacillus thuringiensis*. J Appl Microbiol 86:660–672

Lee MK, You TH, Young BA, Cotrill JA, Valaitis AP, Dean DH (1996) Aminopeptidase N purified from gypsy moth brush border membrane vesicles is a specific receptor for *Bacillus thuringiensis* Cry1Ac toxin. Appl Environ Microbiol 62:2845–2849

Lee MK, Walters FS, Hart H, Palekar N, Chen J-S (2003) The mode of action of the *Bacillus thuringiensis* vegetative insecticidal protein Vip3A differs from that of Cry1Ab δ-endotoxin. Appl Environ Microbiol 69:4648–4657

Li J, Koni PA, Ellar DJ (1996) Structure of the mosquitocidal delta-endotoxin CytB from *Bacillus thuringiensis* ssp. *kyushuensis* and implications for membrane pore formation. J Mol Biol 257:129–152

Liu YB, Tabashnik BE (1997) Experimental evidence that refuges delay insect adaptation to *Bacillus thuringiensis*. Proc R Soc Lond B 264:605–610
Margalith Y, Ben-Dov E (2000) Biological control by *Bacillus thuringiensis* subp. *israelensis*. In: Rechcigl JE, Rechcigl NA (eds) Insect pest management, techniques for environmental protection. Lewis Publishers, Boca Raton
O'Callaghan M, Glare TR, Burgess EPJ, Malone LA (2005) Effects of plants geneticallymodified for insect resistance on non-target organisms. Annu Rev Entomol 50:271–292
Petras SF, Casida LE Jr (1985) Survival of *Bacillus thuringiensis* spores in soil. Appl Environ Microbiol 50:1496–1501
Promdonkoy B, Ellar DJ (2000) Membrane pore architecture of a cytolytic toxin from *Bacillus thuringiensis*. Biochem J 350:275–282
Rasko DA, Altherr MR, Han CS, Ravel J (2005) Genomics of the *Bacillus cereus* group organisms. FEMS Microbiol Rev 29:303–310
Rosas-García NM (2009) Biopesticide production from *Bacillus thuringiensis*: an environmentally friendly alternative. Recent Pat Biotechnol 3:28–36
Silo-Suh LA, Lethbridge BJ, Raffel SJ, He H, Clardy J, Handelsman J (1994) Biological activities of two fungistatic antibiotics produced by *Bacillus cereus* UW85. Appl Environ Microbiol 60:2023–2030
Steinhaus EA (1951) Possible use of *Bacillus thuringiensis* Berliner as an aid in the biological control of the alfalfa caterpillar. Hilgardia 20:359–381
Steinhaus EA (1956) Potentialities for microbial control of insects. Agric Food Chem 4:676–680
Steinhaus EA, Jerrel EA (1954) Further observations on *Bacillus thuringiensis* Berliner and other sporeforming bacteria. Hilgardia 23:1–23
Talalaev EV (1956) Septicemia of the caterpillars of the Siberian silkworm. Mikrobiologiya 25:99
van der Geest LPS van der, Laan PA (1971) Sources of special materials. In: Burges HD, Hussey NW (eds) Microbial control of insects and mites. Academic Press, NY, pp 741–749
van der Laan PA (1967) Insect pathology and microbial control. In: van der Laan PA (ed) Proc Intern Colloq on Insect Path & Microbial Contr. North-Holland Publishers Co., Amsterdam, pp 252–286
Van Frankenhuyzen K (1993) The challenge of *Bacillus thuringiensis*. In: Entwistle PE, Cory JS, Bailey MJ, Higgs S (eds) *Bacillus thuringiensis*, an environmental biopesticide: theory and practice. Wiley, Chichester, pp 1–35
Vouk V, Klas Z (1931) Conditions influencing the growth of the insecticidal fungus *Metarrhizium anisopliae* (Metsch.) Sor. Int Corn Borer Invest Sci Rept 4:24–45
Weiser J (1986) Impact of *Bacillus thuringiensis* on applied entomology in Eastern Europe and in the Soviet Union. In: Krieg A, Huger AM (eds) Mitt. Biol. Bundesanst. Land Forstwirtsch Berl Dahlem, vol 233. Paul Parey, Berlin, pp 37–49
Wirth MC, Park H-W, Walton WE, Federici BA (2005) Cyt1A of *Bacillus thuringiensis* delays evolution of resistance to Cry11A in the mosquito, *Culex quinquefasciatus*. Appl Environ Microbiol 71:185–189
Wu D, Johnson JJ, Federici BA (1994) Synergism of mosquitocidal toxicity between CytA and CryIVD proteins using inclusions produced from clones genes. Mol Microbiol 13:965–972
Zakharyan RA, Agabalyan AS, Chil-Akopyan LA, Gasparyan NS, Bakunts KA, Tatevosyan PE, Afrikyan EK (1976) About the possibility of extrachromosomal DNA in creation of the entomocidal endotoxin of *B. thuringiensis.* Dokl Akad Nauk Arrn SSR 63:42–47 [in Russian.]

Chapter 2
Bacillus thuringiensis Applications in Agriculture

Zenas George and Neil Crickmore

Abstract *Bacillus thuringiensis* (*Bt*) and its insecticidal toxins have been used in agronomical pest control for decades. The mechanism of action of *Bt* toxins on insect pest involves specific molecular interactions which makes *Bt* a popular choice for pest control. The specificity of action of *Bt* toxins reduces the concern of adverse effects on non-target species, a concern which remains with chemical insecticides. Different strains of *Bt* are known to express different classes of toxins which in turn target different insects. *Bt* and its toxins can be formulated into powder or liquid sprays or expressed in transgenic plants. To maximize the effect of *Bt* toxins, multiple toxins are often combined when making *Bt* formulations or expressed in transgenic plants. Though *Bt* is a very effective biological control agent, there are concerns over the development of resistance by insect species and also the narrow spectrum of activity of individual toxins. To address these concerns, new strains of *Bt* expressing novel toxins are actively sought and existing toxins are genetically modified for improved activity.

Keywords Parasporal crystal proteins · Mode of action · Resistance · Use of Bt products · Synergism

2.1 Introduction

Bacillus thuringiensis is a Gram positive spore forming bacteria grouped into the *Bacillus cereus* group of *Bacilli* which produces proteinaceous insecticidal crystals during sporulation which is the distinctive feature between it and other members of the *Bacillus cereus* group (Read et al. 2003; Rasko et al. 2005). *Bacillus thuringiensis* was originally discovered in 1902 by a Japanese biologist Shigetane Ishiwatari who isolated it from diseased silkworm, *Bombyx mori* but it was formally characterised in 1915 by Ernst Berliner of Germany who isolated it from diseased larva of *Ephestia kuhniella* (flour moth caterpillars) in Thuringia province and linked it to the cause of a disease called Schlaffsucht (Milner 1994).

Z. George (✉) · N. Crickmore
Department of Biochemistry, School of Life Sciences, University of Sussex, Falmer, BN1 9QG Brighton, UK
e-mail: n.crickmore@sussex.ac.uk

E. Sansinenea (ed.), *Bacillus thuringiensis Biotechnology*,
DOI 10.1007/978-94-007-3021-2_2, © Springer Science+Business Media B.V. 2012

Bacillus thuringiensis readily proliferates when environmental conditions such as temperature and nutrient availability are favourable whilst the formation of spores have been shown to be triggered by internal and external factors including signals for nutrient starvation, cell density and cell cycle progression (Hilbert and Piggot 2004). The life cycle of *Bt* can be divided for convenience into phases and these are Phase I: vegetative growth; Phase II: transition to sporulation; Phase III: sporulation; and Phase IV: spore maturation and cell lysis (Hilbert and Piggot 2004; Berbert-Molina et al. 2008). The production of the characteristic insecticidal (Cry) proteins deposited in crystals in the mother cell have been shown to mainly start from the onset of sporulation (Sedlak et al. 2000; Xia et al. 2005; Guidelli-Thuler et al. 2009; Pérez-García et al. 2010). A number of *cry*-genes have been shown to be transcribed from two overlapping promoters BtI and BtII by RNA polymerases that contain sporulation dependent sigma factors σ^E and σ^K (Sedlak et al. 2000; Hilbert and Piggot 2004) and a mutation in the consensus region of σ^E has been shown to inhibit transcription from BtI and BtII promoters (Sedlak et al. 2000). It has also been shown that some *Bt* insecticidal proteins are produced and secreted into the culture medium during vegetative growth (Estruch et al. 1996; Donovan et al. 2001; Shi et al. 2004; Bhalla et al. 2005; Leuber et al. 2006; Milne et al. 2008; Singh et al. 2010; Abdelkefi-Mesrati et al. 2011).

As well as the Cry toxins *Bt* produces additional virulence factors including phospholipase C (Palvannan and Boopathy 2005; Martin et al. 2010), proteases (Hajaij-Ellouze et al. 2006; Brar et al. 2009; Infante et al. 2010) and hemolysins (Gominet et al. 2001; Nisnevitch et al. 2010). The virulence factors are controlled by the pleiotropic regulator PlcR and it has been demonstrated that cytotoxicity of *Bt* is PlcR dependent (Ramarao and Lereclus 2006). Deletion of the *plcR* gene has been shown to result in a drastic reduction in the virulence of *Bt* in orally infected insects (Salamitou et al. 2000). The production of virulence factors by *Bt* is necessary but not enough for *Bt* to be called a pathogen (Fedhila et al. 2003) but its production of proteins that have been proved beyond doubt to be independently insecticidal justifies it's name as an insect pathogen (Frankenhuyzen 2009).

The insecticidal proteins in the crystalline bodies produced during sporulation have been shown to contain two types of insecticidal proteins namely Cry toxins and Cyt-toxins and there are one or more toxins produced and packaged into a single crystal or multiple crystals by a *Bt* strain (de Maagd et al. 2001). The Cry toxins acquired the mnemonic Cry from the fact that they are found in the crystal while the Cyt-toxins acquired the mnemonic Cyt because of their in vitro cytolytic activity (Crickmore et al. 1998).

Schnepf and Whiteley (1981) confirmed that the insecticidal ability of *Bt* is as a result of the proteins that it produces by first cloning and heterologously expressing a toxin gene in *E. coli* which showed insecticidal activity to *Manduca sexta* just as the wild type *Bacillus thuringiensis* var. kurstali HD-1 from which it was cloned did. Since this discovery, a great number of other genes have cloned and expressed and the process of *Bt* toxin gene discovery is still ongoing. In order to differentiate between one *Bt* insecticidal gene and the other, the discoverers of the genes gave them arbitrary names like 4.5, 5.3 and 6.6-kb-class genes (Kronstad and Whiteley 1986), *bta* gene (Sanchis et al. 1989), *cry* gene (Donovan et al. 1988) and Type A and

Type B (Hofte et al. 1988) among others. With a steady growth in the number of cloned and characterised novel insecticidal genes coming through, an attempt was made to organise the ever growing data. The first attempt to produce an organised systematic nomenclature of *Bt* insecticidal genes was dependent on the insecticidal activity of the protein they code for to assign a primary rank to the gene and with this system, genes that encode proteins toxic to lepidopteran insects were called *cryI* genes, while lepidopteran and dipteran protein genes were called *cryII* genes, *cryIII* genes were those ones that encoded proteins toxic to coleopterans and *cryIV* genes encoded proteins toxic to dipterans alone (Hofte and Whiteley 1989). Though this system provided a framework for naming newly cloned novel toxins, it was short of a robust system of nomenclature that is able to accommodate new genes without ambiguity. The discovery of wild type gene like *cryIB* that codes for a toxin toxic to both lepidoptera and coleoptera (Bradley et al. 1995) threw the system off balance as it did not have room to accommodate a toxin with such spectrum of activity. Also, toxins like CryIC that had toxicity to both diptera and lepidoptera (Smith and Ellar 1994) did not have a place in the (Hofte and Whiteley 1989) system of nomenclature.

With the difficulty of accommodating newly discovered genes in the (Hofte and Whiteley 1989) nomenclature system arising, there was a need to come up with a robust system of nomenclature and (Crickmore et al. 1998) came up with a system that is based on sequence similarity rather than function based. In the Crickmore et al. system, the mnemonic root was combined with a series of numerals and letters assigned in a hierarchical fashion to indicate degrees of phylogenetic divergence which was estimated by phylogenetic tree algorithms. The mnemonic Cyt was used for parasporal crystal proteins from *Bacillus thuringiensis* that exhibits hemolytic activity or any protein that has obvious sequence similarity to a known Cyt protein and mnemonic Cry was assigned to a parasporal crystal proteins from *Bacillus thuringiensis* that exhibits some experimentally verifiable toxic effect to a target organism, or any protein that has obvious sequence similarity to a known Cry protein. With this system for naming Cry and Cyt proteins widely accepted, it has also been adopted for the naming of the vegetatively produced *Bacillus thuringiensis* toxins and this family of proteins has been given the mnemonic Vip. A website which hosts all the *cry, cyt and vip* cloned genes has been established and is frequently updated as new genes are discovered.

The 3-D crystal structure of Cry-proteins including coleopteran specific Cry8Ea1 (Guo et al. 2009), Cry3Aa (Li et al. 1991) and Cry3Bb (Galitsky et al. 2001), dipteran specific Cry4Aa (Boonserm et al. 2006) and Cry4Ba (Boonserm et al. 2005), lepidopteran specific Cry1Aa (Grochulski et al. 1995), lepidopteran/dipteran specific Cry2Aa (Morse et al. 2001) have been resolved through X-ray crystallographic methods of their activated forms. Also, the 3-D structure of Cyt-proteins have been resolved including activated Cyt2Ba (Cohen et al. 2008) and unprocessed Cyt2Aa (Li et al. 1996). Figure 2.1 is the crystal structure of Cry8Ea1 (Guo et al. 2009) showing the three domain organisation typical of all resolved 3-D structures of Cry toxins while Fig. 2.2 is the 3-D structure of Cyt2Ba which shows overall similarity to 3-D structure of Cyt2Aa that had previously been resolved.

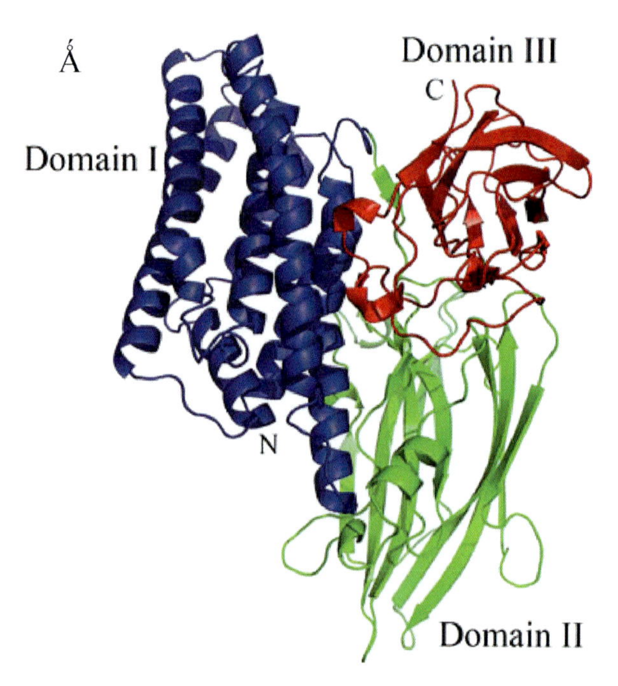

Fig. 2.1 3-D crystal structure of Cry8Ea determined at 2.20 Å (PBD code: 3EB7). The three domains of the protein are represented with different colours with domain I coloured *blue*, domain II coloured *green* while domain III is coloured *red*. (Guo et al. 2009)

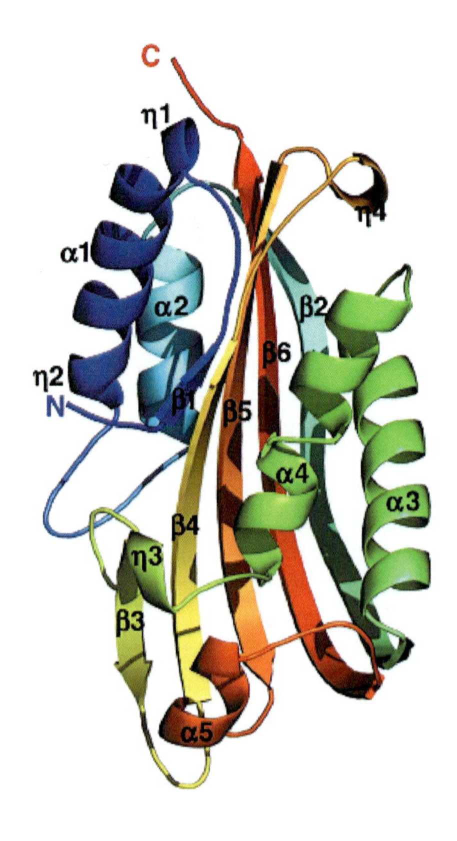

Fig. 2.2 Crystal structure of a monomer of Cyt2Ba determined at 1.80 Å (PDB code: 2RCI). The 'rainbow' colouring scheme is used in colouring the molecule starting with *blue* at the N-terminal and ending with *red* at the C-terminal. (Cohen et al. 2008)

Though different Cry toxins have been shown to have specific targets in their insecticidal activity, the overall 3-D fold of many of them has been shown to be the same (see Fig. 2.1), comprising of three domains (de Maagd et al. 2003). Domain I has been shown to compose of seven α-helices in which the central helix-α5 is hydrophobic and is encircled by six other amphipathic helices. The helical domain I has been shown to share structural similarities with other pore forming bacterial toxins like cytolycin A (Mueller et al. 2009), diphtheria toxin and colicin A (Parker and Pattus 1993). Each of the outer helices of domain I is known to be amphipathic in nature and most of the helices are longer than 30 Å in length (Pigott and Ellar 2007). Domain II is made up of three antiparallel β-sheets packed together to form a β-prism with pseudo threefold symmetry (Li et al. 1991). Two of the sheets are composed of four strands in a Greek key motif and are solvent exposed (Boonserm et al. 2006). The third sheet packs against domain I and is arranged in a Greek-key-like motif with three strands and a short alpha-helix (Pigott and Ellar 2007). The structure of domain II has been compared to those of other β-prism proteins with carbohydrate-binding properties (de Maagd et al. 2003), including vitelline (Shimizu et al. 1994) and *Maclura pomifera* agglutinin (Lee et al. 1998) and it reveals a great topological similarity. Domain III has been shown to contain two antiparallel β-sheets that adopt a β-sandwich fold with the jelly roll topology (Boonserm et al. 2006). Both sheets are composed of five strands, with the outer sheet facing the solvent and the inner sheet packing against domain II. Two long loops extend from one end of the domain and interact with domain I (Grochulski et al. 1995). Domain III shows less structural variability than domain II, and the main differences are found in the lengths, orientations, and sequences of the loops (Boonserm et al. 2005). Domain III has been compared to other carbohydrate-binding protein domains and it shows great degree of similarity (de Maagd et al. 2003) and similarity of domain III was also found with those of domain 4 of the pore-forming toxin aerolysin which is involved in maintenance and stability of the heptameric toxin complex (Lesieur et al. 1999).

Proposed mechanisms of action of Cry and Cyt toxins from *Bt* include pore formation in which *Bt* toxins induce cell death by forming ionic pores following insertion into the membrane, causing osmotic lysis of midgut epithelial cells in their target insect (Knowles and Ellar 1987; Haider and Ellar 1989; Grochulski et al. 1995; Schnepf et al. 1998; Bravo et al. 2004; Rausell et al. 2004). Also, a relatively new mechanism of action of Cry toxins have been proposed which involves the activation of Mg^{2+}-dependent signal cascade pathway that is triggered by the interaction of the monomeric 3-domain Cry toxin with the primary receptor, the cadherin protein BT-R_1 (Zhang et al. 2005, 2006; Soberón et al. 2009). The triggering of the Mg^{2+}-dependent pathway has a knock-on effect and initiates a series of cytological events that include membrane blebbing, appearance of nuclear ghosts, and cell swelling followed by cell lysis (Zhang et al. 2006). The Mg^{2+}-dependent signal cascade pathway activation by Cry toxins have been shown to be analogous to similar effect imposed by other pore forming toxins on their host cells when they are applied at subnanomolar concentration (Parker and Pattus 1993; Nelson et al. 1999; Menzies and Kourteva 2000; Soberón et al. 2009; Porta et al. 2011).

Though the two mechanisms of action seem to differ, with series of downstream events following on from toxin binding to receptors on target cell membranes, there is a degree of commonality in that initially the crystals have to be solubilised in vivo (Aronson et al. 1991; de Maagd et al. 2001; Soberón et al. 2009) or in vitro (Lambert et al. 1992; Bradley et al. 1995; Zhang et al. 2005, 2006) and activated by proteases before (Zhang et al. 2005, 2006) and/or after binding (Gómez et al. 2002; Bravo et al. 2004; Jiménez-Juárez et al. 2007; Soberón et al. 2009) to receptors such as cadherin. The midgut of lepidopteran and dipteran insects have been shown to be alkaline (Berebaum 1980; Gringorten et al. 1993) and this enhances the solubility of Cry toxins (Bravo et al. 2004; Soberón et al. 2009). Those of coleoptera are neutral or slightly acidic and in vitro solubilisation of Cry1Ba (Bradley et al. 1995) and Cry7Aa (Lambert et al. 1992) has been shown to enhance the activity of these toxins towards *Leptinotarsa decemlineata*.

With the pore forming model (Knowles and Ellar 1987; Haider and Ellar 1989; Grochulski et al. 1995; Schnepf et al. 1998; Bravo et al. 2004; Rausell et al. 2004), an ingested crystal toxin is solubilised in the alkaline environment of the insects midgut releasing protoxins which are initially processed by midgut proteases. The initial cleavage of a Cry1A protoxin by the gut proteases results in the removal of the C-terminal half and about 30 amino acid residues from the N-terminal thus releasing active toxin monomers which bind to receptors such as cadherin (Atsumi et al. 2008; Bel et al. 2009; Chen et al. 2009; Fabrick et al. 2009a; Muñóz-Garay et al. 2009; Obata et al. 2009; Pacheco et al. 2009a; Arenas et al. 2010) or proteins anchored to the membrane by GPI-anchored proteins such as aminopeptidase N (Arenas et al. 2010). The initial binding of the activated toxins to receptors is proposed to result in a conformational change which facilitates a second cleavage that removes the N-terminal helix α-1, by a membrane-bound protease. The removal of helix α-1 results in the formation of oligomers that are membrane insertion competent (Bravo et al. 2004). The binding of Cry toxins to the cadherin-like receptors have been shown to involve specific interactions of the variable loop regions in domain II and III with cadherin epitopes (Nair et al. 2008; Chen et al. 2009; Pacheco et al. 2009a; Soberón et al. 2009).

The oligomerised activated toxin that is bound to membrane receptors then inserts the central hydrophobic helix α-4 and 5 (Nair et al. 2008) into the apical membrane of midgut cells causing osmotic shock, bursting of the midgut cells and finally ending in the insect death (Knowles and Ellar 1987; Haider and Ellar 1989; Grochulski et al. 1995; Schnepf et al. 1998; Bravo et al. 2004; Rausell et al. 2004). The pore formation model as proposed by Bravo et al. (2004) for Cry1A toxins is presented in Fig. 2.3.

Cyt-toxins have also been shown to effect killing of its insect targets through unspecific binding to midgut membrane lipids followed by membrane insertion which leads to pore formation and insect death (Li et al. 1996; Cohen et al. 2008; Zhao et al. 2009; Rodriguez-Almazan et al. 2011).

The activation of Mg^{2+}-dependent signal cascade pathway that is triggered by the interaction of the monomeric 3-domain Cry toxin with the primary receptor, the cadherin protein (Zhang et al. 2005, 2006; Soberón et al. 2009) has been shown to trigger a pathway involving stimulation of the stimulatory G protein α-subunit and adenylyl cyclase (AC), increased cyclic adenosine monophosphate (cAMP) levels, and activation of protein kinase A (PKA). Activation of the AC/PKA signalling pathway initiates a series of cytological events that include membrane blebbing, appearance of nuclear ghosts, and cell swelling followed by cell lysis (Zhang et al. 2005, 2006).

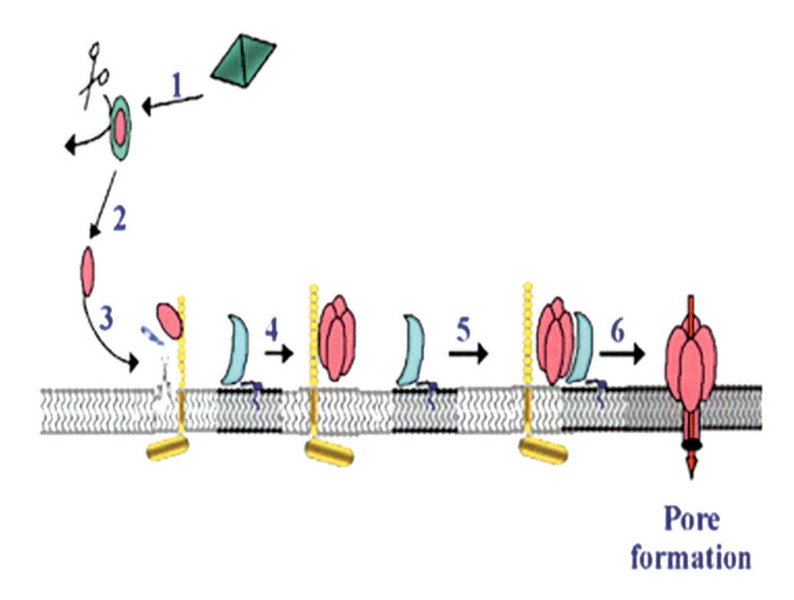

Fig. 2.3 Model of the mode of action of Cry1A toxins. *1* Crystal toxin solubilisation, *2* Initial cleavage by gut proteases, *3* Toxin monomer binding to receptors and second cleavage bymembrane bound protease, *4* Membrane insertion-competent oligomer formation, *5* Binding of oligomeric toxin to receptors, *6* Lytic pore formation. (Bravo et al. 2004)

2.2 Methods of Application of *Bt* and Its Products in Agriculture

Bacillus thuringiensis and its products have been formulated into various forms for application as biological control agents. Such formulations could be solid (powdery or granulated) or liquid. Presently there are over 400 of *Bt* based formulations that has been registered in the market and most of them contain insecticidal proteins and viable spores though the spores are inactivated in some products (Ahmedani et al. 2008). Formulated *Bt* products are applied directly in the form of sprays (Ali et al. 2010). An alternative, and highly successful, method for delivering the toxins to the target insect has been to express the toxin-encoding genes in transgenic plants (Barton et al. 1987; Vaeck et al. 1987; Qaim and Zilberman 2003; Walter et al. 2010; Chen et al. 2011).

2.3 Advantages of Using *Bt* Products Over Chemical Agents in Agricultural Practices

With their specific insecticidal effect on insect pests in the orders coleoptera (beetles and weevils) (López-Pazos et al. 2010; Sharma et al. 2010), diptera (flies and mosquitoes) (Pérez et al. 2007; Roh et al. 2010), hymenoptera (bees and wasps)

(Garcia-Robles et al. 2001; Sharma et al. 2008) and lepidoptera (butterflies and moths) (Baig et al. 2010; Darsi et al. 2010) and to non-insect species such as nematodes (Cappello et al. 2006; Hu et al. 2010), *Bt* toxins have taken centre stage as the major biological control agent and widely preferred to chemical insecticides.

Various assessments have been carried out to check for the safety of *Bt* toxins from sprays or transgenic plants to non-target species in the environment and it has been shown to be mostly environmentally friendly without significant adverse effects (Kapur et al. 2010; Walter et al. 2010; Chen et al. 2011; Randhawa et al. 2011) though there has recently been a laboratory observation that seemed to implicate a commercial *Bt aizawai* strain in the reduction of reproduction in bumblebee (*Bombus terrestris*) workers when applied at a concentration of 0.1% through sugar water and pollen (Mommaerts et al. 2010).

The increased popularity of biological control agents over synthetic chemicals is because of the non-selective lethal effect of the latter agents (Moser and Obrycki 2009; Kristoff et al. 2010; Shah and Iqbal 2010; Eriksson and Wiktelius 2011; Stevens et al. 2011) and the rapid development of resistance by insect pests to synthetic insecticides (Ahmad et al. 2008).

2.4 Threat to Continuous Use of *Bt* as Biological Control Agent

2.4.1 Development of Resistance and Cross Resistance

The continued relevance of *Bt* toxins in the control of insect and non-insect pests is threatened by the development of resistance by the pests in the field (Sayyed et al. 2004) and laboratory reared populations (Pereira et al. 2008; Fabrick et al. 2009b). There have been reports of insect populations resistant to a particular toxin showing resistance to other toxins to which they have not previously been exposed, a term known as 'cross-resistance' (Pereira et al. 2008; Sayyed et al. 2008; Gong et al. 2010; Xu et al. 2010).

There have been a number of proposed modes of resistance of insect pests to *Bt* toxins including reduction of binding of toxins to receptors in the midgut of insects, reduced solubilisation of protoxin, alteration of proteolytic processing of protoxins and toxin degration and or precipitation by proteases (Bruce et al. 2007). The understanding of the mechanism of action of *Bt* toxins (Knowles and Ellar 1987; Haider and Ellar 1989; Grochulski et al. 1995; Schnepf et al. 1998; Bravo et al. 2004; Rausell et al. 2004) have enhanced the experimental verification of some of the modes.

The most studied and experimentally verified mode of resistance is 'mode 1' which is characterized by recessive inheritance, reduced binding by at least one Cry1A toxin, and negligible cross-resistance to Cry1C (Tabashnik et al. 1998; Heckel et al. 2007).

Alteration of protease profile in the midgut of Cry1Ac resistant *Helicoverpa armigera* affected the proteolytic processing of Cry1Ac resulting in the production of 95 and 68 kDa toxin as opposed to the active 65 kDa toxin produced by midgut protease from susceptible population (Rajagopal et al. 2009) suggesting a linkage between improper processing of *Bt* toxin and development of resistance. Sayyed et al. (2005) also demonstrated that a field collected resistant population of *Plutella xylostella* (SERD4) which was subsequently selected in the laboratory using Cry-1Ab and named Cry1Ab-SEL was more sensitive to trypsin-activated Cry1Ab compared to Cry1Ab protoxins again suggesting a defect in processing although no such defect could be identified.

Brush border membrane vesicles from a laboratory selected population of *Ostrinia nubilalis* resistant to Cry1F were found binding the toxin as well as those from a susceptible population and furthermore no differences in activity of luminal gut proteases or proteolytic processing of the toxin were observed (Pereira et al. 2010). This failure to implicate defects in binding or toxin processing in the resistant strain indicates either alternative resistance mechanisms or limitations in the assays used.

2.4.2 *Narrow Spectrum of Activity*

Apart from resistance by pests being a major threat to the future of *Bt* products, the problem of efficacy and spectrum of activity (Regev et al. 1996; de Maagd et al. 2001) remain. In contrast to many synthetic insecticides most *Bt* toxins cloned have a narrow spectrum of activity (Kao et al. 2003; Shu et al. 2009). Only a small minority of toxins (such as Cry1Ba) show activity that spans two to three insect orders (Zhong et al. 2000).

2.5 Strategies to Ensure Continuing Use of *Bt* and Its Products in Agriculture

2.5.1 *Continuous Search for Bt Strains Expressing Toxins with Improved Activity*

The inability of many existing *Bacillus thuringiensis* toxins to overcome the resistance developed by insect species in field (Sayyed et al. 2004) and in laboratory reared populations (Pereira et al. 2008; Fabrick et al. 2009b) is problematic and efforts are continuing to search for *Bt* strains expressing novel toxins with improved activity.

Currently, about 600 insecticidal genes have been cloned from various *Bt* strains and their sequences deposited at the *Bt* toxin nomenclature website (http://www.lifesci.sussex.ac.uk/home/Neil_Crickmore/Bt/). Of these a large number have been heterologously expressed and found to be either independently (Song et al. 2003; Gonzalez-Cabrera et al. 2006; Ibargutxi et al. 2008; Xue et al. 2008; Hu et al. 2010) or in combination (Sharma et al. 2010) toxic to specific insect species in one or more orders.

2.5.2 Use of Synergism Between Bt Products or Between Bt Products and Other Substances

Most of the toxins cloned have narrow spectrum of activity (Kao et al. 2003; Shu et al. 2009) while some expressed toxins like Cyt1Aa show a weak toxicity to mosquitoes on their own but show synergistic activity when combined with other toxins like Cry4Ba and Cry11Aa (Fernandez-Luna et al. 2010).

To boost the efficacy of *Bt* insecticidal toxins and overcome resistance posed by insect pests, the use of other proteins like cadherin fragments have been shown to be a successful strategy (Chen et al. 2007; Abdullah et al. 2009; Pacheco et al. 2009b; Park et al. 2009; Peng et al. 2010). The use of a toxic compound, gossypol derived from the cotton plant has also been used in combination with Cry1Ac to boost its efficacy against a resistant population of *Helicoverpa zea* (Anilkumar et al. 2009). Co-expression of chitinase, an enzyme that is known to disrupt chitin present in the peritrophic membrane in the midgut of insects, has been shown to have an enhanced effect on the efficacy of Cry1Ac against *Helicoverpa armigera* (Ding et al. 2008) and Cry1C against *Spodoptera littoralis* (Regev et al. 1996). Also combinations of Cry toxins have proven to be a very useful strategy employed in boosting efficacy and fighting resistance (Jurat-Fuentes et al. 2003; Kaur 2006; Avisar et al. 2009). The combination of Cry1Ac and Cry2Ab showed a synergistic effect to *Helicoverpa armigera* (Ibargutxi et al. 2008).

It has also been shown that a mixture of crystal protein and spores from the same strain can result in a synergistic insecticidal activity (Johnson and McGaughey 1996; Tang et al. 1996; Johnson et al. 1998).

2.5.3 Genetic Manipulation of Existing Bt Strains and Its Toxin Genes for Improved Activity

With insights gained into the structure of Cry toxins (Li et al. 1991; Grochulski et al. 1995; Galitsky et al. 2001; Morse et al. 2001; Boonserm et al. 2005, 2006) and their mechanism of action (Zhang et al. 2005, 2006; Bravo et al. 2007; Gómez et al. 2007; Pacheco et al. 2009a) molecular genetics has been employed in an attempt to alter or broaden the activity of a given toxin. Herrero et al. (2004) demonstrated that replacing single residues in loops 2 and 3 of domain II with and residues 541–544 in domain III of Cry1Ca with alanine resulted in lower toxicity to *Spodoptera exigua* while its toxicity to *Manduca sexta* was not affected. Swapping domain III of three toxins with little or no activity against *Spodoptera exigua* (Cry1Ac, Cry1Ba and Cry1Ea) with domain III of Cry1Ca resulted in an improved toxicity towards this pest (de Maagd et al. 2000).

Abdullah et al. (2003) remodeled domain II loops of Cry4Ba to resemble that of Cry4Aa and generated mutants that showed improved toxicity to *Culex quinquefasciatus* and *Culex pipiens* of >700-fold and >285-fold respectively. In a similar study, Cry19Aa, a mosquitocidal toxin with specificity toward *Anopheles stephensi* and *Culex pipiens* but with no measurable activity against *Aedes aegypti*, was made more than 42,000-fold more toxic to *Aedes aegypti* by engineering domain II loops 1 and 2 to resemble that of Cry4Ba (Abdullah and Dean 2004). Also, Liu and Dean (2006) introduced mosquitocidal activity to Cry1Aa through rational design to the sequence of loops 1 and 2 based on a sequence alignment with Cry4Ba, a naturally occurring mosquitocidal toxin.

Replacement of domain III of Cry1Ba with Cry1Ac resulted in an improved activity to *Heliothis virescens* (Karlova et al. 2005). Naimov et al. (2001) created a hybrid protein by replacing domain II of Cry1Ba with that of Cry1I which resulted in activity against *Leptinotarsa decemlineata* that was comparable to that of Cry3Aa.

Site directed mutagenesis has great potential to alter toxin-encoding genes particularly when sufficient structural information is available to inform the choice of mutations (Cammack et al. 2006; Fleming et al. 2010; Moustafa et al. 2010; de Maagd et al. 2001, 2003). The understanding of the domain structure and function of Cry1Ac enhanced the use of site directed mutagenesis by Kim et al. (2008) to effect changes to domain I and II that resulted in mutants that showed improve activity to *Ostrinia furnacalis* and *Plutella xylostella*. In another example a triple mutation (N372A, A282G and L283S) in domain II loop of Cry1Ab resulted in a 36-fold increase in toxicity to *Lymantria dispar* and this correlated with an increased binding affinity of greater than 18-fold to brush border membrane vesicles which also resulted in higher toxin concentration at the binding site (Rajamohan et al. 1996).

Natural evolutionary trends have been used in analysing the divergence and host specificity in Cry toxins (de Maagd et al. 2001) and biotechnological techniques like gene shuffling has been used in artificially directing the evolution of new genes with novel characteristics (Stemmer 1994a, b; Zhao and Arnold 1997; Lassner and Bedbrook 2001; Craveiro et al. 2010). Craveiro et al. used DNA shuffling technique to produce four variants of Cry11A12synth and Cry11A12 that, unlike the parent toxins, had toxicity against *Telchin licus licus*. Shan et al. (2011) used error-prone PCR and staggered extension process (StEP) shuffling combined with Red/ET homologous recombination to investigate the insecticidal activity of Cry1Ac and isolated a toxin variant designated as T524N which has increased insecticidal activity against *Spodoptera exigua* larvae while its original insecticidal activity against *Helicoverpa armigera* larvae was retained.

The understanding of the regions that are bound to receptors and how protoxins are processed to a functional form has led to creation of manipulated toxins that mimic the in vivo processing of toxins (Pardo-López et al. 2009). Bravo et al. (2004) demonstrated that the processing of a protoxin to an active toxin in the midgut of a susceptible insect involves the initial cleavage of the protoxin by soluble proteases followed by a second cleavage by membrane bound proteases that removes helix α-1 while the toxin is bound to its receptor. Deletion of the amino-terminal region including helix α-1 of Cry1A toxins resulted in variants that formed oligomers in the absence of cadherin receptor and which killed insects that had developed resistance to Cry1A toxins through mutations in the cadherin gene. The modified toxins were also effective against insects which had acquired reduced susceptibility to native *Bt* toxins due to diminished expression of cadherin protein by cadherin gene silencing through RNA interference (Soberón et al. 2007). Mandal et al. (2007) demonstrated that in vitro truncation of Cry2Aa at the N-terminal of 42-amino acid residues resulted in an improved toxicity against *Spodoptera littoralis* and *Agrotis ipsilon* which was consistent with an observation made by Morse et al. (2001) that the structure of the Cry2Aa protoxin revealed a 49-amino acid N-terminal section preceding helix α-1 that was cleaved in vivo to generate an active toxin and which masked other parts of the toxin believed to interact with the surface of the target cell.

Apart from truncation of the N-terminal residues of domain I to generate toxins with improved activity, Wu et al. (2000) also created mutants (R345A, ΔY350, ΔY351) that involved deletions and specific mutations in loop I of Cry3A domain II which resulted in improved activity against *Tenebrio molitor*.

2.5.4 Use of Gene Stacking

This approach involves the expression of two or more *Bt* toxins with differing spectrum of activities and/or mechanism of action in transgenic plants to control insect pests which has the advantage of controlling pests from many orders as opposed to the narrow spectrum that a single toxin can control. It also has the advantage of reducing development of resistance because if the toxins used are such that there is little potential for cross-resistance between them, therefore there has to be resistance alleles at independent loci before an insect can develop resistance to the stacked toxins, which is a rare event (Gould 1998).

The effectiveness of this method relies on the fact that development of resistance by insect to the stacked toxins would be through similar mechanism e.g. reduced toxin binding and therefore if stacked toxins have different binding sites, it would be difficult for an insect to develop resistance to all the stacked toxins. This assumption has been challenged though as two toxins, Cry1Ac and Cry2Aa that are believed to have different binding sites in *Heliothis virescens* (tobacco budworm) have been shown to have cross-resistance (Jurat-Fuentes et al. 2003) which provokes a rethink on advantages of gene stacking.

References

Abdelkefi-Mesrati L, Boukedi H, Dammak-Karray M, Sellami-Boudawara T, Jaoua S, Tounsi S (2011) Study of the *Bacillus thuringiensis* Vip3Aa16 histopathological effects and determination of its putative binding proteins in the midgut of Spodoptera littoralis. J Invertebr Pathol 106(2):250–254

Abdullah MAF, Dean DH (2004) Enhancement of Cry19Aa mosquitocidal activity against *Aedes aegypti* by mutations in the putative loop regions of domain II. Appl Environ Microbiol 70(6):3769–3771

Abdullah MAF, Alzate O, Mohammad M, McNall RJ, Adang MJ, Dean DH (2003) Introduction of *Culex* toxicity into *Bacillus thuringiensis* Cry4Ba by protein engineering. Appl Environ Microbiol 69(9):5343–5353

Abdullah MAF, Moussa S, Taylor MD, Adang MJ (2009) Manduca sexta (Lepidoptera: Sphingidae) cadherin fragments function as synergists for Cry1A and Cry1C *Bacillus thuringiensis* toxins against noctuid moths *Helicoverpa zea*, *Agrotis ipsilon* and *Spodoptera exigua*. Pest Manag Sci 65(10):1097–1103

Ahmad M, Sayyed AH, Saleem MA (2008) Evidence for field evolved resistance to newer insecticides in *Spodoptera litura* (Lepidoptera: Noctuidae) from Pakistan. Crop Prot 27(10):1367–1372

Ahmedani MS, Haque MI, Afzal SN, Iqbal U, Naz S (2008) Scope of commercial formulations of *Bacillus thuringiensis berliner* as an alternative to methyl bromide against *Tribolium castaneum* adults. Pak J Bot 40(5):2149–2156

Ali S, Zafar Y, Ali GM, Nazir F (2010) *Bacillus thuringiensis* and its application in agriculture. Afr J Biotechnol 9(14):2022–2031

Anilkumar KJ, Sivasupramaniam S, Head G, Orth R, Van Santen E, Moar WJ (2009) Synergistic interactions between Cry1Ac and natural cotton defenses limit survival of Cry1Ac-resistant *Helicoverpa Zea* (Lepidoptera: Noctuidae) on *Bt* cotton. J Chem Ecol 35(7):785–795

Arenas I, Bravo A, Soberón M, Gómez I (2010) Role of alkaline phosphatase from *Manduca sexta* in the mechanism of action of *Bacillus thuringiensis* Cry1Ab toxin. J Biol Chem 285(17):12497–12503

Aronson AI, Han ES, McGaughey W, Johnson D (1991) The solubility of inclusion proteins from *Bacillus thuringiensis* is dependent upon protoxin composition and is a factor in toxicity to insects. Appl Environ Microbiol 57(4):981–986

Atsumi S, Inoue Y, Ishizaka T, Mizuno E, Yoshizawa Y, Kitami M, Sato R (2008) Location of the *Bombyx mori* 175 kDa cadherin-like protein-binding site on *Bacillus thuringiensis* Cry1Aa toxin. FEBS J 275(19):4913–4926

Avisar D, Eilenberg H, Keller M, Reznik N, Segal M, Sneh B, Zilberstein A (2009) The *Bacillus thuringiensis* delta-endotoxin Cry1C as a potential bioinsecticide in plants. Plant Sci 176(3):315–324

Baig DN, Bukhari DA, Shakoori AR (2010) Cry Genes profiling and the toxicity of isolates of *Bacillus thuringiensis* from soil samples against American bollworm, *Helicoverpa armigera*. J Appl Microbiol 109(6):1967–1978

Barton KA, Whiteley HR, Yang NS (1987) *Bacillus-thuringiensis* delta-endotoxin expressed in transgenic *Nicotiana tabacum* provides resistance to lepidopteran insects. Plant Physiol 85(4):1103–1109

Bel Y, Siqueira HAA, Siegfried BD, Ferré J, Escriche B (2009) Variability in the cadherin gene in an *Ostrinia nubilalis* strain selected for Cry1Ab resistance. Insect Biochem Mol Biol 39(3):218–223

Berbert-Molina MA, Prata AMR, Pessanha LG, Silveira MM (2008) Kinetics of *Bacillus thuringiensis* var. *israelensis* growth on high glucose concentrations. J Ind Microbiol Biotechnol 35(11):1397–1404

Berebaum M (1980) Adaptive significance of midgut pH in larval lepidoptera. Am Nat 115(1): 138–146

Bhalla R, Dalal M, Panguluri SK, Jagadish B, Mandaokar AD, Singh AK, Kumar PA (2005) Isolation, characterization and expression of a novel vegetative insecticidal protein gene of *Bacillus thuringiensis*. FEMS Microbiol Lett 243(2):467–472

Boonserm P, Davis P, Ellar DJ, Li J (2005) Crystal structure of the mosquito-iarvicidal toxin Cry-4Ba and its biological implications. J Mol Biol 348(2):363–382

Boonserm P, Mo M, Angsuthanasombat C, Lescar J (2006) Structure of the functional form of the mosquito larvicidal Cry4Aa toxin from *Bacillus thuringiensis* at a 2.8-Angstrom resolution. J Bacteriol 188(9):3391–3401

Bradley D, Harkey MA, Kim MK, Biever KD, Bauer LS (1995) The insecticidal cryIB crystal protein of *Bacillus thuringiensis* ssp. *thuringiensis* has dual specificity to coleopteran and lepidopteran larvae. J Invertebr Pathol 65(2):162–173

Brar SK, Verma M, Tyagi RD, Valéro JR, Surampalli RY (2009) Entomotoxicity, protease and chitinase activity of *Bacillus thuringiensis* fermented wastewater sludge with a high solids content. Bioresour Technol 100(19):4317–4325

Bravo A, Gómez I, Conde J, Muñoz-Garay C, Sánchez J, Miranda R, Zhuang M, Gill SS, Soberón M (2004) Oligomerization triggers binding of a *Bacillus thuringiensis* Cry1Ab pore-forming toxin to aminopeptidase N receptor leading to insertion into membrane microdomains. Biochim Biophys Acta 1667(1):38–46

Bravo A, Gill SS, Soberón M (2007) Mode of action of *Bacillus thuringiensis* Cry and Cyt toxins and their potential for insect control. Toxicon 49(4):423–435

Bruce MJ, Gatsi R, Crickmore N, Sayyed AH (2007) Mechanisms of resistance to *Bacillus thuringiensis* in the Diamondback Moth. Biopestic Int 3(1):1–12

Cammack R, Attwood TK, Campbell PN, Parish JH, Smith AD, Stirling JL, Vella F (2006) Oxford dictionary of biochemistry and molecular biology, 2nd edn. Oxford University Press, New York

Cappello M, Bungiro RD, Harrison LM, Bischof LJ, Griffitts JS, Barrows BD, Aroian RV (2006) A purified *Bacillus thuringiensis* crystal protein with therapeutic activity against the hookworm parasite Ancylostoma ceylanicum. Proc Natl Acad Sci USA 103(41):15154–15159

Chen J, Hua G, Jurat-Fuentes JL, Abdullah MA, Adang MJ (2007) Synergism of *Bacillus thuringiensis* toxins by a fragment of a toxin-binding cadherin. Proc Natl Acad Sci USA 104(35):13901–13906

Chen J, Aimanova KG, Fernandez LE, Bravo A, Soberón M, Gill SS (2009) *Aedes aegypti* cadherin serves as a putative receptor of the Cry11Aa toxin from *Bacillus thuringiensis* subsp. *israelensis*. Biochem J 424(2):191–200

Chen M, Shelton A, Ye GY (2011) Insect-resistant genetically modified rice in china: from research to commercialization. Annu Rev Entomol 56:81–101

Cohen S, Dym O, Albeck S, Ben-Dov E, Cahan R, Firer M, Zaritsky A (2008) High-resolution crystal structure of activated Cyt2Ba monomer from *Bacillus thuringiensis* subsp. israelensis. J Mol Biol 380(5):820–827

Craveiro KIC, Júnior JEG, Silva MCM, Macedo LLP, Lucena WA, Silva MS, Júnior JDADS, Oliveira GR, Magalhães MTQD, Santiago AD, Grossi-De-Sa MF (2010) Variant Cry1Ia toxins generated by DNA shuffling are active against sugarcane giant borer. J Biotechnol 145(3): 215–221

Crickmore N, Zeigler DR, Feitelson J, Schnepf E, Van Rie J, Lereclus D, Baum J, Dean DH (1998) Revision of the nomenclature for the *Bacillus thuringiensis* pesticidal crystal proteins. Microbiol Mol Biol Rev 62(3):807–813

Darsi S, Prakash GD, Udayasuriyan V (2010) Cloning and characterization of truncated cry1Ab gene from a new indigenous isolate of *Bacillus thuringiensis*. Biotechnol Lett 32(9):1311–1315

de Maagd RA, Weemen-Hendriks M, Stiekema W, Bosch D (2000) *Bacillus thuringiensis* delta-endotoxin Cry1C domain III can function as a specificity determinant for Spodoptera exigua in different, but not all, Cry1-Cry1C hybrids. Appl Environ Microbiol 66(4):1559–1563

de Maagd RA, Bravo A, Crickmore N (2001) How *Bacillus thuringiensis* has evolved specific toxins to colonize the insect world. Trends Genet 17(4):193–199 de Maagd RA, Bravo A, Berry C, Crickmore N, Schnepf HE (2003) Structure, diversity, and evolution of protein toxins from spore-forming entomopathogenic bacteria. Annu Rev Genet 37:409–433

Ding XZ, Luo ZH, Xia LQ, Gao B, Sun YJ, Zhang YM (2008) Improving the insecticidal activity by expression of a recombinant cry1Ac gene with chitinase-encoding gene in acrystalliferous *Bacillus thuringiensis*. Curr Microbiol 56(5):442–446

Donovan WP, Gonzalez JM Jr, Gilbert MP, Dankocsik C (1988) Isolation and characterization of EG2158, a new strain of *Bacillus thuringiensis* toxic to coleopteran larvae, and nucleotide sequence of the toxin gene. Mol Gen Genet 214(3):365–372

Donovan WP, Donovan JC, Engleman JT (2001) Gene knockout demonstrates that vip3A contributes to the pathogenesis of *Bacillus thuringiensis* toward *Agrotis ipsilon* and *Spodoptera exigua*. J Invertebr Pathol 78(1):45–51

Eriksson H, Wiktelius S (2011) Impact of chlorpyrifos used for desert locust control on non-target organisms in the vicinity of mangrove, an ecologically sensitive area. Int J Pest Manag 57(1):23–34

Estruch JJ, Warren GW, Mullins MA, Nye GJ, Craig JA, Koziel MG (1996) Vip3A, a novel *Bacillus thuringiensis* vegetative insecticidal protein with a wide spectrum of activities against lepidopteran insects. Proc Natl Acad Sci USA 93(11):5389–5394

Fabrick J, Oppert C, Lorenzen MD, Morris K, Oppert B, Jurat-Fuentes JL (2009a) A novel *Tenebrio molitor* cadherin is a functional receptor for *Bacillus thuringiensis* Cry3Aa toxin. J Biol Chem 284(27):18401–18410

Fabrick JA, Jech LF, Henneberry TJ (2009b) Novel pink bollworm resistance to the Bt toxin Cry-1Ac: effects on mating, oviposition, larval development and survival. J Insect Sci 9(24):1–8

Fedhila S, Gohar M, Slamti L, Nel P, Lereclus D (2003) The *Bacillus thuringiensis* PlcR-regulated gene inhA2 is necessary, but not sufficient, for virulence. J Bacteriol 185(9):2820–2825

Fernandez-Luna MT, Tabashnik BE, Lanz-Mendoza H, Bravo A, Soberón M, Miranda-Rios J (2010) Single concentration tests show synergism among *Bacillus thuringiensis* subsp *israelensis* toxins against the malaria vector mosquito *Anopheles albimanus*. J Invertebr Pathol 104(3):231–233

Fleming NI, Knower KC, Lazarus KA, Fuller PJ, Simpson ER, Clyne CD (2010) Aromatase is a direct target of FOXl2: C134W in granulosa cell tumors via a single highly conserved binding site in the ovarian specific promoter. PLoS One 5(12):e14389

Frankenhuyzen KV (2009) Insecticidal activity of *Bacillus thuringiensis* crystal proteins. J Invertebr Pathol 101(1):1–16

Galitsky N, Cody V, Wojtczak A, Ghosh D, Luft JR, Pangborn W, English L (2001) Structure of the insecticidal bacterial delta-endotoxin Cry3Bb1 of *Bacillus thuringiensis*. Acta Crystallogr D Biol Crystallogr 57:1101–1109

Garcia-Robles I, Sanchez J, Gruppe A, Martinez-Ramirez AC, Rausell C, Real MD, Bravo A (2001) Mode of action of *Bacillus thuringiensis* PS86Q3 strain in hymenopteran forest pests. Insect Biochem Mol Biol 31(9):849–856

Gómez I, Sánchez J, Miranda R, Bravo A, Soberón M (2002) Cadherin-like receptor binding facilitates proteolytic cleavage of helix α-1 in domain I and oligomer pre-pore formation of *Bacillus thuringiensis* Cry1Ab toxin. FEBS Lett 513(2–3):242–246

Gómez I, Pardo-Lopez L, Munoz-Garay C, Fernandez LE, Perez C, Sanchez J, Soberón M, Bravo A (2007) Role of receptor interaction in the mode of action of insecticidal Cry and Cyt toxins produced by *Bacillus thuringiensis*. Peptides 28(1):169–173

Gominet M, Slamti L, Gilois N, Rose M, Lereclus D (2001) Oligopeptide permease is required for expression of the *Bacillus thuringiensis* plcR regulon and for virulence. Mol Microbiol 40(4):963–975

Gong YJ, Wang CL, Yang YH, Wu SW, Wu YD (2010) Characterization of resistance to *Bacillus thuringiensis* toxin Cry1Ac in *Plutella xylostella* from China. J Invertebr Pathol 104(2):90–96

Gonzalez-Cabrera J, Farinos GP, Caccia S, Diaz-Mendoza M, Castanera P, Leonardi MG, Giordana B, Ferre J (2006) Toxicity and mode of action of *Bacillus thuringiensis* cry proteins in the Mediterranean corn borer, *Sesamia nonagrioides* (Lefebvre). Appl Environ Microbiol 72(4):2594–2600

Gould F (1998) Sustainability of transgenic insecticidal cultivars: integrating pest genetics and ecology. Annu Rev Entomol 43:701–726

Gringorten JL, Crawford DN, Harvey WR (1993) High pH in the ectoperitrophic space of the larval lepidopteran midgut. J Exp Biol 183:353–359

Grochulski P, Masson L, Borisova S, Pusztaicarey M, Schwartz JL, Brousseau R, Cygler M (1995) *Bacillus-thuringiensis* CryIA(a) insecticidal toxin crystal structure and channel formation. J Mol Biol 254(3):447–464

Guidelli-Thuler AM, De Abreu IL, Lemos MVF (2009) Expression of the sigma35 and *cry2ab* genes involved in *Bacillus thuringiensis* virulence. Sci Agric (Piracicaba, Braz) 66(3):403–409

Guo S, Ye S, Liu Y, Wei L, Xue J, Wu H, Song F, Zhang J, Wu X, Huang D, Rao Z (2009) Crystal structure of *Bacillus thuringiensis* Cry8Ea1: an insecticidal toxin toxic to underground pests, the larvae of *Holotrichia parallela*. J Struct Biol 168(2):259–266

Haider MZ, Ellar DJ (1989) Mechanism of action of *Bacillus thuringiensis* insecticidal δ-endotoxin: interaction with phospholipid vesicles. Biochim Biophys Acta Biomembr 978(2):216–222

Hajaij-Ellouze M, Fedhila S, Lereclus D, Nielsen-Leroux C (2006) The enhancin-like metalloprotease from the *Bacillus cereus* group is regulated by the pleiotropic transcriptional activator PlcR but is not essential for larvicidal activity. FEMS Microbiol Lett 260(1):9–16

Heckel DG, Gahan LJ, Baxter SW, Zhao JZ, Shelton AM, Gould F, Tabashnik BE (2007) The diversity of *Bt* resistance genes in species of Lepidoptera. J Invertebr Pathol 95(3):192–197

Herrero S, Gonzalez-Cabrera J, Ferre J, Bakker PL, Maagd RA de (2004) Mutations in the *Bacillus thuringiensis* Cry1Ca toxin demonstrate the role of domains II and III in specificity towards *Spodoptera exigua* larvae. Biochem J 384:507–513

Hilbert DW, Piggot PJ (2004) Compartmentalization of gene expression during *Bacillus subtilis* spore formation. Microbiol Mol Biol Rev 68(2):234–262

Hofte H, Whiteley HR (1989) Insecticidal crystal proteins of *Bacillus thuringiensis*. Microbiol Rev 53(2):242–255

Hofte H, Van Rie J, Jansens S, Van Houtven A, Vanderbruggen H, Vaeck M (1988) Monoclonal antibody analysis and insecticidal spectrum of three types of lepidopteran-specific insecticidal crystal proteins of *Bacillus thuringiensis*. Appl Environ Microbiol 54(8):2010–2017

Hu Y, Georghiou SB, Kelleher AJ, Aroian RV (2010) *Bacillus thuringiensis* Cry5B protein is highly efficacious as a single-dose therapy against an intestinal roundworm infection in mice. PLoS Negl Trop Dis 4(3):e614

Ibargutxi MA, Muñoz D, Escudero IRD, Caballero P (2008) Interactions between Cry1Ac, Cry2Ab, and Cry1Fa *Bacillus thuringiensis* toxins in the cotton pests *Helicoverpa armigera* (Hübner) and *Earias insulana* (Boisduval). Biol Control 47(1):89–96

Infante I, Morel MA, Ubalde MC, Martínez-Rosales C, Belvisi S, Castro-Sowinski S (2010) Wool-degrading Bacillus isolates: extracellular protease production for microbial processing of fabrics. World J Microbiol Biotechnol 26(6):1047–1052

Jiménez-Juárez N, Muñoz-Garay C, Gómez I, Saab-Rincon G, Damian-Almazo JY, Gill SS, Soberón M, Bravo A (2007) *Bacillus thuringiensis* Cry1Ab mutants affecting oligomer formation are non-toxic to *Manduca sexta* larvae. J Biol Chem 282(29):21222–21229

Johnson DE, McGaughey WH (1996) Contribution of *Bacillus thuringiensis* spores to toxicity of purified cry proteins towards indianmeal moth larvae. Curr Microbiol 33(1):54–59

Johnson DE, Oppert B, McGaughey WH (1998) Spore coat protein synergizes *Bacillus thuringiensis* crystal toxicity for the Indianmeal moth (*Plodia interpunctella*). Curr Microbiol 36(5): 278–282

Jurat-Fuentes JL, Gould FL, Adang MJ (2003) Dual resistance to *Bacillus thuringiensis* Cry1Ac and Cry2Aa toxins in heliothis virescens suggests multiple mechanisms of resistance. Appl Environ Microbiol 69(10):5898–5906

Kao SS, Hsieh FC, Tzeng CC, Tsai YS (2003) Cloning and expression of the insecticidal crystal protein gene *cry1Ca9* of *Bacillus thuringiensis* G10-01A from Taiwan granaries. Curr Microbiol 47(4):295–299

Kapur M, Bhatia R, Pandey G, Pandey J, Paul D, Jain RK (2010) A case study for assessment of microbial community dynamics in genetically modified Bt cotton crop fields. Curr Microbiol 61(2):118–124

Karlova R, Weemen-Hendriks M, Naimov S, Ceron J, Dukiandjiev S, Maagd RA de (2005) *Bacillus thuringiensis* δ-endotoxin Cry1Ac domain III enhances activity against *Heliothis virescens* in some, but not all Cry1-Cry1Ac hybrids. J Invertebr Pathol 88(2):169–172

Kaur S (2006) Molecular approaches for identification and construction of novel insecticidal genes for crop protection. World J Microbiol Biotechnol 22(3):233–253

Kim YS, Roh JY, Kang JN, Wang Y, Shim HJ, Li MS, Choi JY, Je YH (2008) Mutagenesis of *Bacillus thuringiensis cry1Ac* gene and its insecticidal activity against *Plutella xylostella* and *Ostrinia furnacalis*. Biol Control 47(2):222–227

Knowles BH, Ellar DJ (1987) Colloid-osmotic lysis is a general feature of the mechanism of action of *Bacillus thuringiensis* δ-endotoxins with different insect specificity. Biochim Biophys Acta Gen Subj 924(3):509–518

Kristoff G, Guerrero NRV, Cochón AC (2010) Inhibition of cholinesterases and carboxylesterases of two invertebrate species, *Biomphalaria glabrata* and *Lumbriculus variegatus*, by the carbamate pesticide carbaryl. Aquat Toxicol 96(2):115–123

Kronstad JW, Whiteley HR (1986) Three classes of homologous *Bacillus thuringiensis* crystal-protein genes. Gene 43(1–2):29–40

Lambert B, Hofte H, Annys K, Jansens S, Soetaert P, Peferoen M (1992) Novel *Bacillus thuringiensis* insecticidal crystal protein with a silent activity against coleopteran larvae. Appl Environ Microbiol 58(8):2536–2542

Lassner M, Bedbrook J (2001) Directed molecular evolution in plant improvement. Curr Opin Plant Biol 4(2):152–156

Lee X, Thompson A, Zhang Z, Ton-That H, Biesterfeldt J, Ogata C, Xu L, Johnston RAZ, Young NM (1998) Structure of the complex of *Maclura pomifera* agglutinin and the T- antigen disaccharide, Galβ1,3GalNAc. J Biol Chem 273(11):6312–6318

Lesieur C, Frutiger S, Hughes G, Kellner R, Pattus F, Van Der Goott FG (1999) Increased stability upon heptamerization of the pore-forming toxin aerolysin. J Biol Chem 274(51):36722–36728

Leuber M, Orlik F, Schiffler B, Sickmann A, Benz R (2006) Vegetative insecticidal protein (Vip1Ac) of *Bacillus thuringiensis* HD201: evidence for oligomer and channel formation. Biochemistry 45(1):283–288

Li JD, Carroll J, Ellar DJ (1991) Crystal-structure of insecticidal delta-endotoxin from *Bacillus-thuringiensis* at 2.5-A resolution. Nature 353(6347):815–821

Li J, Koni PA, Ellar DJ (1996) Structure of the mosquitocidal δ-endotoxin CytB from *Bacillus thuringiensis* sp. *kyushuensis* and implications for membrane pore formation. J Mol Biol 257(1):129–152

Liu XS, Dean DH (2006) Redesigning *Bacillus thuringiensis* Cry1Aa toxin into a mosquito toxin. Protein Eng Des Sel 19(3):107–111

López-Pazos SA, Rojas Arias AC, Ospina SA, Cerón J (2010) Activity of *Bacillus thuringiensis* hybrid protein against a lepidopteran and a coleopteran pest. FEMS Microbiol Lett 302(2): 93–98

Mandal CC, Gayen S, Basu A, Ghosh KS, Dasgupta S, Maiti MK, Sen SK (2007) Prediction-based protein engineering of domain I of Cry2A entomocidal toxin of *Bacillus thuringiensis* for the enhancement of toxicity against lepidopteran insects. Protein Eng Des Sel 20(12):599–606

Martin PAW, Gundersen-Rindal DE, Blackburn MB (2010) Distribution of phenotypes among *Bacillus thuringiensis* strains. Syst Appl Microbiol 33(4):204–208

Menzies BE, Kourteva I (2000) Staphylococcus aureus alpha-toxin induces apoptosis in endothelial cells. FEMS Immunol Med Microbiol 29(1):39–45

Milne R, Liu Y, Gauthier D, Frankenhuyzen KV (2008) Purification of Vip3Aa from *Bacillus thuringiensis* HD-1 and its contribution to toxicity of HD-1 to spruce budworm (*Choristoneura fumiferana*) and gypsy moth (*Lymantria dispar*) (Lepidoptera). J Invertebr Pathol 99(2): 166–172

Milner RJ (1994) History of *Bacillus thuringiensis*. Agric Ecosyst Environ 49(1):9–13

Mommaerts V, Jans K, Smagghe G (2010) Impact of *Bacillus thuringiensis* strains on survival, reproduction and foraging behaviour in bumblebees (*Bombus terrestris*). Pest Manag Sci 66(5):520–525

Morse RJ, Yamamoto T, Stroud RM (2001) Structure of Cry2Aa suggests an unexpected receptor binding epitope. Structure 9(5):409–417

Moser SE, Obrycki JJ (2009) Non-target effects of neonicotinoid seed treatments; mortality of coccinellid larvae related to zoophytophagy. Biol Control 51(3):487–492

Moustafa DA, Jain N, Sriranganathan N, Vemulapalli R (2010) Identification of a single-nucleotide insertion in the promoter region affecting the sodc promoter activity in *Brucella neotomae*. PLoS One 5(11):e14112

Mueller M, Grauschopf U, Maier T, Glockshuber R, Ban N (2009) The structure of a cytolytic α-helical toxin pore reveals its assembly mechanism. Nature 459(7247):726–730

Muñóz-Garay C, Portugal L, Pardo-López L, Jiménez-Juárez N, Arenas I, Gómez I, Sánchez-López R, Arroyo R, Holzenburg A, Savva CG, Soberón M, Bravo A (2009) Characterization of the mechanism of action of the genetically modified Cry1AbMod toxin that is active against Cry1Ab-resistant insects. Biochim Biophys Acta Biomembr 1788(10):2229–2237

Naimov S, Weemen-Hendriks M, Dukiandjiev S, Maagd RA de (2001) *Bacillus thuringiensis* delta-endotoxin Cry1 hybrid proteins with increased activity against the Colorado potato beetle. Appl Environ Microbiol 67(11):5328–5330

Nair MS, Xinyan SL, Dean DH (2008) Membrane insertion of the *Bacillus thuringiensis* Cry1Ab toxin: single mutation in domain II block partitioning of the toxin into the brush border membrane. Biochemistry 47(21):5814–5822

Nelson KL, Brodsky RA, Buckley JT (1999) Channels formed by subnanomolar concentrations of the toxin aerolysin trigger apoptosis of T lymphomas. Cell Microbiol 1(1):69–74

Nisnevitch M, Sigawi S, Cahan R, Nitzan Y (2010) Isolation, characterization and biological role of camelysin from *Bacillus thuringiensis* subsp. *israelensis*. Curr Microbiol 61(3):176–183

Obata F, Kitami M, Inoue Y, Atsumi S, Yoshizawa Y, Sato R (2009) Analysis of the region for receptor binding and triggering of oligomerization on *Bacillus thuringiensis* Cry1Aa toxin. FEBS J 276(20):5949–5959

Pacheco S, Gómez I, Arenas I, Saab-Rincon G, Rodríguez-Almazán C, Gill SS, Bravo A, Soberón M (2009a) Domain II loop 3 of *Bacillus thuringiensis* Cry1Ab toxin is involved in a "Ping Pong" binding mechanism with *Manduca sexta* aminopeptidase-N and cadherin receptors. J Biol Chem 284(47):32750–32757

Pacheco S, Gómez I, Gill SS, Bravo A, Soberón M (2009b) Enhancement of insecticidal activity of *Bacillus thuringiensis* Cry1A toxins by fragments of a toxin-binding cadherin correlates with oligomer formation. Peptides 30(3):583–588

Palvannan T, Boopathy R (2005) Phosphatidylinositol-specific phospholipase C production from *Bacillus thuringiensis* serovar. *kurstaki* using potato-based media. World J Microbiol Biotechnol 21(6–7):1153–1155

Pardo-López L, Muñoz-Garay C, Porta H, Rodríguez-Almazán C, Soberón M, Bravo A (2009) Strategies to improve the insecticidal activity of Cry toxins from *Bacillus thuringiensis*. Peptides 30(3):589–595

Park Y, Abdullah MAF, Taylor MD, Rahman K, Adang MJ (2009) Enhancement of *Bacillus thuringiensis* Cry3Aa and Cry3Bb toxicities to coleopteran larvae by a toxin-binding fragment of an insect cadherin. Appl Environ Microbiol 75(10):3086–3092

Parker MW, Pattus F (1993) Rendering a membrane protein soluble in water: a common packing motif in bacterial protein toxins. Trends Biochem Sci 18(10):391–395

Peng DH, Xu XH, Ruan LF, Yu ZN, Sun M (2010) Enhancing Cry1Ac toxicity by expression of the *Helicoverpa armigera* cadherin fragment in *Bacillus thuringiensis*. Res Microbiol 161(5):383–389

Pereira EJG, Lang BA, Storer NP, Siegfried BD (2008) Selection for Cry1F resistance in the European corn borer and cross-resistance to other Cry toxins. Entomol Exp Appl 126(2):115–121

Pereira EJG, Siqueira HAA, Zhuang M, Storer NP, Siegfried BD (2010) Measurements of Cry1F binding and activity of luminal gut proteases in susceptible and Cry1F resistant *Ostrinia nubilalis* larvae (Lepidoptera: Crambidae). J Invertebr Pathol 103(1):1–7

Pérez C, Muñoz-Garay C, Portugal LC, Sánchez J, Gill SS, Soberón M, Bravo A (2007) *Bacillus thuringiensis* ssp. *israelensis* Cyt1Aa enhances activity of Cry11Aa toxin by facilitating the formation of a pre-pore oligomeric structure. Cell Microbiol 9(12):2931–2937

Pérez-García G, Basurto-Ríos R, Ibarra JE (2010) Potential effect of a putative σH-driven promoter on the over expression of the Cry1Ac toxin of *Bacillus thuringiensis*. J Invertebr Pathol 104(2):140–146

Pigott CR, Ellar DJ (2007) Role of receptors in *Bacillus thuringiensis* crystal toxin activity. Microbiol Mol Biol Rev 71(2):255–281

Porta H, Cancino-Rodezno A, Soberón M, Bravo A (2011) Role of MAPK p38 in the cellular responses to pore-forming toxins. Peptides 32(3):601–606

Qaim M, Zilberman D (2003) Yield effects of genetically modified crops in developing countries. Science 299(5608):900–902

Rajagopal R, Arora N, Sivakumar S, Rao NGV, Nimbalkar SA, Bhatnagar RK (2009) Resistance of *Helicoverpa armigera* to Cry1Ac toxin from *Bacillus thuringiensis* is due to improper processing of the protoxin. Biochem J 419 309–316

Rajamohan F, Alzate O, Cotrill JA, Curtiss A, Dean DH (1996) Protein engineering of *Bacillus thuringiensis* delta-endotoxin: mutations at domain II of CryIAb enhance receptor affinity and toxicity toward gypsy moth larvae. Proc Natl Acad Sci USA 93(25):14338–14343

Ramarao N, Lereclus D (2006) Adhesion and cytotoxicity of *Bacillus cereus* and *Bacillus thuringiensis* to epithelial cells are FlhA and PlcR dependent, respectively. Microbes Infect 8(6): 1483–1491

Randhawa GJ, Singh M, Grover M (2011) Bioinformatic analysis for allergenicity assessment of *Bacillus thuringiensis* Cry proteins expressed in insect-resistant food crops. Food Chem Toxicol 49(2):356–362

Rasko DA, Altherr MR, Han CS, Ravel J (2005) Genomics of the *Bacillus cereus* group of organisms. FEMS Microbiol Rev 29(2):303–329

Rausell C, Pardo-López L, Sánchez J, Muñoz-Garay C, Morera C, Soberón M, Bravo A (2004) Unfolding events in the water-soluble monomeric Cry1Ab toxin during transition to oligomeric pre-pore and membrane-inserted pore channel. J Biol Chem 279(53):55168–55175

Read TD, Peterson SN, Tourasse N, Baillie LW, Paulsen IT, Nelson KE, Tettelin H, Fouts DE, Eisen JA, Gill SR, Holtzapple EK, Okstad OA, Helgason E, Rilstone J, Wu M, Kolonay JF, Beanan MJ, Dodson RJ, Brinkac LM, Gwinn M, Deboy RT, Madpu R, Daugherty SC, Durkin AS, Haft DH, Nelson WC, Peterson JD, Pop M, Khouri HM, Radune D, Benton JL, Mahamoud Y, Jiang LX, Hance IR, Weidman JF, Berry KJ, Plaut RD, Wolf AM, Watkins KL, Nierman WC, Hazen A, Cline R, Redmond C, Thwaite JE, White O, Salzberg SL, Thomason B, Friedlander AM, Koehler TM, Hanna PC, Kolsto AB, Fraser CM (2003) The genome sequence of *Bacillus anthracis* Ames and comparison to closely related bacteria. Nature 423(6935):81–86

Regev A, Keller M, Strizhov N, Sneh B, Prudovsky E, Chet I, Ginzberg I, Konczkalman Z, Koncz C, Schell J, Zilberstein A (1996) Synergistic activity of a *Bacillus thuringiensis* delta-endotoxin and a bacterial endochitinase against *Spodoptera littoralis* larvae. Appl Environ Microbiol 62(10):3581–3586

Rodriguez-Almazan C, Ruiz De Escudero I, Cantón PE, Muñoz-Garay C, Pérez C, Gill SS, Soberón M, Bravo A (2011) The amino- and carboxyl-terminal fragments of the *Bacillus thuringensis* Cyt1Aa toxin have differential roles in toxin oligomerization and pore formation. Biochemistry 50(3):388–396

Roh JY, Kim YS, Wang Y, Liu Q, Tao X, Xu HG, Shim HJ, Choi JY, Lee KS, Jin BR, Je YH (2010) Expression of *Bacillus thuringiensis* mosquitocidal toxin in an antimicrobial *Bacillus brevis* strain. J Asia Pac Entomol 13(1):61–64

Salamitou S, Ramisse F, Brehelin M, Bourguet D, Gilois N, Gominet M, Hernandez E, Lereclus D (2000) The plcR regulon is involved in the opportunistic properties of *Bacillus thuringiensis* and *Bacillus cereus* in mice and insects. Microbiology 146(11):2825–2832

Sanchis V, Lereclus D, Menou G, Chaufaux J, Guo S, Lecadet MM (1989) Nucleotide sequence and analysis of the N-terminal coding region of the Spodoptera-active delta-endotoxin gene of *Bacillus thuringiensis aizawai* 7.29. Mol Microbiol 3(2):229–238

Sayyed AH, Raymond B, Ibiza-Palacios MS, Escriche B, Wright DJ (2004) Genetic and biochemical characterization of field-evolved resistance to *Bacillus thuringiensis* toxin Cry1Ac in the diamondback moth, *Plutella xylostella*. Appl Environ Microbiol 70(12):7010–7017

Sayyed AH, Gatsi R, Sales Ibiza-Palacios M, Escriche B, Wright DJ,Crickmore N (2005) Common, but complex, mode of resistance of *Plutella xylostella* to *Bacillus thuringiensis* toxins Cry1Ab and Cry1Ac. Appl Environ Microbiol 71(11):6863–6869

Sayyed AH, Moores G, Crickmore N, Wright DJ (2008) Cross-resistance between a *Bacillus thuringiensis* Cry toxin and non-Bt insecticides in the diamondback moth. Pest Manag Sci 64(8):813–819

Schnepf HE, Whiteley HR (1981) Cloning and expression of the *Bacillus thuringiensis* crystal protein gene in *Escherichia coli*. Proc Natl Acad Sci USA 78(5):2893–2897

Schnepf E, Crickmore N, Van Rie J, Lereclus D, Baum J, Feitelson J, Zeigler DR, Dean DH (1998) *Bacillus thuringiensis* and its pesticidal crystal proteins. Microbiol Mol Biol Rev 62(3): 775–806

Sedlak M, Walter T, Aronson A (2000) Regulation by overlapping promoters of the rate of synthesis and deposition into crystalline inclusions of *Bacillus thuringiensis* δ-endotoxins. J Bacteriol 182(3):734–741

Shah MD, Iqbal M (2010) Diazinon-induced oxidative stress and renal dysfunction in rats. Food Chem Toxicol 48(12):3345–3353

Shan S, Zhang Y, Ding X, Hu S, Sun Y, Yu Z, Liu S, Zhu Z, Xia L (2011) A Cry1Ac toxin variant generated by directed evolution has enhanced toxicity against lepidopteran insects. Curr Microbiol 62(2):358–365

Sharma HC, Dhillon MK, Arora R (2008) Effects of *Bacillus thuringiensis* delta-endotoxin-fed *Helicoverpa armigera* on the survival and development of the parasitoid *Campoletis chlorideae*. Entomol Exp Appl 126(1):1–8

Sharma P, Nain V, Lakhanpaul S, Kumar PA (2010) Synergistic activity between *Bacillus thuringiensis* Cry1Ab and Cry1Ac toxins against maize stem borer (*Chilo partellus Swinhoe*). Lett Appl Microbiol 51(1):42–47

Shi Y, Xu W, Yuan M, Tang M, Chen J, Pang Y (2004) Expression of vipl/vip2 genes in *Escherichia coli* and *Bacillus thuringiensis* and the analysis of their signal peptides. J Appl Microbiol 97(4):757–765

Shimizu T, Vassylyev DG, Kido S, Doi Y, Morikawa K (1994) Crystal structure of vitelline membrane outer layer protein I (VMO-I): a folding motif with homologous Greek key structures related by an internal three-fold symmetry. EMBO J 13(5):1003–1010

Shu C, Yan G, Wang R, Zhang J, Feng S, Huang D, Song F (2009) Characterization of a novel cry8 gene specific to Melolonthidae pests: *Holotrichia oblita* and *Holotrichia parallela*. Appl Microbiol Biotechnol 84(4):701–707

Singh G, Sachdev B, Sharma N, Seth R, Bhatnagar RK (2010) Interaction of *Bacillus thuringiensis* vegetative insecticidal protein with ribosomal S2 protein triggers larvicidal activity in *Spodoptera frugiperda*. Appl Environ Microbiol 76(21):7202–7209

Smith GP, Ellar DJ (1994) Mutagenesis of two surface-exposed loops of the *Bacillus thuringiensis* CryIC δ-endotoxin affects insecticidal specificity. Biochem J 302(2):611–616

Soberón M, Pardo-López L, López I, Gómez I, Tabashnik BE, Bravo A (2007) Engineering modified Bt toxins to counter insect resistance. Science 318(5856):1640–1642

Soberón M, Gill SS, Bravo A (2009) Signaling versus punching hole: how do *Bacillus thuringiensis* toxins kill insect midgut cells? Cell Mol Life Sci 66(8):1337–1349

Song FP, Zhang J, Gu AX, Wu Y, Han LL, He KL, Chen ZY, Yao J, Hu YQ, Li GX, Huang DF (2003) Identification of cry1I-type genes from *Bacillus thuringiensis* strains and characterization of a novel cry1I-type gene. Appl Environ Microbiol 69(9):5207–5211

Stemmer WPC (1994a) DNA shuffling by random fragmentation and reassembly: in vitro recombination for molecular evolution. Proc Natl Acad Sci USA 91(22):10747–10751

Stemmer WPC (1994b) Rapid evolution of a protein in vitro by DNA shuffling. Nature 370(6488):389–391

Stevens MM, Burdett AS, Mudford EM, Helliwell S, Doran G (2011) The acute toxicity of fipronil to two non-target invertebrates associated with mosquito breeding sites in Australia. Acta Trop 117(2):125–130

Tabashnik BE, Liu YB, Malvar T, Heckel DG, Masson L, Ferré J (1998) Insect resistance to *Bacillus thuringiensis*: uniform or diverse? Philos Trans R Soc B Biol Sci 353(1376):1751–1756

Tang JD, Shelton AM, Van Rie J, De Roeck S, Moar WJ, Roush RT, Peferoen M (1996) Toxicity of *Bacillus thuringiensis* spore and crystal protein to resistant diamondback moth (*Plutella xylostella)*. Appl Environ Microbiol 62(2):564–569

Vaeck M, Reynaerts A, Hofte H, Jansens S, Debeuckeleer M, Dean C, Zabeau M, Vanmontagu M, Leemans J (1987) Transgenic plants protected from insect attack. Nature 328(6125):33–37

Walter C, Fladung M, Boerjan W (2010) The 20-year environmental safety record of GM trees. Nat Biotechnol 28(7):656–658

Wu SJ, Koller CN, Miller DL, Bauer LS, Dean DH (2000) Enhanced toxicity of *Bacillus thuringiensis* Cry3A δ-endotoxin in coleopterans by mutagenesis in a receptor binding loop. FEBS Lett 473(2):227–232

Xia L, Sun Y, Ding X, Fu Z, Mo X, Zhang H, Yuan Z (2005) Identification of cry-type genes on 20-kb DNA associated with Cry1 crystal proteins from *Bacillus thuringiensis*. Curr Microbiol 51(1):53–58

Xu L, Wang Z, Zhang J, He K, Ferry N, Gatehouse AMR (2010) Cross-resistance of Cry1Ab-selected Asian corn borer to other Cry toxins. J Appl Entomol 134(5):429–438

Xue J, Liang G, Crickmore N, Li H, He K, Song F, Feng X, Huang D, Zhang J (2008) Cloning and characterization of a novel Cry1A toxin from *Bacillus thuringiensis* with high toxicity to the Asian corn borer and other lepidopteran insects. FEMS Microbiol Lett 280(1):95–101

Zhang X, Candas M, Griko NB, Rose-Young L, Bulla LA (2005) Cytotoxicity of *Bacillus thuringiensis* Cry1Ab toxin depends on specific binding of the toxin to the cadherin receptor BT-R-1 expressed in insect cells. Cell Death Differ 12(11):1407–1416

Zhang XB, Candas M, Griko NB, Taussig R, Bulla LA (2006) A mechanism of cell death involving an adenylyl cyclase/PKA signaling pathway is induced by the Cry1Ab toxin of *Bacillus thuringiensis*. Proc Natl Acad Sci U S A 103(26):9897–9902

Zhao HM, Arnold FH (1997) Optimization of DNA shuffling for high fidelity recombination. Nucleic Acids Res 25(6):1307–1308

Zhao X, Xia L, Ding X, Yu Z, Lü Y, Tao W (2009) Homology modeling of Cyt2Ca1 of *Bacillus thuringiensis* and its molecular docking with inositol monophosphate. Chin J Chem 27(10):2085–2089

Zhong CH, Ellar DJ, Bishop A, Johnson C, Lin SS, Hart ER (2000) Characterization of a *Bacillus thuringiensis* delta-endotoxin which is toxic to insects in three orders. J Invertebr Pathol 76(2):131–139

Chapter 3
Risk Assessment of Bt Transgenic Crops

Sarvjeet Kaur

Abstract The global demand, likely to escalate for at least another 40 years, requires a multifaceted strategy to ensure food security. Augmentation of crop yields in a sustainable manner is essential. Pests destroy on an average 14–25% of the total global agricultural production. Biopesticides are an important component of integrated pest management (IPM) strategies employed for minimization of insect pests-incurred crop yield losses as also for reduction of use of environmentally harmful chemical pesticides. *Bacillus thuringiensis* (Bt) is an aerobic, gram-positive, spore-forming bacterium producing crystal proteins (Cry), which are selectively toxic to target insects. Cry proteins act by insertion into the microvillar brush-border membranes in the midgut of susceptible insects, leading to disruption of osmotic balance, lysis of epithelial cells and eventually death of insect. Some Bt strains additionally secrete vegetative insecticidal proteins (Vips), which cause toxicity in susceptible insects by midgut epithelial cell lysis and gut paralysis. Bt has been used as a microbial biopesticide for the past five decades because of the advantages of specific toxicity against target insects, lack of polluting residues and safety to non-target organisms, and accounts for 95% of the 1% market share of biopesticides in the total pesticide market. However, the use of Bt microbial biopesticide formulations has been rather limited due to the problems of narrow host range, low persistence on plants and inability of foliar application to reach the insects feeding inside the plants, notwithstanding several biotechnological approaches for the development of improved Bt biopesticides. Bt transgenic crops have been developed to overcome the problems of Bt biopesticides and for more effective insect control. Risk assessment in relation to certain concerns raised about environmental and food safety of Bt transgenic crops has been addressed in this chapter.

Keywords Risk assessment · Effects · Persistence of Bt · Degradation of Bt · Resistance

---

S. Kaur (✉)
National Research Centre on Plant Biotechnology, Indian Agricultural Research Institute Campus, New Delhi 110012, India
e-mail: dr_sarvjeetkaur@yahoo.com

E. Sansinenea (ed.), *Bacillus thuringiensis Biotechnology*,
DOI 10.1007/978-94-007-3021-2_3, © Springer Science+Business Media B.V. 2012

3.1 Introduction

The global demand, likely to escalate for at least another 40 years, requires a multifaceted strategy to ensure food security (Godfray et al. 2010). Augmentation of crop yields in a sustainable manner is essential. Pests destroy on an average 14–25% of the total global agricultural production (DeVilliers and Hoisington 2011). Biopesticides are an important component of integrated pest management (IPM) strategies employed for minimization of insect pests-incurred crop yield losses as also for reduction of use of environmentally harmful chemical pesticides. *Bacillus thuringiensis* (Bt) is an aerobic, gram-positive, spore-forming bacterium producing crystal proteins (Cry), which are selectively toxic to target insects (Feitelson et al. 1992; García-Robles et al. 2001; de Escudero et al. 2006). Cry proteins act by insertion into the microvillar brush-border membranes in the midgut of susceptible insects, leading to disruption of osmotic balance, lysis of epithelial cells and eventually death of insect (Schnepf et al. 1998). Some Bt strains additionally secrete vegetative insecticidal proteins (Vips), which cause toxicity in susceptible insects by midgut epithelial cell lysis and gut paralysis (Lee et al. 2003). Bt has been used as a microbial biopesticide for the past five decades because of the advantages of specific toxicity against target insects, lack of polluting residues and safety to non-target organisms, and accounts for 95% of the 1% market share of biopesticides in the total pesticide market (Flexner and Belnavis 1999). However, the use of Bt microbial biopesticide formulations has been rather limited due to the problems of narrow host range, low persistence on plants and inability of foliar application to reach the insects feeding inside the plants, notwithstanding several biotechnological approaches for the development of improved Bt biopesticides (Kaur 2000).

The problems of field application of Bt biopesticides have been overcome by Bt transgenic crops which were first developed using native *cry*1Ac gene (Barton et al. 1987). However, due to premature polyadenylation of bacterial gene in plants, the *in planta* expression of wild type *cry* gene was low (Diehn et al. 1998). Subsequently, synthetic *cry* genes with plant-preferred codon usage have been used to increase the expression of *cry* genes *in planta*. Bt transgenic corn, cotton, tomato, brinjal, rice, potato and soybean have been developed. Transgenic potatoes producing Cry3A protein for the control of Colorado potato beetle in 1995; and maize and cotton producing Cry1A protein for the control of various lepidopteran pests in 1996 were first cultivated in the United States of America. Thereafter, transgenic crops expressing Cry1F and Vip3a for the control of Lepidoptera in maize and cotton, Cry2Ab for the control of Lepidoptera in cotton, and Cry3Bb, Cry34/35 and Cry3A for the control of corn rootworm in maize were given regulatory approvals for cultivation (Gatehouse 2008).

The total acreage of transgenic crops has been steadily increasing with countries including 10 developed and 19 developing countries having adopted commercial cultivation of transgenic crops on 140 million hectares in 2010 (James 2010). The most widely grown Bt crop is cotton (*Gossypium hirsutum L.*), accounting for 64%

of global cotton area devoted to Bt crops, followed by corn (*Zea mays L.*) accounting for 29% of global corn area. Bt cotton has been planted on an increasingly large scale in India and China (Liu 2009; Choudhary and Gaur 2010). The use of Bt crops has resulted in increased yields and significant reductions of insecticide application, thus providing economic and environmental benefits (Shelton et al. 2008; Brookes and Barfoot 2008; Carpenter 2010).

However, while there are undoubted advantages of deployment of Bt transgenic crops for effective insect pest control, certain concerns have been raised about the environmental safety of Bt transgenic crops. Large-scale commercial cultivation of transgenic plants is fraught with complex ecological, economic and social ramifications (Shelton et al. 2002; Kaur 2004). A thorough and objective evaluation of possible risks and potential benefits is essential for biosafety assessment of transgenic crops. Therefore, the cultivation of transgenic crops is under regulatory controls worldwide (Jaffe 2004).

3.2 Perceived Risks with Bt Transgenic Crops

Risk signifies the probability of harmful consequences of an activity. The professed ecological risks of Bt transgenic crops include safety of transgene towards non-target organisms, the effects of introduced gene on the ecologically relevant phenotypic traits, environmental fitness of the crop, potential for invasiveness or weediness, unpredicted effects such as changes in the levels of secondary metabolites and food safety. Several reviews have been published on the effect of transgenic crops on the environment (Conner et al. 2003; Clark et al. 2005; Andow and Zwahlen 2006; Carpenter 2011).

3.2.1 Risk to Non-target Organisms

The non-target organisms are not the intended target of Bt toxins, but come into contact with Bt transgenic crops under field conditions. The non-target organisms include the beneficial insects which are valuable natural enemies of the pests, pollinators, herbivores which feed on the transgenic crop, soil fauna, and species of biodiversity and conservation significance. For ecological benefit over the use of chemical pesticides, the deployment of Bt transgenic plants needs to be compatible with the biological control.

There are three primary routes of exposure to Bt toxin: exposure to Bt pollen, exposure to Bt transgenic crop residue in the soil and trophic exposure via feeding on herbivores which feed on the Bt transgenic crop. The simplest route of exposure is through direct consumption of leaf, stalk, root, seed or pollen of transgenic plant (Clark et al. 2005). Natural enemies are most likely to ingest Bt toxin upon consumption of Bt transgenic plant parts or upon preying on or parasitizing herbivores which feed on Bt plant.

A major concern has been expressed that Bt toxin may be harmful to the desirable non-target organisms that consume the transgenic crop parts and the beneficial natural predators of the pests that ingest toxin-containing plant parts. The potential impact of Bt crops on non-target organisms has been studied in considerable detail in both laboratory and field studies (O'Callaghan et al. 2005; Clark et al. 2005; Romeis et al. 2009; Duan et al. 2010; Yu et al. 2011). The impact ranges from no detrimental effect, to some degree of adverse effect, or actual improvement of abundance of beneficial insects.

3.2.1.1 Studies Reporting No Adverse Effects to Non-target Organisms

Many studies have reported no adverse effects of Bt on non-target organisms which include beneficial pests such as the predators, parasitoids, insects of value such as the pollinators and butterflies and other non-target pests such as the aphids which may be the hosts for parasitoids.

3.2.1.1.1 Pest Predators and Parasitoids

Non-target predators and parasitoids are important components of integrated pest management (IPM) since they prey on both the target and the secondary pests. While insect predators may feed on a variety of hosts during its lifetime, a parasitoid usually completes its entire life in a single host. Parasitoids are important natural enemies of many pest species and are used extensively in biological and integrated control programs.

Several laboratory and field studies have shown that Bt crops due to their narrow spectrum of activity have no detrimental effects on predators and parasitoids. In studies examining effects of Cry protein on lady beetles, which are ecologically important insect predators, no adverse effects on their development or survival were observed (Clark et al. 2005). Bt transgenic plants had no adverse effect on the lady birds feeding on the aphids, which were raised on transgenic plants (Dogan et al. 1996). The effects of Bt potato expressing coleopteran-specific Cry3A on the ladybird beetle *Harmonia axyridis* and the carabid beetle *Nebria brevicollis* were investigated via the bitrophic interaction of the adult ladybird exposed to potato floral tissue and the tritrophic interaction of the carabid consuming a non-target potato pest. The Cry3A protein expressed in floral tissues had no overall significant acute effects on survival, body mass change, fecundity or egg viability of *H. axyridis* and *N. brevicollis*, indicating low risk to coleopteran insects other than the targeted chrysomelid larvae of Colorado potato beetle (*Leptinotarsa decemlineata*), due to high specificity of the Cry3A protein (Ferry et al. 2007).

A laboratory study, which has generated much controversy, has reported mortality of two-spotted ladybird (*Adalia bipuntata*), a polyphagus predator, which was fed with flour moth (*Ephestia kuehniella*) that had been sprayed with solutions of Cry1Ab and Cry3Bb protein produced in *E. coli* (Schmidt et al. 2009). However, this study has been strongly refuted on the basis of lack of proper quantification of degree of exposure of test insects, statistical analyses of differences in observed mortality and absence of negative impact on the larval developmental time and bodyweight of adult beetles (Rauschen 2010; Ricroch et al. 2010). Alvarez-Alfageme et al. (2011) have used tri-trophic and bi-trophic experimental systems and reported that *A. bipunctata* is not sensitive to Cry1Ab and Cry3Bb1, and that the harmful effects observed by Schmidt et al. (2009) were artifacts.

Parasitoids have close relationships with their hosts, possess a relatively narrow host range and are more likely than predators to suffer adverse impacts if their Bt susceptible hosts are wakened or killed with Bt toxin (Romeis et al. 2006). Some studies to examine the effects of Cry proteins on parasitoids that utilize herbivorous hosts feeding on Bt transgenic plants have been carried out. No detrimental effects on the ability of the hymenopteran parasitoid wasp (*Diaeretiella rapae*) to control its green peach aphid host (*Myzus persicae*) was observed in two *Brassica napus* transgenic lines expressing lepidopteran-toxic *cry*1Ac gene or coleopteran-toxic proteinase inhibitor oryzastatin I (*Oz-I*) gene from rice (Schuler et al. 2001).

Plutella xylostella (Diamondback moth) is the most destructive of the insect pests of *Brasssica* crops worldwide. *Cotesia plutellae*, an important endoparasitoid of *P. xylostella*, was unable to survive in Bt-susceptible *P. xylostella* larvae on Bt oilseed rape plants due to premature host mortality, but completed its larval development in Bt-resistant *P. xylostella* larvae (Schuler et al. 2004). Experiments of parasitoid flight and foraging behaviour showed that adult *C. plutellae* females did not distinguish between Bt and wild type oilseed rape plants, and were more attracted to Bt plants damaged by Bt-resistant hosts than by susceptible hosts due to more extensive feeding damage. Furthermore, population scale experiments with mixtures of Bt and wild type plants demonstrated that *C. plutellae* was as effective in controlling resistant *P. xylostella* on transgenic as on wild type plants.

Populations of *P. xylostella* that were resistant to Cry1C protein, and were fed on Cry1C-expressing transgenic plants, became parasitized by *Diadegma insulare*, an important endoparasitoid and biological control agent of *P. xylostella*, without any harmful effect of Cry1C protein on the parasitoid, while parasitism rates on *P. xylostella* strains resistant to four commonly used insecticides were reduced (Chen et al. 2008a). Furthermore, *D. insulare* did not discriminate between Bt or non-Bt Broccoli plants, or Cry1Ac-resistant or susceptible *P. xylostella* genotypes, after two generations of exposure (Liu et al. 2010). The life parameters of *D. insulare* and its subsequent generation from *P. xylostella* reared on Bt broccoli were not significantly different from those of non Bt broccoli.

3.2.1.1.2 Pollinators and Butterflies

The toxicity of Bt to non-target lepidopteran pests has also been examined. Honeybees, which are widespread pollinators, have been used as test insects in feeding tests with Bt plant pollen extensively and found to have no adverse effects on their longevity and development (Hofs et al. 2008; Liu et al. 2009a; Han et al. 2010) Meta-analysis of 25 studies on the effects of lepidopteran- and coleopteran-toxic Cry proteins incorporated in Bt transgenic crops showed that there was no negative impact on the survival of larvae or adults of honey bee (Duan et al. 2008).

Initial laboratory studies indicated a risk to Monarch butterfly (*Danaus plexippus*) larvae that consumed pollen containing high levels of Cry protein deposited on milkweed leaves (Losey 1999). However, the impact of pollen from commercial cultivation of Bt transgenic corn hybrids on Monarch butterfly population was found to be negligible in a 2 year study in Canada (Sears et al. 2001). Long-term low level exposure to Bt pollen resulted in 0.6% additional mortality of Monarch butterfly larvae (Dively et al. 2004). In field studies to examine the potential toxicity of Cry protein to another butterfly, the black swallowtail (*Papilio polyxenes*), no adverse effects were observed (Zangerl et al. 2001).

3.2.1.1.3 Aphids

Aphids are abundant herbivores in most crops and are attacked by a range of specialist predators and parasitoids. Romeis and Meissle (2011) have reviewed the available data and concluded that aphids do not ingest considerable amounts of Cry proteins since they feed exclusively on the phloem sap of Bt transgenic crops. Concentrations of Cry proteins in aphids were found to be below the limits of detection (LOD) in a majority of studies with Bt transgenic plants. Thus, the natural pest enemies that predate on aphids are unlikely to be at risk because of limited exposure. However, if phloem-specific promoters are used for the development of future Bt transgenic crops targeted for sap-sucking insect pests, the presence of Cry proteins in the phloem sap and consequently in aphids feeding on such plants cannot be ruled out.

3.2.1.2 Studies Reporting Adverse Effects to Non-target Organisms

Adverse effects of Bt toxins on beneficial insects have also been reported in some studies. A review has expressed concerns about effects of Bt on non-target insects (Hilbeck and Schmidt 2006). The biological control of the host pest *Helicoverpa armigera* (Hubner) by its parasitoid was said to be not fully compatible with the use of Bt toxin due to early and rapid mortality of the pest (Blumberg et al. 1997). Bt corn-fed prey was found to have some toxicity towards green lacewing (*Chrysoperla carnea*), a beneficial predator insect which is an important biological control agent for many pests, resulting in its reduced fitness (Hilbeck et al. 1998; Dutton et al. 2002). It was claimed that Cry1Ab protein is toxic to *C. carnea*. Andow and Hilbeck (2004) analyzed the experiments performed by different researchers and concluded that transgenic Bt maize might cause increased mortality to *C. carnea* in the field and recommended additional testing by tritrophic studies in the field.

However, subsequent studies in which *C. carnea* larvae were directly fed Cry1Ab protein, no direct toxic effect was observed on *C. carnea* larvae, indicating that the effects observed by Hilbeck et al. (1998) could have been due to *C. carnea* feeding on poor, suboptimal quality lepidopteran prey (Romeis et al. 2004; Lawo and Romeis 2008). An adverse effect was seen on *C. carnea* larvae fed on Bt cotton-fed caterpillars, while no adverse effect was observed on larvae fed on Bt cotton-fed caterpillars that were resistant to Cry1Ac (Lawo et al. 2010). Thus, this study also reiterated that the adverse effects observed with the susceptible strain of caterpillars were due to the quality of the prey and not due to direct toxicity of the Cry1Ac protein to the predator. This underscores the usefulness of Cry protein-resistant strains for testing the sensitivity of beneficial non-target insects (Li and Romeis 2010). Similarly, no negative impact of Bt broccoli expressing Cry1Ac, Cry1Ac and Cry1C or Cry1C proteins, on the herbivore imported cabbageworm (*Pieris rapae*) and its pupal endoparasitoid *Pteromalus puparum* was observed in terms of developmental time, parasitism rate and longevity (Chen et al. 2008b). The previously observed deleterious effect on *P. puparum* developed on Bt plant-fed host was concluded to be due to reduced quality of the host rather than direct toxicity.

Meta-analysis of effects of Bt transgenic plants on non-target arthropods grouped by functional guilds (Predator, Parasitoid, Mixed, Herbivore, Omnivore and Detritivore) revealed significant reduction in predators on Bt cotton as compared with control plants not sprayed with any chemical insecticide (Wolfenberger et al. 2008). The reduction has, however, been shown to be not of consequence for biological control (Naranjo 2005; Naranjo et al. 2005).

In another review based on an analysis of laboratory studies on impact on transgenic plants on natural enemies, it was concluded that Cry proteins often have non-neutral effects on natural enemies, with parasitoids being more susceptible than predators (Lövei et al. 2009). This paper has been severely criticized by Shelton et al. (2009) citing inappropriate methods of analysis, lack of ecological context and properly formulated hypothesis and a negatively biased and incorrect interpretation of published data.

It is pertinent that while most studies have been performed with Cry proteins produced in the bacterium *Escherichia coli* in the laboratory, many others have used Cry protein produced in the transgenic plant as the test material (Hilbeck et al. 2011). The amount of Cry protein expressed in the transgenic plant is dependent on the specific transformation event that resulted in insertion of transgene at a particular location in the plant genome and the promoter used for expression of transgene. Under field conditions, the non-target organisms may be exposed to different, possibly lower-than-laboratory-test concentrations of Cry protein, due to change in expression levels impacted by various environmental factors during the course of growing season of the transgenic plant (Clark et al. 2005). Tritrophic studies using spider mites (*Tetranchycus urticae*) that had fed on Bt maize as the carrier were conducted to assess the effect of Cry proteins on *C. carnea* under more realistic routes of exposure (Alvarez-Alfageme et al. 2011). It was concluded that ingestion of Cry1Ac and Cry3Bb proteins had no effect on the larval mortality, weight or development time. Further data on the toxicity of Bt towards beneficial predators will be help in detailed analysis of any impact on non-target organisms.

3.2.1.3 Studies Reporting Beneficial Effects to Non-target Organisms

Many studies have indicated Bt transgenic crops to be more protective of beneficial insects and secondary pests as compared with the conventional insecticides (Reed et al. 2001). Bt crops have been shown to augment natural control of pests by improving the abundance of some beneficial pests (Yu et al. 2011). The abundance and activity of parasitoids and predators was found to be similar on Bt and non-Bt crops in field studies, whereas the use of conventional insecticides usually resulted in deleterious impact on biological control organisms (Romeis et al. 2006). Meta-analysis of 42 field experiments indicated reduction in abundance of non-target arthropods on Bt transgenic plants as compared with that on control plants, which were not treated with any chemical insecticide. However, when control plants were treated with chemical insecticides, non-target arthropod abundance was significantly higher on Bt transgenic cotton and maize plants (Marvier et al. 2007).

3.2.1.4 Effects on Viruses, Aquatic Animals, Birds and Mammals

No significant differences were observed between the infection rates of Maize dwarf mosaic virus (MDMV) and Maize rough dwarf virus (MRDV) on two generations of Bt maize transgenic varieties and the non-transformed isogenic varieties, suggesting that differences in virus distribution are linked to the genetic background of the maize varieties and distribution of virus reservoirs rather than to Bt-maize cultivation (Achon and Alonso-Dueñas 2009).

Non-target effects of Cry proteins on aquatic animals, birds and mammals have also been investigated in several studies (Flachowsky et al. 2005; Finamore et al. 2008; Trabalza-Marinucci et al. 2008; Chambers et al. 2010; Jensen et al. 2010; Yu et al. 2011; Mancebo et al. 2011). The current data suggests that Bt has no toxic effects on mammalian development and health.

Some studies with the Vip protein have also been performed. The effects of Vip3A protein expressed in transgenic maize on non-target organisms have been assessed in laboratory studies with Vip3A proteins produced in *E. coli* and found to have negligible risk (Raybould and Vlachos 2011).

3.2.2 Persistence of Bt Residue in Soil

Concern has also been expressed about presence of Bt toxin in the soil due to the possible post-harvest plant residue, sloughing of root cells and release of exudates from roots (Ahmad et al. 2005; Clark et al. 2005). The Cry protein in Bt corn tissue in soil was found to persist in the field for several months (Zwahlen et al. 2003). Bt toxin can persist in the soil after release from root exudates of the growing transgenic crops, by binding to the surface-active particles in the soil and may contribute to the selection of resistant soil insects (Saxena and Stotzky 2000).

3.2.2.1 Rate of Bt Degradation in Soil

Variation in the rate of Bt degradation in the soil has been observed (O'Callaghan et al. 2005). This variation could be due to different experimental procedures such as use of dried plant material versus fresh samples and the complexity of the soil ecosystem. Differences in the persistence of Cry proteins observed in various studies have been attributed to different microbial activity due to variations in soil types, growing season, crop management practices and other environmental factors specific to climate and geographic zones.

Most studies have suggested rapid break down of Cry proteins from transgenic plant residues in the soil (Ahmad et al. 2005; Li et al. 2007; Icoz and Stotzky 2008a, b; Zurbrügg et al. 2010). Cry3Bb1 protein released in root exudates and from decaying plant residues of Bt corn was found to degrade rapidly and did not persist in the soil (Icoz and Stotzky 2008a). Decomposition pattern of leaf residues from three Bt transgenic crops (two expressing the Cry1Ab, one the Cry3Bb1 protein) and six non-transgenic hybrids (the three corresponding non-transformed near-isolines and three conventional hybrids) was investigated. Leaf residue decomposition in transgenic and non-transgenic plants was found to be similar, and plant components among conventional hybrids differed more than between transgenic and non-transgenic hybrids. While the concentration of Cry3Bb1 protein was higher in senescent maize leaves than that of Cry1Ab, degradation of Cry3Bb1 protein was faster, indicating that it has a shorter persistence in plant residues.

Decomposition patterns of Bt maize were found to be well within the range of common conventional hybrids, indicating lack of deleterious effects of Bt maize on the activity of the decomposer community (Zurbrügg et al. 2010). Rapid degradation of Cry1Ab protein from plant residues remaining on the field after harvest and no persistence of Cry1Ab protein in different soils at four South German field sites under long-term Bt Maize cultivation spanning over nine growing seasons has been reported (Gruber et al. 2011).

3.2.2.2 Effects on Macroorganisms and Microorganisms in Soil

Few or no toxic effects of Cry proteins on soil macroorganisms such as woodlice, collembolans, mites, earthworms, snails, nematodes, protozoa and the activity of various enzymes in soil have been reported in lab and field experiments (Ahmad et al. 2005; Icoz and Stotzky 2008a, b; Liu et al. 2009b; Bai et al. 2010;

Yu et al. 2011). In a 2 year study to evaluate the effects of Bt corn in comparison of that of its parent isoline, crop management practices and environmental conditions were found to have greater effect on the species diversity and abundance of soil-dwelling Collembola rather than the crop itself (Priestley and Brownbridge 2009).

The release of Cry proteins from Bt plant residues in the soil has raised the concern that these could likely be transferred to higher trophic levels and may adversely affect organisms of the decomposer food web. The Cry1Ab and Cry3Bb1 protein from Bt corn could potentially be transferred to two common agricultural slug pests *Arion lusitanicus* and *Deroceras reticulatum* and their feces by ingestion in a laboratory study (Zurbrügg and Nentwig 2009). After slugs had ceased feeding on Bt corn, Cry1Ab was detectable in fresh slug faeces for a significantly longer time and often in higher amounts than the Cry3Bb1, probably due to different degradation rates. Most of the Cry proteins seemed to be digested or adsorbed by the slugs, but there was still insecticidal-active Cry1Ab protein left in the intestines and feces, indicating that slug predators and organisms associated with the feces may ingest active protein. The potential adverse effects on these organisms and the situation in the field, however, need further investigation.

Hu et al. (2009) have observed that multiple year cultivation of Bt cotton might not affect the soil microbial population. The effects on microbial communities in the soil have been reported to range from no effect, to minor, to significant effects. The microbial communities are assayed by various techniques such as fatty acid methyl ester profiles, analysis of ribosomal DNA sequences and population counts. The effect of transgenic plants has been reported to be negligible in most cases and not any greater compared with the effects of crop variety and species, and change in weather conditions (Raybould 2007).

No consistent conclusions have been obtained on effects of Bt crops on soil ecosystems with respect to soil persistence and dynamics of Bt toxins, effects of Bt cotton on soil microorganisms, and soil biochemical properties in the research data from China (Liu 2009). This could be due to diverse ecological regions, varieties and *cry* genes (single or bivalent) and the inherent complexity of soil ecosystems having multifarious, dynamic and interacting components. The dynamics of soil organic matter, nutrients, hormones, and soil enzymes depend on the plants and soil microorganisms and their interaction, especially in the rhizosphere.

3.2.3 Gene Flow from Bt Transgenic Crops

A major concern of widespread commercial cultivation of Bt transgenic crops has been the perceived, eventual gene flow, that is, the movement of genes from one organism to another. Gene flow can be horizontal or vertical, depending upon the sexual compatibility of the involved organisms.

3.2.3.1 Horizontal Transfer of Genes

Horizontal gene transfer (HGT) is the movement of genes or genetic elements between unrelated or sexually incompatible organisms. Horizontal gene transfer in respect of transgenic crops is likely to be a rare event (Thomson 2001).

There is very little possibility of the horizontal gene transfer from transgenic plants to the terrestrial bacteria as its frequency is extremely low (Nielsen et al. 1998). The cry genes are widespread amongst *B. thuringiensis* strains, which commonly occur in soil and as such, it is far more likely that gene transfer will occur from naturally occurring *B. thuringiensis* to other soil microorganisms than from Bt transgenic plant residues to bacteria. Furthermore, in the unlikely event of gene transfer, the introduced cry genes in Bt transgenic crops having been modified for plant codon usage, only low levels of gene expression in recipient microbes would be expected.

HGT from transgenic plants to Algae, fungi and protists is also expected to be exceedingly rare (Keese 2008). HGT from Bt transgenic plants to other organisms therefore, presents negligible risk to human health and safety or the environment due to the rarity of such events, relative to those HGT events that occur in nature.

3.2.3.2 Vertical Transfer of Genes

Vertical gene transfer refers to the movement of genes or genetic elements by sexual reproduction between related or sexually compatible organisms. The fear of transgene escape from a transgenic plant to its non-transgenic counterpart or wild relatives has been a matter of concern.

It is feared that the escape of transgene to related species or weeds growing near the transgenic crop by pollen dispersal may create ‘superweeds’ which may eventually invade new habitats. Nevertheless, the wild type genotypes are generally more weedy and fitter as most genetic changes tend to reduce rather than increase fitness. Moreover, plants differ in their out-crossing potential, either intra-specific or inter-specific. The potential for a crop to hybridize with a weed is highly dependent on sexual compatibility and relatedness between parent species (Conner et al. 2003). Inter-specific hybrids are generally sterile and less adapted to survival. Transgenic crops are no more likely to transfer transgenes or any other gene to other species than the crop cultivars.

The potential impact of introgression of transgene into landraces or wild relatives would depend on the consequences, that is, whether or not the transgene confers any fitness advantage. Hydroponic, glasshouse, and field experiments performed to evaluate the effects of introgression of cry1Ac and green fluorescent protein (GFP) transgenes on hybrid productivity and competitiveness in experimental *Brassica rapa* x transgenic *Brassica napus* hybrids revealed that hybridization, with or without transgene introgression, resulted in less productive and competitive populations (Halfhill et al. 2005). Furthermore, consideration of the potential impacts should be made in the context of the background genetics of the transgenic crop, which may have a greater impact than the accompanying transgene (Carpenter 2011).

3.2.4 *Acquisition of Weediness Potential*

A concern is expressed over acquisition of weediness potential by transgenic crops due to the presence of the transgene in the new genetic background. The plant may be endowed with putative weediness characteristics such as seed dormancy, phenotypic plasticity, indeterminate growth, continuous flowering and seed production, and seed dispersal and competitiveness, making it more persistent so that the plant may itself become an invasive weed. However, introduction of a transgene into a crop is unlikely to convert it into a significantly greater weed. The weediness characteristics are several, each being controlled by one or more genes and the crop may contain only some of these genes. These attributes are unlikely to arise from a single or few gene transfers of genetic modification (Conner et al. 2003). Genetically modified plants behaved substantially the same as non-transgenic parental lines in over 400 field trials worldwide (Miller and Powell 1994).

Besides, weediness is not an inherent trait of a species and would depend upon the type of habitat, vegetative composition and edaphic conditions (Crawley et al. 1995). Breeding for insect resistance, drawn from the gene pool of wild relatives, has been historically, a fairly common activity without any significant ecological consequences. Yet, the ecological risks need to be considered in the context of large-scale cultivation of transgenics in the proximity of sexually compatible relatives (Bergelson et al. 1999).

3.2.5 *Monoculture and Eventual Loss of Biodiversity*

Another concern raised over the large scale cultivation of Bt transgenic crops, is that it may lead to monoculture and eventual erosion of the genetic diversity of crops because breeding programs will concentrate on a small number of high value cultivars. Preferred cultivation of these crops will eventuate into decreased crop diversity by way of invasion of natural ecosystems. This is especially relevant in the context of developing countries, which are often, the centers of crop diversity. In many of these places, the cultivars are grown in the proximity of the wild relatives.

However, it is important to consider that agriculture has long been an endeavor of creating new improved crops by domestication and natural ecosystems have been intentionally converted into agricultural land. Plant breeding methods such as artificial manipulation of chromosome number, development and addition of substitution lines for specific chromosomes and chemical and radiation treatments to induce mutations and chromosome rearrangement, as well as cell and tissue culture approaches such as embryo rescue, in vitro fertilization and protoplast fusion for generation of inter-specific and inter-generic hybrids for varietal improvement have been employed (Conner et al. 2003). A meta-analysis of genetic diversity trends in crop cultivars released in the last century found no clear general trends in diversity (van de Wouw et al. 2010).

3.2.6 Unintended Effects of Gene Transfer on Plant Metabolism

A concern is expressed that introduction of transgene into the recipient plant may lead to some unintended effects on plant metabolism. Such pleiotropic effects may include altered expression of an unrelated gene either at the site of insertion or distant to the site of insertion, novel traits arising from interaction of the transgenic gene product with endogenous non-target molecules and secondary effects due to altered substrate or product levels in the biochemical pathways. These effects may result in toxicity, allergenicity, weediness, or compositional changes.

Integration of foreign DNA in the vicinity of active genes may lead to co-suppression or changes in the level of expression of these genes by affecting methylation patterns or regulating signal transduction and transcription. Intra-locus instability or ectopic recombination may occur in case of integration of transgene repeats (Andow and Zwahlen 2006). However, unintended effects may also occur otherwise in non-transgenic plants due to chromosomal recombination or under specific environmental conditions. Moreover, the experience with development of transgenic plants indicates that there is little potential for unexpected outcomes that may go undetected during selection (Bradford et al. 2005).

3.2.7 Food Safety of Bt Transgenic Crops

The food safety of Bt transgenic crops has been a contentious issue and more of this topic will be discussed in depth in Chap. 16. The concern over Bt transgenic plants with regard to food safety and human health is whether the transgenic plant is likely to pose a greater risk than the non-transgenic plant variety it is derived from. The food safety concerns expressed in relation to Bt plants are potential toxicity and allergenicity of Cry proteins, changes in nutrient composition of the plants, unintended effects giving rise to toxicity or allergenicity and the safety of antibiotic resistance-marker-encoded proteins included in the transgene cassettes (Shelton et al. 2002; EFSA 2008).

These concerns, evaluated in the context of the predicted range of dietary exposures to establish a reasonable likelihood of safety, do not pose questions about safety of transgenic crops (Chassy 2002). There is no credible evidence of toxicity or allergenicity to Bt crops in humans (Mendelsohn et al. 2003). The concentration of Cry proteins in transgenic plants is usually well below 0.1% of the plant's total protein, and the Cry proteins have not been demonstrated to be toxic to humans

(Chassy 2002). Furthermore, Cry proteins do not contain sequences resembling relevant allergen epitopes and have not been implicated to be allergens.

There has, however, been serious concern over the Bt transgenic 'StarLink' corn carrying Cry9C gene. The registration of this product was approved by the U.S. Environmental Protection Agency (EPA) in 1998 only as limited use for animal feed. The restriction was imposed due to resistance of Cry9C protein to pepsin digestion and heat. The Cry1Ab and Cry1Ac proteins rapidly became inactive in processed corn and cottonseed meal, but Cry9C remained stable when exposed to simulated gastric digestion and heat and was, therefore, not permitted for human consumption (Shelton et al. 2002). Due to uncontrolled cross-pollination with cultivars used for food production, it became commingled with products for human consumption. Upon more detailed evaluation for allergenicity, it was reported to have 'medium likelihood' to be an allergen. However, the combination of the expression level of the protein and the amount of corn found to be commingled posed only 'low probability' of sensitizing individuals to Cry9C (Shelton et al. 2002). Nonetheless, due to the controversy, the registration for Starlink corn variety was voluntarily withdrawn by Aventis Crop Sciences in October 2000.

A concern has been expressed that the antibiotic resistance gene, if used as a selection marker for transgenic crop development, may get transferred to bacterial pathogens in the human gut, thereby spreading drug resistance in mammalian pathogens. Inarguably, antibiotic resistance genes are already widespread in the environment, occurring naturally in bacterial species and are readily transferable by conjugation and transduction among bacteria (Keese 2008). Furthermore, gene transfer from ingested transgenic crop to intestinal microflora is dependent upon survival of the transgene DNA in the gastrointestinal tract. In a study to assess survival of transgene in the gut of human volunteers who were fed transgenic soybean, only a small proportion of transgenic DNA survived passage through the stomach and small intestine, while all the transgenic DNA was degraded in the colon (Netherwood et al. 2004). Gene transfer to intestinal microflora did not occur during the feeding experiment.

Validated animal models have been used for food safety assessment of transgenic crops (Flachowsky et al. 2007). In animal feeding experiments, most plant DNA was found to be degraded in the gastrointestinal tract, transfer of plant DNA from feed into the animal body was not detected and no residues of plant DNA were found in any organ or tissue (Flachowsky et al. 2007). It is also noteworthy that the impact of free DNA of transgenic origin is likely to be negligible as compared to the total amount of free DNA present in the environment from a variety of sources such as pollen, leaves, fruits, decaying plants and animals, and release from microbes (Dale et al. 2002).

The food consumed by humans has had transgenomic recombinations and selective breeding over centuries. Most crop plants are the product of artificial selection. The absence of any negative reports of compromised biosafety indicates that genetic modification of crops for food uses may pose no immediate or significant risks as compared with the conventional crops (Stewart et al. 2000). There are no compelling scientific arguments to demonstrate that Bt transgenic crops are innately different from non-transgenic crops (Dale et al. 2002). Nutritional composition of cottonseed remained unchanged in the Bt transgenic cotton as compared with the non-transgenic counterpart (Tang et al. 2006).

3.2.8 Development of Resistant Pests

A key ecological concern over large scale cultivation of Bt crops is the development of resistance in insects exposed to Bt (Kaur 2004). Transgenic plants expressing a single cry gene throughout the season and in all plant parts can accelerate the onset of resistance in pests by increasing the selection pressure (Gould 1998). For the development of first generation transgenic crops only a few *cry* genes, such as *cry*1Ab, *cry*1Ac, *cry*3A, *cry*3B and *cry*9A have been employed. Fortunately, except for *P. xylostella* (diamondback moth), none of the other insect species have developed resistance to Bt biopesticides in field conditions so far. A resistant strain of *P. xylostella* was able to complete without any adverse effects, its life cycle on Bt transgenic canola which produced high levels of Cry1Ac protein (Ramachandran et al. 1998). There is a threat of development of resistance in *H. armigera* to Cry-1Ac protein upon large-scale cultivation of Bt cotton. An important issue is whether large-scale commercial cultivation of Bt crops can be sustained in view of threat of development of resistance in pests to the currently deployed *cry* genes.

3.3 Risk Analysis of Bt Transgenic Crops

Risk analysis encompasses risk assessment, risk management and risk communication (Johnson et al. 2007; Talas-Og˘ras 2011). Risk analysis is carried out by application of legislation, policies and procedures in a defined framework. Risk analysis thus, entails a wider context than risk assessment to incorporate more extensive concerns, including some 'non-scientific' concerns, in order to make decisions regarding regulatory aspects. Environmental risk assessment (ERA) is essential for decision making and regulatory approval regarding cultivation and commercialization of Bt transgenic crops.

3.3.1 Risk Assessment of Bt Transgenic Crops

Risk assessment is the first stage in predictive risk assessment and entails hazard identification and risk estimation. Hazard identification involves consideration of factors such as population, species and ecosystem characteristics; definition of measurable endpoints and terms of cultivation of transgenic plant such as location, area and time of release. Risk estimation involves measurement of risk in terms of characterization, probability and level of effects. Quantification of risk of transgenic plants could be complex due to the many ways in which a transgene may influence the ecology of the recipient and associated organisms (Wilkinson et al. 2003).

3.3.1.1 Risk Characterization

Risk characterization involves ascertaining the likelihood as well as the consequences of exposure. It can be complex and forms the basis of evaluation and mitigation measures to be adopted. Therefore, environmental studies must have properly formulated hypotheses, experimental designs and testing methods, failing which the interpretation of results is unreliable (Shelton et al. 2009). The risk hypotheses formed to facilitate risk characterization should take into account the case-specific scenario comprising of anticipated pathways of exposure and scientifically testable postulates.

A correct depiction of exposure scenario enables precise characterization of risk. The possible routes, duration and intensity of exposure of Cry protein to the beneficial species of interest are considered in the exposure scenario. For example, the risk to Monarch butterfly from feeding on Milkweed growing in the vicinity of Bt maize due to dispersal of pollen from Bt maize should take note of the timing and intensity of pollen shed and occurrence and distribution of Milkweed and Monarch butterfly in a particular habitat (Sears et al. 2001). Thus, risk hypotheses often comprise of complex interconnected multiple pathways of exposure that could lead to specific adverse effects.

3.3.1.2 Problem Formulation for Environmental Risk Assessment

Problem formulation is the first step during which risk assessors work in collaboration with the risk managers to arrive at a feasible plan for conduct of a case-specific risk assessment taking into consideration the relevant exposure scenarios and potential consequences arising from those scenarios (Wolt et al. 2010). Risk hypotheses are defined in the problem formulation phase. Therefore, problem formulation is an important aspect of ERA to provide the correct science-based framework for evaluating the need for transgenic crop and comparing with other possible solutions to the problem. Problems parameters and constraints are defined in a tractable manner and contextualized with respect to the case-specific environmental concerns, methods to be deployed for risk assessment and measurable endpoints of risk assessment. Problem formulation should be well-designed and consistent to provide guidance to researchers, regulatory and decision-making agencies.

The focus of risk assessment has shifted from assessment of specific indicators of potential adverse environmental effects to a direct assessment of defined risk endpoints taking into account the receiving ecosystem and the relevant ecological functional groups and ecological processes in that ecosystem. Assessment endpoints are ecological entities that need to be protected e.g. the abundance of beneficial insects or potential harms posed to the non-target organisms and biodiversity. Assessment endpoints should be selected carefully, easy to evaluate in the laboratory and likely to indicate the possible adverse effects. Some assessment variables that are used are: mortality levels (LD50), effect concentrations (EC), no observed effect concentration (NOEC), fitness measures such as fecundity, development duration, body mass or the percentage of individuals that reach a certain life-stage and genetic diversity in a specific habitat. Risk assessments for transgenic crops are often contentious and are more effective when assessment endpoints are explicit and clearly defined (Raybould et al. 2010).

3.3.2 Methods for Risk Assessment of Bt Transgenic Crops

Risk is assessed in a stepwise manner by performing laboratory tests, followed by small scale and subsequently large scale field trials to ascertain environmental safety before commercial release of a transgenic crop. In the step-by-step procedure, the assessment of potential adverse impacts of transgenic plants is carried out in four stages of decreased safety measures- (i) research under containment in laboratory and greenhouse, (ii) small-scale confined trials, (iii) large-scale unconfined trials and (iv) commercial release tests, which are mainly concerned with agronomic performance under cultivation (Kjellsson 1997).

The maximum potential hazards and impact of transgenic plant on key non-target organisms as well as the ecosystem are determined. Selection of non-target organisms to be tested is done on the basis of criteria of co-occurrence, abundance, significance and trophic interactions between Bt transgenic plant and other organisms. There are potentially three trophic levels among transgenic plant, and herbivores and their predators. The first trophic level is the Bt plant carrying the *cry* gene(s). The second trophic level is the Bt plant and herbivore feeding on that plant. The third trophic level is the Bt plant-herbivore-predator or the parasitoid. The tritrophic interaction is between the plant and the predator or parasitoid.

Furthermore, risk assessment is done on a case-to-case basis rather than generalization of results from one case to the next, taking into consideration the biology and environment of the plant. The performance of the transgenic crop can vary in different environments due to differences in pollination or cross-pollination between transgenic crop and related species, which may vary in different environments.

3.3.2.1 The Tiered Approach to Risk Assessment

Risk assessment methodologies are structured into specific tiers of testing, relying on available information and may employ surrogate species as test organisms. The tiered toxicity approach originates from the methods applied in chemical pesticide testing and is based on the assumption that lower tier laboratory tests, which expose surrogate non-target organisms to high doses of Cry proteins can detect harmful effects that might be manifested in the field. First-tier experiments are conducted to identify hazards in a 'worst-case scenario' and subsequent tiers are used to quantify risk by introducing more realistic probability and exposure levels (Wilkinson et al. 2003). First tier experiments comprise of laboratory studies at a considerably high dose to identify the worst-case scenario. It is however, expected that the concentration of Cry protein encountered in the field situation is likely to be much less. Therefore, second tier laboratory studies take into account the actual concentrations likely to occur in the field and are also known as semi-field studies. The third tier studies are the experiments conducted in the field for a realistic quantification of the exposure levels and the ensuing risk if any.

First tier assessments are conducted in the laboratory where surrogate non-target organisms representing particular taxonomic or functional guilds are subjected to Bt toxin or Bt plant tissues at typically greater than 10 times the expected exposure concentration (Romeis et al. 2008). The transgenic plant may be judged to have minimal risk if no toxicity to non-target organisms is observed in first tier experiments. However, if potential toxicity is observed, higher tier testing is carried out.

A scientifically rigorous tiered approach to risk assessment of transgenic crops in relation to impact on non-target arthropods, focusing on testing of clearly stated risk hypothesis, making maximum use of available data and using formal decision guidelines to progress from one testing tier to the next, has been described (Romeis et al. 2008). Duan et al. (2010) performed meta-analyses to test the validity of tiered approach towards risk assessment of Bt transgenic plants. Laboratory studies of Cry proteins were found to predict effects that were either consistent with or more conservative than those found in field studies, provided laboratory studies encompassed the full range of ecological contexts in which non-target organisms could get directly or indirectly (via intervening trophic levels) exposed to Cry protein.

3.3.2.2 Issues with Risk Assessment Methodologies

Some issues in risk assessment of Bt transgenic crops are complex and interdisciplinary. A crucial component for a proper risk assessment of transgenic crops is defining the appropriate baseline for comparison and decision. More often than not, the best and most appropriately defined reference point is the environmental impact of plants developed through traditional breeding methods (Conner et al. 2003). Therefore, risk assessment methodologies are based on comparative analysis of substantial equivalence to determine similarities and differences between the transgenic and its unmodified isogenic counterpart plant with respect to expression of new and native proteins and assessment of potential unintended effects (Talas-Og˘ras (2011).

Another issue involves short-term versus long-term experiments. Risk assessment experiments are methodically challenging due to several sources of variation affecting the potential risk. Risk assessment encompasses evaluation of both the anticipated effects as well as the cumulative long-term effects. It entails estimation of future risks based on existing knowledge. The effects on the environment could be immediate-observed during the period of release, or delayed. Therefore, the experiments need to be performed over a period of time for consistent monitoring. Moreover, risk assessment on a larger geographic scale requires coordinated efforts for integration and interpretation of data from diverse disciplines.

There is may be an element of uncertainty present in risk assessment experiments, arising from inherent biological variability and knowledge gaps in respect of perceived risks. Hilbeck et al. (2011) have proposed some shortcomings in the current risk assessment methodologies and proposed a hierarchical scheme from laboratory studies to field trials using test organisms that do occur in the receiving environment.

3.3.3 Risk Management of Bt Transgenic Crops

The deployment of Bt transgenic plants should keep in view the important goal of sustainability. Therefore, appropriate strategies for risk management of Bt crops need to be devised. The risk management process builds upon the risk assessment to determine whether measures are required in order to protect people and/or the environment. Risk management includes evaluation of identified risks to determine what specific treatments are required to mitigate harm to human health and environment. These risk management measures are important in the context of decision-making and regulatory approval of transgenic crops. Thus, risk management encompasses risk evaluation, cost-benefit, public concern, monitoring, risk reduction and decision making to minimize undesired effects of a hazard. It involves practical precautions and restoration measures to manage potential adverse impacts.

3.3.3.1 Risk Assessment of the Effects on Non-target Organisms

Ecological risk assessment requires a thorough understanding of the factors governing the abundance and distribution of the species affected directly by Bt transgenic crop cultivation as well as the indirect influence of these species on the communities they inhabit. A comprehensive set of studies are performed in order to assess the potential non-target effects.

3.3.3.1.1 Early Tier Lab Testing of Non-target Organisms

The initial set of tests are done in the laboratory at possible maximum concentrations of the pure protein or plant tissues on a series of non-target species including typically insect predators, predators, parasitoids and soil dwelling and aquatic invertebrates, chosen on the basis of region but to be representative of the different taxa and ecological guilds. It is appropriate to especially consider those groups of non-target organisms that (i) are widespread and occur regularly, (ii) have important ecological functions in the ecosystem, (iii) serve valued purposes and (iv) can be tested both in the laboratory as well as in simulation studies in the field.

The different routes of exposure to Bt toxin may include direct consumption of plant tissue by herbivores, feeding on the pollen or ingestion of soil-incorporated plant residues. It is however, expected that the exposure to pollen would be limited in view of generally low levels of *cry* gene expression in transgenic pollen and decrease upon increasing distance from the transgenic plant. The fate of Cry protein under field conditions and its movement through non-target insect food-webs has been reviewed (Romeis et al. 2009). Bitrophic and tritrophic exposures are assessed by ecotoxicological testing on a variety of species for regulatory submissions. Ecotoxicological methods have been advocated to yield more useful information for risk assessment as compared to the ecological methods of testing (Raybould 2007). The ecotoxicological principles for regulatory decision making ensure that the research problems are selected by policy relevance, testing of risk hypotheses that predict no harm of transgenic plant to things of value, a preference for tests of ecosystem function, simple comparative models, and accurate and relevant predictions and favouring tests of hypotheses under conditions that provide most rigor.

Confidence in the robustness and reproducibility of the data from early tier laboratory studies based on clearly defined risk hypotheses is a must for acceptance of results for regulatory decision making. Early tier studies comprise of highly controlled laboratory experiments where test species are exposed to high concentrations of Cry toxins. Romeis et al. (2011) have provided recommendations on experimental design for early tier studies to evaluate potential adverse impacts on non-target insects. The surrogate species and their life stages employed in laboratory tests should be appropriate and representative of the species in different habitats likely to be exposed in the field based on the predicted exposure pathway. Laboratory tests are generally shorter in duration than the field tests but are performed at higher concentrations of Cry protein. The duration should though, be sufficient for the response of the measurement endpoints. If any adverse effects are observed in early testing, or if the testing is inconclusive, additional testing at higher tiers is carried out. Andow and Hilbeck (2004) have argued that the chemical composition of transgenic plant may be altered by expression of the transgene, which could have effects on the non-target species, independent of the transgenic product itself. Therefore, the initial laboratory assessments should be broadened to include toxicity tests using both the purified transgene product as well as the whole plant tests using intact transgenic plants.

While risk assessment needs to be thorough, it does not have to be redundant. The risk assessment may not be required for species and life-stages that are at negligible risk because of limited exposure. For example, the predators and parasitoids that specifically attack aphids, which have been reported to contain no or negligible amounts of Cry protein, may be at negligible risk under field conditions (Romeis and Meissle 2011).

3.3.3.1.2 Field Studies for Assessment of Impact on Non-target Organisms

Field studies need to be undertaken when potentially adverse effects are detected in lower tier testing or lower tier tests are not possible either due to lack of testable surrogate or validated test protocol (Romeis et al. 2011). Both the mean density and variability of the test insects in the field need to be considered. If mean densities of non-target insects are low, such as of foliage foraging predators, testing risk hypotheses in the field using mean densities as assessment endpoints requires a large number of replicates and long experimental time.

Furthermore, the natural variability of the conventional plant lines as compared with and its influence on the field densities of the non-target pests should also be taken into account. This reiterates the necessity and usefulness of thoroughly and comprehensively conducting early tier laboratory tests with probable non-target pests likely to be encountered in the field. Further semi-field (under containment using live transgenic plants) or open field tests should be conducted if some degree of hazard is identified in laboratory tests. These tests help confirm whether an effect can still be detected under more realistic rates and routes of exposure. Also, the necessity of conducting field experiments for risk assessment should be examined in a specific scenario. Rauschen et al. (2010) have highlighted the limitations of using higher-tier field release experiments for assessing the potential impact of Bt maize on beetles, a diverse and important group of insects in agricultural systems, in a 6 year field-release study.

The results of field studies on effects of Bt toxin on non-target organisms have been reviewed (Romeis et al. 2006; Chen et al. 2009). The major non-target pests in Bt cotton fields, the sucking pests such as cotton aphid, thrip, whitefly, leafhopper, *etc* and spider mites are not susceptible to Bt proteins used currently and in general exhibit the same pest status and continue to be managed identically in Bt and conventional cotton system (Chen et al. 2009; Yu et al. 2011). However, these secondary pest populations have increased and gradually evolving into key pests in many countries due to the reduced use of insecticides and changes in pest management regimes (Mann et al. 2010; Zhao et al. 2011).

Thus, the potential non-target effects of the transgenic crop cultivation appear to be manageable (Wraight et al. 2000). Further, to avoid any residual build up of Bt toxin in the food chain, possible precautions include tissue-specific expression of the transgene in only the plant parts targeted by the insect pests such as the leaves and stems.

3.3.3.2 Risk Assessment of the Effects of Persistence of Bt in Soil

The process of degradation of plant-produced Cry proteins is different from that of purified Cry proteins due to macrofaunal and microbial decomposition of plant material before Cry proteins interact directly with the soil (Clark et al. 2005). It is currently difficult to accurately ascertain the environmental fate of plant-produced Cry proteins in soil due to lack of sophisticated analytical techniques. Many studies have employed the ecologically unrealistic exposure scenario of using relatively high concentrations of the purified Cry protein expressed in *E. coli* as the test material while the toxin likely to be encountered, if at all, by the soil organisms would be the plant-produced Cry protein at the amount expressed in the plant. It is further likely that Bt toxin, being biodegradable, may not pose a severe threat of environmental contamination.

The effect of Bt toxin on soil organisms can be difficult to predict and may have a cascading effect on the ecological dynamics of soil communities. A thorough evaluation of the persistence of Bt toxin and its effect on soil organisms is warranted (Stotzky 2001). Multisite and long-term experiments are needed to understand the interaction among Bt toxin and soil organisms for risk assessment of Bt crops to soil ecosystems.

3.3.3.3 Risk Assessment of Gene Flow from Bt Transgenic Crops

Gene flow from Bt transgenic crops to wild and weedy relatives needs to be quantitatively assessed (Poppy 2004). The hybridization potential of the transgenic crop with cultivated and wild species, as well as the gene pool of the plant and its related species, should be assessed. Cross fertilization between transgenic and non-transgenic plants is an issue that needs to be addressed. Various factors affecting cross pollination such as isolation distance, geographical conditions, wind direction and velocity, weather conditions and crop biology also need to be considered. Other factors which may influence cross fertilization rates are pollen viability and longevity, male fertility or sterility and synchrony between anthesis of the pollen donor and silking in the recipient field (Sanvido et al. 2008).

3.3.3.3.1 Physical Containment of Gene Flow

The pollen-mediated movement of transgene to sexually compatible species is dependent upon physical proximity. The pollen dispersal pattern, as determined by spatial models taking into account the distance and direction as applied to wind-pollinated species, has been found to range from limited to extensive at different sites, depending upon local conditions such as exposure to wind (Meagher et al. 2003). In a field experiment in China with four plot-size treatments of adjacent Bt transgenic and non-transgenic rice, a dramatic reduction in transgene frequencies with increasing distance from the transgenic crop was observed (Rong et al. 2007). Furthermore, different plot sizes did not significantly affect the frequencies of gene flow. Thus, the pollen-mediated crop-to-crop transgene flow could be maintained at negligible levels with short spatial isolation.

Determination of gene flow in centres of origin or diversity is especially important. Dispersal of pollen and seed from transgenic crop can be reduced by maintaining adequate isolation distance (Chandler and Dunwell 2008). Potential gene flow to wild species can be prevented by establishing exclusion zones and restriction of planting near wild relatives. Multi-location field trials should be conducted to determine the impact of transgenic plant on the wild relatives and on the agronomic and natural environment. While multi-location trials would include location-specific bioassays, global monitoring and bioassays, done in different countries, will help in relating the assessment data properly.

3.3.3.3.2 Biological Containment of Gene Flow

Some of the ecological risks of transgenic escape can be addressed by the transgenic design itself. Genetic manipulation to avoid paternal transmittance of a transgene by pollen is possible by targeting the transgene to the chloroplast or mitochondria, which are maternally inherited. Thus, outcrossing of transgene via pollen can be prevented (Daniell 2007). Chloroplast transformation is an effective strategy for transgene containment and offers the additional advantage of high level of expression of the transgene (Rawat et al. 2011). There could be occasional low level paternal chloroplast transmission in some species, which can be controlled by stringent selection protocols (Ruf et al. 2007).

Gene flow from transgenic crops can be prevented, by using fertility constructs that disable the pollen or ovule function in the plants hemizygous for the transgene (Bergelson et al. 1999). Pollen flow from such crops to wild relatives can produce only sterile hybrids. However, in the event of a mutation disrupting the effectiveness of such fertility constructs, or recombination, which breaks the linkage of the transgene and the infertility gene, a small number of pollen and ovules lacking the fertility construct can also result. The containment strategy for transgenes can be further enhanced through male sterility (Talas-Og˘ras 2011). Another strategy for tackling gene escape is using a chemically inducible promoter to allow expression of transgene only upon exogenous application of a chemical (Cao et al. 2006). There would, thus, be no risk of transgene escape to natural populations in non-agricultural areas due to the absence of the chemical.

There has been concern over escape into environment of the commonly used selectable marker genes (antibiotic and herbicide resistance) for transgenic development. Methods, which rely on elimination of marker gene following transformation, have been devised for the development of marker free transgenic crops such as Cre/lox P recombinase mediated marker gene excision, intrachromosomal recombination based excision, post-transcriptional gene silencing, RNA interference hairpin technology and deactivation strategies such as genetic use restriction technology (GURT) (Lutz and Maliga 2007).

Inclusion of suicide genes into the chromosomal and Ti plasmid DNA of *Agrobacterium tumefaciens* strains used in the development of transgenic plants can control, to some extent, the unintended transfer of disarmed *A. tumefaciens* plasmid harboring foreign genes to wild plants or soil bacteria (Molin et al. 1993). Suicide constructs are based on the lethal genes from *E. coli* and are triggered upon plasmid transfer. Regarding the dissemination of *A. tumefaciens* through transgenic plants, the fears are rather unfounded, for, *A. tumefaciens* is eliminated during tissue culture and is not propagated through seed. Further studies on gene transfer among bacteria in natural environments and its ecological impact will be useful in allaying the concerns regarding transgene escape (Brookes and Barfoot 2008).

3.3.3.4 Risk Assessment of Weediness Potential of Bt Transgenic Crops

Assessment of weediness attributes of a transgenic plant at the time of introduction into a new environment is necessary. It is important to take into consideration the biology of host plant to determine if it has any species-specific characters which might endow "weediness" and to assess its potential to establish itself outside domestic cultivation. Statistically significant differences in quantitative phenotypic data such as plant height and percent seed germination as well as qualitative phenotypic data such as disease susceptibility have been reported between Bt transgenic and control plants but these differences fall within the reported range for the crop species (CERA, ILSI Research Foundation 2010).

3.3.3.5 Risk Assessment of Monoculture and Eventual Loss of Biodiversity due to Bt Transgenic Crops

Impact of transgenic crops in cultivated and natural ecosystems needs to be thoroughly assessed. All the relevant data, regarding important crops targeted for transgenic development, is required in order to successfully predict if the transgene could enable the plant to extend beyond its geographical range. Transgenic plants possessing traits such as high growth rate, large biomass accumulation and high reproductive capacity may represent potential hazard of invasion (Kjellsson 1997). Spatial models of vegetation dynamics could prove valuable for prediction of invasion. Carpenter (2011) has argued that from a broader perspective, transgenic GM crops may actually increase crop diversity by enhancing underutilized alternative crops, the so called orphan crops such as sweet potato making them more suitable for widespread domestication.

3.3.3.6 Risk Assessment of Unintended Effects of Gene Transfer on Plant Metabolism

Many innovations in the evaluation of unintended secondary metabolic changes in transgenic crops such as multi-component analysis of low molecular weight compounds and chemical fingerprinting of transgenic crops, to compare any possible compositional alterations with respect to non-transgenic isogenic parental and closely related species bred at identical as well as multiple sites, have been achieved (Noteborn et al. 2000). The differences, if any, in the concentration of important nutritional components in the transgenic crop and those of a near isogenic non-transgenic line can be detected by plant composition analysis (Raybould et al. 2010). It is pertinent to consider that while the lack of statistically significant differences in the concentration of the analytes is indicative of no harmful effects due to the transgene, detection of statistically significant differences does not necessary indicate deleterious effects of transformation.

Detailed compositional analysis of Bt transgenic Bollgard II cotton was carried out to measure proximate, fiber, amino acid, fatty acid, gossypol, and mineral contents of cottonseed from a total of 14 U.S. field sites over 2 years. The results showed that Bollgard II cotton was compositionally and nutritionally equivalent to conventional cotton varieties (Hamilton et al. 2004). In addition to comparison of chromatographic protein profiles, the new techniques such as genomics for DNA sequence analysis, transcriptomics to study gene expression profiles, proteomics for investigation of novel proteins and metabolomics-based investigation of changes in metabolic pathways are of help in detection of any unexpected changes in transgenic crops vis a vis their isogenic conventional counterparts (Baker et al. 2006; Garcia et al. 2009).

An isogenic control is genetically identical to the transgenic plant, except at the transgene locus. However, for virtually all commercial varieties, such ideal isogenic controls do not exist and the available near-isogenic control varieties may differ from the transgenic variety by up to 4% of genome (Andow and Hilbeck 2004). These authors have argued that the results from single experiments are not very convincing as the observed result could be due either to the transgene and its associated genetic material, or by the correlated systematic differences between the transgenic variety and the near isogenic control, or else by a balancing of both the effects. Assessment of whole plant effects by performing multiple comparisons between several pairs of transgenic varieties and controls has been recommended.

3.3.3.7 Risk Assessment of Food Safety of Bt Transgenic Crops

Risk Assessment of food safety of Bt transgenic crops takes into consideration the source of transgene, *in silico* homology comparison of deduced amino acid sequence of transgene to that of known allergens, physicochemical properties, abundance of Cry protein in the transgenic crop, and specific immunoglobin E binding studies. A comprehensive food safety assessment approach involves structure/activity analysis, serum screening, development of appropriate animal models, and generation of quantitative proteomics data as well as elucidation of basic mechanisms of food allergy (Barber et al. 2008; Selgrade et al. 2009). Sensitive toxicological tests need to be developed to rule out any eventual negative health impact. Toxicity tests with transgenic crops for detection of any unintended effects should comply with the international guidelines taking into consideration all important parameters such as duration of exposure, biochemical and pathological observations and sufficiently large sample size of test animals (Cellini et al. 2004). However, there does not appear to be a consensus on details of methods of performance of serum studies for regulatory agencies (Holzhauser et al. 2008). Not all tests currently being applied to assessment of allergenicity of transgenic crops have a sound scientific basis.

Bioinformatics analysis may assist in the assessment process by prediction, on a purely theoretical basis, of the toxic or allergenic potential of a protein. However, the results of such analyses are not definitive and should be used to identify proteins requiring more rigorous testing (Goodman et al. 2008). Since short-term toxicological studies can provide only limited information on effects of consumption of transgenic food and feed, long-term case-by-case studies are required for nutritional safety assessment of transgenic crops.

Public awareness on the merits of Bt transgenic crops is necessary for garnering consumer acceptance. Also, procedures for regulating the transgenic food, such as labeling, may followed so that the public is assured of no health risks or suspected allergies. There can be adventitious presence of genetically modified (GM) material in non-genetically modified material due to accidental seed impurity, seed planting equipment and practices, cross pollination between GM and non-GM crops, the presence of GM volunteers and product mixing during harvesting, transport and storage processes (Paludelmas et al. 2009). The European Union has defined a 0.9% threshold as the maximum percentage of GM material that may be contained in food and feed without the need to be specifically labeled (European Union 2003). Qualitative and quantitative PCR detection methods for identification and quantification of Bt cotton have been developed (Yang et al. 2005). High throughput detection methods such as quadruplex real time PCR have also been devised for rapid screening of food for the presence of GM material (Gaudron et al. 2009). A construct-specific comprehensive multiplex based PCR detection assay has been reported for identification and validation of Bt transgenic cotton lines (Kamle et al. 2011). However, it is impossible to provide consumers assurance of absolute-zero risk, largely owing to the inadequacy of methods to screen for novel and previously unreported toxicity or allergenicity. Moreover, the zero-risk standard applied to Bt transgenic crops far exceeds the standard used for novel crops produced through traditional methods (Chassy 2002).

3.3.3.8 Management of Development of Resistance

Detailed understanding of resistance mechanisms, increased knowledge of pest biology and evaluation of resistance management strategies is necessary for ensuring the sustainability of Bt transgenic crops. Analysis of more than a decade of global monitoring data on Bt cotton has revealed that the refuges have helped to delay resistance in pests (Tabashnik et al. 2008).

3.3.3.8.1 Elucidation of Mechanism of Resistance Development

Improvement over existing strategies of management of resistance requires elucidation of biochemical mechanism and genetics of resistance development (Ferre and Van Rie 2002). Apart from development of resistance in laboratory assays under high selection pressure, some instances of resistance to Bt crops in field populations of insect pests have also been reported, although the genetic basis of resistance has not been identified in any of these cases (Tabashnik et al. 2009; Storer et al. 2010). Identification of genes associated with resistance can enable efficient detection of recessive alleles in the resistant heterozygotes.

Cadherin has been proposed to be a receptor for Cry1Ac protein (Fabrick and Tabashnik 2007). The development of resistance has been linked to the cadherin locus, though some other studies did not find a correlation between the cadherin locus and the development of resistance. Resistance to Cry1Ac cotton was associated with mutations altering cadherin proteins in several major cotton pests. The resistance to Cry1Ac was found to occur due to retrotransposon-mediated disruption of a Cadherin superfamily gene, encoding for protein involved in larval development and differentiation (Gahan et al. 2001). In a laboratory strain of *H. armigera* resistant to Cry1Ac, resistance was found to be tightly linked to the Cadherin locus in a back cross analysis and the Cadherin gene was found to be disrupted by a premature stop codon (Xu et al. 2005). A resistant allele in pink bollworm, *Pectinophora gossypiella*, was found to have a deletion of eight amino acids in the binding region of Cadherin, indicating the role of binding epitopes in toxicity (Morin et al. 2003). The concentration of gossypol, a cotton defensive chemical, was found to be higher in pink bollworm larvae with cadherin resistance alleles than in larvae lacking such alleles (Williams et al. 2011). Adding gossypol to the larval diet decreased larval weight and survival, and increased the fitness cost affecting larval growth, but not survival, suggesting that increased accumulation of plant defensive chemicals may contribute to fitness costs.

However, in two field-selected strains of diamondback moth, *P. xylostella*, resistance to Cry1A toxins was not linked with an orthologous Cadherin gene suggesting a possibly different genetic basis of resistance in the field-evolved strains (Baxter et al. 2005). Resistance to Cry1Ac toxin has been correlated with the loss of high-affinity, irreversible binding to the mid-gut membrane (Ferre and Van Rie 2002). Reduced binding may though, not be the only mechanism of development of resistance (Li et al. 2004). Mutations in a 12-cadherin-domain protein conferred some Cry1Ac resistance but did not block this toxin binding in *in vitro* assays (Gahan et al. 2010). Loss of glycosyl transferase genes, which may be involved in assembly of the receptor having glycosyl anchor, has also been proposed as a mechanism of development of resistance (Griffitts et al. 2003). Homologuos resistance loci are predicted in *P. xylostella* and *H. virescens* based on the linkage of Cry1A resistance to mannose phosphate isoenzymes (Herrero et al. 2001). Sequestration of Cry protein by esterases in the insect midgut has also been proposed as a mechanism by which insect may develop resistance (Gunning et al. 2005). A map-based cloning approach was employed to identify the resistance gene in *H. virescens* that segregated independently from the cadherin mutation. An inactivating mutation of the ABC transporter ABCC2, that was genetically linked to Cry1Ac resistance and was correlated with loss of Cry1Ac binding to membrane vesicles, was observed (Gahan et al. 2010). ABC proteins are integral membrane proteins with many functions, including export of toxic molecules from the cell, but have not been implicated in the mode of action of Bt toxins. The reduction in toxin binding due to the inactivating mutation suggested that ABCC2 was involved in membrane integration of the toxin pore.

3.3.3.8.2 Monitoring of Development of Resistance

Monitoring the early phases of resistance is crucial. The monitoring of early phase of onset of resistance is difficult due to the generally recessive nature of resistance. Bioassays usually cannot detect heterozygotes for resistance; therefore, DNA screening methods have been devised for monitoring of resistance. Resistance to cry1Ac expressing Bt cotton in four resistant strains of pink bollworm was associated with recessive alleles at the Cadherin locus, suggesting that this locus may be considered as a candidate for developing DNA-based molecular monitoring for resistance in field populations of pests (Tabashnik et al. 2005). Development of resistance may not turn out to be a rapid process in the field conditions. No resistance alleles were detected even after a decade of exposure to Bt cotton in a 4 year survey from 2001–2005 of development of resistance as assessed by presence of mutant alleles of cadherin gene linked to resistance in pink bollworm by PCR amplification using specific primers (Tabashnik et al. 2006). Fitness costs of development of resistance to Bt crops through pleiotropic effects that reduce fitness of pests carrying resistance alleles as compared to susceptible pests lacking such alleles are observed in the absence of Bt toxins (Gassmann et al. 2009). The fitness costs associated with the evolution of resistance to Bt toxins are substantial (Carriere et al. 2001).

However, the threat to development of resistance has to be viewed in a location and crop-pest specific scenario. In diet-incorporation bioassays of measurement of susceptibilities of bollworm *Helicoverpa zea* (Boddie) and Tobacco budworm *H. virescens* (F.) to Cry1Ac among laboratory and field-collected populations, although the ratio of susceptibility of laboratory and field-collected populations was found to be similar, the colonies coming from Bt cotton had lower susceptibility as compared to those from non-Bt cotton as well as Bt corn (Ali et al. 2006).

3.3.3.8.3 Strategies for Resistance Management

Resistance management is a mandated requirement for cultivation of Bt transgenic crops. Each insect-Bt crop system may have a unique management requirement because of the different associated pests of each area and biology of the insects (Shelton et al. 2000; Zhao et al. 2002). The recommendations, emerging from the ecological simulation studies carried out to evolve management practices for appropriate deployment of Bt transgenic crops for delaying the onset of resistance development advocate as-high-as-possible toxin dose to minimize the chances of survival of the resistant heterozygotes, planting non-transgenic refuges to sustain the homozygous susceptible insect population and deployment of multiple toxin genes acting on different receptor sites in the insect midgut (Roush 1997; Kaur 2006).

Refuges of non-Bt host plants can delay resistance by providing susceptible individuals to mate with resistant individuals and by selecting against resistance. Large refuges, low initial resistance allele frequency, incomplete resistance and density dependent population growth in refuges are the factors favoring reversal of inheritance of resistance (Carriere and Tabashnik 2001). Refuges designed to increase the dominance or magnitude of fitness costs could be especially useful for delaying pest resistance. The sustainability of this strategy requires gene flow between insect population feeding on transgenic and that feeding on non-transgenic plants (Bourguet et al. 2000).

Successful commercialization of Bt transgenic crops is dependent upon compliance of recommended resistance management practices. Bt transgenic plants must only be deployed in an integrated pest management (IPM) strategy. Monitoring is required for pest densities and evaluation of economic injury levels so that the pesticides are applied only when necessary. Development of resistant insects usually occurs as localized outbreaks. Some insects are highly mobile and can affect large areas. Studies on the long-range movement of resistant insects to other crops are required to further improve the resistance management strategies. Development of resistance can be delayed by employing some molecular strategies as well.

3.3.3.8.4 Molecular Strategies for Delay of Development of Resistance

The molecular strategies for delaying the onset of resistance include tissue-specific expression of *cry* genes or expression induced via a chemical spray such as that of salicylic acid, overexpression of *cry* gene in chloroplasts, insecticidal gene pyramiding, novel hybrid genes or synthetic modification of hybrid *cry* genes for broader range of insecticidal activity (Christov et al. 1999; Kota et al. 1999; De Cosa et al. 2001). Transgenic broccoli (*Brassica oleracea L.*) plants developed using PR-1a promoter inducible with acibenzolar-s-methyl ester (ASM) caused full mortality to all instars of both SS (susceptible) and RS (resistant) genotypes of diamondback moth, *P. xylostella*, when sprayed with ASM (Bates et al. 2005; Cao et al. 2006). As differential expression of *cry* gene in various plant parts may affect larval survival and pest population dynamics, it is also a critical factor in resistance management (Adamczyk et al. 2001).

The survival of resistant genotypes of pests can be minimized by use of toxic proteins that bind to different sites or receptor and use of toxic protein of such high toxicity that any resistant population is unable to survive. Sequence modification of *cry* genes and construction of hybrid *cry* genes having domains for binding to more than one receptor in the target insect midgut may lead to a higher level of toxicity. Construction of novel Cry proteins, through protein engineering, requires a detailed knowledge of structure and mode of action of Cry proteins (Kaur 2006).

Development of second-generation Bt transgenic crops requires either new insecticidal genes or stacking or pyramiding of genes, wherein more than one insecticidal gene having different binding sites, are used in combination as a resistance management strategy (van der Slam et al. 1994). A laboratory strain of *H. armigera*, resistant to Cry1Ac, was also found to have cross resistance for Cry1Aa, Cry1Ab but not to Cry2Aa protein in diet assays (Xu et al. 2005). Gene pyramiding in transgenic crops can increase effectiveness as well as delay the development of resistance in pests.

Second generation Bt crops with stacked genes are being developed to improve the level of control and to broaden the spectrum of insecticidal protection. Two genes, *cry*1Ab and *cry*1Ac have been fused and transferred in a single transformation event to develop GK cotton in China (Shelton et al. 2002). Bollgard II cotton was developed by inserting *cry*2Ab into Bollgard I cotton variety, DP50B, which was already expressing *cry*1Ac gene (Monsanto Company 2002). The efficacy of Bt cotton expressing both *cry*1Ac and *cry*2Ab2 genes was enhanced as few *cry*1Ac-resistant pink bollworms could survive on such plants (Tabashnik et al. 2003). However, 5 year data from Australia showed a significant exponential increase in the frequency of alleles conferring Cry2Ab resistance in Australian field populations of *Helicoverpa punctigera* since the adoption of a second generation Bt cotton expressing Cry2Ab protein (Downes et al. 2010). Furthermore, the frequency of *cry*2Ab resistance alleles in populations from cropping areas was eightfold higher than that found for populations from non-cropping regions. The field-evolved resistance in a dual-toxin Bt-crop emphasizes the importance of monitoring for resistance.

Development of resistance in *P. xylostella* (diamondback moth) could be delayed by pyramiding of two genes, *cry*1Ac and *cry*1C, in broccoli (Zhao et al. 2003). Bt corn with fused *cry*34 and *cry*35Ab1 (formerly *vip*1 and *vip*2) genes and WideStrike cotton having *cry*1Ac and *cry*1F genes have been developed by Dow AgroSciences (Moellenbeck et al. 2001; Storer et al. 2006). Two cry genes (out of *cry*1Ab, *cry*1Ac, *cry2*A and *cry*1C) pyramided rice lines and their hybrids exhibited excellent efficacy against stem borers and leaf folders in field evaluation (Yang et al. 2011).

Transgenic crops by gene pyramiding with other genes, in addition to, cry genes have also been developed. Since Cry and Vip proteins bind to different receptors in the midgut of susceptible insects, they can be used together in transgenic crops for resistance management. Vip cotton with *vip*3A gene, which provided control from lepidopteran pests has been developed (Benedict and Ring 2004). The threat of cross-resistance between Cry1Ac and Vip3A was observed to be low in field populations of H. armigera indicating that Bt cotton expressing *cry*1Ac and vip3A genes could delay resistance evolution in H. armigera in Bt cotton planting area of China (An et al. 2010). A fusion protein, engineered by combining Cry1Ac with the galactose-binding domain of the nontoxic ricin B-chain, was expressed in transgenic rice and maize (Mehlo et al. 2005). It was found to be significantly more toxic and exhibited a wider spectrum of activity as compared with Cry1Ac alone. However, in Bt cotton expressing *cry*1Ac gene and a trypsin inhibitor gene from cowpea (CpTI), enhanced control of H. armigera under field conditions, due to expression of the CpTI gene, was not demonstrated (Cui et al. 2011).

Detailed information on Cry protein structure and mechanism of action will help in design of appropriate gene pyramids for resistance management. Advances in methods of identification of candidate receptors for novel insecticidal proteins will help in devising suitable gene combinations for pyramiding. An expressed sequence tag (EST) strategy was followed for the identification of candidate receptors in the *Diabrotica virgifera virgifera* (Western corn rootworm) midgut and a cadherin gene encoding a putative Bt receptor and many diverse types of cathepsins (cysteine-proteases) were identified as potential targets of protease inhibitors in the insect midgut (Siegfried et al. 2005).

3.3.4 Risk Communication

Risk communication involves communication and consultations among risk assessors, risk managers and other potential stakeholders who may be affected by the perceived risk. Risk communication is an important component of risk assessment and successful cultivation of Bt transgenic crops.

The release of transgenic crops into the environment has raised public sensitivity in some countries, though not with the same intensity. Public opposition is vociferous in Europe while the transgenics have largely gained approval in USA. Much of public antipathy is due to the lack of knowledge and propaganda by anti-GMO lobbies. Perception of a risk may differ between the scientific experts and lay public. Some articles published in scientific journals have also raised false alarms about risks from transgenic plants (Miller 2010).

The research in genetic engineering of crops is moving at a tremendous pace. It could make the lay public worried about the safeguards of this technology. Gaining confidence and support of consumers is important for success of any technology. Prudent caution in assessment of the profitable and predictable effects vis-a-vis the unintended, inadvertent and accidental effects of transgenic crop cultivation is required. Public should be informed about the potential risks and benefits of Bt transgenic crops.

3.4 Oversight of Bt Transgenic Crops

Implementation of guidelines for resistance management strategies and follow-up on on-farm monitoring is required for large scale cultivation of Bt crops. The resistance management strategies can fail because of inadequate information, planning and implementation, thereby rendering the transgenic crops ineffective against the intended target insects. There are some inherent limitations of risk management in actual practice, particularly in the resource-scarce scenario of the developing countries. The recommendation of growing a buffer zone of non-Bt crops may not be stringently adhered to by farmers. Farmers should be provided adequate information and guidance in cultivation of Bt transgenic crops so that the guidelines for resistance management are strictly implemented. Continued research into the management and implementation strategies of the transgenic crops, regularly updated regulations and increased collaboration between stockholders are needed to maintain the efficiency of Bt in transgenic crops.

Constraints on resistance management practices may vary in different regions. Therefore, the adoption of resistance management tactics can be on a location-specific basis, incorporating regionally based modifications in some agricultural practices.

3.4.1 Post Release Environmental Monitoring (PREM)

Post-market environmental monitoring (PREM) of Bt transgenic crops to observe environmental changes, which may represent environmental harm, is an essential part of regulatory frameworks governing commercial release of transgenic crops. PREM enables decision on, whether monitoring be continued, stopped or the release of transgenic variety for commercial cultivation be permitted or withheld. Such decisions are necessary to initiate remedial measures or to sustain claims of environmental liability. However, there are some challenges in interpretation of transgenic crop monitoring data for environmental decision-making due to methodological limitations of data collection and analysis, and data evaluation. There is controversy among the different stakeholders on what constitutes environmental harm and consequently, what corrective actions need to be taken. The uncertainties can be assessed in a more rigorous way during pre-market risk assessment (Sanvido et al. 2011).

At the same time, it is important for the risk managers to acknowledge the limits of environmental monitoring programs as a tool for decision-making. A challenging aspect of assessment of environmental impact of transgenics is the scale-dependent impact. Risk assessment entails gathering data by making well-designed and carefully monitored tests of increasing scale and complexity. Although in a small scale trial also, the consideration is given to the impact which would be relevant in the case of a large-scale commercial production; the probability of a rare event of inter-specific hybridization with a weed, which may not occur in the limited field trial but may happen when the transgenic is sown under widespread commercial production, cannot be completely negated. Individual complexities of farm and farmers can hinder the performance of whole-farm studies.

Since it is unrealistic to assume that all facets of environmental impact can be forethought, predicted and assessed, it is necessary to undertake sustained monitoring of transgenics post-commercialization. A marker gene such as the green fluorescent protein, can be introduced into the host plant along with the intended transgene for the in-field monitoring of the expression of the transgene and to facilitate the detection of gene flow (Harper et al. 1999; Lei et al. 2011). Expression under different agronomic conditions can be checked with Expressed Sequence Tags (ESTs). Transcriptome analysis-based databases have been developed which serve as comprehensive bioinformatics tool for analysis of differences in gene expression based on gene functional annotations of ESTs (Cheng and Strömvik 2008).

3.4.2 Environmental Risk Assessment (ERA) of Bt Crops in Different Countries

Bt crops have been adopted for commercial cultivation in several countries, including many developing countries. Internationally accepted risk assessment strategies have been followed before commercialization of Bt transgenic crops in different countries.

Environmental risk assessment is required for the regulatory approval of GM organisms in all countries signatories to the Cartagena Biosafety Protocol (CBP 2000). The Cartagena protocol has ratifications from 157 countries; however, the effectiveness of regulatory regimes may not be foolproof particularly in poorer economies (Kothamasi and Vermeylen 2011). Many developing countries have lack of scientific infrastructure and financial constraints. Accessibility of transgenic products may be relatively restricted in developing countries due to vocal opposition to transgenic technology and lack of regulatory mechanisms within which to deploy them (Devilliers and Hoisington 2011). A common framework for risk assessment and terminology is desirable. The key concepts need to be harmonized for broad application of problem formulation applied to different countries.

Public sector has an important role in development of Bt transgenic crops in the developing countries. Therefore, problem formulation should be in the context of public policy. Furthermore, given the higher crop genetic diversity of these countries, the local genetic background, into which the transgene might introgress, needs to be assessed for any adverse impacts. Hence, the developing countries need more pragmatic risk assessment for addressing the biosafety issues in their own particular environment. Because of lesser crop diversity, the risk ratio to transgenics is smaller for the developed world, than that for the developing countries. Relatively few wild relatives of cultivated crops exist in the developed countries. In USA and Australia, the release of transgenic plants has been kept out of the area where the wild relatives exist.

Additionally, the feasibility of follow up on on-farm monitoring of all the agricultural practices required for delaying resistance needs to be facilitated for successful large scale cultivation of the Bt transgenic crops in the developing countries (Kaur 2007). Enforcement of the practice to include non-transgenic refugia in transgenic plantation could be a problem in developing countries. Therefore, these countries need to devise and implement internationally acceptable norms and standards of monitoring and resistance management. Providing incentives for cultivation of Bt transgenic crops with agronomic practices that minimize the impact on environment and nurture crop diversity, sound crop rotation, soil fertility and wildlife biodiversity will be helpful (Dale et al. 2002).

A strong regulatory system should be established and properly implemented in the countries undertaking cultivation of Bt crops. To engender public trust, the key components of such a regulatory system should include: (i) mandatory pre-market approval, (ii) established safety standards, (iii) transparency, (iv) public participation, (v) use of outside scientists for expert scientific advice, (vi) independent agency decisions, (vii) post-approval activities and (viii) enforcement authority (Jaffe 2004). Regulatory costs limit the number of insect-resistant crops that may be developed (Raybould and Quemada 2010). Therefore, formulation of a well defined policy as to how the benefits and harms balance each other is important. There should be consensus among diverse groups of scientists involved in ERA. Hilbeck et al. (2011) have presented an improved ERA concept over that of the EU Directive 2001/18/EC which lays out the main provisions of ERA of GM organisms in Europe. The regulatory oversight should be sufficiently strict without being oppressive to research.

3.5 Conclusion and Future Prospects

Bt transgenic plants offer the benefits of improvement in agricultural production, food, nutrition and human health, but may present risks to the environment, if not monitored and managed properly. Ecological, economic and social concerns expressed about large scale cultivation of transgenic crops have been extensively addressed (Shelton et al. 2002; Kaur 2004). Success of transgenic technology will require a steady and conscientious assessment and management of risks. Effective risk assessment and monitoring entails integrated and multidisciplinary participation from different scientific teams to provide a holistic approach. Further studies on the effect of estimated and realistic Cry protein concentrations in the environment on the non-target organisms, which are representative of different agricultural ecosystems, based on standardized, long duration bioassays and improved biochemical methods, will be helpful.

Since ensuring complete biosafety of transgenic crops is a formidable task, the risk assessment strategies should focus on high probability risk rather than hypothetical or unrecognizable risk. Though some of the perceived eventual ecological effects of transgenics may be cognizable, it may not be feasible to estimate the consequences of these effects in entirety. Furthermore, the perception of a risk may differ between the scientific experts and lay public. For example, the public perception that Bt crops could harm Monarch butterflies has persisted as had been suggested by preliminary incomplete studies even though further studies found only minimal risk (Chapotin and Wolt 2007). A science-based approach and involvement of scientific experts is needed for improving the understanding of public with regard to the transgenic technology and the perceived risks thereof. Other social and ethical concerns should also be assessed for gaining the confidence of public (Kaur, 2004; de Melo- Martín and Meghani 2008).

Some uncertainties may merit continued caution. Therefore, long term studies are necessary. The new techniques of genomics, proteomics and metabolomics will also be useful for improvement of risk assessment methodologies. Management strategies for delay of development of resistance in insect pests to Cry proteins are a must for maintaining the efficacy of Bt crops. Elucidation of biochemical mechanism and genetics of resistance development will help in developing efficient strategies for the monitoring and management. Also, other potential solutions to the insect pest control by way of biological control, changes in farm management practices may be adopted based on their efficacy and sustainability. Accurate and adequate risk assessment and management will dictate the regulations that need to be adopted and guide strategies for development and deployment of future Bt transgenic crops.

References

Achon MA, Alonso-Dueñas N (2009) Impact of 9 years of Bt-maize cultivation on the distribution of maize viruses. Transgenic Res 18:387–397

Adamczyk JJ, Hardee DD, Adams LC, Sumerford DV (2001) Correlating differences in larval survival and development of bollworm (Lepidoptera: Noctuidae) and fall armyworm (Lepidoptera: Noctuidae) to differential expression of Cry1A(c) delta-endotoxin in various plant parts among commercial cultivars of transgenic *Bacillus thuringiensis* cotton. J Econ Entomol 94:284–290

Ahmad A, Wilde GE, Zhu KY (2005) Detectability of coleopteran-specific Cry3Bb1 protein in soil and its effect on nontarget surface and below-ground arthropods. Environ Entomol 34:385–394

Ali MI, Luttrell RG, Young SY (2006) Susceptibilities of *Helicoverpa zea* and *Heliothis virescens* (Lepidoptera: Noctuidae) populations to Cry1ac insecticidal protein. J Econ Entomol 99:164–175

Alvarez-Alfageme F, Bigler F, Romeis J (2011) Laboratory toxicity studies demonstrate no adverse effects of Cry1Ab and Cry3Bb1 to larvae of *Adalia bipunctata* (Coleoptera: Coccinellidae): the importance of study design. Transgenic Res 20:467–479

An J, Gao Y, Wu K, Gould F, Gao J, Shen Z, Lei C (2010) Vip3Aa tolerance response of *Helicoverpa armigera* populations from a Cry1Ac cotton planting region. J Econ Entomol 103:2169–2173

Andow DA, Hilbeck A (2004) Science-based risk assessment for nontarget effects of transgenic crops. Bioscience 54:637–649

Andow DA, Zwahlen C (2006) Assessing environmental risks of transgenic plants. Ecol Lett 9:196–214

Bai YY, Yan RH, Ye GY, Huang FN, Cheng JA (2010) Effects of transgenic rice expressing *Bacillus thuringiensis* Cry1Ab protein on ground-dwelling collembolan community in postharvest seasons. Environ Entomol 39:243–251

Baker JM, Hawkins ND, Ward JL, Lovegrove A, Napie JA, Shewry PR, Beale MH (2006) A metabolomic study of substantial equivalance of field-grown GM wheat. Plant Biotechnol J 4:381–392

Barber D, Rodríguez R, Salcedo G (2008) Molecular profiles: a new tool to substantiate serum banks for evaluation of potential allergenicity of GMO. Food Chem Toxicol 46:35–40

Barton KA, Whiteley HR, Yang NS (1987) *Bacillus thuringiensis* delta-endotoxin expressed in transgenic *Nicotiana tabacum* provides resistance to lepidopteran insects. Plant Physiol 85:1103–1109

Bates SL, Cao J, Zhao JZ, Earle ED, Roush RT, Shelton AM (2005) Evaluation of a chemically inducible promoter for developing a within-plant refuge for resistance management. J Econ Entomol 98:2188–2194

Baxter SW, Zhao J-Z, Gahan LJ, Shelton AM, Tabashnik BE, Heckel DG (2005) Novel genetic basis of field-evolved resistance to Bt toxins in *Plutella xylostella*. Insect Mol Biol 14:327–334

Benedict JH, Ring DR (2004) Transgenic crops expressing Bt proteins: current status, challenges and outlook. In: Koul O, Dhaliwal DS (eds) Transgenic crop protection: concepts and strategies. Science Publishers, Enfield, pp 15–84. ISBN 1-57808-302-8

Bergelson J, Winterer J, Purrington CB (1999) Ecological impacts of transgenic crops. In: Chopra VL, Malik VS, Bhat SR (eds) Applied plant biotechnology. Oxford IBH Publishing Company Pvt. Ltd., New Delhi, pp 325–343

Blumberg D, Navon A, Keren S, Goldenberg S, Ferkovich SM (1997) Interactions among *Helicoverpa armigera* (Lepidoptera: Noctuidae), its larval endoparasitoid *Microplitis croceipes* (Hymenoptera: Braconidae) and *Bacillus thuringiensis*. J Econ Entomol 90:1181–1186

Bondzio A, Stumpff F, Schön J, Martens H, Einspanier R (2008) Impact of *Bacillus thuringiensis* toxin Cry1Ab on rumen epithelial cells (REC)—a new in vitro model for safety assessment of recombinant food compounds. Food Chem Toxicol 46:1976–1984

Bourguet D, Bethenod MT, Pasteur N, Viard F (2000) Gene flow in the European corn borer *Ostrinia nubilalis*: implications for the sustainability of transgenic insecticidal maize. Proc R Soc Lond B Biol Sci 267:117–122

Bradford KJ, Deynze A van, Gutterson N, Parrot W, Strauss SH (2005) Regulating transgenic crops sensibly: lessons from plant breeding, biotechnology and genomics. Nat Biotechnol 23:439–444

Brookes G, Barfoot P (2008) Global impact of biotech crops: socio-economic and environmental effects 1996–2006. AgBioForum 11:21–38

Cao J, Bates SL, Zhao JZ, Shelton AM, Earle ED (2006) *Bacillus thuringiensis* protein production, signal transduction, and insect control in chemically inducible PR-1a/cry1Ab broccoli plants. Plant Cell Rep 25:554–560

Carpenter JE (2010) Peer-reviewed surveys indicate positive impact of commercialized GM crops. Nat Biotechnol 28:319–321

Carpenter JE (2011) Impacts of GM crops on biodiversity. GM Crops 2:1–17

Carriere Y, Tabashnik BE (2001) Reversing insect adaptation to transgenic insecticidal plants. Proc R Soc Lond B Biol Sci 268:1475–1480

Carriere Y, Ellers-Kirk C, Liu YB, Sims MA, Patin AL, Dennehy TJ, Tabashnik BE (2001) Fitness costs and maternal effects associated with resistance to transgenic cotton in the pink bollworm (Lepidoptera: Gelechiidae). J Econ Entomol 96:1571–1576

Cartagena Protocol on Biosafety to the Convention on Biological Diversity (CBP) (2000) Secretariat of the convention on biological diversity, Montreal, Canada. http://www.biodiv.org/biosafety/protocol.asp

Cellini F, Chesson A, Colquhoun I, Constable A, Davies HV, Engel KH, Gatehouse AM, Karenlampi S, Kok EJ, Leguay JJ, Lehesranta S, Noteborn H, Pedersen J, Smith M (2004) Unintended effects and their detection in genetically modified crops. Food Chem Toxicol 42:1089–1125

Chambers CP, Whiles MR, Rosi-marshall EJ, Tank JL, Royer TV, Griffiths NA, Evans-White MA, Stojak AR (2010) Responses of stream macroinvertebrates to Bt maize leaf detritus. Ecol Appl 20:1949–1960

Chandler S, Dunwell JM (2008) Gene flow, risk assessment and the environmental release of transgenic plants. Crit Rev Plant Sci 27:25–49

Chapotin S, Wolt J (2007) Genetically modified crops for the bioeconomy: meeting public and regulatory expectations. Transgenic Res 16:675–688

Chassy B (2002) Food safety evaluation of crops produced through biotechnology. J Am Coll Nutr 21:166S–173S

Chen M, Zhao J-Z, Collins HL, Earle ED, Cao J, Shelton AM (2008a) A critical assessment of the effects of Bt transgenic plants on parasitoids. PLoS One 3:e2284

Chen M, Zhao JZ, Shelton AM, Cao J, Earle ED (2008b) Impact of single-gene and dual-gene Bt broccoli on the herbivore *Pieris rapae* (Lepidoptera: Pieridae) and its pupal endoparasitoid *Pteromalus puparum* (Hymenoptera: Pteromalidae). Transgenic Res 17:545–555

Chen M, Ye G-Y, Liu Z-C, Fang Q, Hu C, Peng Y-F, Shelton AM (2009) Analysis of Cry1Ab toxin bioaccumulation in a food chain of Bt rice, an herbivore and a predator. Ecotoxicology 18:230–238

Cheng KC, Strömvik MV (2008) SoyXpress: a database for exploring the soybean transcriptome. BMC Genomics 9:368

Choudhary B, Gaur K (2010) Bt cotton in India: a country profile. ISAAA series of biotech crop profiles. ISAAA, Ithaca. ISBN 978-1-892456-46-X
Christov NK, Imaishi H, Ohkawa H (1999) Green-tissue-specific expression of a reconstructed cry1C gene encoding the active fragment of *Bacillus thuringiensis* delta-endotoxin in haploid tobacco plants conferring resistance to *Spodoptera litura*. Biosci Biotechnol Biochem 63:1433–1444
Clark BW, Phillips TA, Coats JR (2005) Environmental fate and effects of *Bacillus thuringiensis* (Bt) proteins from transgenic crops: a review. J Agric Food Chem 53:4643–4653
Conner AJ, Glare TR, Nap J-P (2003) The release of genetically modified crops into the environment. Part II. Overview of ecological risk assessment. Plant J 33:19–46
Crawley MJ (1995) Long term ecological effects of the release of genetically modified organisms. In: CEP, Pan-European conference on the potential long-term ecological impact of the genetically modified organisms. Proceedings Strasbourg, 24–26 Nov. 1993. Council of Europe Press, Strasbourg, pp 43–66
Cui JJ, Luo JY, van der Werf W, Ma Y, Xia JY (2011) Effect of pyramiding Bt and CpTI genes on resistance of cotton to *Helicoverpa armigera* (Lepidoptera: Noctuidae) under laboratory and field conditions. J Econ Entomol 104:673–684
Dale PJ, Clarke B, Fontes EMG (2002) Potential for the environmental impact of transgenic crops. Nat Biotechnol 20:567–574
Daniell H (2007) Transgene containment by maternal inheritance: effective or elusive? Proc Natl Acad Sci U S A 104:6879–6880
De Cosa B, Moar W, Lee SB, Miller M, Daniell H (2001) Overexpression of the Bt cry2Aa2 operon in chloroplasts leads to formation of insecticidal crystals. Nat Biotechnol 19:71–74
de Escudero R, Estela A, Porcar M, Martinez C, Oguiza JA, Escriche B, Ferre J, Cadallero P (2006) Molecular and insecticidal characterization of a Cry1I protein toxic to insects of the families Noctuidae, Tortricidae, Plutellidae and Chrysomelidae. Appl Environ Microbiol 72:4796–4804
de Melo-Martín I, Meghani Z (2008) Beyond risk: a more realistic risk-benefit analysis of agricultural biotechnologies. EMBO Rep 9:303–307
DeVilliers SM, Hoisington DA (2011) The trends and future of biotechnology crops for insect pest control. Afr J Biotechnol 10:4677–4681
Dichn SH, Chiu EL, De Rocher EJ, Green PJ (1998) Premature adenylation at multiple sites within a *Bacillus thuringiensis* toxin gene coding region. Plant Physiol 117:143–1443
Dively GP, Rose R, Sears MK, Hellmich RL, Stanley-Horn DE, Calvin DD, Russoe JM, Anderson PL (2004) Effects on monarch butterfly larvae (Lepidoptera: Danaidae) after continuous exposure to Cry1Ab-expresing corn during anthesis. Environ Entomol 33:1116–1125
Dogan EB, Berry RE, Reed GL, Rossignol PA (1996) Biological parameters of convergent lady beetle (Coleoptera: Coccinellidae) feeding on aphids (Homiptera: Aphididae) on transgenic potatoes. J Econ Entomol 89:1105–1108
Downes S, Parker T, Mahon R (2010) Incipient resistance of *Helicoverpa punctigera* to the Cry-2Ab Bt toxin in Bollgard II cotton. PLoS One 5:e12567
Duan JJ, Marvier M, Huesing J, Dively G, Huang ZY (2008) A Meta-analysis of effects of Bt crops on honey bees (Hymenoptera: Apidae). PLoS One 3:1–6
Duan JJ, Lundgren JG, Naranjo S, Marvier M (2010) Extrapolating non-target risk of Bt crops from laboratory to field. Biol Lett 6:74–77
Dutton A, Klein H, Romeis J, Bigler F (2002) Uptake of Bt-toxin by herbivores feeding on transgenic maize and consequences for the predator Chrysoperla carnea. Ecol Entomol 27:441–447
EFSA (2008) European Food Safety Authority. Draft updated guidance document for the risk assessment of genetically modified plants and derived food and feed. EFSA J 727:1–135
European Union (2003) Regulation (EC) No. 1829/2003 of the European Parliament and of the Council of 22 September 2003 on genetically modified food and feed. European Parliament and the Council of the European Union, Brussels
Fabrick JA, Tabashnik BE (2007) Binding of *Bacillus thuringiensis* toxin Cry1Ac to multiple sites of cadherin in pink bollworm. Insect Biochem Mol Biol 37:97–106

Feitelson JS, Payne J, Kim L (1992) *Bacillus thuringiensis* insects and beyond. Biotechnology 10:271–275
Ferre J, Van Rie J (2002) Biochemistry and genetics of insect resistance to *Bacillus thuringiensis*. Annu Rev Entomol 47:501–533
Ferry N, Mulligan EA, Majerus MEN, Gatehouse AMR (2007) Bitrophic and tritrophic effects of Bt Cry3A transgenic potato on beneficial, non-target, beetles. Transgenic Res 16:795–812
Finamore A, Roseelli M, Britti S, Monstra G, Ambra R, Turrini A, Mengheri E (2008) Intestinal and peripheral immune response to MON810 maize ingestion in weaning and old mice. J Agric Food Chem 56:11533–11539
Flachowsky G, Halle I, Aulrich K (2005) Long term feeding of Bt corn-a 10 generation study with quails. Arch Anim Nutr 59:449–451
Flachowsky G, Aulrich K, Böhme H, Halle I (2007) Studies on feeds from genetically modified plants (GMP): contributions to nutritional and safety assessment. Anim Feed Sci Technol 133:2–30
Flexner JL, Belnavis DL (1999) Microbial insecticides. In: Rechcigl JE, Rechcigl NA (eds) Biological and biotechnological control of insect pests. Lewis Publishers, Boca Raton, pp 35–62. ISBN 1-56670-479-0
Gahan LJ, Gould F, Heckel DG (2001) Identification of a gene associated with Bt resistance in *Heliothis virescens*. Science 293:857–860
Gahan LJ, Pauchet Y, Vogel H, Heckel DG (2010) An ABC transporter mutation is correlated with insect resistance to *Bacillus thuringiensis* Cry1Ac toxin. PLoS Genet 6:e1001248. doi:10.1371/journal.pgen.1001248
García MC, García B, García-Ruiz C, Gómez A, Cifuentes A, Marina ML (2009) Rapid characterisation of (glyphosate tolerant) transgenic and non-transgenic soybeans using chromatographic protein profiles. Food Chem 113:1212–1217
García-Robles I, Sanchez J, Gruppe A, Martinez-Ramirez AC, Rausell C, Real MD, Bravo A (2001) Mode of action of *Bacillus thuringiensis* PS86Q3 strain in hymenopteran forest pests. Insect Biochem Mol Biol 31:849–856
Gassmann AJ, Carrière Y, Tabashnik BE (2009) Fitness costs of insect resistance to *Bacillus thuringiensis*. Annu Rev Entomol 54:147–163
Gatehouse J (2008) Biotechnological prospects for engineering insect-resistant plants. Plant Physiol 146:881–887
Gaudron T, Peters C, Boland E, Steinmetz A, Moris G (2009) Development of a quadruplex-real-time-PCR for screening food for genetically modiWed organisms. Eur Food Res Technol 229:295–305
Godfray HCJ, Beddington JR, Crute IR, Haddad L, Lawrence D, Muir JF, Pretty J, Robinson S, Thomas SM, Toulmin C (2010) Food security: the challenge of feeding 9 billion people. Science 327:812–818
Goodman RE, Vieths S, Sampson HA, Hill D, Ebisawa M, Taylor SL, van Ree R (2008) Allergenicity assessment of genetically modified crops—what makes sense? Nat Biotechnol 26:73–81
Gould F (1998) Sustainability of transgenic insecticidal cultivars: integrating pest genetics and ecology. Annu Rev Entomol 43:701–726
Griffitts JS, Huffman DL, Whitacre JL, Barrows BD, Marroquin LD, Muller R, Brown JR, Hennet T, Esko JD, Aroian RV (2003) Resistance to a bacterial toxin is mediated by removal of a conserved glycosylation pathway required for toxin-host interactions. J Biol Chem 278:45594–45602
Gruber H, Paul V, Meyer HH, Müller M (2011) Determination of insecticidal Cry1Ab protein in soil collected in the final growing seasons of a nine-year field trial of Bt-maize MON810. Transgenic Res. doi:10.1007/s11248-011-9509-7
Gunning RV, Dang HT, Kemp FC, Nicholson IC, Moores GD (2005) New resistance mechanism in *Helicoverpa armigera threatens transgenic crops expressing Bacillus thuringiensis* Cry1Ac toxin. Appl Environ Microbiol 71:2558–2563
Halfhill MD, Sutherland JP, Moon HS, Poppy GM, Warwick SI, Weissinger AK, Rufty TW, Raymer PL, Stewart CN Jr (2005) Growth, productivity, and competitiveness of introgressed

weedy *Brassica rapa* hybrids selected for the presence of Bt *cry*1Ac and *gfp* transgenes. Mol Ecol 14:3177–3189
Hamilton KA, Pyla PD, Breeze M, Olson T, Li M, Robinson E, Gallghar SP, Sorbet SP, Chen Y (2004) Bollgard II cotton: compositional analysis and feeding studies of cottonseed from insect-protested cotton (*Gossypium hirsutum L.*) producing the Cry1Ac and Cry2Ab2 proteins. J Agric Food Chem 52:6969–6976
Han P, Niu CY, Lei CL, Cui JJ, Desneux N (2010) Use of an innovative T-tube maze assay and the proboscis extension response assay to assess sublethal effects of GM products and pesticides on learning capacity of the honey bee *Apis mellifera L.* Ecotoxicology 19:1612–1619
Harper BK, Mabon SA, Leffel SM, Halfhill MD, Richards HA, Moyer KA, Stewart CN Jr (1999) Green fluorescent protein as a marker for expression of a second gene in transgenic plants. Nat Biotechnol 17:1125–1129
Herrero S, Ferre J, Escriche B (2001) Mannose phosphate isomerase isoenzymes in *Plutella xylostella* support common genetic bases of resistance to *Bacillus thuringiensis* toxins in Lepidopteran species. Appl Environ Microbiol 67:979–981
Hilbeck A, Schmidt JEU (2006) Another view on Bt proteins- How specific they are and what else they might do? Biopestic Int 2:1–50
Hilbeck A, Baumgartner M, Fried PM, Bigler F (1998) Effects of transgenic *Bacillus thuringiensis* corn-fed prey on mortality and development time of immature *Chrysoperla carnea* (Neuroptera: Chrysopidae). Environ Entomol 27:480–487
Hilbeck A, Meier M, Römbke J, Jänsch S, Teichmann H, Tappeser B (2011) Environmental risk assessment of genetically modified plants—concepts and controversies. Environ Sci Eur 23:13. doi:10.1186/2190-4715-23-13
Hofs JL, Schoeman AS, Pierre J (2008) Diversity and abundance of flower-visiting insects in Bt and non-Bt cotton fields of Maputaland (KwaZulu Natal Province, South Africa). Int J Trop Insect Sci 28:211–219
Holzhauser T, van Ree R, Poulsen LK, Bannon GA (2008) Analytical criteria for performance characteristics of IgE binding methods for evaluating safety of biotech food products. Food Chem Toxicol 46:15–192
Hu HY, Liu XX, Zha ZW, Sun JG, Zhang QW, Liu Z, Yong Y (2009) Effects of repeated cultivation of transgenic Bt cotton on functional bacterial populations in rhizosphere soil. World J Microbiol Biotechnol 25:357–366
Icoz I, Stotzky G (2008a) Cry3Bb1 protein from *Bacillus thuringiensis* in root exudates and biomass of transgenic corn does not persist in soil. Transgenic Res 17:609–620
Icoz I, Stotzky G (2008b) Fate and effects of insect-resistant Bt crops in soil ecosystems. Soil Biol Biochem 40:559–586
ILSI Research Foundation (2010) A review of the environmental safety of the Cry1Ac protein. Center for environmental risk assessment, 26 May 2010. ILSI Research Foundation, Washington DC, p 18
Jaffe G (2004) Regulating transgenic crops: a comparative analysis of different regulatory processes. Transgenic Res 13:5–19
James C (2010) Global view of commercialized transgenic crops: 2010. Brief no. 42. ISAAA (International Service for Acquisition of Agri-biotech Applications), Ithaca, NY, USA. http://www.isaaa.org/publications/briefs/Breif_.htm
Jensen PD, Dively GP, Swan CM, Lamp WO (2010) Exposure and non-target effects of transgenic Bt corn debris in streams. Environ Entomol 39:707–714
Johnson KL, Raybould AJ, Hudson MD, Poppy GM (2007) How does scientific risk assessment of GM crops fit within the wider risk analysis? Trends Plant Sci 12:1–5
Kamle S, Kumar A, Bhatnagar RK (2011) Development of multiplex and construct specific PCR assay for detection of *cry*2Ab transgene in genetically modified crops and product. GM Crops 2:1–8
Kaur S (2000) Molecular approaches towards development of novel *Bacillus thuringiensis* biopesticides. World J Microbiol Biotechnol 16:781–793

Kaur S (2004) Ecological, economic and social perspectives on transgenic crop protection: path for the developing world. In: Koul O, Dhaliwal DS (eds) Transgenic crop protection: concepts and strategies. Science Publishers, Enfield, pp 373–405
Kaur S (2006) Molecular approaches for identification and construction of novel insecticidal genes for crop protection. World J Microbiol Biotechnol 22:233–253
Kaur S (2007) Deployment of Bt transgenic crops: development of resistance and management strategies in the Indian scenario. Biopestic Int 3:23–42
Keese P (2008) Risks from GMOs due to horizontal gene transfer. Environ Biosafety Res 7:123–149
Kjellsson G (1997) Principles and procedures for ecological risk assessment of transgenic plants. In: Kjellsson G, Simonsen V, Ammann K (eds) Methods for risk assessment of transgenic plants. II. Pollination, gene transfer and population impacts. Birkhauser, Basel, pp 221–236. ISBN 3-7643-5696-0
Kota M, Daniell H, Varma S, Garczynski SF, Gould F, Moar WJ (1999) Overexpression of the *Bacillus thuringiensis* (Bt) Cry2Aa2 protein in chloroplasts confers resistance to plants against susceptible and Bt-resistant insects. Proc Natl Acad Sci U S A 96:1840–1845
Kothamasi D, Vermeylen S (2011) Genetically modified organisms in agriculture: can regulations work? Environ Dev Sustain 13:535–546
Lawo NC, Romeis J (2008) Assessing the utilization of a carbohydrate food source and the impact of insecticidal proteins on larvae of the green lacewing, *Chrysoperla carnea*. Biol Control 44:389–398
Lawo NC, Wäckers FL, Romeis J (2010) Characterizing indirect prey-quality mediated effects of a Bt crop on predatory larvae of the green lacewing, *Chrysoperla carnea*. J Insect Physiol 56:1702–1710
Lee MK, Walters FS, Hart H, Palekar N, Chen J-S (2003) The mode of action of the *Bacillus thuringiensis* vegetative insecticidal protein Vip3A differs from that of Cry1Ab δ-endotoxin. Appl Environ Microbiol 69:4648–4657
Lei L, Stewart CN Jr, Tang Z-X, Wei W (2011) Dynamic expression of green fluorescent protein and *Bacillus thuringiensis* Cry1Ac endotoxin in interspecific hybrids and successive backcross generations (BC1 and BC2) between transgenic *Brassica napus* crop and wild *Brassica juncea*. Ann Appl Biol. doi:10.1111/j.1744-7348.2011.00486.x
Li Y, Romeis J (2010) Bt maize expressing Cry3Bb1 does not harm the spider mite, *Tetranychus urticae* or its ladybird beetle predator, *Stethorus punctillum*. Biol Control 53:337–344
Li H, Gonzalez-Cabrera J, Opert B, Ferre J, Higgins RA, Suschman LL, Radke GA, Zhu KY, Huang F (2004) Binding analysis of Cry1Ab and Cry1Ac with membrane vesicles from Bt resistant and susceptible *Ostrinia nubilalis*. Biochem Biophys Res Commun 323:52–57
Li YH, Wu KM, Zhang YJ, Yuan GH (2007) Degradation of Cry1Ac protein within transgenic *Bacillus thuringiensis* rice tissues under field and laboratory conditions. Environ Entomol 36:1275–1282
Liu W (2009) Effects of Bt transgenic crops on soil ecosystems: a review of a ten-year research in China. Front Agric China 3:190–198
Liu B, Shu C, Xue K, Zhou KX, Li XG, Liu DD, Zheng YP, Xu CR (2009a) The oral toxicity of the transgenic Bt+CpTI cotton pollen to honeybees (*Apis mellifera*). Ecotoxicol Environ Safe 72:1163–1169
Liu B, Wang L, Zeng Q, Meng J, Hu WJ, Li XG, Zhou KX, Xu K, Liu DD, Zheng YP (2009b) Assessing effects of transgenic Cry1Ac cotton on the earthworm *Eisenia fetida*. Soil Biol Biochem 41:1841–1846
Liu X, Chen M, Onstad D, Roush R, Shelton AM (2010) Effect of Bt broccoli and resistant genotype of *Plutella xylostella* (Lepidoptera: Plutellidae) on development and host acceptance of the parasitoid Diadegma insulare (Hymenoptera: Ichneumonidae). Transgenic Res 20:887–897
Losey JE, Rayor LS, Carter ME (1999) Transgenic pollen harms monarch larvae. Nature 399:214
Lövei GL, Andow DA, Arpaia S (2009) Transgenic insecticidal crops and natural enemies: a detailed review of laboratory studies. Environ Entomol 38:293–306

Lutz KA, Maliga P (2007) Construction of marker-free transplastomic plants. Curr Opin Biotechnol 18:107–114

Mancebo A, Molier T, González B, Lugo S, Riera L, Arteaga ME, Bada AM, González Y, Pupo M, Hernández Y, González C, Rojas NM, Rodríguez G (2011) Acute oral, pulmonary and intravenous toxicity/pathogenicity testing of a new formulation of *Bacillus thuringiensis* var *israelensis* SH-14 in rats. Regul Toxicol Pharmacol 59:184–190

Mann RS, Gill RS, Dhawan AK, Shera PS (2010) Relative abundance and damage by target and non-target insects on Bollgard and Bollgard II cotton cultivars. Crop Prot 29:793–801

Marvier M, McCreedy C, Regetz J, Kareiva P (2007) A meta-analysis of effects of Bt cotton and maize on nontarget invertebrates. Science 316:1475–1477

Meagher TR, Belanmger FC, Day PR (2003) Using empirical data to model transgene dispersal. Philos Trans R Soc Lond B 358:1157–1162

Mehlo L, Gahakwa D, Nghia PT, Loc NT, Capell T, Gatehouse JA, Gatehouse AM, Christou P (2005) An alternative strategy for sustainable pest resistance in genetically enhanced crops. Proc Natl Acad Sci U S A 102:7812–7816

Mendelsohn M, Kough J, Vaituzis Z, Matthews K (2003) Are Bt crops safe? Nat Biotechnol 21:1003–1009

Miller HI (2010) The tarnished gold standard for GM risk assessment. GM Crops 1:59–61

Miller MC, Powell W (1994) A commercial view of biotechnology in crop production. In: Marshall G, Walters D (eds) Molecular biology in crop protection. Chapman and Hall, London, pp 225–245

Moellenbeck DJ, Peters ML, Bing JW, Rouse JR, Higgins SL, Nevshemal T, Marshall L, Ellis RT, Bystrak PG, Lang BA, Stewart JL, Kouba K, Sondag B, Gustafson B, Nour K, Xu D, Swenson J, Zhang J, Czapla T, Schwab G, Jayne S, Stockhoff BA, Narva K, Schnepf HE, Stelman SJ, Poutre C, Koziel M, Duck N (2001) Insecticidal proteins from *Bacillus thuringiensis* protected corn from corn rootworms. Nat Biotechnol 19:668–672

Molin S, Boe L, Jensen LB, Kristensen CS, Givskov M, Ramos JL, Bej AK (1993) Suicidal genetic elements and their use in biological containment of bacteria. Annu Rev Microbiol 47:139–166

Monsanto Company (2002) Request for additional information on the Direct Grower and Environmental Benefits of Bollgard II cotton to support the registration of the plant-incorporated protectant, Cry2Ab2 insect control protein in cotton. Letter to U.S. Environmental Protection Agency dated October 25, 2002. Contained in MRID# 455588-00

Morin S, Biggs RW, Sisterson MS, Shriver L, Ellere-Kirk C, Higginson D, Holley D, Gahan LJ, Heckel DG, Carriere Y, Dennehy TJ, Brown JK, Tabashnik BE (2003) Three cadherin alleles associated with resistance to *Bacillus thuringiensis* in pink bollworm. Proc Natl Acad Sci U S A 100:5004–5009

Naranjo SE (2005) Long-term assessment of the effects of transgenic Bt cotton on the function of the natural enemy community. Environ Entomol 34:1211–1223

Naranjo SE, Head G, Dively G (2005) Field studies assessing arthropod non-target effects in Bt transgenic crops: introduction. Environ Entomol 34:1178–1180

Netherwood T, Martin-Orue SM, O'Donnell AG, Gockling S, Graham J, Mathers JC, Gilbert HJ (2004) Assessing the survival of transgenic plant DNA in the human gastrointestinal tract. Nat Biotechnol 22:204–209

Nielsen KM, Bones AM, Smalla K, van Elsas JD (1998) Horizontal gene transfer from transgenic plants to terrestrial bacteria—a rare event? FEMS Microbiol Rev 22:79–103

Noteborn HP, Lommen A, van der Jagt RC, Weseman JM (2000) Chemical fingerprinting for the evaluation of unintended secondary metabolic changes in transgenic food crops. J Biotechnol 77:103–114

O'Callaghan M, Glare TR, Burgess EPJ, Malone LA (2005) Effects of plants genetically modified for insect resistance on nontarget organisms. Annu Rev Entomol 50:271–292

Palaudelma`s M, Peñas G, Mele' E, Serra J, Salvia J, Pla M, Nadal A, Messeguer J (2009) Effect of volunteers on maize gene flow. Transgenic Res 18:583–594

Poppy GM (2004) Gene flow from GM plants: towards a more quantitative risk assessment. Trends Biotechnol 22:436–438

Priestley AL, Brownbridge M (2009) Field trials to evaluate effects of Bt-transgenic silage corn expressing the Cry1Ab insecticidal toxin on non-target soil arthropods in northern New England, USA. Transgenic Res 18:425–443
Ramachandran S, Buntin GD, All JN, Tabashnik BE, Reymer PL, Adang MJ, Pulliam DA, Steward CN Jr (1998) Survival, development and oviposition of resistant diamondback moth (Lepidoptera: Plutellidae) on transgenic canolatoxin. J Econ Entomol 91:1239–1244
Rauschen S (2010) A case of "pseudo science"? A study claiming effects of the Cry1Ab protein on larvae of the two spotted ladybird is reminiscent of the case of the green lacewing. Transgenic Res 19:13–16
Rauschen S, Schaarschmidt F, Gathmann A (2010) Occurrence and field densities of Coleoptera in the maize herb layer: implications for environmental risk assessment of genetically modified Bt-maize. Transgenic Res 19:727–744
Rawat P, Singh AK, Ray K, Chaudhary B, Kumar S, Gautam T, Kanoria S, Kaur G, Kumar P, Pental D, Burma PK (2011) Detrimental effect of expression of Bt endotoxin Cry1Ac on in vitro regeneration, in vivo growth and development of tobacco and cotton transgenics. J Biosci 36:363–376. doi:10.1007/s12038-011-9074-5
Raybould A (2007) Ecological versus ecotoxicological methods for assessing the environmental risks of transgenic crops. Plant Sci 173:589–602
Raybould A, Quemada H (2010) Bt crops and food security in developing countries: realised benefits, sustainable use and lowering barriers to adoption. Food Sec 2:247–259
Raybould A, Vlachos D (2011) Non-target organism effects tests on Vip3A and their application to the ecological risk assessment for cultivation of MIR162 maize. Transgenic Res 20:599–611
Raybould A, Tuttle A, Shore S, Stone T (2010) Environmental risk assessments for transgenic crops producing output trait enzymes. Transgenic Res 19:595–609
Reed GL, Jensen AS, Riebe J, Head G, Duan JJ (2001) Transgenic Bt potato and conventional insecticides for Colorado potato beetle management: comparative efficacy and non-target impacts. Entomol Exp Appl 100:89–100
Ricroch A, Bergé JB, Kuntz M (2010) Is the German suspension of MON810 maize cultivation scientifically justified? Transgenic Res 19:1–12
Romeis J, Meissle M (2011) Non-target risk assessment of Bt crops-Cry protein uptake by aphids. J Appl Entomol 135:1–6
Romeis J, Dutton A, Bigler F (2004) *Bacillus thuringiensis* toxin has no direct effect on larvae of the green lacewing *Chrysoperla carnea* (Stephens) (Neuroptera: Chrysopidae). J Insect Physiol 50:175–183
Romeis J, Meissle M, Bigler F (2006) Transgenic crops expressing *Bacillus thuringiensis* toxins and biological control. Nat Biotechnol 24:63–71
Romeis J, Bartsch D, Bigler F, Candolfi MP, Gielkens MMC, Hartley SE, Hellmich RL, Huesing JE, Jepson PC, Layton R, Quemada H, Raybould A, Rose RI, Schiemann J, Sear MK, Shelton AM, Sweet J, Vaituzis Z, Wolt JD (2008) Assessment of risk of insect-resistant transgenic crops to non-target arthropods. Nat Biotechnol 26:203–208
Romeis J, Meissle M, Raybould A, Hellmich RL (2009) Impact of insect-resistant transgenic crops on above-ground nontarget arthropods. In: Ferry N, Gatehouse AMR (eds) Environmental impact of genetically modified crops. CAB International, Wallingford, pp 165–198
Romeis J, Hellmich RL, Candolfi MP, Carstens K, De Schrijver A, Gatehouse AMR, Herman RA, Huesing JE, McLean MA, Raybould A, Shelton AM, Waggoner A (2011) Recommendations for the design of laboratory studies on non-target arthropods for risk assessment of genetically engineered plants. Transgenic Res 20:1–22
Rong J, Lu BR, Song Z, Su J, Snow AA, Zhang X, Sun S, Chen R, Wang F (2007) Dramatic reduction of crop-to-crop gene flow within a short distance from transgenic rice fields. New Phytol 173:346–353
Roush R (1997) Managing resistance to transgenic crops. In: Carozzi N, Koziel MG (eds) Advances in insect control: the role of transgenic plants. Taylor and Francis, London, pp 271–294
Ruf S, Karcher D, Bock R (2007) Determining the transgene containment level provided by chloroplast transformation. Proc Natl Acad Sci U S A 104:6998–7002

Sanvido O, Widmer F, Winzeler M, Streit B, Szerencsits E, Bigler F (2008) Definition and feasibility of isolation distances for transgenic maize cultivation. Transgenic Res 17:317–335
Sanvido O, Romeis J, Bigler F (2011) Environmental change challenges decision-making during post-market environmental monitoring of transgenic crops. Transgenic Res. doi:10.1007/s11248-011-9524-8
Saxena D, Stotzky G (2000) Insecticidal toxin from *Bacillus thuringiensis* is released from roots of transgenic Bt corn in vitro and in situ. FEMS Microbiol Ecol 33:35–39
Schmidt J, Braun C, Whitehouse L, Hilbeck A (2009) Effects of activated Bt transgene products (Cry1Ab, Cry3Bb) on immature stages of the ladybird *Adalia bipunctata* in laboratory ecotoxicity testing. Arch Environ Contam Toxicol 56:221–228
Schnepf E, Crickmore N, Van Rie J, Lereclus D, Baum J, Feitelson J, Zeigler DR, Dean DH (1998) *Bacillus thuringiensis* and its pesticidal crystal proteins. Microbiol Mol Biol Rev 62:775–806
Schuler TH, Denholm I, Jouanin L, Clark AJ, Poppy GM (2001) Population scale laboratory studies of the effect of transgenic plant on non-target insects. Mol Ecol 10:1845–1853
Schuler TH, Denholm I, Clark SJ, Stewart CN, Poppy GM (2004) Effects of Bt plants on the development and survival of the parasitoid *Cotesia plutellae* (Hymenoptera: Braconidae) in susceptible and Bt-resistant larvae of the diamondback moth, *Plutella xylostella* (Lepidoptera: Plutellidae). J Insect Physiol 50:435–443
Sears MK, Hellmich RL, Stanley-Horn DE, Oberhauser KS, Pleasants JM, Mattila HR, Siegfried BD, Dively GP (2001) Impact of Bt corn pollen on monarch butterfly populations: a risk assessment. Proc Natl Acad Sci U S A 98:11937–11942
Selgrade MK, Bowman CC, Ladics GS, Privalle L, Laessig SA (2009) Safety assessment of biotechnology products for potential risk of food allergy: implications of new research. Toxicol Sci 110:31–39
Shelton AM, Tang JD, Roush RT, Metz TD, Earle ED (2000) Field tests on managing resistance to Bt-engineered plants. Nat Biotechnol 18:339–342
Shelton AM, Zhao JZ, Roush RT (2002) Economic, ecological, food safety and social consequences of the deployment of the Bt transgenic plants. Annu Rev Entomol 47:845–881
Shelton AM, Romeis J, Kennedy GG (2008) IPM and insect protected transgenic plants: thoughts for the future. In: Romeis J, Shelton AM, Kennedy GG (eds) Integration of insect-resistant, genetically modified crops within IPM programs. Springer, Dordrecht, pp 419–429
Shelton A, Naranjo S, Romeis J, Hellmich R, Wolt J, Federici B et al (2009) Setting the record straight: a rebuttal to an erroneous analysis on transgenic insecticidal crops and natural enemies. Transgenic Res 18:317–322
Siegfried BD, Waterfield N, ffrench-Constant RH (2005) Expressed sequence tags from Diabrotica virgifera virgifera midgut identify a coleopteran cadherin and a diversity of cathepsins. Insect Mol Biol 14:137–143
Stewart CN Jr, Richards HA, Halfhill MD (2000) Transgenic plants and biosafety: science, misconceptions and public perceptions. Biotechniques 29:832–836, 838–843
Storer NP, Babcock JM, Edwards JM (2006) Field measures of western corn root worm (Coleoptera: Chrysomelidae) mortality caused by Cry34/35Ab1 proteins expressed in maize event 59122 and implications for trait durability. J Econ Entomol 99:1381–1387
Storer NP, Babcock JM, Schlenz M, Meade T, Thompson GD et al (2010) Discovery and characterization of field resistance to Bt maize: *Spodoptera frugiperda* (Lepidoptera: Noctuidae) in Puerto Rico. J Econ Entomol 103:1031–1038
Stotzky G (2001) Release, persistence and biological activity in soil of insecticidal proteins from *Bacillus thuringiensis*. In: Letourneau DK, Burrows BE (eds) Genetically engineered organisms. CRC Press, Boca Raton, pp 187–222
Tabashnik BE, Carriere Y, Dennehy TJ, Morin S, Sisterson MS, Roush RE, Shelton AM, Zhao JZ (2003) Insect resistance to transgenic Bt crops: lessons from the laboratory and the field. J Econ Entomol 96:1031–1038
Tabashnik BE, Biggs RW, Higginson DM, Henderson S, Unnithan DC, Unnithan GC, Elers-Kirk C, Sisterson MS, Dennehy TJ, Carriere Y, Shai M (2005) Association between resistance to Bt cotton and cadherin genotype in pink bollworm. J Econ Entomol 98:635–644

Tabashnik BE, Fabrick JA, Henderson S, Biggs RW, Yafuso CM, Nyboer ME, Manhardt NM, Coughlin LA, Sollome J, Carriere Y, Dennehy TJ, Morin S (2006) DNA screening reveals pink bollworm resistance to Bt cotton remains rare after a decade of exposure. J Econ Entomol 99:1525–1530

Tabashnik BE, Gassmann AJ, Crowder DW, Carrière Y (2008) Insect resistance to Bt crops: evidence versus theory. Nat Biotechnol 26:199–202

Tabashnik BE, Van Rensburg JBJ, Carrière Y (2009) Field-evolved insect resistance to Bt crops: definition, theory, and data. J Econ Entomol 102:2011–2025

Talas-Og˘ras T (2011) Risk assessment strategies for transgenic plants. Acta Physiol Plant 33:647–657

Tang M, Huang K, Li W, Zhou K, He X, Lu Y (2006) Absence of effect after introducing *Bacillus thuringiensis* gene on nutritional composition in cottonseed. J Food Sci 71:S38–S41

Thomson JA (2001) Horizontal transfer of DNA from GM crops to bacteria and to mammalian cells. J Food Sci 66:188–193

Trabalza-Marinucci M, Brandi G, Rondini C, Avellini L, Giammarini C, Costarelli S, Acuti G, Orlandi C, Filippini G, Chiaradia E, Malatesta M, Crotti S, Antonini C, Amaglianib G, Manuali E, Mastrogiacomo AR, Moscati L, Haouet MN, Gaiti, A, Magnani M (2008) A three-year longitudinal study on the effects of a diet containing genetically modified Bt176 maize on the health status and performance of sheep. Livestock Sci 113:178–190

van de Wouw M, van Hintum T, Kik C, van Treuren R, Visser B (2010) Genetic diversity trends in twentieth century crop cultivars: a meta analysis. Theor Appl Genet 120:1241–1252

Van Der Salm T, Bosch D, Honee G, Feng L, Munsteman E, Bakker P, Steikema WJ, Visser B (1994) Insect resistance of transgenic plants that express modified *Bacillus thuringiensis cry1*A(b) and *cry*IC genes: a resistance management strategy. Plant Mol Biol 26:51–59

Wilkinson MJ, Sweet J, Poppy GM (2003) Risk assessment of GM plants: avoiding gridlock. Trends Plant Sci 8:208–213

Williams JL, Ellers-Kirk C, Orth RG, Gassmann AJ, Head G et al (2011) Fitness cost of resistance to Bt Cotton linked with increased gossypol content in Pink Bollworm Larvae. PLoS ONE 6(6):e21863

Wolfenbarger LL, Naranjo SE, Lundgren JG, Bitzer RJ, Watrud LS (2008) Bt crop effects on functional guilds of non-target arthropods: a meta-analysis. PLoS One 3:1–11

Wolt JD, Keese P, Raybould A, Fitzpatrick JW, Burachik M, Gray A, Olin SS, Schiemann J, Sears M, Wu F (2010) Problem formulation in the environmental risk assessment for genetically modified plants. Transgenic Res 19:425–436

Wraight CL, Zangerl AR, Carroll MJ, Berenbaum MR (2000) Absence of toxicity of *Bacillus thuringiensis* pollen to black swallowtails under field conditions. Proc Natl Acad Sci U S A 97:7700–7703

Xu X, Yu L, Wu Y (2005) Disruption of a Cadherin gene associated with resistance to Cry1Ac-endotoxin of *Bacillus thuringiensis* in *Helicoverpa armigera*. Appl Environ Microbiol 71:948–954

Yang L, Pan A, Zhang K, Yin C, Qian B, Chen J, Huang C, Zhang D (2005) Qualitative and quantitative PCR methods for event-specific detection of genetically modified cotton Mon1445 and Mon531. Transgenic Res 14:817–831

Yang Z, Chen H, Tang W, Hua H, Lin Y (2011) Development and characterization of transgenic rice expressing two *Bacillus thuringiensis* genes. Pest Manag Sci 67:414–422

Yu H-L, Li Y-H, Wu K-M (2011) Risk assessment and ecological effects of transgenic bt crops on non-target organisms. J Integr Plant Biol 53:520–538

Zangerl AR, McKenna D, Wraight CL, Carroll M, Ficarello P, Warner R, Berenbaum MR (2001) Effects of exposure to event 176 *Bacillus thuringiensis* corn pollen on monarch and black swallowtail caterpillars under field conditions. Proc Natl Acad Sci U S A 98:11908–11912

Zhao JZ, Li YX, Collins HL, Shelton AM (2002) Examination of the F2 screen for rare resistance alleles to *Bacillus thuringiensis* toxins in the diamondback moth (Lepidoptera: Plutellidae). J Econ Entomol 95:14–21

Zhao JH, Ho P, Azadi H (2011) Benefits of Bt cotton counterbalanced by secondary pests? Perceptions of ecological change in China. Environ Monit Assess 173:985–994
Zhao JZ, Cao J, Li YX, Collins HL, Roush RT, Earle ED, Shelton AM (2003) Transgenic plants expressing two *Bacillus thuringiensis* toxins delay insect resistance evolution. Nat Biotechnol 21:1493–1497
Zurbrügg C, Nentwig W (2009) Ingestion and excretion of two transgenic Bt corn varieties by slugs. Transgenic Res 18:215–225
Zurbrügg C, Hönemann L, Meissle M, Romeis J, Nentwig W (2010) Decomposition dynamics and structural plant components of genetically modified Bt maize leaves do not differ from leaves of conventional hybrids. Transgenic Res 19:257–267
Zwahlen C, Hilbeck A, Gugerli P, Nentwig W (2003) Degradation of the Cry1Ab protein within the transgenic *Bacillus thuringiensis* corn tissue in the field. Mol Ecol 12:765–775

Chapter 4
Use and Efficacy of Bt Compared to Less Environmentally Safe Alternatives

Mohd Amir Fursan Abdullah

Abstract Bt is classified as a biopesticide. Market surveys from several states in the United States of America suggest that the main reason for using biopesticides is it is considered safe for the environment. Public perception and organic farming have risen in importance as reasons for using biopesticides. However, the main reason for not using biopesticides continues to be perceptions of low efficacy, while higher costs and lack of awareness appeared to have less significance. Even though limited by its narrow activity, Bt usage has increased significantly over the years. The narrow activity spectrum of Bt is to its advantage for it is well suited for IPM programs that preserve large segments of indigenous invertebrate predator and parasitoid populations. On the other hand, usage of broad spectrum pesticides that also kill non-target beneficial insects, may lead to pest outbreak when the pesticide loses efficacy due to pest resistance. Less environmentally safe pesticides also have their roles to play in IRM and IPM programs. When combined with practices in IPM, they will have even less effect on the environment while reducing the probability of resistance. Substantial economic returns and environmental safety will drive the usage of safe pest control measures, which Bt may play a significant role.

Keywords Use of Bt · Efficacy of Bt · Biopesticide · Safety · Pest resistance

4.1 Biopesticide Classification

According to the United States Environmental Protection Agency (EPA), biopesticides are certain types of pesticides derived from such natural materials as animals, plants, bacteria, and certain minerals. There are many materials currently available as biopesticides. There were approximately 195 registered biopesticide active ingredients and 780 products at the end of 2001.

Biopesticides fall into three major classes: Microbials, biochemicals, and plant-incorporated-protectants.

M. A. F. Abdullah (✉)
InsectiGen, Inc., ADS Rm. 458, 425 River Road, 30602-2771 Athens, GA, USA
e-mail: mamir@insectigen.com

E. Sansinenea (ed.), *Bacillus thuringiensis Biotechnology*,
DOI 10.1007/978-94-007-3021-2_4, © Springer Science+Business Media B.V. 2012

Microbial pesticides consist of a microorganism as the active ingredient. The most widely used microbial pesticides are subspecies and strains of Bt. Each strain of this bacterium produces a different composition of insecticidal proteins, and is active against a narrow range of related species of insect larvae. A Bt toxin is normally active against some species in an insect Order, however, in some cases, a Bt toxin may be active against insects in two different Orders, for example, Cry1I is toxic against coleopteran and lepidopteran insects (reviewed in (van Frankenhuyzen 2009)).

Biochemical pesticides are naturally occurring substances that control pests by non-toxic mechanisms. Conventional pesticides, by contrast, are generally synthetic materials that directly kill or inactivate the pest. Thus, natural products such as spinosad (produced by *Saccharopolyspora spinosa* bacterium) and pyrethrum (produced by *Chrysanthemum* plants) are not considered as biopesticides because they act on nervous system of the pest to kill the pest. Biochemical pesticides include substances, such as insect sex pheromones, which interfere with mating, as well as various scented plant extracts that attract insect pests to traps. The EPA has established a special committee to determine whether a substance is classified as biochemical pesticide.

Plant-Incorporated-Protectants are pesticidal substances that plants produce from genetic material that has been added to the plant. Crops such as corn and cotton have been genetically engineered by introducing genes for the Bt pesticidal protein into the plant's own genetic material. The plant, instead of the Bt bacterium, produces the substance that destroys the pest. The protein and its genetic material, but not the plant itself, are regulated by EPA.

4.2 Biopesticide Market Survey

Market surveys on biopesticide usage are included here since Bt is a biopesticide. Surveys (Melnick et al. 2009) were conducted by Biopesticide Industry Alliance (BPIA) in 2002 and 2008 across several market segments (California Pest Control Advisors (PCAs), Florida Distributors, Florida Growers, California Growers, and Golf Course Superintendents in California, Florida, Illinois, Michigan, North Carolina, New York, Ohio, Pennsylvania, Texas, Wisconsin) with the following objectives:

1. Assessing the level of industry awareness of biological-based pesticide products,
2. Obtaining personal experience levels with biopesticide products,
3. Identifying perceptions of biopesticide technology and the sources of information used in industry in the usage decision process,
4. Exploring the rationale behind biopesticide usage decision process,
5. Quantifying the reaction of targeted groups to the prospects of certification standards,
6. Estimating the potential impact of a certification program on the adoption of biopesticides,
7. Determining perceptions on whether BPIA activities have had impact on the perception of biopesticides among the targeted groups.

The surveys suggest that the main reason biopesticides are used continue to be environmental safety. While public perception rose in importance for California PCAs, California Growers and Florida Distributors, organic farming also rose as a reason for using biopesticides. On the other hand, the reason for not using biopesticides continues to be perceptions of low efficacy. Other reasons include higher costs, and lack of awareness. The main conclusions from the 2008 survey are that the overall "concept" of using a biopesticide seems to be a positive idea to most audiences. However, there is an over-riding perception of "risk" that is an important obstacle to increased biopesticide use. The perceived "risks" include:

1. Credibility and reputation of the PCA and/or Distributor may be compromised,
2. Peer pressure from industry colleagues,
3. Loss of income to PCAs and growers.

Although the survey was conducted in the US, it may reflect public perception of biopesticides in other countries as well. The survey results may suggest ways for the industry to increase public acceptance of biopesticides in general and Bt specifically.

4.3 Use of Bt in IPM

According to the EPA, integrated pest management (IPM) is an approach to pest management that relies on a combination of common-sense practices, which use current, comprehensive information on the life cycles of pests and their interaction with the environment. Then, this information, in combination with available pest control methods, is used to control pest damage by the most economical means, and with the least possible hazard to people, property, and the environment.

Biocontrol via insect parasitoids, fungal pathogens, mating disruption, crop rotation, adjustments in date of planting, polyculture, and conventional chemical sprays are compatible with Bt use. Indigenous insect populations and introduced biocontrol agents that are harmed by broad spectrum chemical pesticides are not generally harmed by Bt, thus making Bt well suited to IPM programs that preserve large segments of indigenous invertebrate predator and parasitoid populations (Navon 2000). The early driving forces on implementing IPM was the emerging and growing threat of insecticide resistance in crop pests, which later includes reduction of pesticide use and protection of the environment.

4.4 Use and Efficacy of Bt

The efficacy of Bt is dependent on its narrow spectrum of toxicity. Most of the Bt products in agriculture are targeted against lepidopteran insects. Also, since it is only active when ingested by the specific target pests, topically-applied Bt is not effective against insects that normally bore into plant tissues such as stem and fruit borers (i.e. *Ostrinia nubilalis*, and *Helicoverpa zea*) and leaf-miners (i.e. *Tuta absoluta* and *Phyllocnistis citrella*). However, these limitation could be overcome; *O. nubilalis* is effectively controlled by genetically modified corn that expresses Bt toxin in its tissues (reviewed in (Sanahuja et al. 2011; Meissle et al. 2011)), while addition of surfactant appears to increase the toxicity of topically-applied Bt against leafminer *P. citrella* by increasing Bt penetration into the mines (Shapiro et al. 1998).

Bt has been used successfully to control insect pests in agriculture (i.e. vegetable cultivation, tree fruit and nut crops), forestry, and mosquito control (Marrone 1994). According to California's pesticide use report (2009), Bt applications (which include aizawai and kurstaki strains) rose significantly from 2002–2009 for crops such as broccoli, cabbage, cauliflower, corn, leaf lettuce, tomatoes, and strawberry. The three crops that received the highest Bt applications (which include aizawai and kurstaki strains) in 2009 in California were strawberry, tomato, and pomegranate; about 32,000 gross pounds of Bt was applied for strawberry crops, followed by about 30,000 gross pounds for tomatoes (fresh and for processing), and about 27,000 gross pounds for pomegranate. For pomegranate, only the use of Bt strain kurstaki was reported and interestingly the application of the Bt increased substantially from 2006–2009, while from 1991–2005 the usage was almost negligible. Globally, sales of Bt in 2005 was estimated at US$ 159 million (CPL Business Consultants, October 2006). Research and development activity on Bt such as improving strains and formulations, and development of synergists have led to improved efficacy of Bt (Sanahuja et al. 2011).

Bt also plays an important role in management of insecticide resistance. Effective insecticide and miticide resistance management (IRM) strategies attempt to minimize the selection for resistance from any one type of insecticide or miticide. In practice, alternations, sequences or rotations of compounds from groups with different modes of action can provide a sustainable and effective approach to IRM (i.e., where resistance develops from altered target sites in the insect).

Pests that were once effectively controlled by pesticides are on the rebound. Some examples of pesticide-resistant insects include *Spodoptera litura* (also known as the oriental leafworm moth, cluster caterpillar, cotton leafworm, tobacco cutworm, tropical armyworm, taro caterpillar, tobacco budworm, rice cutworm, and cotton cutworm) (Kranthi et al. 2001; Ahmad et al. 2007; Armes et al. 1997), *Helicoverpa armigera* (cotton bollworm) (Wu and Guo 2005), *Plutella xylostella* (diamondback moth) (Bommarco et al. 2011), and *Tuta absoluta* (tomato leafminer) (Vacas et al. 2011). The applications of broad spectrum pesticides also remove natural predators and other beneficial insects. With the absence of natural predators, resistant pest population can recover quickly when pesticide control becomes ineffective.

Heightened public concern of adverse environmental effects associated with the heavy use of chemical insecticides has fuelled examination of alternative methods for insect pest control. One of the potential alternatives is the use of entomopathogenic microorganisms such as Bt. Insect pests that are resistant to chemical insecticides are not cross-resistant to Bt due to different modes of toxicity. Thus, Bt products have been applied to control the resistant insects. However, over reliance on Bt alone have also led to development of resistance. Laboratory selections with single Bt toxins have demonstrated that more than a dozen insect species have the potential to develop resistance against Bt (Tabashnik et al. 1998). However, to date only field-evolved resistance of *P. xylostella* (Tabashnik et al. 1997; Baxter et al. 2005) and greenhouse-evolved resistance of *Trichoplusia ni* (Janmaat and Myers 2003) have been reported for Bt formulations. This could be due to the multiple toxins (with affinity for different receptors) present in commercial Bt products. In order to prolong the utility of Bt, it is best to include Bt in IRM and IPM programs.

The less environmentally safe pesticides also have their roles to play in IRM and IPM. Most of the highly toxic pesticides based on older chemistries have been discontinued in their production and usage. Newer pesticides that have less effect on the environment, combined with practices in IPM, will have even less effect on the environment while reducing the probability of resistance. In the end, high economic returns and safety to the environment will motivate the usage of safe pest control measures, of which Bt may play a significant role.

References

Ahmad M, Arif I, Ahmad M (2007) Occurence of insecticide resistance in field populations of *Spodoptera litura* (Lepidoptera: Noctuidae) in Pakistan. Crop Prot 26:807–809

Armes NJ, Wightman JA, Jadhav DR, Ranga Rao GV (1997) Status of insecticide resistance in *Spodoptera litura* in Andhra Pradesh India. Pestic Sci 50:240–248

Baxter SW, Zhao JZ, Gahan LJ, Shelton AM, Tabashnik BE, Heckel DG (2005) Novel genetic basis of field-evolved resistance to Bt toxins in *Plutella xylostella*. Insect Mol Biol 14(3):327–334. doi:10.1111/j.1365-2583.2005.00563.x

Bommarco R, Miranda F, Bylund H, Bjorkman C (2011) Insecticides suppress natural enemies and increase pest damage in cabbage. J Econ Entomol 104(3):782–791

California's pesticide use report (2009) California department of pesticide regulation. http://www.cdpr.ca.gov/docs/pur/purmain.htm. Accessed 25 May 2011

Janmaat AF, Myers J (2003) Rapid evolution and the cost of resistance to *Bacillus thuringiensis* in greenhouse populations of cabbage loopers, *Trichoplusia ni*. Proc Biol Sci 270(1530):2263–2270. doi:10.1098/rspb.2003.2497

Kranthi KR, Jadhav DR, Wanjari RR, Ali SS, Russell D (2001) Carbamate and organophosphate resistance in cotton pests in India, 1995–1999. Bull Entomol Res 91(1):37–46

Marrone PG (1994) Present and future use of *Bacillus thuringiensis* in Integrated Pest Management systems: an industrial perspective. Biocontrol Sci Technol 4:517–526

Meissle M, Romeis J, Bigler F (2011) Bt maize and integrated pest management-a European perspective. Pest Manag Sci 67(9):1049–1058. doi:10.1002/ps.2221

Melnick R, Stoneman B, Marrone PG, Messerschmidt O, Donaldson M (2009) Biopesticide use and attitudes survey results. Biopesticide Industry Alliance. http://www.ipmcenters.org/ipm-symposium09/37-2_Stoneman.pdf. Accessed 25 July 2011

Navon A (2000) *Bacillus thuringiensis* insecticides in crop protection—reality and prospects. Crop Prot 19:669–676

Sanahuja G, Banakar R, Twyman RM, Capell T, Christou P (2011) *Bacillus thuringiensis*: a century of research, development and commercial applications. Plant Biotechnol J 9:283–300

Shapiro JP, Schroeder WJ, Stansly PA (1998) Bioassay and efficacy of *Bacillus thuringiensis* and organosilicone surfactant against the citrus leafminer (Lepidoptera: Phyllocnistidae). Fla Entomol 81(2):201–210

Tabashnik BE, Liu YB, Malvar T, Heckel DG, Masson L, Ballester V, Granero F, Mensua JL, Ferre J (1997) Global variation in the genetic and biochemical basis of diamondback moth resistance to *Bacillus thuringiensis*. Proc Natl Acad Sci U S A 94(24):12780–12785
Tabashnik BE, Liu YB, Malvar T, Heckel DG, Masson L, Ferre J (1998) Insect resistance to *Bacillus thuringiensis*: uniform or diverse? Philos Trans R Soc Lond B 353:1751–1756
Vacas S, Alfaro C, Primo J, Navarro-Llopis V (2011) Studies on the development of a mating disruption system to control the tomato leafminer, *Tuta absoluta* Povolny (Lepidoptera: Gelechiidae). Pest Manag Sci 67(11):1473–1480. doi:10.1002/ps.2202
van Frankenhuyzen K (2009) Insecticidal activity of *Bacillus thuringiensis* crystal proteins. J Invertebr Pathol 101(1):1–16. doi:10.1016/j.jip.2009.02.009
Wu KM, Guo YY (2005) The evolution of cotton pest management practices in China. Annu Rev Entomol 50:31–52. doi:10.1146/annurev.ento.50.071803.130349

Chapter 5
Protein Engineering of *Bacillus thuringiensis* δ-Endotoxins

Alvaro M. Florez, Cristina Osorio and Oscar Alzate

Abstract Protein engineering of insecticidal Bt δ-endotoxins is a powerful tool for designing novel Cry toxins with altered properties, including changing the toxin's specificity. By following some elementary rules governing the structure/function relationship, it has been possible to create new toxins with modified properties including increased toxicity and binding affinity, enhanced ion-transport activity, and changes in insect specificity. These methods have also produced valuable information and have led to an improved understanding of the mode of action of these important biopesticides. The results discussed in this chapter derive from rational molecular design where protein structure is modified by incorporating single or multiple amino acid substitutions aimed at modifying specific protein functions. In this review, we analyze several protein modifications that have been successfully used for creating stable, functional proteins with minimal structural alterations. The understanding and proper use of protein engineering approaches may help in implementing appropriate pest management strategies by improving the efficacy of these toxins against insect pests.

Keywords Protein engineering · δ-endotoxins · Protein domains · Protein structure · Designing novel toxins

---

O. Alzate (✉)
Department of Cell and Developmental Biology, School of Medicine, University of North Carolina, 438A Taylor Hall, CB# 7090, Mason Farm Road, 27599 Chapel Hill, NC, USA
e-mail: alzate@med.unc.edu

C. Osorio
Systems Proteomics Center, Program in Molecular Biology and Biotechnology, School of Medicine, University of North Carolina, 438A Taylor Hall, CB# 7090, Mason Farm Road, 27599 Chapel Hill, NC, USA

Alvaro M. Florez
Laboratorio de Biología Molecular y Biotecnología, Facultad de Medicina, Universidad de Santander, Bucaramanga, Colombia

E. Sansinenea (ed.), *Bacillus thuringiensis Biotechnology*,
DOI 10.1007/978-94-007-3021-2_5, © Springer Science+Business Media B.V. 2012

5.1 Introduction

The toxic action of many bacterial toxins is based on their ability to interact with cell membranes by inducing ion pores that disrupt the membrane potential and leads to cell lysis. In many cases, this process is mediated by recognition of specific receptors on membranes of the target organism. A family of these toxins is produced as crystalline inclusions by the gram-positive bacterium *Bacillus thuringiensis* (*Bt*) during the stationary growth phase. These toxins are known as Cry and Cyt δ-endotoxins (See Chap. 1 and 2 in this part for a review of the *Bt* toxins). These δ-endotoxins were classified for the first time by Höfte and Whiteley (Höfte and Whiteley 1989), who proposed classifying the *Bt* crystal proteins into four main pathotypes depending on their specificities: against Lepidoptera (CryI, now Cry1), against Coleoptera (CryIII, now Cry3), against Diptera (CryIV, now Cry4, 10, 11; and CytA, now Cyt1A), and against both Lepidoptera and Coleoptera (CryII now Cry2) (Höfte and Whiteley 1989). Then, in 1993 a *Bt* δ-endotoxin nomenclature committee was created in order to update the nomenclature based on amino acid identity and, in 1998, a new nomenclature was published assigning an unique name to holotype sequences based on the degree of divergence (Crickmore et al. 1998). *Bt* δ-endotoxins are active against a diverse range of insects including the orders Lepidoptera, Diptera, Coleoptera, Hymenoptera, Homoptera, Orthoptera and Mallophaga and also show activity against Nematodes, Mites and Protozoa (Schnepf et al. 1998). To date, 229 sequences have been grouped into 218 holotypes for 69 *cry* families and 11 holotypes for three *cyt* families (Crickmore et al. 2011).

The δ-endotoxins are produced by around 100 *Bt* subspecies (Sanahuja et al. 2011); with more toxins and more species being assayed every day searching for toxicity against insects of almost every order (van Frankenhuyzen 2009) (to learn more about the *Bt* toxin specificity database, the reader should visit the web site http://www.glfc.forestry.ca/bacillus/). The specific activity of all the *Bt* toxins is summarized in the dendogram shown in Fig. 5.1, modified from http://www.lifesci.sussex.ac.uk/home/Neil_Crickmore/Bt/ with permission (Crickmore et al. 1998, 2011), which also includes activity against cancer cells (van Frankenhuyzen 2009).

Since 1938, several *Bt*-based insecticidal products have been introduced, including improved formulations, new bacterial strains, and new toxins. In general, the new toxins were developed to reduce their susceptibility to UV light, heat, extreme pH, and proteolytic degradation, and to increase their stability and their ability to form ion-channels (Kaur 2000; Sanahuja et al. 2011).

The family of three-domain *Bt* δ-endotoxins consists of 69 different groups encoded by more than 500 different *cry* gene sequences and clustered into four phylogenetically non-related protein families. Sequence alignments of the activated toxins reveal up to five conserved blocks and three conserved domains (Fig. 5.1, Crickmore et al. 2011). The activated toxins (Fig. 5.2), with molecular weights ranging from 70–130 kDa, result from proteolytic degradation of the protoxins by specific proteases in the insect midgut (Bietlot et al. 1990; Höfte and Whiteley 1989; Schnepf et al. 1998).

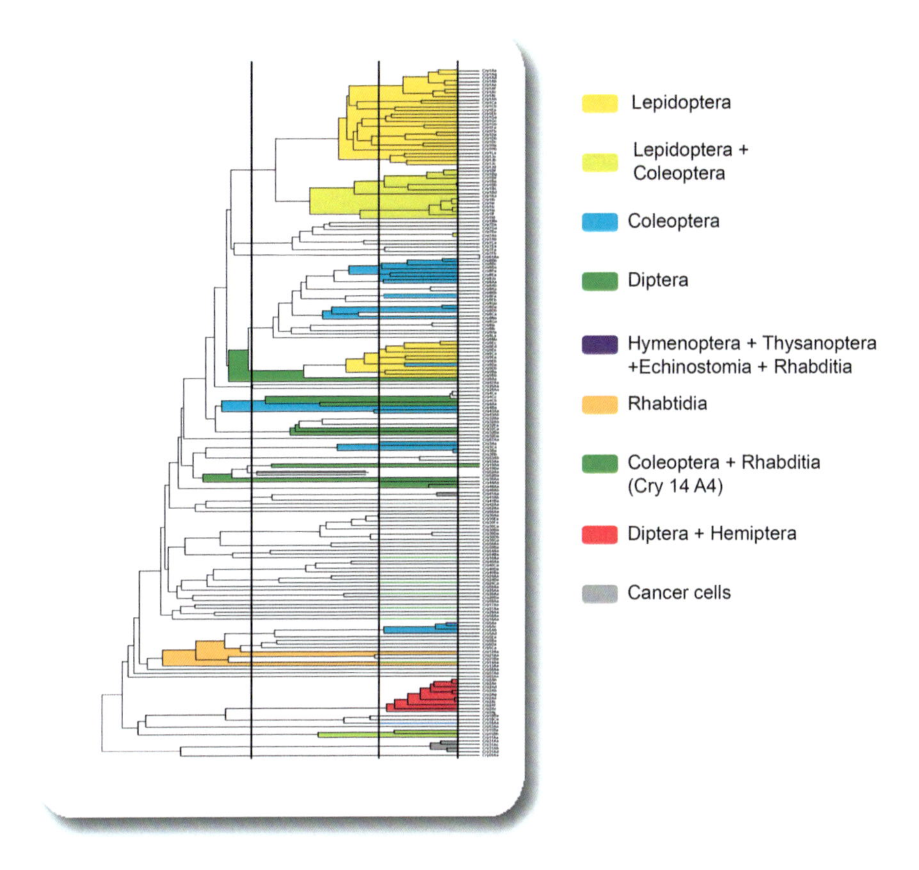

Fig. 5.1 Classification of *Bt* δ-endotoxins. The dendogram was downloaded from http://www.lifesci.sussex.ac.uk/Home/Neil_Crickmore/Bt/, with permission from Dr. Neil Crickmore and then colored to indicate the specificity of the δ-endotoxin groups. (Crickmore et al. 2011; van Frankenhuyzen 2009)

The structure of the *Bt* δ-endotoxins in their active conformation is composed of three distinct structural domains. Domain I is a bundle of 7–8 α-helices characterized by the presence of a centrally located hydrophobic α-helix (α-helix 5) (Grochulski et al. 1995; Li et al. 1991). Two major functions have been associated with domain I: insertion into the midgut membrane and ion channel formation (Fig. 5.2). Domain II is involved in receptor binding and may play a role in membrane partitioning (Nair and Dean 2008). Exchanging domains between δ-endotoxins active against different insects has been used to demonstrate that the portion of the protein that is responsible for insect specificity resides in domains II and III (Ge et al. 1989). Domain III participates in receptor binding and insect specificity and ion channel activity (reviewed in (Schnepf et al. 1998)).

Understanding the structure-function relationship in Cry δ-endotoxins is fundamental for structural-based rational design of modified molecules to improve their

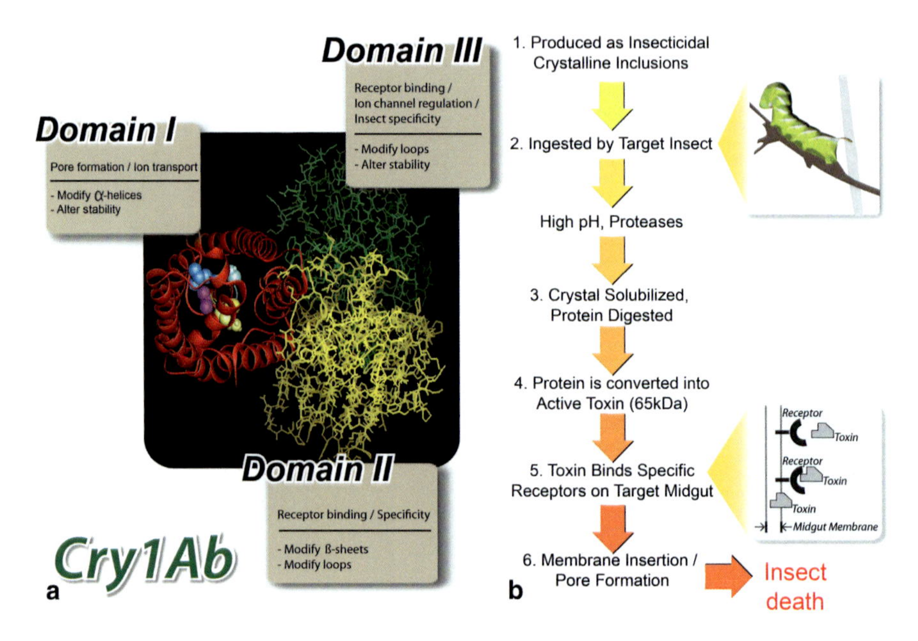

Fig. 5.2 The structure of the Cry1 δ-endotoxins, their mode of action, and the potential targets for protein engineering are perfectly interconnected. **A** Domain I is involved in pore formation, domain II is involved in receptor binding, and domain III has influence on receptor binding, ion channel formation, and insect specificity; all three domains have also specific functions in the structural stability of the protein. **B** Major aspects of the mode of action of the 3-domain δ-endotoxins. The 3D-structure corresponds to Cry1Ab created by molecular modeling with the program Quanta using the coordinates for Cry1Aa (Pdb code 1CIY), and energy minimized with MM^{+}. (Alcantara et al. 2001)

function. To understand such principles, we will briefly review what is known about the mode of action of Cry toxins and how this mode of action depends on the protein structure (Fig. 5.2): (1) proteins are produced by *Bt* as protoxins encapsulated in crystalline inclusion bodies; (2) the crystals dissolve in the alkaline environment of the insect midgut and the protoxins are digested by specific proteases leading to activated toxins; (3) activated toxins bind specific receptors on the insect midgut; (4) receptor-bound toxins undergo conformational changes—in some cases resulting in oligomeric structures that are able to modify the target membrane; and (5) the protein alters the structural stability of the membrane producing channels or pores that alter the membrane potential leading to cell lysis and insect death.

The initial interaction between δ-endotoxins and the target membrane is mediated by domains II and III (Arenas et al. 2010; Pacheco et al. 2009; Schnepf et al. 1998) which are able to bind to amino peptidase N (APN) and alkaline phosphatase (ALP) receptors followed by binding to a cadherin-like protein (in some cases this process leads to the cleavage of α-helix 1 (Aronson 2000)), leading to protein oligomerization (Bravo et al. 2004). In some instances, the specific binding of some of the *Bt* toxins to glycosylphosphatidylinositol (GPI)-anchored proteins in conjunction with APN and ALP results in the insertion of the oligomeric toxins into the target membranes (Likitvivatanavong et al. 2011). For some Cry toxins, oligomers insert into the membrane creating pores which causes osmotic shock and finally results in insect death (Zhuang et al. 2002). Another model proposed to explain the mode of action of Cry1Ab suggests that the toxin binds to a cadherin-like protein receptor thereby activating signaling cascades leading to increased cyclic-AMP and protein kinase, which eventually results in cell death (Zhang et al. 2006). An alternative mode of action, specific for insect cells and not insect larvae, has been proposed in which toxin-receptor interaction is followed by toxicity *without* the formation of ion pores (Zhang et al. 2006). Regardless of the mechanism by which a toxin kills the target insect, this chapter will show that protein engineering can be used to modify the function of a protein by manipulating its structural properties.

The study of δ-endotoxins has been driven by modifications for the purpose of altering their major functions (Table 5.1). These mutant proteins (called by some "muteins") result in specific phenotypes which are subsequently analyzed in regards to: (i) structural stability, (ii) binding affinity, (iii) pore forming capacity, (iv) ion channel conductance, and (v) toxicity (Table 5.1).

In this review we describe several approaches for modulating the function of Cry toxins by modifying their structural characteristics and we describe how these modified proteins should be evaluated in order to determine if these modifications have, in fact, led to improvements to the structure/function relationship. The Table 5.1 presents a summary of the functions that will be modified, the structural characteristics that *should* be modified in order to obtain an altered protein with a modified function, without significant alteration of the protein's structure, and methods to validate the effects of such modification.

5.2 Protein Engineering of *Bt* δ-endotoxins

5.2.1 Manipulating the Protein-activation Mechanisms

Most *Bt* δ-endotoxins are activated by trypsin-like proteases in the insect midgut, resulting in a protease-resistant protein of approximately 60–70 kDa (Höfte and Whiteley 1989). This proteolytic activation is an important factor in the resistance mechanisms developed by some insect species. To counteract this defense mechanism, some researchers have used protein engineering to develop inhibitors of digestive enzymes (Gatehouse 2011; Sanahuja et al. 2011).

After proteolytic cleavage, most δ-endotoxins show modifications including loss of α-helix 1 (Aronson 2000) and nicking of some regions on α-helices 3 and 4 (Carroll et al. 1997). The latter modifications result in higher binding affinity to the midgut columnar epithelial cells and increased ion channel formation; however, removal of some amino acids from the N-terminus results in altered toxicity (Schnepf 1995). In other cases, such as in the case of Cry2A, sequential cleavage of the

Table 5.1 Function targeted for modification, target structure to achieve a modified molecule, and methods for protein modification and monitoring

Function targeted for modification	Structure targeted to modify function	Assay[a]	Monitoring method	References
Stability of crystal inclusion	The crystal Little is known about the crystal properties. In principle, the disulfide bridges conferring stability to the crystal can be modified by SDM. This will allow the inclusions to last longer after spreading	Expose purified crystal to environmental conditions, test for crystal stability	IR spectroscopy	Choma et al. (1990)
Resistance to protease activity Protein stability	Any amino acid residue susceptible to proteolytic degradation	Expose purified proteins to different levels of proteases and determine fragment formation	1D-PAGE, Thermal stability (monitored by CD or 1D-PAGE) Fragment analysis by mass spectrometry, Calorimetry	Ge et al. (1991) Hussain et al. (1996, 2010) Audtho et al. (1999) Lightwood et al. (2000) Coux et al. (2001) Gomez et al. (2002) Kirouac et al. (2006) Fortier et al. (2007) Alzate et al. (2010)
Resistance to UV inactivation	Mostly surface exposed motifs containing Trp	Expose purified proteins to UV, and assay for protein stability and toxicity	Mass spectrometry, Raman spectroscopy, Circular dichroism spectroscopy, "Whole tissue" Voltage clamping	Pusztai et al. (1991) Pozsgay et al. (1987)
Binding to specific receptors	Motifs identified in receptor binding; specifically loops of domain II (in particular the region containing Phe371)	Binding assays—accompanied of structural analysis to demonstrate that the amino acid substitution do not alter the protein structure	Binding assays, Circular dichroism, Thermal stability analysis, Surface Plasmon Resonance, Calorimetry, FRETb	Rajamohan et al. (1995, 1996a, b) Lee et al. (1995) Rajamohan and Dean (1996) Masson et al. (1999) Gomez et al. (2003)

Table 5.1 (continued)

Function targeted for modification	Structure targeted to modify function	Assay[a]	Monitoring method	References
Membrane partitioning	α-Helices of domain I	Ion channel formation, Protein protection assays, Protein stability, Conductivity on lipid bilayers	Circular dichroism, Fluorescence, "Whole tissue" Voltage clamping, Calorimetry, Light scattering, SDSLc, Osmotic Swelling	Lorence et al. (1995) Schwartz et al. (1997b, c) Vachon et al. (2004) Alzate et al. (2006, 2009, 2010) Girard et al. (2008) Nair and Dean (2008) Lebel et al. (2009) Hussain et al. 2010) Zavala et al. (2011) Alcantara et al. (2001)
Oligomerization	Amino acid residues involved in oligomerization	Determine formation of large molecules and test for receptor binding, ion conductance and toxicity	IR spectroscopy, Calorimetry, Fluorescence spectrocopy, Ultra-centrifugation, "Whole tissue" Voltage clamping	Rausell et al. (2004a) Munoz-Garay et al. (2006) Jimenez-Juarez et al. ((2007, 2008) Soberon et al. (2007) Rodriguez-Almazan et al. (2009) Arenas et al. (2010)
Specificity	Amino acid residues involved in receptor binding	Toxicity against insects other than target	Bioassays	Abdullah et al. (2003) Liu and Dean (2006)

[a] All modifications should be accompanied by toxicity assays against the target insect.
[b] Fluorescence Resonance Energy Transfer.
[c] Site-directed Spin Labeling

protoxin only at very specific locations results in activated toxins (Audtho et al. 1999). As a result of this cleavage-dependent activation process, it is possible to create protease-resistant δ-endotoxins by protein engineering. For instance, several modified Cry2Aa1 proteins were created by replacing Leu144 with His to mimic the corresponding position (His161) from Cry3A. These changes resulted in mutant proteins with higher stability and higher resistance to proteolytic cleavage, albeit showing decreased toxicity (Audtho et al. 1999).

5.2.2 Modulating the Membrane Partitioning Mechanisms

Membrane-partitioning mechanisms are modified through structural manipulation of the α-helical domain I. Six of the seven α-helices of this domain surround the hydrophobic α 5-helix. The external α-helices are amphipathic in nature with a hydrophobic side oriented towards α-helix 5 creating a hydrophobic core. Most of these α-helices are large enough (~30 Å) so that, in principle, they should be able to span hydrophobic membranes (Li et al. 1991). Multiple polar residues are involved in the formation of salt bridges or hydrogen bonds (Li et al. 1991). These properties may be modified by site-directed mutagenesis designed to modulate the membrane partitioning properties of the δ-endotoxins. For instance, specific amino acid substitutions A92D, R93D (Wu and Aronson 1992) and A92E and Y153D (Chen et al. 1995) in the Cry1Ab δ-endotoxin have shown that the introduction of negatively-charged amino acids residues have detrimental effects on membrane insertion and result in loss of toxic activity towards *Manduca sexta*. Other analyses, as will be described later, have revealed that, in some cases, the participation of all domains is necessary for proper conductance and pore formation, and that relative movement of domain I favors ion channel formation. Ion conductance may be modulated by restricting the relative movements of the α-helices with disulfide bridges (Alzate et al. 2006; Schwartz et al. 1997a).

Several mutations have been introduced into different helices of domain I which do not have significant effects on protein toxicity. Some exceptions are mutations in α-helix 5, in which a significant reduction in toxicity has been observed (Aronson 1995; Gazit and Shai 1993; Wu and Aronson 1992). There are also important modifications to α-helix 2, including the mutations of Pro70 that result in "muteins" with impaired ion transport (Arnold et al. 2001), as well as mutations to α-helix 7 (Alcantara et al. 2001; Chandra et al. 1999). A summary of specific mutations targeting amino acid residues in domain I is presented in Table 5.2.

5.2.2.1 Modulating the Function of α-Helix 2

Domain I of the Cry1A family of *Bt* δ-endotoxins contains an interesting motif in which an α-helix—in this case α-helix 2, is broken down into two α-helices by the presence of a proline residue. This motif is common to many membrane-inserting

Table 5.2 Mutations targeted at modifying the function of Domain I

Toxin	Mutations	Effect	Reference
Cry1Ab	F50K	Fail pore formation; loss of toxicity	Ahmad and Ellar (1990)
	P70A	Loss toxicity	Arnold et al. (2001)
	P70G		
	L157C	Decreased toxicity to *M. sexta*	Alzate et al. (2009)
	S176C		
	S170C	No effect in toxicity	Alzate et al. (2009)
	S170R_1*		
	L157R_1*		
	S176R_1*	Decreased toxicity to *M. sexta*	Alzate et al. (2009)
	V171C	Increased toxicity to *L. dispar*	Alzate et al. (2010)
	L157C		
	D225A	Decreased toxicity to *M. sexta*	Alcantara et al. (2001)
	W226A		
	D242A		
	N230A		
	R234A		
	R233A		
	R224A		
	R228A		
	Y229A		
	E235A		
	F247A		
	R228K	Reduce toxin's current inhibition	Alcantara et al. (2001)
	F232Y		
	E235A		
	E235Q		
Cry1Ac	A92D	Increase toxicity	Wu and Aronson (1992)
Cry1Ac	R93H	Decrease toxicity	Wu and Aronson (1992)
	R93G		
	R93A		
	R93S		
	R209A	Same toxicity as wild type	Aronson (1995)
	R209P		
	T213A		
	W210L		
	V218N		
	Y211N		
	Y211R		
	Y211D		
	W210C	Same toxicity as wild type	Aronson (1995)
	Y211D		
	Y211C		
	G214E		
	I132S	Decrease toxicity	Kumar and Aronson (1999)
	I132L		
	I132V		
	I132N		
	M130T	Same toxicity as wild type	Kumar and Aronson (1999)
	M131I		
Cry1Ac1	N135Q	Pore formation affected	Tigue et al. (2001)

*Spin-labeled mutant toxin

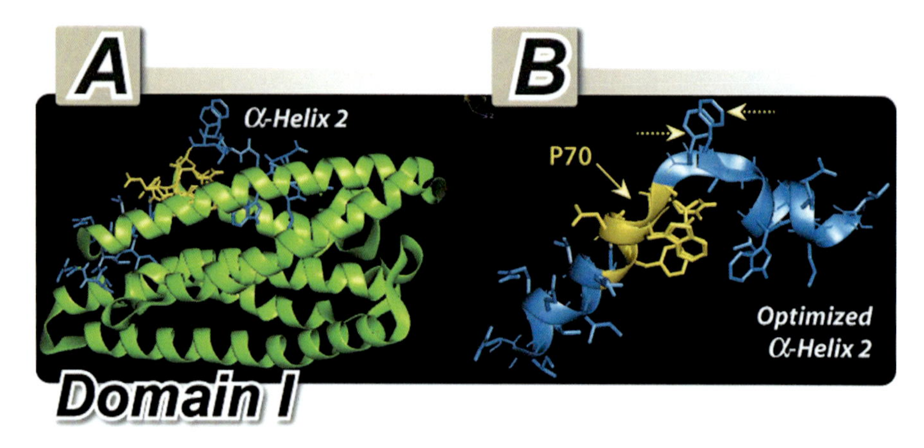

Fig. 5.3 Optimized structure of α-helix 2. **A** α-Helical domain I from Cry1Ab with α-helix 2 shown in *blue* in a "stick" representation, the remaining helices of the domain are displayed as "ribbons". **B** Energy optimized structure for the ^{69}GPS^{71} deletion. The *dashed arrows* indicate how the deletion induced changes in the 3D localization of all the amino acid residues in the α-helix

proteins, and is highly conserved in the Cry1A δ-endotoxin family. In the activated Cry1Ab toxin this broken-helix motif results from the presence of Pro70 (Fig. 5.3) (Arnold et al. 2001).

This broken-helix motif was used to generate three individual Cry1Ab mutant proteins in which the proline residue was substituted for alanine (P70A), and glycine (P70G); finally the "elbow" region ^{69}GPS^{71} was eliminated. The rationale for selecting these amino acid substitutions was as follows:first, the P-to-A mutation was expected to produce a *longer* α-helix 2, by using Ala, which is known to induce α-helix formation; second, the P-to-G substitution was created to generate a *more* flexible "elbow" thereby producing an α-helix with similar length and conformation as the original helix, but with more mobility; and third, the GPS deletion was expected to eliminate the "kink" produced by the Pro residue at position 70. The results of these mutations were analyzed by biophysical techniques, and molecular modeling as shown in Fig. 5.3 (Arnold et al. 2001).

Two types of analysis were performed to validate the effects of these mutations, a structural analysis and a functional analysis. The structural analysis utilized Circular Dichroism (CD). Polarized light is very sensitive to the orientation of α-helices in a protein; therefore CD, a technique that measures the degree of rotation of a polarized beam, can be used to determine the "helicity" of a protein. In addition, using the proper calibration and the proper standards, the percentage of β-sheets and random coils can be determined. Thus CD gives a representative structural signature from which one can determine the secondary structure of the protein (i.e., by providing an estimation of the fraction of a molecule that is in the α-helix, β-sheet, β-turn, or random coil conformations) (Berova et al. 2000; Greenfield 2006). The functional analysis was performed with "whole tissue" voltage clamping (VC), a technique in which ion currents are measured while the membrane potential is "fixed" by applying a specific voltage (Wolfersberger et al. 1987; Wolfersberger 1990; Wolfersberger and Spaeth 1987).

Based on the CD and VC analyses, P70A and P70G exhibited similar *ellipticity* compared to the wild type toxin, suggesting that no changes to the α-helical content resulted from the amino acid substitutions; and that both mutant toxins were able to disrupt the ion transport. This is in agreement with the toxicity assays, and explains why these toxins exhibit poor toxicity (Arnold et al. 2001). It is also observed that when the conserved broken-helix motif is deleted, the α-helical content is increased and the protein exhibits higher thermal stability than the wild type protein. These experiments suggest an approach for developing a mutant toxin with similar structural and functional properties as the wild type, but with a higher stability against thermal denaturation. In addition these findings suggest that α-helix 2 is required for proper ion transport and that recovery of α-helical content restores wild type-like function.

5.2.2.2 Manipulating the Ion-transport Properties of α-Helix 5

Countless reports have shown the importance of α-helix 5 in ion channel formation, with several mutations to this region resulting in loss of toxicity (Alzate et al. 2006, 2010; Wu and Aronson 1992). Several substitutions in which single cysteine residues have been introduced at the N-terminal, central, and C-terminal regions of α-helix 5 of Cry1Ab displayed interesting effects on the ion-transport capabilities of this toxin, both when the cysteine residue was unmodified or when it was covalently tagged with N-ethyl-maleimide or methyl methanesulfonothioate (Alzate et al. 2009). As in the previous section, structural and functional analyses were performed with CD and VC. Site-directed spin-labeling (SDSL) was also used (Hubbell et al. 2000). Site-directed spin labeling is a powerful technique in which a modified cysteine residue is covalently tagged with a paramagnetic reagent, and the effects of the mutation are analyzed by Electron Paramagnetic Resonance (EPR, Atherton 1993). This technique is particularly useful for studying the membrane-bound state and the dynamics of membrane-associated proteins (Altenbach et al. 1990, 1994; Hubbell et al. 2000, 2003).

The thiol groups from the introduced Cys residues are reacted with *S*-(2,2,5,5-tetramethyl-2,5-dihydro-1H-pyrrol-3-yl)methyl methanesulfonothioate (MTSL) creating a covalent bond with the cysteine. This reaction results in the release of the sulfinic acid (CH_3SO_2), leaving the MTSL moiety attached to the protein and adding 186.3 daltons to the protein's mass. This "spin label" acts as a molecular reporter that allows the study of macromolecular structures and their dynamics based on the mobility of the spin label, which is determined by its environment (Altenbach et al. 1994).

Using CD, VC and SDSL monitored by EPR, it was possible to determine that two cysteine substitutions in α-helix 5 of Cry1Ab, L157C and S176C, are important for ion-transport (Alzate et al. 2009). Interestingly, it was observed that these mutations may be used to modulate the time required by the protein to partition into the target membrane—a variable that can be measured by VC; adding one more variable to the list of structural parameters that can be manipulated to design Cry toxins with improved toxicity. These results, which were unique for amino acid residue S176, and were not observed for the other two mutations in the α-helix 5, suggest that this residue is critical for ion transport or for positioning the toxin in the membrane, or both (Alzate et al. 2009).

Another mutation in Cry1Ab, V171C, has been used to show important differences in toxin susceptibilities between *M. sexta* and *L. dispar*. This V171C mutant protein displayed an increased rate of partitioning into *L. dispar* midgut membranes, resulting in a 25-fold increase in toxicity; however, in toxic assays against *M. sexta*, the toxicity level was actually decreased compared to the wild type toxin. These results clearly suggest differences in the host-dependent mechanisms that mediate toxic action (Alzate et al. 2010). These results also suggest that a toxin undergoes conformational changes after translocation into the midgut membrane, a property that should be explored further for designing δ-endotoxins with enhanced toxicity.

5.2.2.3 Manipulating the Ion-discrimination Capabilities of α-Helix 7

α-Helix 7 is located in the core of the toxin between domains I and II (Grochulski et al. 1995). Several studies have shown the possible involvement of α-helix 7 in protein action and its potential function as biosensor (Chandra et al. 1999; Gazit et al. 1998; Grochulski et al. 1995). Protein engineering of Cry1Ab was used to determine the ability of this α-helix to discriminate between K^+ and Rb^+ transport in *M. sexta* midgut membranes (Alcantara et al. 2001).

Six mutations in α-helix 7, R224A, R228A, Y229A, R233A, E235A, and F247A, produced activated toxins similar to the wild type. Interestingly, mutant proteins R224A, R228A, and E235A were able to discriminate between Rb^+ and K^+ transport (Alcantara et al. 2001). CD spectroscopy indicated that none of the mutations resulted in significant changes to the protein's structure, suggesting that changes in the protein capability to discriminate between the two cations is dependent on either the charge of the amino acid, the location of the amino acid, or the conformation of the protein in the target membrane. Although at this time, these suggestions are purely speculative, it will be valuable to explore further the ability of these toxins to discriminate between different ions in order to find alternative ways for improving their toxicity (Alcantara et al. 2001).

5.2.3 Protein Engineering of Domains II and III

Domain II is the most variable region consisting of three anti-parallel β-sheets, and domain III is a β-sandwich (Grochulski et al. 1995; Li et al. 1991). Both domains are important for receptor recognition, particularly the loop regions of domain II, which are involved in receptor binding and toxin specificity (Pigott and Ellar 2007; Rajamohan et al. 1995, 1996b; Schnepf et al. 1998).

The hydrophobicity of some of these amino acid residues is very important in regulating the molecular interactions between the toxins and their receptors on the target membranes, as well as regulation of the translocation process of the toxins. Based on these properties, several studies have shown that modifications to δ-endotoxins resulted in modified proteins with higher potency and a wider spectrum of toxicity towards target insects. Some of these mutant proteins displayed improved receptor-binding interactions, resulting in improved toxicity. For instance, the deletion of Phe371 and mutations F371A and G374A in Cry1Ab resulted in loss of toxicity towards *M. sexta* larvae. Binding assays for these mutant toxins were similar to the wild type toxin, suggesting that mutations in residues 370–375 of loop II do not affect the overall binding but rather the *irreversible* binding of the toxin to the *M. sexta* midgut membranes (Rajamohan et al. 1995). This irreversible binding is considered by many *Bt* investigators to be the translocation of the toxin into the target membrane. In a similar fashion, the N372A and N372G mutant proteins displayed different functional roles in receptor binding and toxicity toward *L. dispar*. These mutant proteins displayed increased toxicity and enhanced binding affinity to midgut brush border membrane vesicles. The improved binding affinity of N372A suggests that this side chain may allow the toxin to fit more efficiently into the receptor's binding pockets (Rajamohan et al. 1996a). The hydrophilic/hydrophobic relationship found in the receptor/toxin interaction motif should be explored further to improve the binding efficiency and membrane partitioning of δ-endotoxins.

Several approaches have been used for the purpose of modifying and manipulating toxins that exhibit different specificities to insect targets, including interchanging segments involved in receptor recognition. Based on domain II mapping it is possible to identify which loops are important for this function, and thus incorporate loop sequences from one toxin into another to modify the receptor recognition properties. For example, the Cry4Aa toxin exhibits low activity against mosquito species *Aedes aegypti*, *Anopheles quadrimaculatus, Culex pipiens* and *Culex quinquefasciatus*, while Cry4Ba is toxic against *A. aegypti* and *A. quadrimaculatus* but not against *C. pipiens* and *C. quinquefasciatus*. The loop sequences 1 and 2 from Cry4Aa have been used to replace the corresponding sequences of Cry4Ba. These modifications resulted in an increase of toxicity of greater than 700-fold against *C. quincefasciatus* and greater than 285-fold against *C. pipiens* (Abdullah et al. 2003).

Rational molecular design has also helped introduce specificity to a toxin that was previously known not to have activity against a different insect species. Using several rounds of deletions and substitutions in domain II it was possible to manipulate the specificity of Cry1Aa. Cry1Aa is a lepidopteran-specific δ-endotoxin without activity against mosquitoes. However, the insertion of three loops from domain II of Cry4B into Cry1Aa results in this toxin becoming active against *C. pipiens* (Liu and Dean 2006).

Table 5.3 Disulfide bridge mutants

Name	Mutations	Disulfide bridge
SS1	(Arg99Cys-Ala-144Cys)*	α helix 3 and 4
SS2	(Val162Cys-Ala207Cys)	α helix 5 and 6
SS3	(Ser176Cys-Ser252Cys)	α helix 5 and 7
SS3	(Arg224Cys-Ser279Cys)	α helix 7 and loop Domain II

*Two single mutants were also created and tested individually: Arg99Cys and Ala144Cys (Alzate et al. 2006)

As mentioned above, domain III is less variable than domain II (Boonserm et al. 2005, 2006; Grochulski et al. 1995; Li et al. 1991). For some toxins, these differences are more evident, as between some related toxins such as Cry1Aa and Cry 1Ac. In Cry1Ac, the hairpin extension created by forming a pocket in which the *N*-acetylgalactosamine (GalNAc) binds (Burton et al. 1999; Li et al. 2001). The creation of novel toxins by *in vivo* recombination of domains I and II from Cry1Ab and domain III from Cry1Ac, resulting in "hybrid" toxins with high insecticidal activity and specificity has been reported (de Maagd et al. 1996). This particular hybrid displayed a 60-fold increase in toxicity to *Spodoptera exigua* larvae compared to Cry1Ab and a sixfold decrease in toxicity compared to Cry1Ac (de Maagd et al. 1996). Other toxin hybrids containing domains I and II from Cry1Ac, Cry1Ba, Cry1Ea, or Cry1Fa and domain III from Cry1Ca have been reported (de Maagd et al. 2000), indicating the overall power of this strategy for designing *Bt*-derived toxins. Similarly, hybrid toxins have been made by exchanging a fragment between amino acid residues 451–623 of Cry1Aa with Cry1Ac, which resulted in altered receptor binding properties (Lee et al. 1995). The specificity of the amino acid residue Thr524—located in the β-16-β17 loop of domain III—for receptor-binding interactions has been used to increase the activity of Cry1Ac towards Lepidopteran insects (Shan et al. 2010).

5.2.4 *Modifying the Conformational Requirements for Membrane Partitioning with Disulfide Bridges*

Cry1A δ-endotoxins are cysteine-free proteins (Grochulski et al. 1995). This property offers a natural advantage that may be exploited by creating more stable, functionally-altered proteins by engineering disulfide bridges into the proteins (Alzate et al. 2006; Schwartz et al. 1997a). The introduction of disulfide bridges at specific locations (Table 5.3) holding together α-helices and some loops resulted in trypsin-resistant toxins with modified ion channel properties in both artificial phospholipids vesicles (Schwartz et al. 1997a) and insect midgut membranes (Alzate et al. 2006).

These mutant proteins were less toxic than the wild-type protein towards insect larvae from *Bombix mori* and *M. sexta*; however, treatment with the reducing agent β-mercaptoethanol that specifically targets the artificially-introduced disulfide bridges, permitted the recovery of toxic activity for all mutant toxins except one (Alzate et al. 2006).

Structural and functional analyses of these mutant proteins with CD, thermal denaturation, binding assays, and electrophysiology in artificial phospholipid vesicles and midgut membranes, indicate that there have been small structural changes in the secondary structure that do not diminish the resistance to proteolysis (Alzate et al. 2006; Schwartz et al. 1997a). Binding experiments on *B. mori* and *M. sexta* did not show significant differences, indicating that these mutants have the same binding parameters as the wild-type toxin. These mutant toxins also conserved their ability to penetrate into membranes, both in the oxidized or reduced states (Alzate et al. 2006). The electrophysiological analysis of mutant toxins SS2 and SS3 (Table 5.3) suggests that these mutant proteins have a more rigid structure and a more tightly-packed α-helical domain I, resulting in improved ability to translocate into biological membranes (Alzate et al. 2006).

We can conclude that there is no requirement for separation of the α-helices for translocation into the midgut membrane and that all possible disulfide bridges are formed. Structural changes are not required for ion-transport activity and changes in protein activity are the result of individual mutations of the residues that form the disulfide bridges.

5.2.5 Altering the Oligomerization Properties of Bt δ-Endotoxins

Protein oligomerization has been observed in several Cry δ-endotoxins (Munoz-Garay et al. 2009; Pardo-Lopez et al. 2006; Rausell et al. 2004b), and this has been postulated to be a crucial step favoring receptor recognition and formation of ion pores.

Several mutations have been described that affect δ-endotoxin oligomerization and correlate with severe disruption in toxicity (Rodriguez-Almazan et al. 2009). Mutants with residues N135C, A140K, T142C, or T143D of Cry1Ab failed to form an oligomeric structure and showed reduced membrane insertion, resulting in a drastic reduction in toxicity (Rodriguez-Almazan et al. 2009). Cry1Ab double-negative mutant (DN) proteins D136N/T143D and E129K/D136N formed hetero-oligomers which failed to form ion channels in black lipid bilayers (Rodriguez-Almazan et al. 2009). It has been suggested that the formation of these hetero-oligomers results in the inability of the proteins to create ion pores (Rodriguez-Almazan et al. 2009). The mechanisms responsible for protein oligomerization and the relationship of these alternative structures with ion channel formation and toxicity should be studied further. This could open new opportunities for protein engineering of δ-endotoxins.

5.3 Conclusions

The structure of *Bt* δ-endotoxins includes regions responsible for finding the target membrane and creating ion pores. A combination of domain structures participates in ion-channel formation and oligomeric transformations. At the present time many features of the δ-endotoxins' mode of action have been elucidated, but many still remain to be uncovered. As presented in the previous sections, all of these structural regions may be targeted to create novel toxins with modified properties.

Multiple techniques can be used to design and develop modified toxins, including molecular modeling, bioinformatics, and site-directed mutagenesis. In a similar fashion, multiple techniques may be used to analyze the results of those specific modifications, including bioassays, circular dichroism (CD), "whole tissue" voltage clamping (VC), fluorescence spectroscopy, Electron Paramagnetic Resonance (EPR), and Site Directed Spin Labeling (SDSL). We hope that this review will provide guidance to many *Bt* researchers in rational protein design and that it will help in the development of modified *Bt* δ-endotoxins with improved biological function, without adversely affecting the structural stability of the toxins.

5.4 Perspectives

The next big step in *Bt* protein engineering will be to understand the membrane bound state of the toxins and to use this information to design proteins with improved structure and function. At the present time there is contradictory evidence from many groups that supports several different membrane-bound models. This clearly indicates that a "final" model able to explain all observations is still needed. As indicated above, the three domains of the *Bt* toxins carry on specific functions that can be manipulated by protein engineering. A major goal in this research should be the development of "chimeric" proteins in which *Bt* δ-endotoxin domains are combined with domains from other unrelated proteins to create molecules with new functions. Just to mention an example, combining the ion-pore forming domain of Cry toxins with the catalytic domain of diphtheria toxin may allow the development of families of proteins carrying this catalytic domain into a new type of target cells.

Acknowledgments We would like to thank Drs. Rehan Hussain (Columbus Children's Hospital) and Carol Parker (University of Victoria Proteome Center) for careful review of the manuscript, and to Mr. Diego Velez for his contribution to Figs. 5.1–5.3. Special thanks to Dr. Donald H. Dean from the Entomology Department, The Ohio State University, for his support and encouragement.

References

Abdullah MA, Alzate O, Mohammad M, McNall RJ, Adang MJ, Dean DH (2003) Introduction of Culex toxicity into *Bacillus thuringiensis* Cry4Ba by protein engineering. Appl Environ Microbiol 69(9):5343–5353

Ahmad W, Ellar DJ (1990) Directed mutagenesis of selected regions of a *Bacillus thuringiensis* entomocidal protein. FEMS Microbiol Lett 56(1–2):97–104

Alcantara EP, Alzate O, Lee MK, Curtiss A, Dean DH (2001) Role of a-helix 7 of *Bacillus thuringiensis* Cry1Ab δ-endotoxin in membrane insertion, structural stability, and ion channel activity. Biochemistry 40(8):2540–2547

Altenbach C, Marti T, Khorana HG, Hubbell WL (1990) Transmembrane protein structure: spin labeling of bacteriorhodopsin mutants. Science 248(4959):1088–1092

Altenbach C, Greenhalgh DA, Khorana HG, Hubbell WL (1994) A collision gradient method to determine the immersion depth of nitroxides in lipid bilayers: application to spin-labeled mutants of bacteriorhodopsin. Proc Natl Acad Sci U S A 91(5):1667–1671

Alzate O, You T, Claybon M, Osorio C, Curtiss A, Dean DH (2006) Effects of disulfide bridges in domain I of *Bacillus thuringiensis* Cry1Aa δ-endotoxin on ion-channel formation in biological membranes. Biochemistry 45(45):13597–13605

Alzate O, Hemann CF, Osorio C, Hille R, Dean DH (2009) Ser170 of *Bacillus thuringiensis* Cry1Ab δ-endotoxin becomes anchored in a hydrophobic moiety upon insertion of this protein into *Manduca sexta* brush border membranes. BMC Biochem 10:25

Alzate O, Osorio C, Florez AM, Dean DH (2010) Participation of valine 171 in α-Helix 5 of *Bacillus thuringiensis* Cry1Ab δ-endotoxin in translocation of toxin into *Lymantria dispar* midgut membranes. Appl Environ Microbiol 76(23):7878–7880

Arenas I, Bravo A, Soberon M, Gomez I (2010) Role of alkaline phosphatase from *Manduca sexta* in the mechanism of action of *Bacillus thuringiensis* Cry1Ab toxin. J Biol Chem 285(17):12497–12503

Arnold S, Curtiss A, Dean DH, Alzate O (2001) The role of a proline-induced broken-helix motif in α-helix 2 of *Bacillus thuringiensis* δ-endotoxins. FEBS Lett 490(1–2):70–74

Aronson A (1995) The protoxin composition of *Bacillus thuringiensis* insecticidal inclusions affects solubility and toxicity. Appl Environ Microbiol 61(11):4057–4060

Aronson A (2000) Incorporation of protease K into larval insect membrane vesicles does not result in disruption of integrity or function of the pore-forming *Bacillus thuringiensis* δ-endotoxin. Appl Environ Microbiol 66(10):4568–4570

Atherton NM (1993) Principles of electron spin resonance. Ellis Horwood, Prentice Hall, London

Audtho M, Valaitis AP, Alzate O, Dean DH (1999) Production of chymotrypsin-resistant *Bacillus thuringiensis* Cry2Aa1 δ-endotoxin by protein engineering. Appl Environ Microbiol 65(10):4601–4605

Berova N, Nakanishi K, Woody R (2000) Circular dichroism: principles and applications, 2nd edn. Wiley-VCH, New York

Bietlot HP, Vishnubhatla I, Carey PR, Pozsgay M, Kaplan H (1990) Characterization of the cysteine residues and disulfide linkages in the protein crystal of *Bacillus thuringiensis*. Biochem J 267:309–315

Boonserm P, Davis P, Ellar DJ, Li J (2005) Crystal structure of the mosquito-larvicidal toxin Cry4Ba and its biological implications. J Mol Biol 348(2):363–382

Boonserm P, Mo M, Angsuthanasombat C, Lescar J (2006) Structure of the functional form of the mosquito larvicidal Cry4Aa toxin from *Bacillus thuringiensis* at a 2.8-angstrom resolution. J Bacteriol 188(9):3391–3401

Bravo A, Gomez I, Conde J, Munoz-Garay C, Sanchez J, Miranda R, Zhuang M, Gill SS, Soberon M (2004) Oligomerization triggers binding of a *Bacillus thuringiensis* Cry1Ab pore-forming toxin to aminopeptidase N receptor leading to insertion into membrane microdomains. Biochim Biophys Acta 1667(1):38–46

Burton SL, Ellar DJ, Li J, Derbyshire DJ (1999) *N*-acetylgalactosamine on the putative insect receptor aminopeptidase N is recognized by a site on the domain III lectin-like fold of a *Bacillus thuringiensis* insecticidal toxin. J Mol Biol 287:1011–1022

Carroll J, Wolfersberger MG, Ellar DJ (1997) The *Bacillus thuringiensis* Cry1Ac toxin-induced permeability change in *Manduca sexta* midgut brush border membrane vesicles proceeds by more than one mechanism. J Cell Sci 110:3099–3104

Chandra A, Ghosh P, Mandaokar AD, Bera AK, Sharma RP, Das S, Kumar PA (1999) Amino acid substitution in α-helix 7 of Cry1Ac δ-endotoxin of *Bacillus thuringiensis* leads to enhanced toxicity to *Helicoverpaarmigera*Hubner. FEBS Lett 458(2):175–179

Chen XJ, Curtiss A, Alcantara E, Dean DH (1995) Mutations in domain I of *Bacillus thuringiensis* δ-endotoxin CryIAbreduce the irreversible binding of toxin to *Manduca sexta* brush border membrane vesicles. J Biol Chem 270(11):6412–6419

Choma CT, Surewicz WK, Carey PR, Pozsgay M, Raynor T, Kaplan H (1990) Unusual proteolysis of the protoxin and toxin from *Bacillus thuringiensis*: structural implications. Eur J Biochem 189:523–527

Coux F, Vachon V, Rang C, Moozar K, Masson L, Royer M, Bes M, Rivest S, Brousseau R, Schwartz JL, Laprade R, Frutos R (2001) Role of interdomain salt bridges in the pore-forming ability of the *Bacillus thuringiensis* toxins Cry1Aa and Cry1Ac. J Biol Chem 276(38):35546–35551

Crickmore N, Zeigler DR, Feitelson J, Schnepf E, Van Rie J, Lereclus D, Baum J, Dean DH (1998) Revision of the nomenclature for the *Bacillus thuringiensis* pesticidal crystal proteins. Microbiol Mol Biol Rev 62(3):807–813

Crickmore N, Zeigler DR, Schnepf E, Van Rie J, Lereclus D, Baum J, Bravo A, Dean DH (2011) *Bacillus thuringiensis* toxin nomenclature. http://www.lifesci.sussex.ac.uk/Home/Neil_Crickmore/Bt/

de Maagd RA, Kwa MS, van der Klei H, Yamamoto T, Schipper B, Vlak JM, Stiekema WJ, Bosch D (1996) Domain III substitution in *Bacillus thuringiensis* δ-endotoxin CryIA(b) results in superior toxicity for *Spodopteraexigua* and altered membrane protein recognition. Appl Environ Microbiol 62(5):1537–1543

de Maagd RA, Weemen-Hendriks M, Stiekema W, Bosch D (2000) *Bacillus thuringiensis* δ-endotoxin Cry1C domain III can function as a specificity determinant for *Spodopteraexigua* in different, but not all, Cry1-Cry1C hybrids. Appl Environ Microbiol 66(4):1559–1563

Fortier M, Vachon V, Frutos R, Schwartz JL, Laprade R (2007) Effect of insect larval midgut proteases on the activity of *Bacillus thuringiensis* Cry toxins. Appl Environ Microbiol 73(19):6208–6213

Gatehouse JA (2011) Prospects for using proteinase inhibitors to protect transgenic plants against attack by herbivorous insects. Curr Protein Pept Sci. (Epub ahead of print)

Gazit E, Shai Y (1993) Structural and functional characterization of the alpha 5 segment of *Bacillus thuringiensis* δ-endotoxin. Biochemistry 32(13):3429–3436

Gazit E, La Rocca P, Sansom MS, Shai Y (1998) The structure and organization within the membrane of the helices composing the pore-forming domain of *Bacillus thuringiensis* δ-endotoxin are consistent with an "umbrella-like" structure of the pore. Proc Natl Acad Sci U S A 95(21):12289–12294

Ge AZ, Shivarova NI, Dean DH (1989) Location of the Bombyxmori specificity domain on a *Bacillus thuringiensis* δ-endotoxin protein. Proc Natl Acad Sci U S A 86(11):4037–4041

Ge AZ, Rivers D, Milne R, Dean DH (1991) Functional domains of *Bacillus thuringiensis* insecticidal crystal proteins: refinement of *Heliothisvirescens* and *Trichoplusiani*specificity domains on CryIA(c). J Biol Chem 266:17954–17958

Girard F, Vachon V, Prefontaine G, Marceau L, Su Y, Larouche G, Vincent C, Schwartz JL, Masson L, Laprade R (2008) Cysteine scanning mutagenesis of alpha4, a putative pore-lining helix of the *Bacillus thuringiensis* insecticidal toxin Cry1Aa. Appl Environ Microbiol 74(9):2565–2572

Gomez I, Sanchez J, Miranda R, Bravo A, Soberon M (2002) Cadherin-like receptor binding facilitates proteolytic cleavage of helix alpha-1 in domain I and oligomer pre-pore formation of *Bacillus thuringiensis* Cry1Ab toxin. FEBS Lett 513(2–3):242–246

Gomez I, Dean DH, Bravo A, Soberon M (2003) Molecular basis for *Bacillus thuringiensis* Cry1Ab toxin specificity: two structural determinants in the *Manduca sexta* Bt-R1 receptor interact with loops alpha-8 and 2 in domain II of Cy1Ab toxin. Biochemistry 42(35):10482–10489

Greenfield NJ (2006) Using circular dichroism spectra to estimate protein secondary structure. Nat Protoc 1(6):2876–2890

Grochulski P, Masson L, Borisova S, Pusztai-Carey M, Schwartz JL, Brousseau R, Cygler M (1995) *Bacillus thuringiensis* CryIA(a) insecticidal toxin: crystal structure and channel formation. J Mol Biol 254(3):447–464

Höfte H, Whiteley HR (1989) Insecticidal crystal proteins of *Bacillus thuringiensis*. Microbiol Rev 53:242–255

Hubbell WL, Cafiso DS, Altenbach C (2000) Identifying conformational changes with site-directed spin labeling. Nat Struct Biol 7(9):735–739

Hubbell WL, Altenbach C, Hubbell CM, Khorana HG (2003) Rhodopsin structure, dynamics, and activation: a perspective from crystallography, site-directed spin labeling, sulfhydryl reactivity, and disulfide cross-linking. Adv Protein Chem 63:243–290

Hussain SR, Aronson AI, Dean DH (1996) Substitution of residues on the proximal side of Cry1A *Bacillus thuringiensis* δ-endotoxins affects irreversible binding to Manduca sexta midgut membrane. Biochem Biophys Res Commun 226(1):8–14

Hussain SA, Flórez AM, Dean DH, Alzate O (2010) Preferential protection of domains II and III of *Bacillus thuringiensis* Cry1Aa toxin by brush border membrane vesicles. Rev Colomb Biotechnol 12(2):14–26

Jimenez-Juarez N, Munoz-Garay C, Gomez I, Saab-Rincon G, Damian-Almazo JY, Gill SS, Soberon M, Bravo A (2007) *Bacillus thuringiensis* Cry1Ab mutants affecting oligomer formation are non-toxic to *Manduca sexta* larvae. J Biol Chem 282(29):21222–21229

Jimenez-Juarez N, Munoz-Garay C, Gomez I, Gill SS, Soberon M, Bravo A (2008) The pre-pore from *Bacillus thuringiensis* Cry1Ab toxin is necessary to induce insect death in *Manduca sexta*. Peptides 29(2):318–323

Kaur S (2000) Molecular approaches towards development of novel *Bacillus thuringiensis* biopesticides. World J Microbiol Biotechnol 16:781–793

Kirouac M, Vachon V, Quievy D, Schwartz JL, Laprade R (2006) Protease inhibitors fail to prevent pore formation by the activated *Bacillus thuringiensis* toxin Cry1Aa in insect brush border membrane vesicles. Appl Environ Microbiol 72(1):506–515

Kumar AS, Aronson AI (1999) Analysis of mutations in the pore-forming region essential for insecticidal activity of a *Bacillus thuringiensis* δ-endotoxin. J Bacteriol 181(19):6103–6107

Lebel G, Vachon V, Prefontaine G, Girard F, Masson L, Juteau M, Bah A, Larouche G, Vincent C, Laprade R, Schwartz JL (2009) Mutations in domain I interhelical loops affect the rate of pore formation by the *Bacillus thuringiensis* Cry1Aa toxin in insect midgut brush border membrane vesicles. Appl Environ Microbiol 75(12):3842–3850

Lee MK, Young BA, Dean DH (1995) Domain III exchanges of *Bacillus thuringiensis* CryIA toxins affect binding to different gypsy moth midgut receptors. Biochem Biophys Res Commun 216:306–312

Li J, Carroll J, Ellar DJ (1991) Crystal structure of insecticidal d-endotoxin from *Bacillus thuringiensis* at 2.5 Å resolution. Nature 353:815–821

Li J, Derbyshire DJ, Promdonkoy B, Ellar DJ (2001) Structural implications for the transformation of the *Bacillus thuringiensis* δ-endotoxins from water-soluble to membrane-inserted forms. Biochem Soc Trans 29(Pt 4):571–577

Lightwood DJ, Ellar DJ, Jarrett P (2000) Role of proteolysis in determining potency of *Bacillus thuringiensis* Cry1Ac δ-endotoxin. Appl Environ Microbiol 66(12):5174–5181

Likitvivatanavong S, Chen J, Evans AM, Bravo A, Soberon M, Gill SS (2011) Multiple receptors as targets of cry toxins in mosquitoes. J Agric Food Chem 59(7):2829–2838

Liu XS, Dean DH (2006) Redesigning *Bacillus thuringiensis* Cry1Aa toxin into a mosquito toxin. Protein Eng Des Sel 19(3):107–111

Lorence A, Darszon A, Diaz C, Lievano A, Quintero R, Bravo A (1995) Delta-endotoxins induce cation channels in *Spodopterafrugiperda* brush border membranes in suspension and in planar lipid bilayers. FEBS Lett 360(3):217–222

Masson L, Tabashnik BE, Liu YB, Brousseau R, Schwartz JL (1999) Helix 4 of the *Bacillus thuringiensis* Cry1Aa toxin lines the lumen of the ion channel. J Biol Chem 274(45):31996–32000

Munoz-Garay C, Sanchez J, Darszon A, de Maagd RA, Bakker P, Soberon M, Bravo A (2006) Permeability changes of *Manduca sexta* midgut brush border membranes induced by oligomeric structures of different Cry toxins. J Membr Biol 212(1):61–68

Munoz-Garay C, Rodriguez-Almazan C, Aguilar JN, Portugal L, Gomez I, Saab-Rincon G, Soberon M, Bravo A (2009) Oligomerization of Cry11Aa from *Bacillus thuringiensis* has an important role in toxicity against *Aedesaegypti*. Appl Environ Microbiol 75(23):7548–7550

Nair MS, Dean DH (2008) All domains of Cry1A toxins insert into insect brush border membranes. J Biol Chem 283(39):26324–26331

Pacheco S, Gomez I, Arenas I, Saab-Rincon G, Rodriguez-Almazan C, Gill SS, Bravo A, Soberon M (2009) Domain II loop 3 of *Bacillus thuringiensis* Cry1Ab toxin is involved in a "ping pong" binding mechanism with *Manduca sexta*aminopeptidase-N and cadherin receptors. J Biol Chem 284(47):32750–32757

Pardo-Lopez L, Gomez I, Rausell C, Sanchez J, Soberon M, Bravo A (2006) Structural changes of the Cry1Ac oligomeric pre-pore from *Bacillus thuringiensis* induced by *N*-Acetylgalactosamine facilitates toxin membrane insertion. Biochemistry 45(34):10329–10336

Pigott CR, Ellar DJ (2007) Role of receptors in *Bacillus thuringiensis* crystal toxin activity. Microbiol Mol Biol Rev 71(2):255–281

Pozsgay M, Fast P, Kaplan H, Carey PR (1987) The effect of sunlight on the protein crystals from *Bacillus thuringiensis* subsp. kurstaki HD1 and NRD12, a Raman spectroscopy study. J Invertebr Pathol 50:246–253

Pusztai M, Fast P, Gringorten L, Kaplan H, Lessard T, Carey PR (1991) The mechanism of sunlight-mediated inactivation of *Bacillus thuringiensis* crystals. Biochem J 273(Pt 1):43–47

Rajamohan F, Dean DH (1996) Molecular biology of bacteria on biological control of insects. In: Gunasekaran M, Weber DJ (eds) Molecular biology of the biological control of pests and diseases of plants. CRC Press, Boca Raton, pp 105–122

Rajamohan F, Alcantara E, Lee MK, Chen XJ, Curtiss A, Dean DH (1995) Single amino acid changes in domain II of *Bacillus thuringiensis* CryIAb δ-endotoxin affect irreversible binding to *Manduca sexta* midgut membrane vesicles. J Bacteriol 177(9):2276–2282

Rajamohan F, Alzate O, Cotrill JA, Curtiss A, Dean DH (1996a) Protein engineering of *Bacillus thuringiensis* δ-endotoxin: mutations at domain II of CryIAb enhance receptor affinity and toxicity toward gypsy moth larvae. Proc Natl Acad Sci U S A 93(25):14338–14343

Rajamohan F, Cotrill JA, Gould F, Dean DH (1996b) Role of domain II, loop 2 residues of *Bacillus thuringiensis* CryIAb δ-endotoxin in reversible and irreversible binding to *Manduca sexta* and *Heliothisvirescens*. J Biol Chem 271(5):2390–2396

Rausell C, Munoz-Garay C, Miranda-CassoLuengo R, Gomez I, Rudino-Pinera E, Soberon M, Bravo A (2004a) Tryptophan spectroscopy studies and black lipid bilayer analysis indicate that the oligomeric structure of Cry1Ab toxin from *Bacillus thuringiensis* is the membrane-insertion intermediate. Biochemistry 43(1):166–174

Rausell C, Pardo-Lopez L, Sanchez J, Munoz-Garay C, Morera C, Soberon M, Bravo A (2004b) Unfolding events in the water-soluble monomeric Cry1Ab toxin during transition to oligomeric pre-pore and membrane-inserted pore channel. J Biol Chem 279(53):55168–55175

Rodriguez-Almazan C, Zavala LE, Munoz-Garay C, Jimenez-Juarez N, Pacheco S, Masson L, Soberon M, Bravo A (2009) Dominant negative mutants of *Bacillus thuringiensis* Cry1Ab toxin function as anti-toxins: demonstration of the role of oligomerization in toxicity. PLoS One 4(5):e5545

Sanahuja G, Banakar R, Twyman RM, Capell T, Christou P (2011) *Bacillus thuringiensis*: a century of research, development and commercial applications. Plant Biotechnol J 9(3):283–300

Schnepf HE (1995) *Bacillus thuringiensis* toxins: regulation, activities and structural diversity. Curr Opin Biotechnol 6:305–312

Schnepf E, Crickmore N, Van Rie J, Lereclus D, Baum J, Feitelson J, Zeigler DR, Dean DH (1998) *Bacillus thuringiensis* and its pesticidal crystal proteins. Microbiol Mol Biol Rev 62(3):775–806

Schwartz JL, Juteau M, Grochulski P, Cygler M, Prefontaine G, Brousseau R, Masson L (1997a) Restriction of intramolecular movements within the Cry1Aa toxin molecule of *Bacillus thuringiensis* through disulfide bond engineering. FEBS Lett 410(2–3):397–402

Schwartz JL, Lu YJ, Sohnlein P, Brousseau R, Laprade R, Masson L, Adang MJ (1997b) Ion channels formed in planar lipid bilayers by *Bacillus thuringiensis* toxins in the presence of *Manduca sexta* midgut receptors. FEBS Lett 412:270–276

Schwartz JL, Potvin L, Chen XJ, Brousseau R, Laprade R, Dean DH (1997c) Single-site mutations in the conserved alternating-arginine region affect ionic channels formed by CryIAa, a *Bacillus thuringiensis* toxin. Appl Environ Microbiol 63(10):3978–3984

Shan S, Zhang Y, Ding X, Hu S, Sun Y, Yu Z, Liu S, Zhu Z, Xia L (2010) A Cry1Ac toxin variant generated by directed evolution has enhanced toxicity against Lepidopteran insects. Curr Microbiol 62(2):358–365

Soberon M, Pardo-Lopez L, Lopez I, Gomez I, Tabashnik BE, Bravo A (2007) Engineering modified *Bt* toxins to counter insect resistance. Science 318(5856):1640–1642

Tigue NJ, Jacoby J, Ellar DJ (2001) The α-helix 4 residue, Asn135, is involved in the oligomerization of Cry1Ac1 and Cry1Ab5 *Bacillus thuringiensis* toxins. Appl Environ Microbiol 67(12):5715–5720

Vachon V, Prefontaine G, Rang C, Coux F, Juteau M, Schwartz JL, Brousseau R, Frutos R, Laprade R, Masson L (2004) Helix 4 mutants of the *Bacillus thuringiensis* insecticidal toxin Cry1Aa display altered pore-forming abilities. Appl Environ Microbiol 70(10):6123–6130

van Frankenhuyzen K (2009) Insecticidal activity of *Bacillus thuringiensis* crystal proteins. J Invertebr Pathol 101(1):1–16

Wolfersberger MG (1990) The toxicity of two *Bacillus thuringiensis* δ-endotoxins to gypsy moth larvae is inversely related to the affinity of binding sites on midgut brush border membranes for the toxins. Experientia 46(5):475–477

Wolfersberger MG, Spaeth DD (1987) Activity of spore-crystal preparations from twenty serotypes of *Bacillus thuringiensis* toward *Manduca sexta* larvae *in vivo* and *in vitro*. Z Angew Entomol 103:138–141

Wolfersberger M, Lüthy P, Maurer A, Parenti P, Sacchi FV, Giordana B, Hanozet GM (1987) Preparation and partial characterization of amino acid transporting brush border membrane vesicles from the larval midgut of the cabbage butterfly (*Pierisbrassicae*). Comp Biochem Physiol 86A:301–308

Wu D, Aronson AI (1992) Localized mutagenesis defines regions of the *Bacillus thuringiensis* δ-endotoxin involved in toxicity and specificity. J Biol Chem 267(4):2311–2317

Zavala LE, Pardo-Lopez L, Canton PE, Gomez I, Soberon M, Bravo A (2011) Domains II and III of *Bacillus thuringiensis* Cry1Ab toxin remain exposed to the solvent after insertion of part of domain I into the membrane. J Biol Chem 286:19109–19117

Zhang X, Candas M, Griko NB, Taussig R, Bulla LA Jr (2006) A mechanism of cell death involving an adenylyl cyclase/PKA signaling pathway is induced by the Cry1Ab toxin of *Bacillus thuringiensis*. Proc Natl Acad Sci U S A 103(26):9897–9902

Zhuang M, Oltean DI, Gomez I, Pullikuth AK, Soberon M, Bravo A, Gill SS (2002) *Heliothisvirescens* and *Manduca sexta* lipid rafts are involved in Cry1A toxin binding to the midgut epithelium and subsequent pore formation. J Biol Chem 277(16):13863–13872

Part II
Genetics of *Bacillus thuringiensis*

Chapter 6
Evolution of the *Bacillus cereus* Group

Ole Andreas Økstad and Anne-Brit Kolstø

Abstract The *Bacillus cereus* group contains six approved bacterial species—*Bacillus thuringiensis, Bacillus cereus* (*sensu stricto*), *Bacillus anthracis, Bacillus weihenstephanensis, Bacillus mycoides*, and *Bacillus pseudomycoides*. In addition, thermotolerant *B. cereus* var. *cytotoxis* could constitute a member of a novel species in the group. Historically, *Bacillus thuringiensis* has been viewed as a species separate from the other members of the group, mainly based on its entomopathogenic properties and production of insecticidal crystal toxins. However, following the sequencing of more than 85 strains from the *B. cereus* group, whole genome phylogenies and MLST analyses of more than 1400 strains based on chromosomal markers clearly show that the *Bacillus cereus* group exhibits a complex population structure, in which strains of *B. thuringiensis* are strongly intermixed with other *B. cereus* group species. From the analysis of genome sequence data, it is clear that the only principal factor differentiating *B. thuringiensis* from *B. cereus* is the presence of insecticidal crystal toxin gene(s), which in the vast majority of cases are carried on plasmid elements that are frequently conjugative or mobilizable, and which occasionally could be lost by the bacterium. Also, strikingly, some *B. thuringiensis* strains are among the closest relatives to the *B. anthracis* clonal complex. Chromosomal factors potentially involved in stabilizing long-term carriage of plasmids in *B. thuringiensis* strains, have not yet been elucidated.

Keywords *Bacillus cereus* group · Phylogeny · Genomes · Sequencing · Plasmids

6.1 The *Bacillus cereus* Group

The *B. cereus* group of bacteria (*B. cereus sensu lato*) is a subgroup of the *B. subtilis* group in the universal rRNA-based *Bacillus* phylogeny (Priest 1993), and includes six approved species: *B. thuringiensis*, *B. cereus* (*sensu stricto*), *B. anthracis*,

O. A. Økstad (✉) · A.-B. Kolstø
Laboratory for Microbial Dynamics (LaMDa), Department of Pharmaceutical Biosciences, University of Oslo, Blindern, 0316 Oslo, Norway
e-mail: aloechen@farmasi.uio.no

A.-B. Kolstø
e-mail: a.b.kolsto@farmasi.uio.no

E. Sansinenea (ed.), *Bacillus thuringiensis Biotechnology*,
DOI 10.1007/978-94-007-3021-2_6, © Springer Science+Business Media B.V. 2012

B. weihenstephanensis, B. mycoides and *B. pseudomycoides*. In addition, three newly identified thermotolerant strains form a phylogenetic outgroup within the *B. cereus* group (*B. cereus* var. *cytotoxis*) (Lund et al. 2000; Fagerlund et al. 2007; Auger et al. 2008), and have been suggested to constitute a separate species (*B. cytotoxicus* or *B. cytoxis*) (Lapidus et al. 2008). Among the six approved species, *B. thuringiensis, B. cereus* and *B. anthracis* are by far the most extensively studied, which is further substantiated by the number of genome sequencing studies carried out; out of the 85 sequenced genomes within the *B. cereus* group, 80 belong to these three species (17 *B. thuringiensis*, 19 *B. anthracis* and 44 *B. cereus*) (Fig. 6.1; Entrez genome project: http://www.ncbi.nlm.nih.gov/genomeprj; Genomes OnLine database: http://www.genomesonline.org).

While *B. thuringiensis* is an entomopathogenic bacterium and the most commonly used commercial biological pesticide worldwide (Soberon et al. 2007), *B. anthracis* is a highly monomorphic species within the *B. cereus* group (Keim et al. 1997; Van Ert et al. 2007) and the cause of anthrax, which is mainly a disease of herbivores but which can present similar acute disease in humans and other mammals. In several areas of the world, including parts of Africa, Australia and Asia, *B. anthracis* is endemic or hyperendemic (http://www.vetmed.lsu.edu/whocc/mp_world.htm). *B. cereus* is a frequent cause of two human food-poisoning syndromes (diarrhoeal and emetic, respectively; Drobniewski 1993; Stenfors Arnesen et al. 2008), as well as of various opportunistic and nosocomial infections (reviewed by Drobniewski 1993; Kotiranta et al. 2000; Bottone 2010). These bacteria can be found in the environment; while *B. anthracis* spores are strongly biased toward sites of previous infections (e.g. earlier carcass sites), *B. cereus* and *B. thuringiensis* are common in soil and soil-associated niches.

6.2 *B. cereus* Group Phylogeny

The phylogeny of the *B. cereus* group has been mapped by various methods, including multilocus enzyme electrophoresis (MLEE; Helgason et al. 1998, 2000a, b), amplified fragment length polymorphism (AFLP; Keim et al. 1997; Jackson et al. 1999; Ticknor et al. 2001; Hill et al. 2004), and multilocus sequence typing (MLST; Helgason et al. 2004; Barker et al. 2005; Tourasse et al. 2006). Altogether five different MLST schemes exist for the group, largely employing non-overlapping strain sets, but with certain strains common to several schemes. This has allowed to combine the five schemes into one phylogeny using supertree methodology (Tourasse and Kolstø 2008), encompassing 1430 isolates (SuperCAT database; http://mlstoslo.uio.no). Furthermore, typing data from MLEE, AFLP and MLST methods have been integrated, resulting in a global multi-scheme and multi-datatype phylogenetic analysis providing the currently most comprehensive view of the *B. cereus* group population (Tourasse et al. 2010), including 2213 isolates (HyperCAT database; http://mlstoslo.uio.no). By MLST analysis, the *B. cereus* group population can be grouped into three main clusters (Helgason et al. 2004), which can be further di-

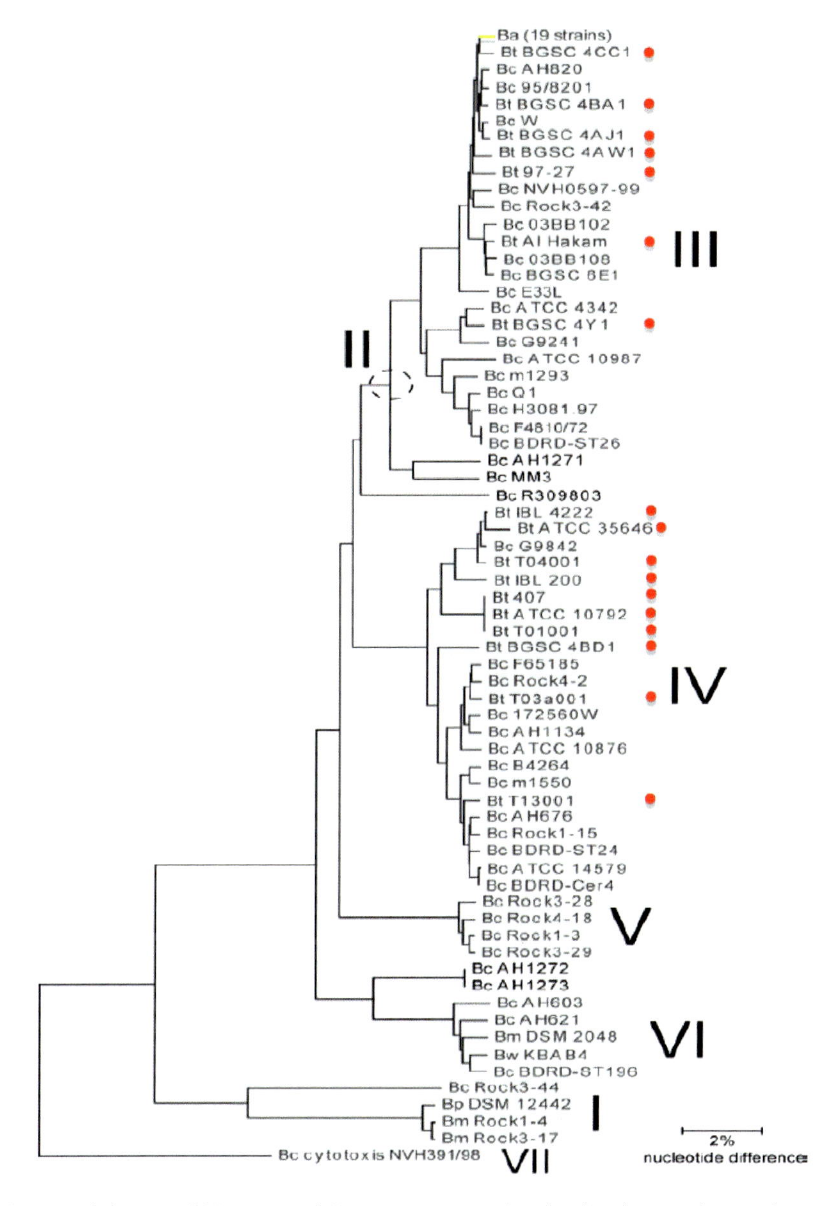

Fig. 6.1 Phylogeny of 85 sequenced *B. cereus* group strains showing the complex species structure of the group, with representatives of cluster I and III-VII from Guinebretière et al. (2008). *Red circles* designate *B. thuringiensis* strains. None of the 85 sequenced strains are assigned to cluster II, but a *dashed circle* indicates where cluster II would emerge based on extrapolation from the *B. cereus* group supertree available at the HyperCAT database (http://mlstoslo.uio.no). The 19 *B. anthracis* isolates form a clonal complex and are represented here as a single lineage (*yellow* labeled branch). The tree was constructed based on concatenation of chromosomal regions (a total of ~1.5 Mbp) conserved in all strains (described in the supplementary file of Tourasse and Kolsto (2008)). Bc, *B. cereus*; Bt, *B. thuringiensis*; Ba, *B. anthracis*; Bm, *B. mycoides*; Bw, *B. weihenstephanensis*; Bp, *B. pseudomycoides*. Tree generated by Nicolas J. Tourasse

vided into a seven cluster phylogeny, as described by Guinebretière and co-workers (2008). *B. anthracis* and emetic *B. cereus* strains form independent clonal structures within main cluster I (Tourasse and Kolstø 2008; equivalent to clade 1 in Didelot et al. 2009). The main clusters in the population are reproduced both in the SuperCAT and HyperCAT phylogenies (Tourasse and Kolstø 2008; Tourasse et al. 2010), as well as in a phylogenetic analysis of the *B. cereus* group, utilizing 157 single-copy genes of the family *Bacillaceae* (Schmidt et al. 2011). While cluster I and II (Tourasse and Kolstø 2008) contain *B. anthracis* and most of the clinical *B. cereus* group isolates (equivalent to group II and III in Helgason et al. 2004; clades 1 and 2 in Didelot et al. 2009), cluster III is heterogeneous and contains *B. cereus* strains from various sources, in addition to multiple outgroups (*B. pseudomycoides* isolates, *B. cereus* var. *cytotoxicus*). Interestingly, Didelot and co-workers, by analysis of 667 strains using the Priest MLST scheme, identified partial barriers to gene flow between the three clades (Didelot et al. 2009).

Although the *B. cereus* group in general is phylogenetically heterogeneous, single strains may be very closely related, when employing genetic markers at the chromosomal level. Furthermore, closely related strains may be of the same, or of different species, and members of the various species belonging to the group are frequently phylogenetically intermixed. In particular, this is the case for *B. cereus* and *B. thuringiensis* isolates (Fig. 6.1). In full *B. cereus* group SuperCAT and HyperCAT phylogenies (http://mlstoslo.uio.no), *B. thuringiensis* strains are distributed to all three main population clusters, and similar intermixing of *B. thuringiensis* with other *B. cereus* group isolates as that shown in Fig. 6.1 is observed (Tourasse and Kolstø 2008; Tourasse et al. 2010).

6.2.1 Complex Genomes

B. cereus group bacteria generally exhibit complex and dynamic genomes, different strains carrying a highly variable number of plasmids (1–> 12) which come in different sizes (2–600 kb; frequently > 80 kb) (reviewed in: Kolstø et al. 2009; He et al. 2011), and some of which may be conjugative or mobilizable. Strains can also host a number of different IS-elements, and in addition typically carry bacteriophage DNA, which may be integrated in the chromosome as prophage(s) or may replicate as independent linear genetic elements (Carlson et al. 1994; Rasko et al. 2005; Verheust et al. 2005; Sozhamannan et al. 2006; Lapidus et al. 2008).

6.3 *Bacillus cereus* Group Genome Sequencing

A multitude of *B. cereus* group strains have been sequenced during the past decade. Following the sequencing of the bioterror attack strain *B. anthracis* Ames Florida (Read et al. 2002), the *B. anthracis* Ames Porton model strain (Read et al. 2003) and the first *B. cereus* strain (ATCC 14579, type strain; Ivanova et al. 2003), 112 additional genomes have been or are being sequenced (Entrez genome project: http://www.ncbi.nlm.nih.gov/genomeprj; Genomes OnLine database: http://www.genomesonline.org), allowing large scale comparative genomics studies. The genome sequences cover all six approved species within the group (http://www.ncbi.nlm.nih.gov/genomeprj; http://www.genomesonline.org; Kolstø et al. 2009), in addition to the *B. cereus* NVH391-98 strain from the remote "*Bacillus cytotoxicus*" cluster, and altogether provide good coverage of the group phylogeny, as mapped by MLST (http://mlstoslo.uio.no).

The first genomes sequenced from the *B. cereus* group, those of *B. anthracis* Ames and of *B. cereus* strains ATCC 14579 and ATCC 10987, were all closed, and serve as reference genomes for later projects. Altogether 22 genomes have now been closed (http://www.genomesonline.org), showing that chromosome sizes for *B. cereus* group members are in the range 5.2–5.4 Mb (with the exception of *B. cereus* NVH 391-98 which is 4.1 Mb), showing that the genomes of the *B. cereus* group members, despite their marked phenotypic differences, have many common features and share a common chromosomal genetic background: Chromosomes have a GC content of around 35.3–35.4%, and strains typically carry a large number of rRNA operons (11 and 13 for *B. anthracis* Ames Porton and *B. cereus* ATCC 14579, respectively). Furthermore, chromosomes are to a large extent syntenic, with common orthologous genes being organized in a conserved order (Ivanova et al. 2003; Rasko et al. 2004). Also, a core gene set has been identified, counting 3000 ± 200 genes (Lapidus et al. 2008) out of the in excess of 5000 genes that make up a typical *B. cereus* group chromosome (excluding *B. cereus* NVH391-98, which has an unusually small genome; Lapidus et al. 2008). Furthermore, each strain characteristically has in the order of 400–800 strain-specific genes (Lapidus et al. 2008), which may potentially be involved in niche adaptation processes, and which are contributing to a large *B. cereus* group pan-genome (20–25000 genes estimated in Lapidus et al. 2008). Another unifying feature for the *B. cereus* group bacteria is the presence of several ubiquitous interspersed repeat elements in the size range 100–400 bp, many of which are unique to the group (Tourasse et al. 2006). These repeats, named *bcr1-bcr18*, are non-protein coding, but several copies seem to be expressed at the RNA level and are predicted to constitute non-autonomous mobile elements belonging to the class 'miniature inverted-repeat transposable elements (MITEs)' (Økstad et al. 2004; Tourasse et al. 2006; Klevan et al. 2007; Delihas 2008; Kristoffersen et al. 2011).

6.4 *Bacillus cereus* Group Bacteria—Pathogens and Symbionts

Both *B. thuringiensis* and *B. cereus* are common bacteria of the soil, but are known to colonize invertebrate guts as endosymbionts, an environment which has been suggested to constitute their natural habitat (Margulis et al. 1998; Jensen et al. 2003).

Given the entomopathogenicity of *B. thuringiensis* and human gastrointestinal, opportunistic and nosocomial infections caused by *B. cereus*, both of these bacteria can on occasion leave their suggested endosymbiotic cycle in invertebrate hosts, for a proposedly less frequent pathogenic life cycle in insects and mammals, respectively (Jensen et al. 2003). *B. anthracis* on the other hand, may in the environment primarily exist as a highly stable, dormant spore in the soil, constituting a reservoir for anthrax disease in warm-blooded animals. The possibility of vegetative cells to grow in nutrient-rich soil environments has however also been suggested (Jensen et al. 2003; Saile and Koehler 2006).

6.4.1 Virulence Determinants

The species-specific virulence of *B. thuringiensis* toward insects is dependent on particular Cry or Cyt insecticidal crystal toxins, while full virulence of *B. anthracis* toward mammals is dependent on two tripartite toxins (lethal toxin and edema toxin) and synthesis of a poly-γ-D-glutamate capsule. The common genetic background shared by these bacteria has however been suggested to confer on them a basal pathogenicity which is associated with chromosomally encoded toxins aiding the colonisation of hosts as diverse as insects and mammals. Both *B. cereus* and *B. thuringiensis* strains produce several common virulence factors, including enterotoxins, phospholipases, haemolysins, proteases, and other degradative enzymes. These are extracellular virulence factors, and are encoded in the chromosome by genes that are under transcriptional control of a global pleiotropic regulator, PlcR. Interestingly, the *plcR* regulon genes are frequently found in strains of several species within the *B. cereus* group, *B. thuringiensis* and *B. anthracis* included (Read et al. 2003; Han et al. 2006; Scarano et al. 2009). The universal presence of *plcR* regulon genes in the group could suggest that *B. cereus* group organisms in general may have the potential to be pathogenic (Read et al. 2003; Han et al. 2006; Hendriksen et al. 2006; McIntyre et al. 2008; Stenfors Arnesen et al. 2008). In B. a*nthracis* however, *plcR* is mutated and non-functional (Agaisse et al. 1999), leading to *B. anthracis* producing a very limited extracellular proteome compared to the other *B. cereus* group species (Gohar et al. 2005) and the chromosomal virulence factors belonging to the PlcR regulon (Gohar et al. 2008) not being synthesized, enterotoxins included. Thus, the PlcR regulon is clearly not necessary for *B. anthracis* virulence (reviewed in Kolstø et al. 2009).

6.5 What Distinguishes the Species in the *B. cereus* Group?

Species distinctions of *B. thuringiensis* and *B. anthracis*, used to separate these species from *B. cereus*, have traditionally been assigned on the basis of phenotypic features—more specifically by the different pathogenic specificities of these bacteria towards insect larvae and vertebrates, respectively. Interestingly, however, the genes coding for these species-specific virulence properties are plasmid-borne, a fact unknown at the time of species designation; in *B. thuringiensis* the crystal protein toxins (Cry or Cyt) causing pathogenicity to insects are almost exclusively encoded by genes present on plasmids of various sizes, and often associated with IS elements (Schnepf et al. 1998). Similarly in *B. anthracis*, the two large plasmids pXO1 (182 kb) and pXO2 (95 kb), which are necessary for full *B. anthracis* virulence (Mock and Fouet 2001; Passalacqua and Bergman 2006), carry the genes encoding the anthrax toxin components (*pag*, *lef* and *cya*; Okinaka et al. 1999) and accompanying regulators (*atxA*, *pagR*) (Uchida et al. 1993; Dai et al. 1995; Mignot et al. 2003; Fouet and Mock 2006), and the poly-γ-D-glutamic acid (polyglutamate) capsule and associated regulators (acpA, acpB) (Pannucci et al. 2002b; Van der Auwera et al. 2005), respectively. The phylogenetic intermixing of strains of the different species, in particular *B. cereus* and *B. thuringiensis* isolates (Fig. 6.1; Tourasse and Kolstø 2008; Tourasse et al. 2011; http://mlstoslo.uio.no), and the fact that the main phenotypical traits classically used to separate *B. thuringiensis* and *B. anthracis* from *B. cereus*, are carried by plasmids, has led to disputes regarding the species definitions within the group (Helgason et al. 2000b; Rasko et al. 2005). Hitherto, no species-specific property other than the plasmid-borne *cry* and *cyt* genes has been identified; *B. thuringiensis* strains can carry the same chromosomally encoded virulence genes as *B. cereus*, including the genes for enterotoxins, phospholipases, haemolysins, and proteases (Han et al. 2006; Scarano et al. 2009), which may in fact be important for the virulence of the bacterium, following its entry into the insect larvae haemocol (Fedhila et al. 2002, 2003, 2004). Thus, a *B. thuringiensis* strain that has lost the *cry*- or *cyt*-containing plasmids will be indistinguishable from *B. cereus*, and will be identified as such. Therefore, although the current species nomenclature is kept, largely based on the well-established differences observed in pathogenicity profiles of the two species toward insects, the two bacterial species are indistuingishable when a chromosomal phylogeny is reconstructed based on a sufficient number of isolates (Tourasse and Kolstø 2008; Tourasse et al. 2011; http://mlstoslo.uio.no). *B. cereus* and *B. thuringiensis* have therefore been suggested to constitute one species in genetic terms (Helgason et al. 2000b).

B. thuringiensis is an insect pathogen; however, isolates may also have the potential to act as opportunistic pathogens in humans and animals, in certain cases possibly causing tissue necrosis, pulmonary infections, or food poisoning (Hernandez et al. 1999; Ghelardi et al. 2007; McIntyre et al. 2008). In line with this, the sequencing of *B. thuringiensis* strains Al-Hakam, 97-27 (*var.* konkukian) and ATCC 35646 (*var.* israelensis) confirmed that *B. thuringiensis* strains can carry the same chromosomal virulence factors that are typical to *B. cereus* (i.e. PlcR-regulated enterotoxins, hemolysins, phospholipases, proteases etc.; Han et al. 2006). With what frequency human infections by *B. thuringiensis* may actually occur is however at present uncertain, since *B. cereus* group strains isolated from human infections (food-poisoning included) are most often not tested for the presence of crystal toxin genes or for the production of such toxins. Also, it is conceivable that *B. thuringiensis*, during an infection in man, might lose the plasmid encoding its entomopatogenic properties. This would make the isolate practically indistinguishable from *B. cereus*. In fact, the ability to cause the *B. cereus* diarrhoeal syndrome may be an inherent feature of all *B. cereus* group species carrying a functional *PlcR* regulator gene. Indeed, *B. thuringiensis* was identified in food or clinical samples from four outbreaks, and *B. mycoides* in one outbreak, sampled in British Columbia, Canada in the period 1991–2005 (McIntyre et al. 2008). Seemingly, contrary to the emetic isolates which group into two clusters phylogenetically, isolates causing the diarrhoeal syndrome are located throughout the *B. cereus* group phylogeny, and may frequently share identical molecular typing data (based on chromosomal markers) with environmental strains isolated from soil or plants (Tourasse et al. 2011; http://mlstoslo.uio.no). Held together with the fact that genes encoding the non-hemolytic enterotoxin complex (Nhe) are found ubiquitously among *B. cereus* group organisms and that *B. cereus* group species other than *B. cereus sensu stricto* have been characterized as the cause of gastroenteritis (McIntyre et al. 2008), this emphasizes the opportunistic nature of *B. cereus* group bacteria, and the potential for other *B. cereus* group species to cause food-borne disease. It should be noted however, that in a study of greenhouse workers with occupational exposure to *B. thuringiensis*, although detecting human gastrointestinal carriage of *B. thuringiensis*, no gastrointestinal symptoms correlated with the presence of *B. thuringiensis* in fecal samples (Jensen et al. 2002).

6.6 Importance of Plasmids to *B. cereus* Group Biology

Plasmids are key elements in encoding several of the phenotypes characteristic of each species in the *B. cereus* group, including *B. thuringiensis* entomopathogenicity and *B. anthracis* virulence, as well as the synthesis of emetic toxin in emetic *B. cereus* strains. In general, *B. cereus* group strains are well-known for having the potential for harbouring one, or often several, plasmids in the cell at the same time (reviewed in Kolstø et al. 2009), however there is frequently a lack of knowledge of what features *B. cereus* and *B. thuringiensis* plasmids encode. The discovery of pXO1 plasmid-like sequences in a large proportion of 19 *B. cereus* and *B. thuringiensis* strains tested for pXO1 gene markers by CGH (Read et al. 2003), was at the time striking, and indicated that plasmids with similarity to pXO1 could potentially exist in other species than *B. anthracis*, including *B. thuringiensis*. This was later confirmed by sequencing of the *B. thuringiensis* pBtoxis plasmid, which carries 4 *cry* and 2 *cyt* toxin genes and has 29 out of its 125 genes in common with plasmid pXO1 (Berry et al. 2002), sequencing of *B. cereus* ATCC 10987 (Rasko et al. 2004) as well as plasmids from emetic and periodontal *B. cereus* isolates (Rasko et al. 2007), and by screening of a large number of *B. cereus* group strains for pXO1 gene markers by PCR (Pannucci et al. 2002a). A number of studies have also identified and functionally described pXO2-like plasmids in different *B. cereus* group species, including *B. thuringiensis* (van der Auwera et al. 2005; Hu et al. 2009a; Amadio et al. 2009). None of the above mentioned studies however revealed a pXO1- or pXO2-like plasmid carrying orthologs to the *B. anthracis* toxin or capsule genes. With the discovery of *B. cereus* strains causing severe anthrax-like disease symptoms, however, and which encode the anthrax toxins from variants of the pXO1 plasmid as well as producing a protective capsule, the principle differences separating the *B. anthracis* and *B. cereus* species are being eroded. Such strains include *B. cereus* G9241, which was isolated from the sputum and blood of a patient with life-threatening pneumonia, and carries a 191 kb plasmid (pBCX01) with 99.6 identity to pXO1 in shared regions, as well as a second 218 kb plasmid (pBC218) which has a gene cluster encoding a polysaccharide capsule (Hoffmaster et al. 2004). Perhaps even more striking, *B. cereus* strains causing anthrax-like disease in great apes have been isolated in the Cote De Ivore and Cameroon (*B. cereus* CI and CA, respectively; Leendertz et al. 2004, 2006; Klee et al. 2006), and been shown to carry full *B. anthracis* pXO1 and pXO2 plasmids, including the pathogenicity islands of both plasmids (Klee et al. 2010). The strains have a frameshift mutation in the chromosomal *plcR* gene (Klee et al. 2006) rendering the protein non-functional, and unable to bind PapR, its cognate peptide pheromone, which is necessary for PlcR regulon activation (Slamti and Lereclus 2002, 2005; Bouillaut et al. 2008). Thus, as is the case for *B. anthracis*, expression of the PlcR regulon is abolished in the CI and CA strains (Klee et al. 2010).

B. cereus group plasmids can be mobile or capable of mobilizing other plasmids in the group, e.g. pXO14 which is efficient in mobilizing pXO1 and pXO2 (Reddy et al. 1987). It is however not known exactly to what extent horizontal transfer of pXO1- and pXO2-like plasmids may occur in the *B. cereus sensu lato* population. Hu and co-workers however, in a study of 80 soil-borne *B. cereus* group isolates from sympiatric communities, detected pXO1- and pXO2-like replicons in 12% and 21% of the isolates, respectively, and showed that strains of the same MLST sequence type had different plasmid contents, while a number of pXO2 replicons were hosted by distinct bacterial isolates, strongly arguing for lateral transfers of genetic material among sympiatric bacteria of the *B. cereus* group (Hu et al. 2009b). Given the fact that newly emerging pathogens (CI/CA strains) seem to arise from transfer of such plasmids, that *B. cereus* strains encoding alternative capsules (G9241) and possibly other virulence traits from plasmid elements exist, and that knowledge of plasmid diversity in the group is rather limiting, a systematic study targeting *B. cereus* group plasmids seems warranted. Also, the chromosomal genetic background for the long-term stability of the plasmids, including the multitude of variants identified in specific *B. thuringiensis* strains (e.g. as recently described in He et al. 2011), is not understood, and should be subject to investigation. In a future perspective, functional analysis of *B. thuringiensis* plasmids, and the availability of a steadily increasing number of *B. thuringiensis* genome sequences, could potentially provide for the engineering of commercial *B. thuringiensis* strains with increased safety to humans.

References

Agaisse H, Gominet M, Økstad OA, Kolstø AB, Lereclus D (1999) PlcR is a pleiotropic regulator of extracellular virulence factor gene expression in *Bacillus thuringiensis*. Mol Microbiol 32:1043–1053

Amadio AF, Benintende GB, Zandomeni RO (2009) Complete sequence of three plasmids from *Bacillus thuringiensis* INTA-FR7-4 environmental isolate and comparison with related plasmids from the *Bacillus cereus* group. Plasmid 62:172–182

Auger S, Galleron N, Bidnenko E, Ehrlich SD, Lapidus A, Sorokin A (2008) The genetically remote pathogenic strain NVH391-98 of the *Bacillus cereus* group is representative of a cluster of thermophilic strains. Appl Environ Microbiol 74:1276–1280

Barker M, Thakker B, Priest FG (2005) Multilocus sequence typing reveals that *Bacillus cereus* strains isolated from clinical infections have distinct phylogenetic origins. FEMS Microbiol Lett 245:179–184

Berry C, O'Neil S, Ben-Dov E, Jones AF, Murphy L, Quail MA, Holden MT, Harris D, Zaritsky A, Parkhill J (2002) Complete sequence and organization of pBtoxis, the toxin-coding plasmid of *Bacillus thuringiensis* subsp. *israelensis*. Appl Environ Microbiol 68:5082–5095

Bottone EJ (2010) *Bacillus cereus*, a volatile human pathogen. Clin Microbiol Rev 23:382–398

Bouillaut L, Perchat S, Arold S, Zorrilla S, Slamti L, Henry C, Gohar M, Declerck N, Lereclus D (2008) Molecular basis for group-specific activation of the virulence regulator PlcR by PapR heptapeptides. Nucleic Acids Res 36:3791–3801

Carlson CR, Caugant DA, Kolstø AB (1994) Genotypic Diversity among *Bacillus cereus* and *Bacillus thuringiensis* Strains. Appl Environ Microbiol 60:1719–1725

Dai Z, Sirard JC, Mock M, Koehler TM (1995) The *atxA* gene product activates transcription of the anthrax toxin genes and is essential for virulence. Mol Microbiol 16:1171–1181

Delihas N (2008) Small mobile sequences in bacteria display diverse structure/function motifs. Mol Microbiol 67:475–481

Didelot X, Barker M, Falush D, Priest FG (2009) Evolution of pathogenicity in the *Bacillus cereus* group. Syst Appl Microbiol 32:81–90

Drobniewski FA (1993) *Bacillus cereus* and related species. Clin Microbiol Rev 6:324–338

Fagerlund A, Brillard J, Furst R, Guinebretiere MH, Granum PE (2007) Toxin production in a rare and genetically remote cluster of strains of the *Bacillus cereus* group. BMC Microbiol 7:43. doi:10.1186/1471-2180-7-43

Fedhila S, Nel P, Lereclus D (2002) The InhA2 metalloprotease of *Bacillus thuringiensis* strain 407 is required for pathogenicity in insects infected via the oral route. J Bacteriol 184:3296–3304

Fedhila S, Gohar M, Slamti L, Nel P, Lereclus D (2003) The *Bacillus thuringiensis* PlcR-regulated gene *inhA2* is necessary, but not sufficient, for virulence. J Bacteriol 185:2820–2825

Fedhila S, Guillemet E, Nel P, Lereclus D (2004) Characterization of two *Bacillus thuringiensis* genes identified by in vivo screening of virulence factors. Appl Environ Microbiol 70:4784–4791

Fouet A, Mock M (2006) Regulatory networks for virulence and persistence of *Bacillus anthracis*. Curr Opin Microbiol 9:160–166

Ghelardi E, Celandroni F, Salvetti S, Fiscarelli E, Senesi S (2007) *Bacillus thuringiensis* pulmonary infection: critical role for bacterial membrane-damaging toxins and host neutrophils. Microbes Infect 9:591–598

Gohar M, Gilois N, Graveline R, Garreau C, Sanchis V, Lereclus D (2005) A comparative study of Bacillus cereus, *Bacillus thuringiensis* and *Bacillus anthracis* extracellular proteomes. Proteomics 5:3696–3711

Gohar M, Faegri K, Perchat S, Ravnum S, Økstad OA, Gominet M et al (2008) The PlcR virulence regulon of *Bacillus cereus*. PLoS One 3:e2793. doi:10.1371/journal.pone.0002793

Guinebretière MH, Thompson FL, Sorokin A, Normand P, Dawyndt P, Ehling-Schulz M, Svensson B, Sanchis V, Nguyen-The C, Heyndrickx M, De Vos P (2008) Ecological diversification in the *Bacillus cereus* group. Environ Microbiol 10:851–865

Han CS, Xie G, Challacombe JF, Altherr MR, Bhotika SS, Brown N et al (2006) Pathogenomic sequence analysis of Bacillus cereus and *Bacillus thuringiensis* isolates closely related to *Bacillus anthracis*. J Bacteriol 188:3382–3390

He J, Wang J, Yin W, Shao X, Zheng H, Li M, Zhao Y, Sun M, Wang S, Yu Z (2011) Complete genome sequence of *Bacillus thuringiensis* subsp. chinensis strain CT-43. J Bacteriol 193:3407–3408

Helgason E, Caugant DA, Lecadet MM, Chen Y, Mahillon J, Lövgren A, Hegna I, Kvaløy K, Kolstø AB (1998) Genetic diversity of *Bacillus cereus/B. thuringiensis* isolates from natural sources. Curr Microbiol 37:80–87

Helgason E, Caugant DA, Olsen I, Kolstø AB (2000a) Genetic structure of population of *Bacillus cereus* and *B. thuringiensis* isolates associated with periodontitis and other human infections. J Clin Microbiol 38:1615–1622

Helgason E, Økstad OA, Caugant DA, Johansen HA, Fouet A, Mock M et al (2000b) *Bacillus anthracis, Bacillus cereus, and Bacillus thuringiensis*—one species on the basis of genetic evidence. Appl Environ Microbiol 66:2627–2630

Helgason E, Tourasse NJ, Meisal R, Caugant DA, Kolstø AB (2004) Multilocus sequence typing scheme for bacteria of the *Bacillus cereus* group. Appl Environ Microbiol 70:191–201

Hendriksen NB, Hansen BM, Johansen JE (2006) Occurrence and pathogenic potential of *Bacillus cereus* group bacteria in a sandy loam. Antonie Van Leeuwenhoek 89:239–249

Hernandez E, Ramisse F, Cruel T, le Vagueresse R, Cavallo JD (1999) *Bacillus thuringiensis* serotype H34 isolated from human and insecticidal strains serotypes 3a3b and H14 can lead to death of immunocompetent mice after pulmonary infection. FEMS Immunol Med Microbiol 24:43–47

Hill KK, Ticknor LO, Okinaka RT, Asay M, Blair H, Bliss KA, Laker M, Pardington PE, Richardson AP, Tonks M, Beecher DJ, Kemp JD, Kolstø AB, Wong AC, Keim P, Jackson PJ (2004) Fluorescent amplified fragment length polymorphism analysis of *Bacillus anthracis, Bacillus cereus*, and *Bacillus thuringiensis* isolates. Appl Environ Microbiol 70:1068–1080

Hoffmaster AR, Ravel J, Rasko DA, Chapman GD, Chute MD, Marston CK et al (2004) Identification of anthrax toxin genes in a *Bacillus cereus* associated with an illness resembling inhalation anthrax. Proc Natl Acad Sci 101:8449–8454 (U S A)

Hu X, Van der Auwera G, Timmery S, Zhu L, Mahillon J (2009a) Distribution, diversity, and potential mobility of extrachromosomal elements related to the *Bacillus anthracis* pXO1 and pXO2 virulence plasmids. Appl Environ Microbiol 75:3016–3028

Hu X, Swiecicka I, Timmery S, Mahillon J (2009b) Sympatric soil communities of *Bacillus cereus* sensu lato: population structure and potential plasmid dynamics of pXO1- and pXO2-like elements. FEMS Microbiol Ecol 70:344–355

Ivanova N, Sorokin A, Anderson I, Galleron N, Candelon B, Kapatral V et al (2003) Genome sequence of *Bacillus cereus* and comparative analysis with *Bacillus anthracis*. Nature 423:87–91

Jackson PJ, Hill KK, Laker MT, Ticknor LO, Keim P (1999) Genetic comparison of *Bacillus anthracis* and its close relatives using amplified fragment length polymorphism and polymerase chain reaction analysis. J Appl Microbiol 87:263–269

Jensen GB, Larsen P, Jacobsen BL, Madsen B, Smidt L, Andrup L (2002) *Bacillus thuringiensis* in fecal samples from greenhouse workers after exposure to *B. thuringiensis*-based pesticides. Appl Environ Microbiol 68:4900–4905

Jensen GB, Hansen BM, Eilenberg J, Mahillon J (2003) The hidden lifestyles of *Bacillus cereus* and relatives. Environ Microbiol 5:631–640

Keim P, Kalif A, Schupp J, Hill K, Travis SE, Richmond K et al (1997) Molecular evolution and diversity in *Bacillus anthracis* as detected by amplified fragment length polymorphism markers. J Bacteriol 179:818–824

Klee SR, Ozel M, Appel B, Boesch C, Ellerbrok H, Jacob D et al (2006) Characterization of *Bacillus anthracis*-like bacteria isolated from wild great apes from Cote d'Ivoire and Cameroon. J Bacteriol 188:5333–5344

Klee SR, Brzuszkiewicz EB, Nattermann H, Brüggemann H, Dupke S, Wollherr A, Franz T, Pauli G, Appel B, Liebl W, Couacy-Hymann E, Boesch C, Meyer FD, Leendertz FH, Ellerbrok

H, Gottschalk G, Grunow R, Liesegang H (2010) The genome of a *Bacillus* isolate causing anthrax in chimpanzees combines chromosomal properties of *B. cereus* with *B. anthracis* virulence plasmids. PLoS One 5:e10986. doi:10.1371/journal.pone.0010986
Klevan A, Tourasse NJ, Stabell FB, Kolstø AB, Økstad OA (2007) Exploring the evolution of the *Bacillus cereus* group repeat element bcr1 by comparative genome analysis of closely related strains. Microbiology 153:3894–3908
Kolstø AB, Tourasse NJ, Økstad OA (2009) What sets *Bacillus anthracis* apart from other *Bacillus* species? Annu Rev Microbiol 63:451–476
Kotiranta A, Lounatmaa K, Haapasalo M (2000) Epidemiology and pathogenesis of *Bacillus cereus* infections. Microbes Infect 2:189–198
Kristoffersen SM, Tourasse NJ, Kolstø AB, Økstad OA (2011) Interspersed DNA repeats bcr1-bcr18 of *Bacillus cereus* group bacteria form three distinct groups with different evolutionary and functional patterns. Mol Biol Evol 28:963–983
Lapidus A, Goltsman E, Auger S, Galleron N, Segurens B, Dossat C et al (2008) Extending the *Bacillus cereus* group genomics to putative food-borne pathogens of different toxicity. Chem Biol Interact 171:236–249
Leendertz FH, Ellerbrok H, Boesch C, Couacy-Hymann E, Matz-Rensing K, Hakenbeck R et al (2004) Anthrax kills wild chimpanzees in a tropical rainforest. Nature 430:451–452
Leendertz FH, Lankester F, Guislain P, Neel C, Drori O, Dupain J et al (2006) Anthrax in Western and Central African great apes. Am J Primatol 68:928–933
Lund T, De Buyser ML, Granum PE (2000) A new cytotoxin from *Bacillus cereus* that may cause necrotic enteritis. Mol Microbiol 38:254–261
Margulis L, Jorgensen JZ, Dolan S, Kolchinsky R, Rainey FA, Lo SC (1998) The Arthromitus stage of *Bacillus cereus*: intestinal symbionts of animals. Proc Natl Acad Sci 95:1236–1241 (U S A)
McIntyre L, Bernard K, Beniac D, Isaac-Renton JL, Naseby DC (2008) Identification of *Bacillus cereus* group species associated with food poisoning outbreaks in British Columbia, Canada. Appl Environ Microbiol 74:7451–7453
Mignot T, Mock M, Fouet A (2003) A plasmid-encoded regulator couples the synthesis of toxins and surface structures in *Bacillus anthracis*. Mol Microbiol 47:917–927
Mock M, Fouet A (2001) Anthrax. Annu Rev Microbiol 55:647–671
Økstad OA, Tourasse NJ, Stabell FB, Sundfaer CK, Egge-Jacobsen W, Risøen PA et al (2004) The bcr1 DNA repeat element is specific to the *Bacillus cereus* group and exhibits mobile element characteristics. J Bacteriol 186:7714–7725
Okinaka RT, Cloud K, Hampton O, Hoffmaster AR, Hill KK, Keim P, Koehler TM, Lamke G, Kumano S, Mahillon J, Manter D, Martinez Y, Ricke D, Svensson R, Jackson PJ (1999) Sequence and organization of pXO1, the large *Bacillus anthracis* plasmid harboring the anthrax toxin genes. J Bacteriol 181:6509–6515
Pannucci J, Okinaka RT, Sabin R, Kuske CR (2002a) *Bacillus anthracis* pXO1 plasmid sequence conservation among closely related bacterial species. J Bacteriol 184:134–141
Pannucci J, Okinaka RT, Williams E, Sabin R, Ticknor LO, Kuske CR (2002b) DNA sequence conservation between the *Bacillus anthracis* pXO2 plasmid and genomic sequence from closely related bacteria. BMC Genomics 3:34. doi:10.1186/1471-2164-3-34
Passalacqua KD, Bergman NH (2006) *Bacillus anthracis*: interactions with the host and establishment of inhalational anthrax. Future Microbiol 1:397–415
Priest F (1993) Systematics and ecology of Bacillus. In: Sonenshein AL, Hoch JA, Losick R (eds) *Bacillus subtilis* and other Gram-positive bacteria: biochemistry, physiology and molecular genetics, 1st edn. American Society for Microbiology, Washington, DC
Rasko DA, Ravel J, Økstad OA, Helgason E, Cer RZ, Jiang L et al (2004) The genome sequence of *Bacillus cereus* ATCC 10987 reveals metabolic adaptations and a large plasmid related to *Bacillus anthracis* pXO1. Nucleic Acids Res 32:977–988
Rasko DA, Altherr MR, Han CS, Ravel J (2005) Genomics of the *Bacillus cereus* group of organisms. FEMS Microbiol Rev 29:303–329

Rasko DA, Rosovitz MJ, Økstad OA, Fouts DE, Jiang L, Cer RZ et al (2007) Complete sequence analysis of novel plasmids from emetic and periodontal *Bacillus cereus* isolates reveals a common evolutionary history among the B. cereus-group plasmids, including *Bacillus anthracis* pXO1. J Bacteriol 189:52–64
Read TD, Salzberg SL, Pop M, Shumway M, Umayam L, Jiang L et al (2002) Comparative genome sequencing for discovery of novel polymorphisms in *Bacillus anthracis*. Science 296:2028–2033
Read TD, Peterson SN, Tourasse N, Baillie LW, Paulsen IT, Nelson KE et al (2003) The genome sequence of *Bacillus anthracis* Ames and comparison to closely related bacteria. Nature 423:81–86
Reddy A, Battisti L, Thorne CB (1987) Identification of self-transmissible plasmids in four *Bacillus thuringiensis* subspecies. J Bacteriol 169:5263–5270
Saile E, Koehler TM (2006) *Bacillus anthracis* multiplication, persistence, and genetic exchange in the rhizosphere of grass plants. Appl Environ Microbiol 72:3168–3174
Scarano C, Virdis S, Cossu F, Frongia R, De Santis EP, Cosseddu AM (2009) The pattern of toxin genes and expression of diarrheal enterotoxins in *Bacillus thuringiensis* strains isolated from commercial bioinsecticides. Vet Res Commun 33(Suppl 1):257–260
Schmidt TR, Scott EJ 2nd, Dyer DW (2011) Whole-genome phylogenies of the Family *Bacillaceae* and expansion of the sigma factor gene family in the *Bacillus cereus* species-group. BMC Genomics 12:430. doi:10.1186/1471-2164-12-430
Schnepf E, Crickmore N, Van Rie J, Lereclus D, Baum J, Feitelson J et al (1998) *Bacillus thuringiensis* and its pesticidal crystal proteins. Microbiol Mol Biol Rev 62:775–806
Slamti L, Lereclus D (2002) A cell-cell signaling peptide activates the PlcR virulence regulon in bacteria of the *Bacillus cereus* group. EMBO J 21:4550–4559
Slamti L, Lereclus D (2005) Specificity and polymorphism of the PlcR-PapR quorum-sensing system in the *Bacillus cereus* group. J Bacteriol 187:1182–1187
Soberon M, Pardo-Lopez L, Lopez I, Gomez I, Tabashnik BE, Bravo A (2007) Engineering modified Bt toxins to counter insect resistance. Science 318:1640–1642
Sozhamannan S, Chute MD, McAfee FD, Fouts DE, Akmal A, Galloway DR et al (2006) The *Bacillus anthracis* chromosome contains four conserved, excision-proficient, putative prophages. BMC Microbiol 6:34. doi:10.1186/1471-2180-6-34
Stenfors Arnesen LP, Fagerlund A, Granum PE (2008) From soil to gut: *Bacillus cereus* and its food poisoning toxins. FEMS Microbiol Rev 32:579–606
Ticknor LO, Kolstø AB, Hill KK, Keim P, Laker MT, Tonks M, Jackson PJ (2001) Fluorescent Amplified Fragment Length Polymorphism Analysis of Norwegian *Bacillus cereus* and *Bacillus thuringiensis* Soil Isolates. Appl Environ Microbiol 67:4863–4873
Tourasse NJ, Kolstø AB (2008) SuperCAT: a supertree database for combined and integrative multilocus sequence typing analysis of the *Bacillus cereus* group of bacteria (including *B. cereus, B. anthracis and B. thuringiensis*). Nucleic Acids Res 36:D461-D468
Tourasse NJ, Helgason E, Økstad OA, Hegna IK, Kolstø AB (2006) The *Bacillus cereus* group: novel aspects of population structure and genome dynamics. J Appl Microbiol 101:579–593
Tourasse NJ, Økstad OA, Kolstø AB (2010) HyperCAT: an extension of the SuperCAT database for global multi-scheme and multi-datatype phylogenetic analysis of the *Bacillus cereus* group population. Database 2010:baq017. doi:10.1093/database/baq017
Tourasse NJ, Helgason E, Klevan A, Sylvestre P, Moya M, Haustant M, Økstad OA, Fouet A, Mock M, Kolstø AB (2011) Extended and global phylogenetic view of the *Bacillus cereus* group population by combination of MLST, AFLP, and MLEE genotyping data. Food Microbiol 28:236–244
Uchida I, Hornung JM, Thorne CB, Klimpel KR, Leppla SH (1993) Cloning and characterization of a gene whose product is a trans-activator of anthrax toxin synthesis. J Bacteriol 175:5329–5338
Van der Auwera GA, Andrup L, Mahillon J (2005) Conjugative plasmid pAW63 brings new insights into the genesis of the *Bacillus anthracis* virulence plasmid pXO2 and of the *Bacillus thuringiensis* plasmid pBT9727. BMC Genomics 6:103. doi:10.1186/1471-2164-6-103

Chapter 7
Bacillus thuringiensis Genetics and Phages—From Transduction and Sequencing to Recombineering

Alexei Sorokin

Abstract Experimental mapping of genes was less intensive in the *Bacillus cereus* group, which includes *B. thuringiensis*, than in *Bacillus subtilis*, generally considered as the model for Gram-positive bacteria. Nevertheless the genomic sequencing equalized densities of available gene maps. Moreover, the genes responsible for such complex phenomena like virulence or psychrotolerance could only be identified using the authentic bacteria prone to possess these properties. The new experimental approaches of post-genomic genetics should therefore be considered. Phage-mediated gene transduction and recombineering perspectives for the *B. cereus* group are reviewed. In combination with new generation sequencing these approaches will constitute the gene identification methodologies in the post-genomics time.

Keywords *Bacillus thuringiensis* · *Bacillus cereus* group · Genetic mapping · Transduction · Recombineering

7.1 Introduction: A Dramatic Story of the *Bacillus cereus* Genetics

In the early days of *Bacillus* genetics the investigators working with *Bacillus cereus* and *Bacillus subtilis* have tried to establish detailed genetic maps of their organism of choice. The *B. cereus* group has applied interest because of two representative species—*B. thuringiensis* and *B. anthracis*. The former is a useful insect pathogen, widely employed to protect crops. The second is animal pathogen, causing problems for veterinary, and notified as a potential biological weapon. The *B. subtilis* group includes also *B. licheniformis* and *B. amyloliquefaciens*, both are attractive because of their high potential for extracellular enzyme production. As

A. Sorokin (✉)
Genome Analysis Team ANALGEN, UMR1319 MICALIS,
Centre de Recherche de Jouy-en-Josas, Institut National de la Recherche Agronomique (INRA) bat. 440, Domaine de Vilvert, 78352 Jouy en Josas cedex, France
e-mail: alexei.sorokine@jouy.inra.fr

E. Sansinenea (ed.), *Bacillus thuringiensis Biotechnology*,
DOI 10.1007/978-94-007-3021-2_7,

long as up to the beginning of 80th the efforts to construct detailed genetic maps of strains from both bacterial groups were equally intensive. These efforts included attempts for developing tools based on bacteriophages (transduction), plasmids (conjugation) and direct transformation of bacteria with native DNA. However the strain *B. subtilis* 168 had a powerful hidden advantage. Genetic mapping of this organism has dramatically intensified after introducing of a very reproductive, efficient and simple transformation protocol (Anagnostopoulos and Spizizen 1961). This resulted in producing of a very detailed genetic map (Piggot and Hoch 1985; Anagnostopoulos et al. 1993). The investigators working with *B. cereus* had only fastidious transduction or protoplast transformation protocols. The constructed maps, very limited in the number of genetic markers, could not be competitive to justify a "model" status for this organism (Heierson et al. 1983; Vary 1993). The interests of academic community were finally shifted to *B. subtilis*, which became the preferred model and paradigm for Bacillus genetics and physiology. Academic and methodological studies using *B. cereus* were abandoned for some time. Interests were not however abandoned for the issues oriented to applications, like technological aspects of crystal protein production and studies of anthrax and protection against it (Koehler 2009; Mock and Fouet 2001; Schnepf et al. 1998; Shelton et al. 2002).

The *B. subtilis* scientific community was winning the rally. The successes of *B. subtilis* genetics prevented further developments of experimental *B. cereus* maps. Moreover, the ease of genetic manipulating with *B. subtilis* 168 has lead to novel unexpected applications. The examples are metabolic engineering of strains able to produce important metabolites like riboflavin or some other (Sauer et al. 1997; Van Arsdell et al. 2005; Wu and Wong 2002). It is remarkable that the genetic studies of riboflavin biosynthesis regulation in *B. subtilis* led to discovery of riboswitches (Mironov et al. 2002; Gusarov et al. 1997). The elegant works performed by A.-B. Kolsto's group on physical mapping of the *B. thuringiensis* and *B. cereus* chromosomes (Carlson et al. 1992, 1996; Carlson and Kolsto 1993; Kolsto et al. 1990) could not fill the gap, competing with massive mapping of *B. subtilis* genomes made by several groups in parallel (Itaya and Tanaka 1991; Azevedo et al. 1993; Amjad et al. 1991). The use of efficient transformation protocols and broad interest of the community were good pre-requisites for success of the latter.

It is not surprising that *B. subtilis* 168 became one of the first model organisms to apply the genomic sequencing (Kunst et al. 1997; Kunst and Devine 1991). This strain could be the first bacterium sequenced due to favorite financing and so-called consortium approach with a large number of teams involved (Devine 1995; Kunst et al. 1995). However the whole-genome shot-gun strategy, adapted by other projects, has appeared to be more efficient compared to clone-by-clone sequencing of the mapped libraries (Fleischmann et al. 1995; Fraser and Fleischmann 1997). By the way, afterwards the *B. subtilis* 168 genome has been entirely re-sequenced at least twice and still remains one of the most important bacterial genomes that have ever been studied (Harwood and Wipat 1996; Tosato and Bruschi 2004; Srivatsan et al. 2008; Barbe et al. 2009). At the beginning of the third millennium genomic sequencing became routine and practical importance of *B. thuringiensis*, as insect pathogen, and of *B. cereus* and *B. anthracis* as animal pathogens have started to play back. The formal nature of genomic sequencing minimizes its dependence of the biological features of studied organism. The first two sequenced genomes of the *B. cereus* group were reported in 2003 (Ivanova et al. 2003; Read et al. 2003) and about 30 completed and 100 of draft sequences are now available in public databases (Kolsto et al. 2009; Markowitz et al. 2010). These numbers can be compared to only five completed and about 20 draft genomes from the *B. subtilis* group (Earl et al. 2008; Alcaraz et al. 2010). The results of multiple locus sequence typing (MLST) of the *B. cereus* group are also impressive, counting more than thousand of characterized strains (Didelot et al. 2009; Tourasse and Kolsto 2007; Tourasse et al. 2010a). These phylogenetic studies provided a good basis to choose the strains for genomic sequencing. A recent excellent review can be recommended for the reader interested to acquire a comprehensive view of the *B. cereus* group genomics and phylogeny (Kolsto et al. 2009).

The genetic maps of bacteria are now presented in the form of formally detected protein coding regions rather than experimentally mapped genetic markers as it was done earlier (Fang et al. 2005; Moszer et al. 1995; Medigue and Moszer 2007). The experimentally established genetic map of *B. subtilis* 168 is much denser than that for any of the *B. cereus* group strains (Biaudet et al. 1996). However, if the gene functions are defined by similarity to known proteins, the maps for the two organisms would look comparable. Still the experimental characterization of gene functions is more complicate in *B. cereus* than in *B. subtilis*. Nevertheless *B. subtilis* 168 cannot replace strains of the *B. cereus* group in studies of properties, to which *B. subtilis* is not pertinent. For example, the virulence genes of *B. cereus* can only be studied using the strains of this species. Another example is the ability of some *B. cereus* group strains to grow at low temperatures. This property of strains, designated together as the *B. weihenstephanensis* species (Lechner et al. 1998), is important considering the problem of refrigerated food contamination by the *B. cereus* group pathogens. The time for novel genetics approaches is therefore coming and the availability of well annotated genomic sequences is only the starting point. It can be called as post-genomic genetics, by analogy to other post-genomic approaches. The next important step is revision of effectiveness of old and new methods in these terms. In this chapter I shall review studies on phage-mediated gene transduction performed with bacteria of the *B. cereus* group. This genetic method, in conjunction with new generation sequencing (NGS) and recombineering, could be a very promising approach in identification of genes involved in formation of complex properties, like virulence or psychrotolerance. Transduction might also be a useful method for constructing of strains with novel desired properties. Recombineering is entirely novel for these bacteria. Some genetic prerequisites of it existing in the *B. cereus* group will be briefly outlined with a hint for the use as a genetic instrument.

7.2 Phage-mediated Gene Transduction

Genetic transduction is the process of genetic marker transfer between different bacterial strains with participation of a phage. This definition necessarily implies that the phage must infect both bacterial strains—the donor and the recipient. An important characteristic of this process is its frequency. The frequency can be defined as the number of transductants that is the number of bacterial colonies showing the desired new phenotype, to the total number of colony forming units (CFU). The absolute number of transductants per milliliter of the recipient culture is also an important parameter, especially if a selection is applied for enrichment of a particular phenotype. For the phages used in *B. thuringiensis* the experimentally measured transduction frequencies usually were not high, of the order 10^{-7}–10^{-5} (Thorne 1968a, b; Landen et al. 1981; Perlak et al. 1979; Lecadet et al. 1980). Transduction can be specialized or generalized. The term **specialized transduction** designate the process when a specific gene is transferred with much higher frequency than the others. Contrary to that the **generalized transduction** is such that many genes can be transferred, by the same phage, with comparable frequencies.

The specialized transduction was actively studied for the phages λ of *Escherichia coli* and SPβ of *B. subtilis* (Zahler 1982; Campbell 2006). In both cases the specialized transduction can be explained by erroneous excision of the phage, residing in the form of prophage in the bacterial chromosome. Instead of the "pure" phage DNA some of phage particles contained DNA, which encodes genes closely located to the phage chromosomal insertion site. If the corresponding phage particle is still infective, the donor bacterium genes could enter into the recipient cell and, after being "saved" by recombination, transfer the corresponding genetic marker to the recipient cell. A confirmation of this model of specialized transduction is that the transduced genetic markers are usually mapped close to the phage integration site (Zahler 1982; Campbell 2006). The specialized transduction was usually observed for the temperate phages with *cos* packaging mechanism (Campbell 2006). This mechanism implies high sequence specificity of the corresponding terminase that produces the ends of the DNA to be packaged into the phage head. Due to this high specificity of the terminase, DNA foreign to the phage cannot serve as a substrate for this enzyme.

The generalized transduction is more valuable for constructing of new strains or genetic mapping. Gene mapping was intensively carried out using the phages P1 or P22 in *E. coli* and *Salmonella enterica*, respectively (Lennox 1955; Thomason et al. 2007; Zinder and Lederberg 1952; Kufer et al. 1982; Tye et al. 1974; Susskind and Botstein 1978) and the phages SPP1, PBS1 and SP10 in *B. subtilis* (Ferrari et al. 1978; Thorne 1962; Lovett and Young 1970; de Lencastre 1979; Takahashi 1961). Several early papers on the physical characterization of plaque forming units and transducing particles have shown that the transducing particles are indistinguishable from the phage particles (de Lencastre 1980, 1981; Yamagishi and Takahashi 1968). The conclusion was that the only difference between them is that transducing units contain bacterial DNA instead of the phage DNA (de Lencastre 1980, 1981). It was also concluded that the transducing particles represent the pieces of bacterial DNA packaged into the phage envelop. The formation of covalently joined hybrids between the phage and bacterial DNA was therefore excluded. The *B. subtilis* phages PBS1 and SP10 have an advantage that they are present in a pseudolysogenic state in the cells (Hemphill and Whiteley 1975). This is similar to the state of the phage P1 of Gram-negative bacteria. The existence of lysogeny and significant size of the phage PBS1 chromosome, estimated to be 76 MDa (Lovett et al. 1974), promoted active use of PBS1 for gene mapping and constructing of detailed genetic map of *B. subtilis*. Subsequent genomic sequence revealed that the constructed map was fairly precise (Rivolta and Pagni 1999).

It is probably the big size of the PBS1 chromosome that prevented its active study as model transducing phage of *B. subtilis*. Instead, the phage SPP1 was intensively studied, even although the practical use of SPP1 for genetic mapping was less common. This can be explained by a relatively moderate size of SPP1 chromosome (44 kb), high virulence and limited host range. Due to these studies, mostly carried out in the laboratory of P. Tavares (Gif-sur-Yvette, France), the phage SPP1 became a paradigm for fundamental research of transducing phages of Bacillus (Alonso et al. 1986) and, in particular, of the DNA packaging mechanism (Cuervo et al. 2007; Oliveira et al. 2010; Isidro et al. 2004; Lebedev et al. 2007; Lhuillier et al. 2009; Orlova et al. 2003). Comprehensive bibliography on the phage DNA packaging can be found in recent reviews (Rao and Feiss 2008; Alonso et al. 2006).

Study of the molecular mechanism of generalized transduction did not attract sufficient attention. In addition to the mentioned above reports on the physical characterization of transducing particles, other studies used genetic approaches. P1 phage mutants with increased transduction frequencies were isolated (Iida et al. 1998; Wall and Harriman 1974). Some mutations were linked to the head assembly process (Iida et al. 1998). Presumably the mutations affecting the availability of chromosomal DNA free ends, the processivity of the packaging reaction, or the stability of transduced DNA molecules in the recipient cell all could result in a changed transduction frequency. Mutants with altered transduction ability of phage P22 were analyzed more intensively (Schmieger and Buch 1975; Raj et al. 1974; Schmieger 1972; Casjens et al. 1992). The data were interpreted as indicating that host DNA to be packaged is cut by the mutants at sites different from those cut by wild-type phage, presumably due to an altered specificity of the responsible nuclease (Schmieger and Backhaus 1976).

Creation of cloning systems that use the *in vitro* packaging of DNA into the phage heads can be considered as direct modeling of the phage-mediated transduction mechanism. Such systems were successfully created and used for gene cloning and genomic physical map constructing. The commonly used systems, so-called cosmids and PAC vectors, are based on the phages λ and P1 of *E.coli* (Sternberg 1990; Collins and Weissman 1984). The common principles in creating of such systems are: (i) using of *in vitro* DNA cleavage at *cos* or *pac* site integrated into the vector; (ii) using *in vitro* packaging of such cleaved DNA into the prepared phage proheads and reconstitution of the infective particles containing this DNA. A similar process of bacterial DNA packaging can be imagined for the fraction of phage particles responsible for the transduction *in vivo*.

7.3 Phage-mediated Gene Transduction in the *B. cereus* Group

7.3.1 Gene Transduction Experiments

Tranduction experiments using bacteria of the *B. cereus* group were done mostly during the years 60–70. Two purposes were primarily targeted. First, before the era of massive application of genetic engineering and genome sequencing, this was one of the most straightforward methods to build genetic map of this microorganism. Second, this was an efficient approach to construct new strains for the practical use. Most experiments were done using *B. thuringiensis* strains, with evident practical purpose for improving of their use as insecticides, or with *B. anthracis*. Since molecular phylogeny of this group was not yet elaborated and routinely used at that time there are some difficulties in interpreting of some interesting transduction data that were obtained. Here I shall compile the main results providing, where possible, their updated interpretation. Some of the studied phages are available in public collections and can be used today. However it appears that the most frequently reported phages are not optimal for many purposes. At present, when the detailed genetic maps based on sequencing data are available, transduction could be helpful for genetic studies of complex phenomena, such as virulence, spore germination or psychrotolerance. Transduction by phages can be as well considered as a promising tool for the synthetic biology manipulations. For these purposes a phage with broad host range, capable to package large DNA molecules would be considered as optimal. Such a phage, similar to the PBS1 of the *B. subtilis* group, was not yet identified and characterized in sufficient details for the *B. cereus* group bacteria.

It is important, for the interpretation of the transduction experiments, to relate the data with the phylogenetic positions of the strains for which such experiments were carried out. The phylogeny of the *B. cereus* group is somewhat ambiguous, which is probably because of the differences in the species concept used by different authors. It is usually considered that the *B. cereus* group contains six officially accepted species (Rasko et al. 2005; Kolsto et al. 2009). A novel well distinct moderately thermophilic species, "*B. cytotoxicus*", was not yet included into this list (Auger et al. 2008; Fagerlund et al. 2007; Lapidus et al. 2008). Some reports also indicate that several accepted species could be genetically considered as only one (Helgason et al. 2000a). It is not entirely clear if the species *B. mycoides*, prone to the filamentous growth, and the psychrotolerant species *B. weihenstephanensis* should not be considered as one, based on the formalized genetic clustering, in spite of the characteristic phenotypes (Lechner et al. 1998; Guinebretiere et al. 2008). Also an ambiguous situation exists for some *B. cereus* strains, including the type strain

B. cereus ATCC14579, clustered with the crystal toxin producers (Sorokin et al. 2006; Priest et al. 2004). In this review I shall follow the simplified presentation of the *B. cereus* group phylogeny based on the MLST data (Auger et al. 2008; Priest et al. 2004; Sorokin et al. 2006). All strains are positioned in four clusters—C, T, W and Y (Auger et al. 2008). *B. anthracis* and most of the *B. cereus* strains belong to the cluster C. Crystal protein producing *B. thuringiensis* strains—T. Psychrotolerant *B. weihenstephanensis*, filamentous *B. mycoides* and *B. pseudomycoides*—W. Finally, the moderately thermophilic strains "*B. cytotoxicus*" will constitute the cluster Y. Such simplified classification permits to identify the phylogenetic affiliation of the barely characterized strains used in transduction experiments long ago. I shall also use the available data for the *B. thuringiensis* strains (cluster T) to affiliate them into the Tolworthi, Kurstaki, Sotto and Thuringiensis sub-clusters as proposed by Priest et al. (2004). Using of such simplified classification mostly based on DNA sequencing data, is partially justified by the view that the phage specificity is usually compatible with the formalized phylogenetic classification of the bacterial strains.

The phage **CP-51** is the most commonly referred transducing phage used for the *B. cereus* group (Thorne 1968a, b). This phage was isolated from soil using *B. cereus* NRRL569 strain as indicator (Thorne 1968b). This strain, synonymous to *B. cereus* ATCC10876 and *B. cereus* BGSC 6A3, is phylogenetically close to the Kurstaki sub-cluster of *B. thuringiensis* strains (Priest et al. 2004). It is not therefore surprising that CP-51 was active on different *B. thuringiensis* strains confirming its broad host range inside of this cluster (Thorne 1978; Yelton and Thorne 1970). Cross-transduction experiments among various strains of *B. thuringiensis* were however largely unsuccessful. This is rather explained by insufficient similarity among the DNA of strains than by defense tools, like restriction. The phage itself did not appear to be restricted when grown on a particular host and assayed with other hosts as indicators. Transduction of plasmids was reported even between *B. thuringiensis* and *B. anthracis* strains (Ruhfel et al. 1984). A partial nucleotide sequence of the original NRRL569 strain is now available (NCBI acc. # NZ_CM000715).

Because of the virulent nature of this phage, difficulties were encountered in selecting and scoring transductants of certain mutants. However, with the use of phage treated with UV light to inactivate most of the PFU, along with appropriately enriched plating medium, reasonable yields of transductants were obtained for most mutants. Temperature sensitive derivative of this phage, CP-51ts, was also used for the studies of germination genes in *B. cereus* NRRL569 (Clements and Moir 1998). In this work generalized transduction with phage CP-51ts was indicating the linkage of germination defects to the resistance marker of transposon, used to generate the mutations. This was necessary to exclude that observed effects were from multiple mutations. The phage CP-51ts is extremely lytic at 30°C. It contained a mutation which prevented phage replication at temperatures above 40°C. Thus the temperature was used to allow the selection of transductants. Only the mutants confirming 100% linkage of germination and transposon mediated antibiotic resistance were further studied.

Several strains, which do not belong to the cluster T, were also tested for CP-51 and, in particular, the strains *B. cereus* ATCC7004, ATCC9592 or ATCC11950 have appeared to be resistant (Yelton and Thorne 1970; Thorne 1968a). The strain *B. cereus* ATCC11950, synonymous to *B. cereus* W, is sensitive to the phages of γ group (Schuch and Fischetti 2006; Fouts et al. 2006). Draft sequence of its genome is available (NCBI acc. # NZ_ABCZ02000000) and it shares 99% of nucleotide sequence identity with *B. anthracis* strains (cluster C). Since identity to the strains of clusters T and C is only 90% it is not surprising that the strain is resistant to the phage CP-51, specific for *B. thuringiesis*. Another resistant strain, *B. cereus* ATCC7004, can be positioned, using MLEE data (Helgason et al. 2000b), to the group III (Guinebretiere et al. 2008; Tourasse et al. 2010a, b) also corresponding to the cluster C. Phylogenetic affiliation of *B. cereus* ATCC9592 is obscure. The strain was recently used in experiments on thermo-resistance of spores (Rice et al. 2004). Based on the cited data it is tempting to conclude that activity of CP-51 is restricted to the cluster T. However reports that the phage is active on *B. anthracis* (Ruhfel et al. 1984) and recent studies where a thermosensitive derivative CP-51ts was used to transfer genetic markers in *B. anthracis* (Sastalla et al. 2010; Moody et al. 2010; Giorno et al. 2009) are contradictory to this conclusion.

Although the nucleotide sequence of the CP-51 genome was not yet published, some important characteristics had been reported (Klumpp et al. 2010a). CP-51 was found to carry 41% of genes that can be regarded as counterparts with the *B. subtilis* phage SP01 (hit-length threshold 100 amino acids, 38–100% identity). The authors also confirmed physical characteristics of this phage obtained earlier by Thorne. The head diameter was approximately 90 nm, and the tail length 160–185 nm. The genome size of CP-51 is 138 kb and it has invariable genome ends. CP-51 DNA contains HMU instead of thymine (Yelton and Thorne 1971). Surprising is the ability of CP-51 for transducing of genetic markers. This feature is not in agreement with the proposed presence of invariable terminal repeats in SP01-related phages (Klumpp et al. 2008; Stewart et al. 2009). To explain this ambiguity the authors propose that the observed infrequent transduction is due to packaging errors by the phage terminase (Klumpp et al. 2010a).

Phage **CP-54** was isolated from soil in a manner similar to that described for CP-51 (Thorne 1968b), except that wild-type *B. thuringiensis subsp allesti* NRRL4041 was the host organism and streptomycin was omitted from the medium. CP-54 is active on all strains of *B. thuringiensis* tested. Electron microscope images of CP-54 are almost identical to that of CP-51 (Yelton and Thorne 1970; Lecadet et al. 1980). The two phages differ in the following: (i) CP-54 has a broader host range than CP-51; (ii) plaque morphologies are different; (iii) CP-54 is more cold labile than CP-51; (iv) the two phages are not serologically identical; (v) CP-54 has DNA of higher molecular weight, >70 MDa compared to 57 MDa in CP-51 (Yelton and Thorne 1971; Lecadet et al. 1980). Generally, strains that were hosts for both phages were more easily lysed with CP-54 than with CP-51. Both phages formed colony-centered plaques on *B. cereus* 569, and both gave rise to mutants which were more virulent and formed clear plaques. Such mutants derived from CP-51 did not have the extended host range characteristic of CP-54. Efforts to isolate host range mutants of CP-51 which were active on strains 4040 and 4041 of *B. thuringiensis* were unsuccessful. Transduction results with CP-54 were analogous to those with CP-51 (Thorne 1978).

CP-54Ber is a phage isolated from lysates of CP-54 grown on *B. cereus* 569 obtained after repeated subculturing and selected for the ability to infect *B. thuringiensis* var. *berliner* 1715. The mutant phage gave turbid plaques on the *berliner* strain and it was able to mediate generalized transduction in this host with frequencies 10^{-7}–10^{-5} (Lecadet et al. 1980). The two phages differed in their inhibition by specific antibodies. Lysates of both phages were extremely labile at low temperatures, as it was reported for CP-54 (Thorne and Holt 1974). Storage during 7 days at 4°C caused 70% inactivation while inactivation only reached 30% at 15°C and 45% at 20°C. The phages were considerably stabilized using dimethyl sulphoxide (10%) and $MgSO_4$ (20 mM). Antiserum was used to improve the frequency of transductants but it decreased their number. Using CP-54Ber the linkage groups of some genetic markers were determined and therefore the phage can be suitable for genetic mapping.

Differences between CP-54 and CP-54Ber were shown with regard to host range. Most of the tested strains, including *B. cereus* 569, were susceptible to CP-54Ber. However, some strains, like *sotto*, *dendrolimus* and *subtoxicus*, did not give a clear response, although they were susceptible to the parental CP-54 phage. Cross-transduction was also much less efficient (10-fold lower) than the homologous transduction. It confirms that high homology is important for efficient recombination. In most cases the numbers of recombinants obtained with CP-54Ber for the *berliner* strain were significantly higher than those reported by Thorne (1978) using phages CP-54 or CP-51 and strains from other serotypes. The stocks of CP54Ber and transduction protocols are available from BGSC (http://www.bgsc.org).

During the work with CP-51 another transducing phage, **CP-53**, was isolated from the lysates of the strain *B. cereus* ATCC6464 (Yelton and Thorne 1970). This phage was supposed to be carried as a prophage by parental *B. cereus* ATCC6464, since the phage generated turbid plaques on *B. cereus* 569 and the cured 6464 UM4 mutant used as indicators. All five tested auxotrophic mutants of *B. cereus* 569 were transduced to prototrophy by CP-53 with frequencies 10^{-6} to 5×10^{-6}. Electron micrographs revealed that whereas CP-51 has a tail core surrounded by a contractile sheath, CP-53 has a long flexible tail without a contractile sheath. Also, contrary to CP-51, CP-53 is stable in the cold. It was also mentioned that UV-induction of CP-53 in *B. cereus* 6464 was reported earlier (Altenbern and Stull 1965).

CP-51 gave greater frequencies of cotransduction for linked markers than did CP-53 (Yelton and Thorne 1971). This is not surprising, since CP-51 is a larger phage which carried more DNA than CP-53. The difference of cotransduction of closely linked markers (80%/30%=2.7) corresponded roughly to the difference of sizes of phage chromosomes (54.3 MDa/17 MDa=3.2) (Yelton and Thorne 1971). CP-51 DNA contained 5-hydroxymethyluracil in place of thymine, whereas CP-53 DNA contained no unusual bases. The host range was not extensively characterized for CP-53 but it was shown that the lysogenized by CP-53 cells became immune to this phage.

Another interesting transducing phage is **TP-13**, similar by some characteristics to SP15 and PBS1 of *B. subtilis*. These included large size, similar morphology, and adsorption specificity to motile cells. TP-13 mediated generalized transduction in several strains of *B. thuringiensis* at frequencies of 10^{-6}–10^{-5} (Perlak et al. 1979). This phage, along with the phage **TP-18** having much smaller genome, was successfully used for constructing of a version of genetic map of *B. thuringiensis* NRRL4042 (Barsomian et al. 1984). TP-13 was shown to be capable of converting an oligosporogenic (Osp), acrystalliferous mutant of *B. thuringiensis* to Spo+Cry+(Perlak et al. 1979). All 17 tested strains of different subspecies of *B. thuringiensis* were sensitive to TP-13 with the exception of strain 4048 (var *aizawai*). Purified TP-13 was as effective as crude lysates in converting Osp Cry- cells to Spo+Cry+, and conversion was independent of the host used for phage propagation. At the same time TP-13 did not appear to form stable lysogens with UM8-13 mutant used as indicator. Upon loss of TP-13, converted cells of UM8-13 returned to the Osp Cry- phenotype. Plaques formed by TP-13 on lawns of non-converted mutants were colony centered, characteristic for temperate phages, but cells within the plaques did not sporulate. All these data indicated that TP-13 was maintained in the converted cells as a non-integrated lysogen. Observations of TP-13 by electron microscopy indicated a large bacteriophage with a head diameter of 120 nm and a tail length of 260 nm, resembling SP15 (Tyeryar et al. 1969) and PBS1 (Eiserling 1967) in morphology and size. The head size indicated that the genome of TP-13 must be similar to that of SP15 estimated to be 250 MDa (Tyeryar et al. 1969). Potentially TP-13 represented one of the best phages for genetic mapping of distant markers or strain constructing in *B. thuringiensis*.

A similar phage although smaller in size, **ϕ63**, was described by Landen et al. (1981). By electron microscopy the phage with the head size of 95 nm and the tail length 200 nm was similar to TP-13. The frequency of transduction was 10^{-7}–10^{-6}. The phage was forming either turbid or clear plaques on 7 of 10 tested *B. thuringiensis* subspecies and one tested *B. cereus* BclI strain. There was a modest restriction on subspecies *finitimus* and no growth at all on subspecies *israelensis*, *aizawai* and on one of the derivatives of *alesti*. There was no growth on *B. subtilis*. Double streptomycin and spectinomycin resistance mutants of the host *B. thuringiensis* subsp *gelechiae*, originated from a commercial powder, were isolated to directly measure the co-transduction frequencies, since the corresponding mutated ribosomal protein genes are clustered. The expected spontaneous double mutation frequency was 10^{-14}. Using additional rifampicin resistance marker permitted to detect the heterologous co-transduction of *rifA* and *strA* in four of seven subspecies of *B. thuringiensis*. Comparison to CP-54 demonstrated greater stability of ϕ63 and 10-fold higher transduction frequencies. The co-transducing capacities of ϕ63 were suggested to be similar to the *B. subtilis* phage SP-10 of the same order of size. SP-10 can co-transduce markers separated by 3–5% of the 4.1 Mbase *B. subtilis* genome (Tyeryar et al. 1969). ϕ63 was subsequently used to construct of a limited marker linkage map of *B. thuringiensis* (Heierson et al. 1983).

The temperate phages ϕHD67, ϕHD130, ϕHD228 and ϕHD248 isolated after UV-induction from the subspecies *aizawai* (Inal et al. 1990) have been shown to mediate both intra-subspecific (*aizawai*) and inter-subspecific (*aizawai-kurstaki*) generalized transduction (Inal et al. 1992). The phage **ϕHD248** was used for genetic analysis of *B. thuringiensis* subsp. *aizawai* (H-serotype 7), which identified the linkage groups *bio-2-his-2-met-1-cit-1-arg-2-lys-3* and *trp-9-purC1-thi-1* (Inal et al. 1996). It was thus demonstrated the possibility for fine genetic mapping using this phage. The phage was also found to have a broad host range, plating on 7 out of 14 H-serotypes tested. However, the phage seems to be specific to the strains of Kurstaki sub-cluster, based on MLST studies (Priest et al. 2004). Characterization of the ϕHD248 genome revealed a linear double-stranded DNA of 47.15 kb, lacking of cohesive ends and having a partial circular permutation (Inal et al. 1996). This feature was proposed to be due to the headful packaging mechanism (Tye et al. 1974), thus explaining the generalized transduction by this and related phages.

The phage **TP21** (to be re-called TP21-T, see below) is interesting because it presents in the *B. thuringiensis* subsp *kurstaki* HD1 cells as autonomously replicating lysogen (Walter and Aronson 1991). However, only two (*cysC* and *trpB/F*) of seven tested markers were transduced into *B. cereus* at a frequency significantly exceeding the reversion rates. Transducing phage DNA hybridized to insertion sequence (IS231-like) probe, thus indicating a region of homology that was indispensable for the transduction. All of transductants contained a 44-kb plasmid, shown to correspond to the transducing phage, designated TP21. DNA of the transducing phage was able to transform *B. subtilis trpF* and *trpB* auxotrophs. Transfer of the particular *cys* and *trp* genes was much greater than that of the other markers tested, consistent with the specialized transduction. The homology between this 44-kb plasmid and the *B. thuringiensis* chromosome was an explanation for the specialized transduction by TP21 (Walter and Aronson 1991). Homologous recombination that occurs is responsible for the specific marker transfer. The phage was proposed to be useful for extensive mapping of particular regions of the chromosome and also for mobilizing additional chromosomal genes or genes on other plasmids, in particular the toxin genes.

Recently the nucleotide sequence of a *B. cereus* phage TP21 was determined (Klumpp et al. 2010b). However it was shown that the sequenced genome does not correspond to the transducing phage described above. In fact three phages with identical name TP21 were described and the authors suggested to rename the transducing phage as TP21-T and the sequenced one as TP21-L (Klumpp et al. 2010b). They have also mentioned that TP21-T was partially sequenced (unpublished). Since a partial sequence of entire genome of one of the *B. thuringiensis subsp kurstaki* strains, T03a001, is available (NCBI acc. #CM000751) it should be possible to identify sequencing contigs corresponding to the phage TP21-T.

The following can be concluded summarizing the studies on the phage-mediated gene transduction in the *B. cereus* group:

(i) Three groups of phages can be identified able to carry out the generalized transduction. The corresponding phages are different in the genome sizes, having the phage chromosomes of approximately 40 kb (CP-53, ϕHD248); 100 kb (CP-51, CP-54, TP-18) and more than 200 kb (TP13, ϕ63).

(ii) The phages of the tree groups are also different in their host ranges. Small phages are relatively strain-specific infecting only a few strains from rather narrow cluster. As an example the phage ϕHD248 is active only on half of the closely related strains of the *kurstaki* lineage (Inal et al. 1996; Priest et al. 2004). These phages presumably represent a family of λ-like temperate phages, but having the *pac* mechanism of DNA packaging and partially permuted genomes. Middle genome size phages belong to the SP01 family (Klumpp et al. 2010a). The ends of their linear chromosome are fixed and the transduction is probably due to DNA packaging errors. Because of this the frequencies are very low. Ironically the phage CP-51 was the most prominent for the generalized transduction studies in the *B. cereus* group (Thorne 1978). The large phages (TP13, ϕ63) are similar in their gene mapping potential to the PBS1 of *B. subtilis* and P1 of *E. coli*. In fact the most extensive gene mapping experiments in *B. thuringiensis* were done using these phages (Barsomian et al. 1984; Heierson et al. 1983). These phages have broader host range, less virulent due to plasmid-like lysogeny state and able to package large pieces of the host genome.

(iii) Most of transduction studies on the bacteria of the *B. cereus* group were carried out using *B. thuringiensis*, or T-cluster strains. No such phages were identified for the W (*B. weinhestephanensis*) or C-cluster strains, neither for the recently described "*B. cytotoxicus*" or Y-cluster (Auger et al. 2008). Presumably it was due to the interest for genetic studies and strain construction in the insecticidal bacteria, although some experiments on *B. anthracis* were also reported (Thorne 1968a). The inter-subspecific transduction is greatly restricted by the low efficiency of recombination between diverged DNA.

The value of transduction for bacterial gene mapping has started to decrease with the advent of genome sequencing methods. The first bacterial genome was sequenced in 1995 (Fleischmann et al. 1995) and afterwards hundreds of sequences became available. The first two genomes of the B. cereus group were reported in 2003 (Ivanova et al. 2003; Read et al. 2003) and since then about 30 complete and 100 of draft sequences appeared in public databases (Kolsto et al. 2009; Rasko et al. 2005). The advent of new generation sequencing (NGS) methodologies made the production of a draft for bacterial genomic sequence a matter of few days (Metzker 2009). However the apparent saturation of genomic information has created new perspectives of research using the old genetic methods. In particular, the bacteriophage mediated transduction can be a very efficient technique for tracing of genes responsible for complex phenotypes. Such phenotypes like spore formation and germination, virulence or ability to grow at extreme temperatures could be mentioned as important for the *B. cereus* group studies. The corresponding genetic systems are rather complex involving tens and hundreds of genes.

The virulence is dependent on many factors in addition to aggressive toxin and physical protection from the host defense. During the "trade-off" evolution the most virulent bacterial strains were eliminated from natural populations (Anderson and May 1979, 1982). Modified strains can therefore be constructed with reduced or increased virulence starting from a model laboratory isolate. The NGS approaches can be then used for identifying the genetic changes influencing the virulence. Detailed studies of the detected genes would elucidate their importance for virulence and mechanisms of their action. Notably the phage mediated gene transduction can be used as an instrument for introducing of virulence related genes from the strains with elevated pathogenicity into the model organism. Study of the latter using NGS would reveal what are the genes that were thus introduced.

Psychrotolerance, designating the ability to grow at low temperatures, is another property of practical importance with complex genetics. The strains possessing this ability in the *B. cereus* group were somehow ecologically secluded from the mesophiles resulting into the genetically distinct strain cluster *B. weihenstephanensis* granted with the species status (Lechner et al. 1998; Sorokin et al. 2006). Representatives of this species are usually less virulent than mesophiles, for as yet unknown reasons, although some pathogenic potential exists also in *B. weihenstephanensis* (Baron et al. 2007; Stenfors Arnesen et al. 2008; Stenfors et al. 2002). Genomic sequencing of a representative strain *B. weihenstephanenesis* KBAB4 did not reveal striking differences in genome organization as compared to the mesophilic strains (Lapidus et al. 2008). Some potentially indicative features could be mentioned. As an example, KBAB4 possesses higher number of fatty acid desaturases types as compared to mesophilic or moderately thermophilic strains (4, 2 and 1, respectively). Since the multiplicity of desaturases potentially increases the fluidity of membranes, this feature is in concordance with supposed higher membrane fluidity of the psychrotolerant strains. However, such analysis does not clarify of how important is the elevated membrane fluidity in comparison to other factors of cold adaptation. Direct experimentation could be the only approach to elucidate this question. The sequenced moderately thermophilic strain "*B. cytotoxicus*" NVH391-98 possesses genome that is 1.5 Mb smaller than those of other *B. cereus* group strains (Lapidus et al. 2008). It is however not clear how to relate this difference with the psychrotolerance or thermoresistance. In addition it is impossible to conclude from genome comparisons if the mesophilic strains already have all necessary genes to develop psychrotolerance under appropriate selective conditions. It appears to be more probable that to acquire this property some genes must be brought there from other strains. In particular such transfer can be offered by the phage-mediated transduction.

Analysis of experiments of gene transduction described above demonstrated that an efficient phage for straightforward gene transfer or genetic mapping was not yet well characterized for the *B. cereus* group. The large phages, like TP13 or ϕ63, appeared to be the most promising, but they were not sufficiently intensively used and characterized as CP-51 or CP-54. Ironically, the transduction mediated by the latter two phages seems to be a rare experimental artifact. Two additional important advantages of the phages TP13 and ϕ63 is their temperate characters and broad host ranges. It appears therefore that the future experiments using gene transduction in the *B. cereus* group should be based on these two phages or similar isolates. Since these two phages can package DNA of a large size it is also possible that their use could help to overcome the problem of inefficient heterologous recombination. All reported experiments detected very important decrease of the transduction frequencies, 10-fold and more, using heterologous donor and recipient. In most of such cases inefficient recombination is supposed to be the problem, although it was not directly demonstrated.

I outline briefly some potentially interesting phages for the future gene transduction experiments. Either they were detected and characterized due to genomic sequencing or identified during recent experimental studies on phage isolating.

7.3.2 Potentially Transducing Phages for the B. cereus Group

I shall review two kinds of data. First, several phages characterized experimentally will be considered for their potential of genetic transduction. Several properties will be taken into account. These are the phage host ranges, virulence, genome sizes and packaging mechanisms. Second type of data will be based on the analysis of bacterial genomic sequences. Multiple potentially inducible phages can be identified, either integrated into the bacterial chromosomes, or present as autonomously replicating elements. Phylogenetic analysis of the terminase genes of such "genomic" phages could provide a hint of the packaging mechanism and therefore their potential for the gene transduction.

Several reports provide basic data on the types of phages that can be found in the *B. cereus* group bacteria. One of the most comprehensive characterizations of phage types was done in the works of Ackerman's group (Ackermann 2003; Ackermann et al. 1995). The phages were divided according to their morphologies from which six can be found in Firmicutes (Ackermann 2007). From these only tailed phages represent an interest as potential vehicles for gene transduction. As for the others, like recently characterized linear genome phages (Ravantti et al. 2003; Sozhamannan et al. 2008; Verheust et al. 2003, 2005), they could not be considered as efficient gene transducers due to their replication mode. Nevertheless, the mentioned above phage CP-51 belongs to the SP01 family recently reviewed (Klumpp et al. 2010a). The authors of this review indicate that the phage could not be considered as a good gene transducer on the basis of its genomic structure. Low transduction frequencies that were seen in experiments could be explained by errors of phage DNA packaging.

The most extensively characterized are the λ-like temperate phages, in particular represented by the γ-phage family of *B. anthracis* (Minakhin et al. 2005; Schuch and Fischetti 2006; Fouts et al. 2006). Such phages having genomes of 35–45 kb are often found integrated in the chromosomes of the sequenced *B. cereus* group strains (Ivanova et al. 2003; Rasko et al. 2005; Read et al. 2003; Sozhamannan et al. 2006). Most of them possess the terminase gene those sequence cluster with enzymes recognizing *cos* site (Casjens et al. 2005). Such phages, like λ of *E. coli* and, presumably, TP-21 of *B. thuringiensis* could be suitable only for specialized transduction. Although more rarely, some other related phages encode headful terminase. The phage phBC391A2 was recently identified in "*B. cytotoxicus*" NVH391-98 strain

(Auger et al. 2008). The phage possesses very narrow host range, presumably due to immunity provided either by similar resident prophages or due to specificity of the phage receptor. phBC391A2 possesses terminase, those similarity corresponds to the headful packaging mechanism, and permuted end redundancy (not published). Although they are rare, such phages can be detected in other sequenced genomes of the *B. cereus* group. Two proteins with identity more than 90% were detected in the current version of the NCBI database. The corresponding genes are encoded by the genomes of *B. thuringiensis* serovar *finitimus YBT-020* (NCBI acc. ## CP002508.1 for the genome and ADY20353 for the protein (Zhu et al. 2011)) and *B. cereus 03BB108* (NCBI acc. ## NZ_ABDM 0200065.1 and ZP_03115627). phBC391A2 and similar phages can be useful for generalized transduction, although each of them apparently has a narrow spectrum of hosts. Other probably similar phage, ϕHD248, is described above. Generalized transduction was demonstrated for the ϕHD248, which confirms that such phages can be useful for limited spectra of hosts (Inal et al. 1990, 1992, 1996).

Large phages, similar to TP13 or ϕ63, can be regarded as the most promising for genetic transduction in *B. cereus*. In addition to large genome size, permitting transfer of significant bacterial genome pieces, such phages also possess broad host spectra. Bcp1 is another experimentally characterized phage with large genome (Schuch and Fischetti 2009). It is different from TP13 or ϕ63 by head morphology. The phage was not used for genetic transduction experiments but it was shown to infect rather broad range of strains, as distinct as *B. anthracis* Sterne (cluster C) and *B. cereus* ATCC14579 (cluster T) (Schuch and Fischetti 2009). Moreover the phage was found to be lysogenic, which is an advantage if it is used for the genetic transduction. It is not known if the lysogeny is due to maintenance of the phage as a low copy plasmid or due to integration into the host chromosome. Usually in the *B. cereus* group the phages of the *Myoviridae* family, morphologically similar to Bcp1, are virulent (Ahmed et al. 1995). However, the same paper reports phage 11, having a different morphology, but with similar head size suggesting large genome. Another feature of the phage 11 is that it infected almost all of tested 142 typeable strains and produced opaque plaques. The future transduction experiments, targeting genetic mapping or novel strain constructing, should take into account the above considerations. The ideal transducing phage with broad specificity can be based on TP13, ϕ63, Bcp1 or 11, or similar phages, tested for the capacity of genetic marker transfer between heterologous strains.

The second possibility for developing of an efficient vehicle for transduction could be based on information available from genomic sequencing. Many of about 100 sequenced *B. cereus* group strains contain episomes resembling non-integrated prophages, rather than metabolic plasmids. The described above phage TP21-T from *B. thuringiensis subsp kurstaki* is a good example of the such phage tested experimentally (Walter and Aronson 1991). Another example of non-integrated prophages, although without a transduction perspective, is the family of phages possessing linear genomes replicating via a protein-priming mechanism (Verheust et al. 2003, 2005; Sozhamannan et al. 2008; Stromsten et al. 2003). These phages, having the genomes of about 15 kb, were found in *B. thuringiensis*, *B. cereus* and

B. anthracis strains and are able to be stably maintained as one copy linear episomes. Upon induction with DNA damaging agent they switch to the lytic development and produce infective phage particles.

Recently another 53 kb *Siphoviridae* temperate phage, designated vB_BceS-IEBH, able to replicate as a plasmid without lysing the host, was characterized in a cereulide-producing strain *B. cereus* CD555 (Smeesters et al. 2011). The induced phage was able to form turbid plaques on one of the 28 tested strains—*B. thuringiensis sv thuringiensis* GBJ085. This result is rather surprising considering the difference of phylogenetic positions of emetic (cluster C) and enthomopathogenic strains (cluster T) (Helgason et al. 2004; Priest et al. 2004; Sorokin et al. 2006). The phage vB_BceS-IEBH encodes terminase that corresponds to the headful packaging mechanism. Headful packaging is compatible with the experimental data on restriction enzyme digestion of the phage DNA, which seems to be partially permuted (Smeesters et al. 2011). The existence of such "phage-like" extrachromosomal elements seems to be rather general in sequenced genomes of the *B. cereus* group bacteria. For example such contigs of 56 kb are present in the genomic sequences of the *B. thuringiensis Al Hakam* (the plasmid designated pALH1) or *B. cereus* F837/76 strains (Challacombe et al. 2007; Lapidus et al. 2008). Psychrotolerant strain *B. weihenstephanensis* KBAB4 has four extrachromosomal elements one of which, pBWB404 of 53 kb, can appear to be an inducible prophage (Lapidus et al. 2008). Another element, pBWB401 of 417 kb, lost most of the phage-related functions, but still keeps extended similarity to the genome of the *B. thuringiensis* phage 0305ϕ8-36 (Hardies et al. 2007).

Scrutiny of genome organization of such extrachromosomal elements provides a simple phenomenological hypothesis of their evolution. On the first step of this evolution a virulent phage becomes temperate having acquired the functions stabilizing it in the form of a low copy plasmid. These are presumably the cases of TP21-T in *B. thuringiensis* subsp *kurstaki*, vB_BceS-IEBH in the emetic strain, pALH1 in *B. thuringiensis Al Hakam* and of pBWB404 in *B. weihenstephanensis* KBAB4. Although biological activity of the prophages was not experimentally demonstrated in the latter two cases, the two episomes carry the genes for cell lysis, DNA packaging and for phage envelope synthesis. The following important step of this evolution would be an acquisition of functions useful for the bacterial host and gradual degradation of phage-related genes, first of all of those destined to kill the host. The 65 kb plasmid pBWB403 of *B. weihenstephanensis* KBAB4 still keeps genes for phage envelope, including tape measure protein. But at least two thirds of the plasmid is occupied by the genes encoding functions useful for the bacterial host rather than for the phage. In particular these are multiple enzymes involved in DNA metabolism, although some of these genes should certainly be related to the maintenance of the plasmid itself. Far evolved 417 kb plasmid pBWB401 keeps traces of similarity to the 218 kb *B. thuringiensis* phage 0305ϕ8-36 (Hardies et al. 2007). It is interesting to note also that several contigs of partially sequenced *B. thuringiensis* subsp *israelensis* ATCC 35646 strain share more extended similarity with the same phage (Hardies et al. 2007). This observation is not surprising since the two *B. thuringiensis* strains (ATCC35646 and the host of 0305ϕ8-36) are more closely related to each other than to the strain KBAB4 of *B. weihenstephanensis*. This hypothesis of gradual evolution of phages into metabolic plasmids in the *B. cereus* group is not new. One previous study described a 48 kb phage J7W-1 integrated into the 69 kb plasmid pAF101 in the strain *B. thuringiensis* AF101 (Kanda et al. 1998). The phage was excised from the plasmid upon induction by DNA damaging agents, temperature or during mating (Kanda et al. 1989, 1998, 2000a, b). It was also shown to be able to lysogenise *B. thuringiensis* subsp *israelensis*. It was not clear, whether in the *israelensis* strain the phage DNA was integrated into the chromosome, although some additional plasmid was reported to appear (Kanda et al. 1998).

7.4 Recombineering Perspectives for the *B. cereus* Group

Recombineering is a relatively new term designating genetic manipulations with bacteria and other organisms using homologous recombination. It is usually based on the use of efficient and relatively relaxed for the DNA homology recombination systems of bacteriophages (Muyrers et al. 1999, 2001; Zhang et al. 1998, 2000; Yu et al. 2000, 2003; Sawitzke et al. 2007). Bacteriophages often contain their own systems of homologous recombination. Such systems compete with the cellular *recA* system in order to provide the phage with recombination functions necessary for the successful replication of its DNA (Kuzminov 1999; Weigel and Seitz 2006). The most studied and apparently rather effective such recombination systems are the *red-gam* pathway of the phage λ and *recET* system of the phage Rac (Stahl 1998; Martinsohn et al. 2008; Li et al. 1998; Muniyappa and Radding 1986; Kolodner et al. 1994; Hall et al. 1993). These systems were adapted for the recombineering technology and successfully used in many applications (Muyrers et al. 2001; Sawitzke et al. 2007; Yu et al. 2000; Zhang et al. 1998; Datsenko and Wanner 2000).

In gram-positive bacteria the most characterized system is that of the phage SPP1 (Alonso et al. 2006; Ayora et al. 2002; Martinez-Jimenez et al. 2005; Vellani and Myers 2003). The functionality of this system was first demonstrated in plasmid transduction experiments (Alonso et al. 1986) and in the studies of formation of high molecular weight DNA of plasmids (Bravo and Alonso 1990). The recombination functions encoded by this phage were shown to be related to three gene products: G34.1P, G35P and G36P. G34.1P is a counterpart of RecE exonuclease and G35P corresponds to the RecT SSA protein. The role of G36P is speculated, although not yet confirmed, to involve protection of ssDNA tails from exonuclease. Although it is expected that the systems similar to the recET of E. coli can be as efficient in gram positive bacteria, the recent study targeting the goal of measuring of their functionality indicated that the situation is more complex. As an example the genes encoding G34.1P and G35P were three orders less efficient in E. coli oligonucleotide recombination test than the red system of phage λ (Datta et al. 2008). A similar result was recently obtained for the recombinases of new families (Lopes et al. 2010). The recombinase of the phage λ was the most efficient in the in vivo recombination test. Most of other tested recombinases provided 10^2–10^3-fold less efficient DNA exchange (Lopes et al. 2010). Apart from the phage-related functions, such systems are attractive for gene engineering methodology because of significantly reduced requirement for the presence of homologous DNA in the recombination driving synapses (Martinsohn et al. 2008; Swingle et al. 2010).

As a supplementary to genomic sequencing the recombineering has a potential to become an efficient experimental method for genetic mapping. Constructing of mutant strains can be done using PCR products or even oligonucleotides (Swingle et al. 2010). Such manipulations can complement gene mapping obtained by whole genome sequencing. It is also an advantage the possibility to modify the genes of interest without leaving any traces in the form of antibiotic resistance genes. This feature permits to collect together many modifications in the bacterial genome and thus to study the multi-gene phenomena. Universalism is another great advantage of recombineering systems. If the frequency of recombination is sufficiently high, a recombineering experiment would only need introducing of a PCR fragment or oligonucleotide into the recipient cell. This can be achieved by electroporation, which is rather effective for many bacteria and was elaborated for the *B. cereus* group (Belliveau and Trevors 1989; Bone and Ellar 1989; Lereclus et al. 1989; Masson et al. 1989; Schurter et al. 1989; Macaluso and Mettus 1991; Peng et al. 2009; Groot et al. 2008; Turgeon et al. 2006). Screening of mutant strains can be done directly by sequencing of the modified chromosomal parts. Constructing of an efficient recombineering system for the *B. cereus* group bacteria is therefore a useful prerequisite for genetic studies of these bacteria.

The bacteriophages detected in the sequenced genomes of the *B. cereus* group contain recombination genes, in particular similar to those of the phage SPP1. As an example the proteins G34.1P and G35P share 34% and 49% of identity with the proteins encoded by Bcer98_2617 and Bcer98_2619 of "*B. cytotoxicus*" NVH391-98. In this bacterium the corresponding region encodes potentially inducible prophage phBC391A1 (Auger et al. 2008). The proteins of "*B. cytotoxicus*" share 80–90% of identity to six other such couples from the sequenced *B. cereus* group genomes and hundreds of similar proteins from other Gram+ bacteria and phages. This recombination system is therefore fairly represented in the *B. cereus* group. The corresponding genes can be expressed and the recombination tested *in vivo* in a way similar to the one described earlier (Datta et al. 2008; Lopes et al. 2010). The expressed genes could also be introduced into the remote strains to increase the frequency of heterologous transduction. As it was demonstrated, decrease of DNA similarity has important impact on the transduction, presumably due to the inefficient recombination (Thorne 1978). Recombineering approach can therefore be used to broaden applications of the phage-mediated transduction for the heterologous hosts.

References

Ackermann HW (2003) Bacteriophage observations and evolution. Res Microbiol 154(4):245–251
Ackermann HW (2007) 5500 Phages examined in the electron microscope. Arch Virol 152(2):227–243

Ackermann HW, Azizbekyan RR, Bernier RL, de Barjac H, Saindoux S, Valero JR, Yu MX (1995) Phage typing of *Bacillus subtilis* and *B. thuringiensis*. Res Microbiol 146(8):643–657

Ahmed R, Sankar-Mistry P, Jackson S, Ackermann HW, Kasatiya SS (1995) *Bacillus cereus* phage typing as an epidemiological tool in outbreaks of food poisoning. J Clin Microbiol 33(3):636–640

Alcaraz LD, Moreno-Hagelsieb G, Eguiarte LE, Souza V, Herrera-Estrella L, Olmedo G (2010) Understanding the evolutionary relationships and major traits of *Bacillus* through comparative genomics. BMC Genomics 11:332

Alonso JC, Luder G, Trautner TA (1986) Requirements for the formation of plasmid-transducing particles of *Bacillus subtilis* bacteriophage SPP1. EMBO J 5(13):3723–3728

Alonso J C, Tavares P, Lurz R, Trautner TA (2006) Bacteriophage SPP1. In: Calendar R (ed) The Bacteriophages, 2nd edn. Oxford University Press, NY, pp 331–349

Altenbern RA, Stull HB (1965) Inducible lytic systems in the genus *Bacillus*. J Gen Microbiol 39:53–62

Amjad M, Castro JM, Sandoval H, Wu JJ, Yang M, Henner DJ, Piggot PJ (1991) An SfiI restriction map of the *Bacillus subtilis* 168 genome. Gene 101(1):15–21

Anagnostopoulos C, Spizizen J (1961) Requirements for transformation in *Bacillus Subtilis*. J Bacteriol 81(5):741–746

Anagnostopoulos C, Piggot PJ, Hoch JA (1993) The genetic map of Bacillus subtilis. In: Sonenshein AL, Hoch JA, Losick R (eds) *Bacillus subtilis* and other Gram-positive bacteria. American Society for Microbiology, Washington, DC, pp 425–461

Anderson RM, May RM (1979) Population biology of infectious diseases: Part I. Nature 280(5721):361–367

Anderson RM, May RM (1982) Coevolution of hosts and parasites. Parasitology 85(Pt 2):411–426

Auger S, Galleron N, Bidnenko E, Ehrlich SD, Lapidus A, Sorokin A (2008) The genetically remote pathogenic strain NVH391-98 of the *Bacillus cereus* group is representative of a cluster of thermophilic strains. Appl Environ Microbiol 74(4):1276–1280

Ayora S, Missich R, Mesa P, Lurz R, Yang S, Egelman EH, Alonso JC (2002) Homologous-pairing activity of the *Bacillus subtilis* bacteriophage SPP1 replication protein G35P. J Biol Chem 277(39):35969–35979

Azevedo V, Alvarez E, Zumstein E, Damiani G, Sgaramella V, Ehrlich SD, Serror P (1993) An ordered collection of *Bacillus subtilis* DNA segments cloned in yeast artificial chromosomes. Proc Natl Acad Sci USA 90(13):6047–6051

Barbe V, Cruveiller S, Kunst F, Lenoble P, Meurice G, Sekowska A, Vallenet D, Wang T, Moszer I, Medigue C, Danchin A (2009) From a consortium sequence to a unified sequence: the *Bacillus subtilis* 168 reference genome a decade later. Microbiology 155(Pt 6):1758–1775

Baron F, Cochet MF, Grosset N, Madec MN, Briandet R, Dessaigne S, Chevalier S, Gautier M, Jan S (2007) Isolation and characterization of a psychrotolerant toxin producer, *Bacillus weihenstephanensis*, in liquid egg products. J Food Prot 70(12):2782–2791

Barsomian GD, Robillard NJ, Thorne CB (1984) Chromosomal mapping of *Bacillus thuringiensis* by transduction. J Bacteriol 157(3):746–750

Belliveau BH, Trevors JT (1989) Transformation of *Bacillus cereus* vegetative cells by electroporation. Appl Environ Microbiol 55(6):1649–1652

Biaudet V, Samson F, Anagnostopoulos C, Ehrlich SD, Bessieres P (1996) Computerized genetic map of *Bacillus subtilis*. Microbiology 142(Pt 10):2669–2729

Bone EJ, Ellar DJ (1989) Transformation of *Bacillus thuringiensis* by electroporation. FEMS Microbiol Lett 49(2–3):171–177

Bravo A, Alonso JC (1990) The generation of concatemeric plasmid DNA in *Bacillus subtilis* as a consequence of bacteriophage SPP1 infection. Nucleic Acids Res 18(16):4651–4657

Campbell A (2006) General aspects of lysogeny. In: Calendar R (ed) The bacteriophages, 2nd edn. Oxford University Press, NY, pp 66–73

Carlson CR, Kolsto AB (1993) A complete physical map of a *Bacillus thuringiensis* chromosome. J Bacteriol 175(4):1053–1060

Carlson CR, Gronstad A, Kolsto AB (1992) Physical maps of the genomes of three *Bacillus cereus* strains. J Bacteriol 174(11):3750–3756
Carlson CR, Johansen T, Kolsto AB (1996) The chromosome map of *Bacillus thuringiensis* subsp. *canadensis* HD224 is highly similar to that of the *Bacillus cereus* type strain ATCC 14579. FEMS Microbiol Lett 141(2–3):163–167
Casjens S, Sampson L, Randall S, Eppler K, Wu H, Petri JB, Schmieger H (1992) Molecular genetic analysis of bacteriophage P22 gene 3 product, a protein involved in the initiation of headful DNA packaging. J Mol Biol 227(4):1086–1099
Casjens SR, Gilcrease EB, Winn-Stapley DA, Schicklmaier P, Schmieger H, Pedulla ML, Ford ME, Houtz JM, Hatfull GF, Hendrix RW (2005) The generalized transducing *Salmonella* bacteriophage ES18: complete genome sequence and DNA packaging strategy. J Bacteriol 187(3):1091–1104
Challacombe JF, Altherr MR, Xie G, Bhotika SS, Brown N, Bruce D, Campbell CS, Campbell ML, Chen J, Chertkov O, Cleland C, Dimitrijevic M, Doggett NA, Fawcett JJ, Glavina T, Goodwin LA, Green LD, Han CS, Hill KK, Hitchcock P, Jackson PJ, Keim P, Kewalramani AR, Longmire J, Lucas S, Malfatti S, Martinez D, McMurry K, Meincke LJ, Misra M, Moseman BL, Mundt M, Munk AC, Okinaka RT, Parson-Quintana B, Reilly LP, Richardson P, Robinson DL, Saunders E, Tapia R, Tesmer JG, Thayer N, Thompson LS, Tice H, Ticknor LO, Wills PL, Gilna P, Brettin TS (2007) The complete genome sequence of *Bacillus thuringiensis Al Hakam*. J Bacteriol 189(9):3680–3681
Clements MO, Moir A (1998) Role of the gerI operon of *Bacillus cereus* 569 in the response of spores to germinants. J Bacteriol 180(24):6729–6735
Collins FS, Weissman SM (1984) Directional cloning of DNA fragments at a large distance from an initial probe: a circularization method. Proc Natl Acad Sci USA 81(21):6812–6816
Cuervo A, Vaney MC, Antson AA, Tavares P, Oliveira L (2007) Structural rearrangements between portal protein subunits are essential for viral DNA translocation. J Biol Chem 282(26):18907–18913
Datsenko KA, Wanner BL (2000) One-step inactivation of chromosomal genes in *Escherichia coli* K-12 using PCR products. Proc Natl Acad Sci USA 97(12):6640–6645
Datta S, Costantino N, Zhou X, Court DL (2008) Identification and analysis of recombineering functions from Gram-negative and Gram-positive bacteria and their phages. Proc Natl Acad Sci USA 105(5):1626–1631
de Lencastre H, Archer LJ (1979) Transducing activity of bacteriophage SPP1. Biochem Biophys Res Commun 86(3):915–919
de Lencastre H, Archer LJ (1980) Characterization of bacteriophage SPP1 transducing particles. J Gen Microbiol 117(2):347–355
de Lencastre H, Archer LJ (1981) Molecular origin of transducing DNA in bacteriophage SPP1. J Gen Microbiol 122(2):345–349
Devine KM (1995) The *Bacillus subtilis* genome project: aims and progress. Trends Biotechnol 13(6):210–216
Didelot X, Barker M, Falush D, Priest FG (2009) Evolution of pathogenicity in the *Bacillus cereus* group. Syst Appl Microbiol 32(2):81–90
Earl AM, Losick R, Kolter R (2008) Ecology and genomics of *Bacillus subtilis*. Trends Microbiol 16(6):269–275
Eiserling FA (1967) The structure of *Bacillus subtilis* bacteriophage PBS 1. J Ultrastruct Res 17(3):342–347
Fagerlund A, Brillard J, Furst R, Guinebretiere MH, Granum PE (2007) Toxin production in a rare and genetically remote cluster of strains of the *Bacillus cereus* group. BMC Microbiol 7:43
Fang G, Ho C, Qiu Y, Cubas V, Yu Z, Cabau C, Cheung F, Moszer I, Danchin A (2005) Specialized microbial databases for inductive exploration of microbial genome sequences. BMC Genomics 6:14
Ferrari E, Canosi U, Galizzi A, Mazza G (1978) Studies on transduction process by SPP1 phage. J Gen Virol 41(3):563–572

Fleischmann RD, Adams MD, White O, Clayton RA, Kirkness EF, Kerlavage AR, Bult CJ, Tomb JF, Dougherty BA, Merrick JM et al (1995) Whole-genome random sequencing and assembly of *Haemophilus influenzae* Rd. Science 269(5223):496–512

Fouts DE, Rasko DA, Cer RZ, Jiang L, Fedorova NB, Shvartsbeyn A, Vamathevan JJ, Tallon L, Althoff R, Arbogast TS, Fadrosh DW, Read TD, Gill SR (2006) Sequencing *Bacillus anthracis* typing phages gamma and cherry reveals a common ancestry. J Bacteriol 188(9):3402–3408

Fraser CM, Fleischmann RD (1997) Strategies for whole microbial genome sequencing and analysis. Electrophoresis 18(8):1207–1216

Giorno R, Mallozzi M, Bozue J, Moody KS, Slack A, Qiu D, Wang R, Friedlander A, Welkos S, Driks A (2009) Localization and assembly of proteins comprising the outer structures of the *Bacillus anthracis* spore. Microbiology 155(Pt 4):1133–1145

Groot MN, Nieboer F, Abee T (2008) Enhanced transformation efficiency of recalcitrant *Bacillus cereus* and *Bacillus weihenstephanensis* isolates upon in vitro methylation of plasmid DNA. Appl Environ Microbiol 74(24):7817–7820

Guinebretiere MH, Thompson FL, Sorokin A, Normand P, Dawyndt P, Ehling-Schulz M, Svensson B, Sanchis V, Nguyen-The C, Heyndrickx M, De Vos P (2008) Ecological diversification in the *Bacillus cereus* Group. Environ Microbiol 10(4):851–865

Gusarov II, Kreneva RA, Rybak KV, Podcherniaev DA, Iomantas Iu V, Kolibaba LG, Polanuer BM, Kozlov Iu I, Perumov DA (1997) Primary structure and functional activity of the *Bacillus subtilis ribC* gene. Mol Biol (Mosk) 31(5):820–825

Hall SD, Kane MF, Kolodner RD (1993) Identification and characterization of the *Escherichia coli* RecT protein, a protein encoded by the recE region that promotes renaturation of homologous single-stranded DNA. J Bacteriol 175(1):277–287

Hardies SC, Thomas JA, Serwer P (2007) Comparative genomics of *Bacillus thuringiensis* phage 0305phi8-36: defining patterns of descent in a novel ancient phage lineage. Virol J 4:97

Harwood CR, Wipat A (1996) Sequencing and functional analysis of the genome of *Bacillus subtilis* strain 168. FEBS Lett 389(1):84–87

Heierson A, Landen R, Boman HG (1983) Transductional mapping of 9 linked chromosomal genes in *Bacillus thuringiensis*. Mol Gen Genet 192:118–123

Helgason E, Okstad OA, Caugant DA, Johansen HA, Fouet A, Mock M, Hegna I, Kolsto AB (2000a) *Bacillus anthracis, Bacillus cereus*, and *Bacillus thuringiensis*—one species on the basis of genetic evidence. Appl Environ Microbiol 66(6):2627–2630

Helgason E, Caugant DA, Olsen I, Kolsto AB (2000b) Genetic structure of population of *Bacillus cereus* and *B. thuringiensis* isolates associated with periodontitis and other human infections. J Clin Microbiol 38(4):1615–1622

Helgason E, Tourasse NJ, Meisal R, Caugant DA, Kolsto AB (2004) Multilocus sequence typing scheme for bacteria of the Bacillus cereus group. Appl Environ Microbiol 70(1):191–201

Hemphill HE, Whiteley HR (1975) Bacteriophages of *Bacillus subtilis*. Bacteriol Rev 39(3):257–315

Iida S, Hiestand-Nauer R, Sandmeier H, Lehnherr H, Arber W (1998) Accessory genes in the darA operon of bacteriophage P1 affect antirestriction function, generalized transduction, head morphogenesis, and host cell lysis. Virology 251(1):49–58

Inal JR, Karunakaran V, Burges HD (1990) Isolation and propagation of phages naturally associated with the aizawai variety of *Bacillus thuringiensis*. J Appl Bacteriol 68(1):17–21

Inal JR, Karunakaran V, Burges HD (1992) Generalised transduction in *Bacillus thuringiensis* var. *aizawai*. J Appl Bacteriol 72:87–90

Inal JM, Karunakaran V, Jones DR (1996) *Bacillus thuringiensis* subsp. *aizawai* generalized transducing phage fHD248: restriction site map and potential for fine-structure chromosomal mapping. Microbiology 142:1409–1416

Isidro A, Henriques AO, Tavares P (2004) The portal protein plays essential roles at different steps of the SPP1 DNA packaging process. Virology 322(2):253–263

Itaya M, Tanaka T (1991) Complete physical map of the *Bacillus subtilis* 168 chromosome constructed by a gene-directed mutagenesis method. J Mol Biol 220(3):631–648

Ivanova N, Sorokin A, Anderson I, Galleron N, Candelon B, Kapatral V, Bhattacharyya A, Reznik G, Mikhailova N, Lapidus A, Chu L, Mazur M, Goltsman E, Larsen N, D'Souza M, Walunas T, Grechkin Y, Pusch G, Haselkorn R, Fonstein M, Ehrlich SD, Overbeek R, Kyrpides N (2003) Genome sequence of *Bacillus cereus* and comparative analysis with *Bacillus anthracis*. Nature 423(6935):87–91

Kanda K, Tan Y, Aizawa K (1989) A novel phage genome integrated into a plasmid in *Bacillus thuringiensis* strain AF101. J Gen Microbiol 135(11):3035–3041

Kanda K, Kitajima Y, Moriyama Y, Kato F, Murata A (1998) Association of plasmid integrative J7W-1 prophage with *Bacillus thuringiensis* strains. Acta Virol 42(5):315–318

Kanda K, Takada Y, Kawasaki F, Kato F, Murata A (2000a) Mating in *Bacillus thuringiensis* can induce plasmid integrative prophage J7W-1. Acta Virol 44(3):189–192

Kanda K, Kayashima T, Kato F, Murata A (2000b) Temperature influences induction of a J7W-1-related phage in *Bacillus thuringiensis* serovar *indiana*. Acta Virol 44(3):183–187

Klumpp J, Dorscht J, Lurz R, Bielmann R, Wieland M, Zimmer M, Calendar R, Loessner MJ (2008) The terminally redundant, nonpermuted genome of Listeria bacteriophage A511: a model for the SPO1-like myoviruses of Gram-positive bacteria. J Bacteriol 190(17):5753–5765

Klumpp J, Lavigne R, Loessner MJ, Ackermann HW (2010a) The SPO1-related bacteriophages. Arch Virol 155(10):1547–1561

Klumpp J, Calendar R, Loessner MJ (2010b) Complete nucleotide sequence and molecular characterization of *Bacillus* phage TP21 and its relatedness to other phages with the same name. Viruses 2:961–971

Koehler TM (2009) *Bacillus anthracis* physiology and genetics. Mol Aspects Med 30(6):386–396

Kolodner R, Hall SD, Luisi-DeLuca C (1994) Homologous pairing proteins encoded by the *Escherichia coli* recE and recT genes. Mol Microbiol 11(1):23–30

Kolsto AB, Gronstad A, Oppegaard H (1990) Physical map of the *Bacillus cereus* chromosome. J Bacteriol 172(7):3821–3825

Kolsto AB, Tourasse NJ, Okstad OA (2009) What sets *Bacillus anthracis* apart from other *Bacillus* species? Annu Rev Microbiol 63:451–476

Kufer B, Backhaus H, Schmieger H (1982) The packaging initiation site of phage P22. Analysis of packaging events by transduction. Mol Gen Genet 187(3):510–515

Kunst F, Devine K (1991) The project of sequencing the entire *Bacillus subtilis* genome. Res Microbiol 142(7–8):905–912

Kunst F, Vassarotti A, Danchin A (1995) Organization of the European *Bacillus subtilis* genome sequencing project. Microbiology 141(Pt 2):249–255

Kunst F, Ogasawara N, Moszer I, Albertini AM, Alloni G, Azevedo V, Bertero MG, Bessieres P, Bolotin A, Borchert S, Borriss R, Boursier L, Brans A, Braun M, Brignell SC, Bron S, Brouillet S, Bruschi CV, Caldwell B, Capuano V, Carter NM, Choi SK, Codani JJ, Connerton IF, Danchin A et al (1997) The complete genome sequence of the gram-positive bacterium *Bacillus subtilis*. Nature 390(6657):249–256

Kuzminov A (1999) Recombinational repair of DNA damage in *Escherichia coli* and bacteriophage lambda. Microbiol Mol Biol Rev 63(4):751–813

Landen R, Heierson A, Boman HG (1981) A phage for generalized transduction in *Bacillus thuringiensis* and mapping of four genes for antibiotic resistance. J Gen Microbiol 123:49–59

Lapidus A, Goltsman E, Auger S, Galleron N, Segurens B, Dossat C, Land ML, Broussolle V, Brillard J, Guinebretiere MH, Sanchis V, Nguen-The C, Lereclus D, Richardson P, Wincker P, Weissenbach J, Ehrlich SD, Sorokin A (2008) Extending the *Bacillus cereus* group genomics to putative food-borne pathogens of different toxicity. Chem Biol Interact 171(2):236–249

Lebedev AA, Krause MH, Isidro AL, Vagin AA, Orlova EV, Turner J, Dodson EJ, Tavares P, Antson AA (2007) Structural framework for DNA translocation via the viral portal protein. EMBO J 26(7):1984–1994

Lecadet MM, Blondel MO, Ribier J (1980) Generalized transduction in *Bacillus thuringiensis* var. *berliner* 1715 using bacteriophage CP-54Ber. J Gen Microbiol 121(1):203–212

Lechner S, Mayr R, Francis KP, Pruss BM, Kaplan T, Wiessner-Gunkel E, Stewart GS, Scherer S (1998) *Bacillus weihenstephanensis* sp. nov. is a new psychrotolerant species of the *Bacillus cereus* group. Int J Syst Bacteriol 48(Pt 4):1373–1382

Lennox ES (1955) Transduction of linked genetic characters of the host by bacteriophage P1. Virology 1(2):190–206

Lereclus D, Arantes O, Chaufaux J, Lecadet M (1989) Transformation and expression of a cloned delta-endotoxin gene in *Bacillus thuringiensis*. FEMS Microbiol Lett 51(1):211–217

Lhuillier S, Gallopin M, Gilquin B, Brasiles S, Lancelot N, Letellier G, Gilles M, Dethan G, Orlova EV, Couprie J, Tavares P, Zinn-Justin S (2009) Structure of bacteriophage SPP1 head-to-tail connection reveals mechanism for viral DNA gating. Proc Natl Acad Sci USA 106(21):8507–8512

Li Z, Karakousis G, Chiu SK, Reddy G, Radding CM (1998) The beta protein of phage lambda promotes strand exchange. J Mol Biol 276(4):733–744

Lopes A, Amarir-Bouhram J, Faure G, Petit MA, Guerois R (2010) Detection of novel recombinases in bacteriophage genomes unveils Rad52, Rad51 and Gp2.5 remote homologs. Nucleic Acids Res 38(12):3952–3962

Lovett PS, Young FE (1970) Genetic analysis in *Bacillus pumilus* by PBSI-mediated transduction. J Bacteriol 101(2):603–608

Lovett PS, Bramucci D, Bramucci MG, Burdick BD (1974) Some properties of the PBP1 transduction system in *Bacillus pumilus*. J Virol 13(1):81–84

Macaluso A, Mettus AM (1991) Efficient transformation of *Bacillus thuringiensis* requires nonmethylated plasmid DNA. J Bacteriol 173(3):1353–1356

Markowitz VM, Chen IM, Palaniappan K, Chu K, Szeto E, Grechkin Y, Ratner A, Anderson I, Lykidis A, Mavromatis K, Ivanova NN, Kyrpides NC (2010) The integrated microbial genomes system: an expanding comparative analysis resource. Nucleic Acids Res 38(Database issue):D382–D390

Martinez-Jimenez MI, Alonso JC, Ayora S (2005) *Bacillus subtilis* bacteriophage SPP1-encoded gene 34.1 product is a recombination-dependent DNA replication protein. J Mol Biol 351(5):1007–1019

Martinsohn JT, Radman M, Petit MA (2008) The lambda red proteins promote efficient recombination between diverged sequences: implications for bacteriophage genome mosaicism. PLoS Genet 4(5):e1000065

Masson L, Prefontaine G, Brousseau R (1989) Transformation of *Bacillus thuringiensis* vegetative cells by electroporation. FEMS Microbiol Lett 51(3):273–277

Medigue C, Moszer I (2007) Annotation, comparison and databases for hundreds of bacterial genomes. Res Microbiol 158(10):724–736

Metzker ML (2009) Sequencing technologies—the next generation. Nat Rev Genet 11(1):31–46

Minakhin L, Semenova E, Liu J, Vasilov A, Severinova E, Gabisonia T, Inman R, Mushegian A, Severinov K (2005) Genome sequence and gene expression of *Bacillus anthracis* bacteriophage Fah. J Mol Biol 354(1):1–15

Mironov AS, Gusarov I, Rafikov R, Lopez LE, Shatalin K, Kreneva RA, Perumov DA, Nudler E (2002) Sensing small molecules by nascent RNA: a mechanism to control transcription in bacteria. Cell 111(5):747–756

Mock M, Fouet A (2001) Anthrax. Annu Rev Microbiol 55:647–671

Moody KL, Driks A, Rother GL, Cote CK, Brueggemann EE, Hines HB, Friedlander AM, Bozue J (2010) Processing, assembly and localization of a *Bacillus anthracis* spore protein. Microbiology 156(Pt 1):174–183

Moszer I, Glaser P, Danchin A (1995) SubtiList: a relational database for the *Bacillus subtilis* genome. Microbiology 141(Pt 2):261–268

Muniyappa K, Radding CM (1986) The homologous recombination system of phage lambda. Pairing activities of beta protein. J Biol Chem 261(16):7472–7478

Muyrers JP, Zhang Y, Testa G, Stewart AF (1999) Rapid modification of bacterial artificial chromosomes by ET-recombination. Nucleic Acids Res 27(6):1555–1557

Muyrers JP, Zhang Y, Stewart AF (2001) Techniques: recombinogenic engineering—new options for cloning and manipulating DNA. Trends Biochem Sci 26(5):325–331

Oliveira L, Cuervo A, Tavares P (2010) Direct interaction of the bacteriophage SPP1 packaging ATPase with the portal protein. J Biol Chem 285(10):7366–7373

Orlova EV, Gowen B, Droge A, Stiege A, Weise F, Lurz R, van Heel M, Tavares P (2003) Structure of a viral DNA gatekeeper at 10 A resolution by cryo-electron microscopy. EMBO J 22(6):1255–1262

Peng D, Luo Y, Guo S, Zeng H, Ju S, Yu Z, Sun M (2009) Elaboration of an electroporation protocol for large plasmids and wild-type strains of *Bacillus thuringiensis*. J Appl Microbiol 106(6):1849–1858

Perlak FJ, Mendelsohn CL, Thorne CB (1979) Converting bacteriophage for sporulation and crystal formation in *Bacillus thuringiensis*. J Bacteriol 140(2):699–706

Piggot PJ, Hoch JA (1985) Revised genetic linkage map of *Bacillus subtilis*. Microbiol Rev 49(2):158–179

Priest FG, Barker M, Baillie LW, Holmes EC, Maiden MC (2004) Population structure and evolution of the *Bacillus cereus* group. J Bacteriol 186(23):7959–7970

Raj AS, Raj AY, Schmieger H (1974) Phage genes involved in the formation generalized transducing particles in *Salmonella*—Phage P22. Mol Gen Genet 135(2):175–184

Rao VB, Feiss M (2008) The bacteriophage DNA packaging motor. Annu Rev Genet 42:647–681

Rasko DA, Altherr MR, Han CS, Ravel J (2005) Genomics of the *Bacillus cereus* group of organisms. FEMS Microbiol Rev 29(2):303–329

Ravantti JJ, Gaidelyte A, Bamford DH, Bamford JK (2003) Comparative analysis of bacterial viruses Bam35, infecting a gram-positive host, and PRD1, infecting gram-negative hosts, demonstrates a viral lineage. Virology 313(2):401–414

Read TD, Peterson SN, Tourasse N, Baillie LW, Paulsen IT, Nelson KE, Tettelin H, Fouts DE, Eisen JA, Gill SR, Holtzapple EK, Okstad OA, Helgason E, Rilstone J, Wu M, Kolonay JF, Beanan MJ, Dodson RJ, Brinkac LM, Gwinn M, DeBoy RT, Madpu R, Daugherty SC, Durkin AS, Haft DH, Nelson WC, Peterson JD, Pop M, Khouri HM, Radune D, Benton JL, Mahamoud Y, Jiang L, Hance IR, Weidman JF, Berry KJ, Plaut RD, Wolf AM, Watkins KL, Nierman WC, Hazen A, Cline R, Redmond C, Thwaite JE, White O, Salzberg SL, Thomason B, Friedlander AM, Koehler TM, Hanna PC, Kolsto AB, Fraser CM (2003) The genome sequence of *Bacillus anthracis Ames* and comparison to closely related bacteria. Nature 423(6935):81–86

Rice EW, Rose LJ, Johnson CH, Boczek LA, Arduino MJ, Reasoner DJ (2004) Boiling and *Bacillus* spores. Emerg Infect Dis 10(10):1887–1888

Rivolta C, Pagni M (1999) Genetic and physical maps of the *Bacillus subtilis* chromosome. Genetics 151(4):1239–1244

Ruhfel RE, Robillard NJ, Thorne CB (1984) Interspecies transduction of plasmids among *Bacillus anthracis, B. cereus*, and *B. thuringiensis*. J Bacteriol 157(3):708–711

Sastalla I, Rosovitz MJ, Leppla SH (2010) Accidental selection and intentional restoration of sporulation-deficient *Bacillus anthracis* mutants. Appl Environ Microbiol 76(18):6318–6321

Sauer U, Hatzimanikatis V, Bailey JE, Hochuli M, Szyperski T, Wuthrich K (1997) Metabolic fluxes in riboflavin-producing *Bacillus subtilis*. Nat Biotechnol 15(5):448–452

Sawitzke JA, Thomason LC, Costantino N, Bubunenko M, Datta S, Court DL (2007) Recombineering: in vivo genetic engineering in *E. coli, S. enterica*, and beyond. Methods Enzymol 421:171–199

Schmieger H (1972) Phage P22-mutants with increased or decreased transduction abilities. Mol Gen Genet 119(1):75–88

Schmieger H, Backhaus H (1976) Altered cotransduction frequencies exhibited by HT-mutants of *Salmonella*-phage P22. Mol Gen Genet 143(3):307–309

Schmieger H, Buch U (1975) Appearance of transducing particles and the fate of host DNA after infection of *Salmonella typhimurium* with P22-mutants with increased transducing ability (HT-mutants). Mol Gen Genet 140(2):111–122

Schnepf E, Crickmore N, Van Rie J, Lereclus D, Baum J, Feitelson J, Zeigler DR, Dean DH (1998) *Bacillus thuringiensis* and its pesticidal crystal proteins. Microbiol Mol Biol Rev 62(3):775–806

Schuch R, Fischetti VA (2006) Detailed genomic analysis of the Wbeta and gamma phages infecting *Bacillus anthracis*: implications for evolution of environmental fitness and antibiotic resistance. J Bacteriol 188(8):3037–3051

Schuch R, Fischetti VA (2009) The secret life of the anthrax agent *Bacillus anthracis*: bacteriophage-mediated ecological adaptations. PLoS One 4(8):e6532

Schurter W, Geiser M, Mathe D (1989) Efficient transformation of *Bacillus thuringiensis* and *B. cereus* via electroporation: transformation of acrystalliferous strains with a cloned delta-endotoxin gene. Mol Gen Genet 218(1):177–181

Shelton AM, Zhao JZ, Roush RT (2002) Economic, ecological, food safety, and social consequences of the deployment of bt transgenic plants. Annu Rev Entomol 47:845–881

Smeesters PR, Dreze PA, Bousbata S, Parikka KJ, Timmery S, Hu X, Perez-Morga D, Deghorain M, Toussaint A, Mahillon J, Van Melderen L (2011) Characterization of a novel temperate phage originating from a cereulide-producing *Bacillus cereus* strain. Res Microbiol 162(4):446–459

Sorokin A, Candelon B, Guilloux K, Galleron N, Wackerow-Kouzova N, Ehrlich SD, Bourguet D, Sanchis V (2006) Multiple-locus sequence typing analysis of *Bacillus cereus* and *Bacillus thuringiensis* reveals separate clustering and a distinct population structure of psychrotrophic strains. Appl Environ Microbiol 72(2):1569–1578

Sozhamannan S, Chute MD, McAfee FD, Fouts DE, Akmal A, Galloway DR, Mateczun A, Baillie LW, Read TD (2006) The *Bacillus anthracis* chromosome contains four conserved, excision-proficient, putative prophages. BMC Microbiol 6:34

Sozhamannan S, McKinstry M, Lentz SM, Jalasvuori M, McAfee F, Smith A, Dabbs J, Ackermann HW, Bamford JK, Mateczun A, Read TD (2008) Molecular characterization of a variant of *Bacillus anthracis*-specific phage AP50 with improved bacteriolytic activity. Appl Environ Microbiol 74(21):6792–6796

Srivatsan A, Han Y, Peng J, Tehranchi AK, Gibbs R, Wang JD, Chen R (2008) High-precision, whole-genome sequencing of laboratory strains facilitates genetic studies. PLoS Genet 4(8):e1000139

Stahl FW (1998) Recombination in phage lambda: one geneticist's historical perspective. Gene 223(1–2):95–102

Stenfors Arnesen LP, Fagerlund A, Granum PE (2008) From soil to gut: *Bacillus cereus* and its food poisoning toxins. FEMS Microbiol Rev 32(4):579–606

Stenfors LP, Mayr R, Scherer S, Granum PE (2002) Pathogenic potential of fifty *Bacillus weihenstephanensis* strains. FEMS Microbiol Lett 215(1):47–51

Sternberg N (1990) Bacteriophage P1 cloning system for the isolation, amplification, and recovery of DNA fragments as large as 100 kilobase pairs. Proc Natl Acad Sci USA 87(1):103–107

Stewart CR, Casjens SR, Cresawn SG, Houtz JM, Smith AL, Ford ME, Peebles CL, Hatfull GF, Hendrix RW, Huang WM, Pedulla ML (2009) The genome of *Bacillus subtilis* bacteriophage SPO1. J Mol Biol 388(1):48–70

Stromsten NJ, Benson SD, Burnett RM, Bamford DH, Bamford JK (2003) The *Bacillus thuringiensis* linear double-stranded DNA phage Bam35, which is highly similar to the *Bacillus cereus* linear plasmid pBClin15, has a prophage state. J Bacteriol 185(23):6985–6989

Susskind MM, Botstein D (1978) Molecular genetics of bacteriophage P22. Microbiol Rev 42(2):385–413

Swingle B, Markel E, Costantino N, Bubunenko MG, Cartinhour S, Court DL (2010) Oligonucleotide recombination in Gram-negative bacteria. Mol Microbiol 75(1):138–148

Takahashi I (1961) Genetic transduction in *Bacillus subtilis*. Biochem Biophys Res Commun 5:171–175

Thomason LC, Costantino N, Court DL (2007) *E. coli* genome manipulation by P1 transduction. In: Ausubel FM et al (eds) Current protocols in molecular biology, Chapter 1:Unit 1 17

Thorne CB (1962) Transduction in *Bacillus subtilis*. J Bacteriol 83:106–111

Thorne CB (1968a) Transduction in *Bacillus cereus* and *Bacillus anthracis*. Bacteriol Rev 32(4 Pt 1):358–361

Thorne CB (1968b) Transducing bacteriophage for *Bacillus cereus*. J Virol 2(7):657–662

Thorne CB (1978) Transduction in *Bacillus thuringiensis*. Appl Environ Microbiol 35(6):1109–1115

Thorne CB, Holt SC (1974) Cold lability of *Bacillus cereus* bacteriophage CP-51. J Virol 14(4):1008–1012

Tosato V, Bruschi CV (2004) Knowledge of the *Bacillus subtilis* genome: impacts on fundamental science and biotechnology. Appl Microbiol Biotechnol 64(1):1–6

Tourasse NJ, Kolsto AB (2007) SuperCAT: a supertree database for combined and integrative multilocus sequence typing analysis of the *Bacillus cereus* group of bacteria (including *B. cereus, B. anthracis* and *B. thuringiensis*). Nucleic Acids Res

Tourasse NJ, Helgason E, Klevan A, Sylvestre P, Moya M, Haustant M, Okstad OA, Fouet A, Mock M, Kolsto AB (2010a) Extended and global phylogenetic view of the *Bacillus cereus* group population by combination of MLST, AFLP, and MLEE genotyping data. Food Microbiol 28(2):236–244

Tourasse NJ, Okstad OA, Kolsto AB (2010b) HyperCAT: an extension of the SuperCAT database for global multi-scheme and multi-datatype phylogenetic analysis of the *Bacillus cereus* group population. Database (Oxford) 2010:baq017

Turgeon N, Laflamme C, Ho J, Duchaine C (2006) Elaboration of an electroporation protocol for *Bacillus cereus* ATCC 14579. J Microbiol Methods 67(3):543–548

Tye BK, Chan RK, Botstein D (1974) Packaging of an oversize transducing genome by *Salmonella* phage P22. J Mol Biol 85(4):485–500

Tye BK, Huberman JA, Botstein D (1974) Non-random circular permutation of phage P22 DNA. J Mol Biol 85(4):501–528

Tyeryar FJ Jr, Taylor MJ, Lawton WD, Goldberg ID (1969) Cotransduction and cotransformation of genetic markers in *Bacillus subtilis* and *Bacillus licheniformis*. J Bacteriol 100(2):1027–1036

Van Arsdell SW, Perkins JB, Yocum RR, Luan L, Howitt CL, Chatterjee NP, Pero JG (2005) Removing a bottleneck in the *Bacillus subtilis* biotin pathway: bioA utilizes lysine rather than S-adenosylmethionine as the amino donor in the KAPA-to-DAPA reaction. Biotechnol Bioeng 91(1):75–83

Vary P (1993) The genetic map of *Bacillus megaterium*. In: Sonenshein AL, Hoch JA, Losick R (eds) *Bacillus subtilis* and other Gram-positive bacteria. American Society for Microbiology, Washington, DC, pp 475–481

Vellani TS, Myers RS (2003) Bacteriophage SPP1 Chu is an alkaline exonuclease in the SynExo family of viral two-component recombinases. J Bacteriol 185(8):2465–2474

Verheust C, Jensen G, Mahillon J (2003) pGIL01, a linear tectiviral plasmid prophage originating from *Bacillus thuringiensis* serovar *israelensis*. Microbiology 149(Pt 8):2083–2092

Verheust C, Fornelos N, Mahillon J (2005) GIL16, a new gram-positive tectiviral phage related to the *Bacillus thuringiensis* GIL01 and the *Bacillus cereus* pBClin15 elements. J Bacteriol 187(6):1966–1973

Wall JD, Harriman PD (1974) Phage P1 mutants with altered transducing abilities for *Escherichia coli*. Virology 59(2):532–544

Walter TM, Aronson AI (1991) Transduction of certain genes by an autonomously replicating *Bacillus thuringiensis* phage. Appl Environ Microbiol 57(4):1000–1005

Weigel C, Seitz H (2006) Bacteriophage replication modules. FEMS Microbiol Rev 30(3):321–381

Wu SC, Wong SL (2002) Engineering of a *Bacillus subtilis* strain with adjustable levels of intracellular biotin for secretory production of functional streptavidin. Appl Environ Microbiol 68(3):1102–1108

Yamagishi H, Takahashi I (1968) Transducing particles of PBS 1. Virology 36(4):639–645

Yelton DB, Thorne CB (1970) Transduction in *Bacillus cereus* by each of two bacteriophages. J Bacteriol 102(2):573–579

Yelton DB, Thorne CB (1971) Comparison of *Bacillus cereus* bacteriophages CP-51 and CP-53. J Virol 8(2):242–253
Yu D, Ellis HM, Lee EC, Jenkins NA, Copeland NG, Court DL (2000) An efficient recombination system for chromosome engineering in *Escherichia coli*. Proc Natl Acad Sci USA 97(11):5978–5983
Yu D, Sawitzke JA, Ellis H, Court DL (2003) Recombineering with overlapping single-stranded DNA oligonucleotides: testing a recombination intermediate. Proc Natl Acad Sci USA 100(12):7207–7212
Zahler SA (ed) (1982) Specialized transduction in *Bacillus subtilis*, vol 1. The molecular biology of the Bacilli. Academic Press, London
Zhang Y, Buchholz F, Muyrers JP, Stewart AF (1998) A new logic for DNA engineering using recombination in *Escherichia coli*. Nat Genet 20(2):123–128
Zhang Y, Muyrers JP, Testa G, Stewart AF (2000) DNA cloning by homologous recombination in *Escherichia coli*. Nat Biotechnol 18(12):1314–1317
Zhu Y, Shang H, Zhu Q, Ji F, Wang P, Fu J, Deng Y, Xu C, Ye W, Zheng J, Zhu L, Ruan L, Peng D, Sun M (2011) Complete genome sequence of *Bacillus thuringiensis* serovar *finitimus* strain YBT-020. J Bacteriol 193(9):2379–2380
Zinder ND, Lederberg J (1952) Genetic exchange in Salmonella. J Bacteriol 64(5):679–699

Chapter 8
Conjugation in *Bacillus thuringiensis*: Insights into the Plasmids Exchange Process

Gislayne T. Vilas-Bôas and Clelton A. Santos

Abstract Since the discovery of the conjugation process in bacteria, many studies focusing on this issue have contributed to a better understanding of the biology, ecology, genetics and consequently to the taxonomy of bacteria. In this chapter, the mechanisms of the conjugation process in Gram-positive species were revised and detailed, including a set of events as the contact between donor and recipient cells, the DNA processing and its inter cellular transport, and the variations of the conjugal mating systems. Studies focusing on conjugative transfer in *Bacillus thuringiensis*, involving the detection of *cry* genes in large conjugative plasmids, the genetic basis of the process, the main plasmids, and methodological variations of mating systems are discussed. Nowadays conjugal mating systems are again prominence and several studies have been conducted to evaluate plasmid exchange both within and between *B. thuringiensis* and closely related species belonging to the *Bacillus cereus* group. Thus, conjugal mating systems became an important tool to understand the role of plasmids in the behavior and in genome evolution of *B. thuringiensis*.

Keywords Conjugation · *Bacillus thuringiensis* · *Bacillus cereus* group · Plasmids exchange

8.1 Introduction

8.1.1 The Nature and the Variations of the Conjugation Process

The discovery of the conjugation process in 1946 by Lederberg and Tatum was told in detail in 1986 by Lederberg (Lederberg 1986). This discovery has the merit of inaugurating the bacterial genetics and, at the time, was considered as the most fundamental advance in the history of bacteriology. Conjugal matings systems are re-

G. T. Vilas-Bôas (✉) · C. A. Santos
Departamento de Biologia Geral, CCB, Universidade Estadual de Londrina,
CP 6001, 86051-990 Londrina/PR, Brazil
e-mail: gvboas@uel.br

E. Sansinenea (ed.), *Bacillus thuringiensis Biotechnology*,
DOI 10.1007/978-94-007-3021-2_8,

markable in mediating DNA transfer between a wide range of bacterial genera, and in some cases, from bacteria to fungal and plant cells (Battisti et al. 1985; Oultram and Young 1985; Koehler and Thorne 1987; Beijersbergen et al. 1992; Hayman and Bolen 1993; Mahmood et al. 1996; Bertolla and Simonet 1999; Heinemann 1999; Christie and Vogel 2000; Chumakov 2000; Chen et al. 2002; Broothaerts et al. 2005). In this sense, bacterial conjugation is a specialized process involving transfer of DNA from a donor to a recipient bacterial cell, which is then referred as transconjugant or exconjugant and posses the ability to start new rounds of conjugation. Studies of conjugational transfer have been reviewed in several articles (Clark and Adelberg 1962; Curtiss 1969; Waters and Guiney 1993; Andrup 1998; Davison 1999; Grohmann et al. 2003; Prozorov 2003; Bahl et al. 2009; Frost and Koraimann 2010).

In the conjugation process there must be the contact between donors and recipients cells, forming specific aggregates. When the contact breaks, exconjugants are formed. In the donor cells, mobilization followed by exit of DNA from the cell occurs. The donor DNA is moved between cells and enters the recipient cells. Thus, the central question in bacterial conjugation is how the DNA is transported through the cell envelopes of the mating cells.

Bacterial conjugation implies a set of events including DNA processing and its transport from one cell to the other. The first event generates a single-stranded DNA (ssDNA) copy. This process begins with a relaxase, which binds to the origin of transfer (*oriT*) sequence and cleaves the DNA strand destined for transfer (T-strand). The relaxase plus one or more auxiliary proteins form a nucleoprotein complex called relaxosome, which remains covalently bound to the 5′ end of the T-strand, resulting in the formation of the relaxase-T-strand transfer intermediate (Grohmann et al. 2003; Chen et al. 2005; Garcillán-Barcia et al. 2009). The second event, inter cellular transport, occurs by a type IV secretion system (T4SS), also known as mating pair formation (MPF) apparatus which is a plasmid-encoded multiprotein complex involved in the traffic of the DNA strand from the donor to the recipient cell (Grohmann et al. 2003). During DNA transfer, signals conferring substrate recognition are carried not by the DNA but by the relaxase. These signals are conserved in different protein substrates and consist of positively charged or hydrophobic clusters of C-terminal residues. Therefore, conjugation systems are thus considered as protein-trafficking systems that have evolved the capacity to recognize and translocate relaxases and, consequently, translocate DNA molecules (Chen et al. 2005). In that way, DNA is actively pumped into the recipient cell by the type IV coupling protein (T4CP) (Llosa et al. 2002; Christie 2004), which is responsible for connecting DNA processing and translocation between cells.

Any DNA molecule can be transferred by conjugation, but the genes that code for proteins that form the conjugative apparatus, necessary for DNA transfer are found mainly in the conjugative plasmids and in the conjugative transposons. Thus, transmissible DNA can be classified according to their mobilization ability, as being conjugative (self-transmissible) or mobilizable (transmissible only in the presence of additional conjugative functions). A plasmid that carries the *tra* genes and encodes for its own set of MPF complex is considered as conjugative and tends to be large (with at least 30 kb) with low copy number, while a naturally occurring mobilizable plasmids carry a mobilization region (*mob*) encoding specific relaxosome components and the its own *oriT*, but lack the functions required for mating pair formation and tend to be small (with up to 15 kb) and have high copy number (Garcillán-Barcia et al. 2009). However, there is a third type of plasmids, denominated nonmobilizable plasmids, which lack both, a *mob* region and an *oriT* site, but can become mobilized by recombination with a conjugative plasmid to form transferable plasmidic DNA molecule (Andrup et al. 1996).

In addition to plasmids and transposons, it is also noteworthy that conjugation systems are also widespread in chromosome-borne mobile genetic elements (MGEs), frequently referred as integrative and conjugative elements (ICEs) and as integrative and mobilizable elements (IMES) (Wozniak and Waldor 2010, for a recent review about this subject).

However, the unidirectional DNA transfer from a donor to a recipient strain is not the only way to transfer of DNA, but there is also a third mechanism related to conjugative DNA transfer that was described and denominated retromobilization (see Ankenbauer 1997 for a review). In this process the DNA from recipient cell is captured by the donor cell harboring the conjugative plasmid. When captured DNA carries chromosomal markers it is denominated retrotransfer, when captured DNA carries markers from mobilizable plasmids, it is named retromobilization. Top et al. (1992) proposed two mechanisms to explain the transfer of a mobilizable plasmid from a recipient to a donor strain. In the bidirectional model (one-step model) transfer is a one step process of bidirectional DNA transfer consisting of a single conjugative event during which DNA flows freely and simultaneously between donor and recipient cells. In the unidirectional model (two-step model) two successive rounds of unidirectional transfer occur. The first step corresponds to the transfer of the conjugative plasmid to the recipient cell and the second step corresponds to the transfer of the mobilizable plasmid from the recipient to the donor cell. Indeed, in Gram-negative species, some studies suggest the two-step model of retromobilization (Heinemann and Ankenbauer 1993; Heinemann et al. 1996; Sia et al. 1996).

Therefore, in conjugal mating systems events of conjugation, mobilization and retromobilization may occur and several variations of mating systems were developed and classified in two categories, the biparental matings and the triparental matings, which use two or three bacterial strains, respectively. The classical mating system is the most known of the biparental matings and involves the transfer of a conjugative plasmid from the donor to the recipient cell, which is then named exconjugant. However, there are variations of the biparental mating systems involving the mobilization and retromobilization of plasmids. In the biparental mobilization three situations can occur:

(i) Transfer of a conjugative plasmid from the donor to the recipient;
(ii) Mobilization of a mobilizable plasmid from the donor to the recipient;
(iii) Transfer of both plasmids to the recipient.

In the same way, in the biparental retromobilization, transfer of a conjugative plasmid from the donor to the recipient and retromobilization of a mobilizable plasmid from the recipient to the donor may occur.

The same events may occur in the triparental matings: transfer of conjugative plasmids, mobilization and retromobilization of plasmids. Two different systems were described: the classical triparental matings are made using a donor, a recipient and a helper strain, which can transfer a mobilizable plasmid or a nonmobilizable plasmid, as detected by Timmery et al. (2009). In this system transfer of a conjugative plasmid from the donor to the recipient and to the helper strain may occur as well as, mobilization of the plasmid from the helper strain to the recipient, transfer of both plasmids to the recipient, and retromobilization of the plasmid from the helper strain to the donor. The second system, denominated Two-recipient mating system uses three strains, a donor and two recipients and is possible to detect, in this system, whether there is preferential transfer of the conjugative plasmid from the donor to one or other recipient (Santos et al. 2010).

Bacterial abilities to transfer plasmid DNA both by conjugation and mobilization and also to capture genetic material from other bacteria were described previously and demonstrated the importance of gene transfer in the evolution of bacterial genomes. In this context, the contribution of horizontal gene transfer to the genetic composition and diversity of *B. thuringiensis* was assayed in order to visualize its ecological importance as a genetic variation resource.

8.1.2 The Conjugation Process in B. thuringiensis

Strains of *B. thuringiensis* usually exhibit a complex plasmid profile, with up to 17 plasmids in sizes ranging from 2–600 kb (Lereclus et al. 1982; McDowell and Mann 1991; Berry et al. 2002; Han et al. 2006; Kashyap and Amla 2007; Amadio et al. 2009; Zhong et al. 2011). Interest has predominantly been focused on the larger and conjugative plasmids (González and Carlton 1980; Battisti et al. 1985; Reddy et al. 1987; Jensen et al. 1996; Wilcks et al. 1998), which frequently carry the *cry* genes (González et al. 1981; Kronstad et al. 1983; González and Carlton 1984), or show the presence of transposon (Lereclus et al. 1986) as well as insertion sequences (Mahillon et al. 1994).

The first report of occurrence of conjugation between strains of *B. thuringiensis* (González and Carlton 1982) is considered as a classic study and allowed the confirmation of the plasmid location of the genes encoding the crystal protein (*cry* genes). In the same year, González et al. (1982) also showed the occurrence of conjugation between strains of *B. thuringiensis* (Cry^{+}, donor) and *Bacillus cereus* (Cry^{-}, recipient), which started to produce the crystals with the same antigenicity of those produced by the donor strain.

However, due to absence of selective markers on natural plasmids of *B. thuringiensis* strains, the first conjugation studies with strains of *B. thuringiensis* used plasmids from other bacterial species. Lereclus et al. (1983) transferred the pAMβ1 plasmid (17 Md) from *Streptococcus faecalis* that confers resistance to erythromycin (Em^R) and lyncomycin (Lm^R) to several strains of *B. thuringiensis* by a filter-mating process. The *B. thuringiensis* exconjugant strains were also used as donors and transferred the plasmid during intraspecific mating, as well as also permitted the transfer of their own mobilizable plasmids.

In the same way, the lack of selective markers on the conjugative plasmids stimulated the use of small non-conjugative but mobilizable antibiotic resistance-encoding plasmids to monitor their transfer. The pBC16 plasmid was the most widely used. This plasmid was originally isolated from *B. cereus* and carry genes encoding resistance to antibiotics. The use of this plasmid allowed characterizing the plasmids pXO11, pXO12, pXO13, pXO14, pXO15 and pXO16, isolated from different *B. thuringiensis* subspecies, as self-transmissible plasmids and capable to promote the transfer of smaller mobilizable plasmids into a variety of recipients including *B. thuringiensis*, *B. cereus* and *Bacillus anthracis* strains (Battisti et al. 1985; Reddy et al. 1987). Andrup et al. (1993) also conducted a study using antibiotic resistance encoding small mobilizable plasmids (pBC16 and pAND006) to monitor transfer between strains of *B. thuringiensis* subsp. *israelensis*. The appearance of macroscopic aggregates when cells of exponentially growing strains of donor and recipient strains were mixed in broth allowing the identification of two aggregation phenotypes, strains Agr^+ which aggregate when combined with an Agr^- strain. The mobilization of the plasmids was unidirectional, i.e. from the Agr^+ cells to the Agr^- cells. In addition, this coaggregation was characterized as non-pheromone-induced and protease-sensitive.

In the following works it was demonstrated that the genetic basis of the aggregation system in *B. thuringiensis* subsp. *israelensis* was located on plasmid pXO16 (Jensen et al. 1995) and that this plasmid can mobilize both rolling-circle replicating plasmids and plasmids based on tetha-replicating origins (Andrup et al. 1996) including small "nonmobilizable" plasmids lacking both a *mob* gene and an *oriT* site.

The pHT73 plasmid from *B. thuringiensis* subsp. *kurstaki* KT0 was the first conjugative plasmid of *B. thuringiensis* marked with a gene conferring resistance to antibiotic (erythromycin) and allowed to monitor conjugation frequencies without the need of using a mobilizable plasmid (Vilas-Bôas et al. 1998) and was denominated pHT73-Em^R. In the same year, Wilcks et al. (1998) demonstrated that *B. thuringiensis* subsp. *kurstaki* pHT73 harbors two self-transmissible plasmids, pHT73 which carries the crystal toxin gene and pAW63 without *cry* genes, and that both are independently able to mobilize pBC16 plasmid. However, under the used conditions, pAW63 plasmid was considered more efficient than pHT73 plasmid, both regarding conjugative transfer and as a mobilizing agent, and pAW63 showed to be self-transmissible to *B. thuringiensis israelensis*, *B. cereus*, *Bacillus licheniformis*, *Bacillus subtilis* and *Bacillus sphaericus*. This study also demonstrated that *B. thuringiensis* subsp. *kurstaki* KT0, contains only one large plasmid, the crystal toxin plasmid and that the two toxin-encoding plasmids from both strains are identical and were named pHT73, which is also able to mobilize smaller non-conjugative plasmids (Lereclus et al. 1985).

All the conjugation studies using *B. thuringiensis* strains presented here showed unilateral gene transfer. However, Timmery et al. (2009) also demonstrated the retromobilization abilities of the conjugative plasmid pXO16 of *B. thuringiensis* subsp. *israelensis*, which were compared with the conjugation and mobilization abilities of this plasmid using mobilizable plasmids and a nonmobilizable element lacking the *mob* gene and *oriT* site. Kinetics experiments showed that retromobilization was delayed when compared to pXO16 conjugation, suggesting that pXO16 retromobilization follows a successive model of transfer, i.e. the two-step model, as already suggested by studies in Gram-negative species (Heinemann and Ankenbauer 1993; Heinemann et al. 1996; Sia et al. 1996). Moreover, it was detected the retromobilization of a nonmobilizable element (pC194 lacking the *mob* and *oriT* features) at frequencies even 100-fold lower than those obtained with the mobilizable plasmids indicating that the presence of mob and *oriT* regions has a positive effect on retromobilization, although they are not essential for the occurrence of the process.

The use of variable and alternative methods to assay conjugal mating allowed the detection of conjugation, mobilization and retromobilization. *In vitro* assays using liquid medium, solid surface, soil samples, water and foodstuffs combined with *in vivo* assays in insect larvae as well as the sequencing of many plasmids, shed light in the gene transfer process and led to increased interest in the role of the plasmids in biology of *B. thuringiensis*. Thus, the horizontal gene transfer both within and between species, has been frequently used to investigate the taxonomic position of *B. thuringiensis* and the other species of the *B. cereus* group.

8.2 Conjugation as a Tool to Understand the Ecology of *B. cereus* Group

The *B. cereus* group includes six very closely related species: *B. cereus*, *B. thuringiensis*, *B. anthracis*, *Bacillus mycoides*, *Bacillus pseudomycoides*, and *Bacillus weihenstephanensis*. Recent studies have suggested that *B. cereus sensu stricto*, *B. anthracis*, and *B. thuringiensis* should be considered as members of a unique species designated *B. cereus sensu lato* (Daffonchio et al. 2000; Helgason et al. 2000a, b; Bavykin et al. 2004). However, other studies have obtained sufficient genetic discrimination between *B. cereus*, *B. thuringiensis*, and *B. anthracis* (Harrell et al. 1995; Keim et al. 1997; Chang et al. 2003; Radnedge et al. 2003) and showed genetic differentiation among *B. cereus* and *B. thuringiensis* (Vilas-Bôas et al. 2002; Cherif et al. 2003; Peruca et al. 2008). Thus, there is still no consensus to whether these bacteria should be classified as separate taxa or that *B. thuringiensis* represents a subspecies of *B. cereus*.

Formerly, these species were classified as distinct because of the great relevance of their phenotypical differences, which formed the basis for their classification (Vilas-Bôas et al. 2007). Thus *B. anthracis* strains are capable of capsule formation and production of toxins that lead to carbuncles in animals and humans, causing the disease known as anthrax. *B. thuringiensis* forms a parasporal crystal (Cry proteins) that is active on larvae of variety of insect orders, allowing the use of products from this bacterium for biological control. Finally, *B. cereus* lacks both of these characteristics and can cause foodborne illness. Interestingly, the determinants of the typical characteristics of each of the three species are present in conjugative megaplasmids as the well-known pHT73 (Vilas-Bôas et al. 1998), pBtoxis (Ben-Dov et al. 1999), pXO1 and pXO2 (Kaspar and Robertson 1987), pCER270 (Rasko et al. 2007), whose horizontal transfer was described to a variety of recipient strains belonging to the *B. cereus* group.

Two main approaches have been used in the resolution of taxonomic issues involving *B. cereus* and *B. thuringiensis*. The first is based in genomic comparisons and uses various techniques such as multi-locus enzyme electrophoresis (Helgason et al. 1998; Vilas-Bôas et al. 2002), sequencing of discrete protein-coding genes (Helgason et al. 2000a), and the multi-locus sequence typing technique (Helgason et al. 2004; Ko et al. 2004; Priest et al. 2004). In this approach the strains from the two species are clustered and the analysis of the results allows genetic discrimination or not between the two species. In the second approach, *in vivo* and *in vitro* assays conducted in controlled conditions in samples of soil, insects, water and foods allow to access the behavior of these bacteria involving spores germination, spores formation and persistence, as well as vegetative cell multiplication and conjugation (Aly et al. 1985; Ohana et al. 1987; Jarrett and Stephenson 1990; Vilas-Bôas et al. 1998, 2000; Takatsuka and Kunimi 2000; Thomas et al. 2000, 2001, 2002; Suzuki et al. 2004; Raymond et al. 2010a, b). Thus, since horizontal gene transfer has been considered as an important process to increase knowledge about the ecology of *B. thuringiensis*, several studies in this field will be detailed in the following paragraphs.

Taking into account that the ecological role of *B. thuringiensis* in the environment is poorly understood, many studies with this subject were conducted aiming to know the behavior of *B. thuringiensis* in soil, water, insects and foodstuffs (Pruett et al. 1980; Petras and Casida 1985; Ohana et al. 1987; Thomas et al. 2000, 2001; Furlaneto et al. 2000; Vilas-Bôas et al. 2000; Ferreira et al. 2003; Van der Auwera et al. 2007; Modrie et al. 2010; Santos et al. 2010). Thus, Aly et al. (1985) were the first authors to demonstrate spore germination, sporulation and toxin production of *B. thuringiensis* subsp. *israelensis* in dead larvae of mosquitoes. Supporting the idea that whether there was germination of spores and multiplication of vegetative cells genetic exchange could occur, Jarrett and Stephenson (1990) infected *Galleria mellonella* and *Spodoptera littoralis* larvae and described dead larvae as hot spots for genetic transfer between *B. thuringiensis* strains, obtaining conjugation frequencies near at 100%. These results suggested that plasmid transfer between *B. thuringiensis* strains could occurs in nature, resulting in the production of new genetic combinations within populations of the bacteria, generating new questions regarding the niche of *B. thuringiensis* and the impact of horizontal gene transfer in the environment.

Subsequently, other studies focusing on mating transfer in environmental conditions began to emerge in order to better understand the evolution of bacteria belonging to *B. cereus* group. However only in the 2000s there was the great explosion of these studies, when the *B. thuringiensis* based products won the world market, beginning to be more popularly and widely used in biological control of agricultural pests and vectors of human disease.

Vilas-Bôas et al. (1998) monitored the conjugative transfer of the pHT73 plasmid from *B. thuringiensis* subsp. *kurstaki* KT0 strain carrying an insecticidal crystal protein gene (*cry1Ac*) and marked with a gene conferring resistance to erythromycin. The assays were conducted with *B. thuringiensis* in broth culture, soil microcosms and infected larvae of the lepidopteran insect *Anticarsia gemmatalis*. The insect larvae were the most favorable environment for the occurrence of conjugation showing the largest conjugation frequencies (10^{-1} exconjugants/recipient).

Thomas et al. (2000) monitored the plasmid transfer between *B. thuringiensis* subsp. *kurstaki* HD1 and *B. thuringiensis* subsp. *tenebrionis* under environmentally relevant laboratory conditions *in vitro*, in soil, and in insects. For *B. thuringiensis* subsp. *kurstaki* HD1, which was used as donor strain, plasmid transfer was detected in dead susceptible lepidopteran insect (*Lacanobia oleracea*) larvae but not in the nonsusceptible coleopteran insect (*Phaedon chocleriae)*. These results confirmed those obtained by Vilas-Bôas et al. (1998) that in susceptible lepidopteran insects there is a greater opportunity for growth of *B. thuringiensis* strains, providing suitable conditions for efficient plasmid transfer in the environment. In another work Thomas et al. (2001) evaluated the plasmid transfer between *B. thuringiensis* subsp. *israelensis* strains in laboratory culture, river water, and dipteran larvae, as well as the mobilization of pBC16 plasmid. The results obtained in the two studies suggested that the conjugative transfer of the plasmids could occur only when there was vegetative cell multiplication inside the coleopteran and dipteran dead larvae, as well as that some susceptible insects could not support the spore germination and the vegetative cell multiplication. This finding was posteriorly discussed by Suzuki et al. (2004), and named as Toxin Specificity (TS), which refers to the ability of a *B. thuringiensis* to produce toxins to kill the insect, and Host Specificity (HS) that refers to the ability of a *B. thuringiensis* strain to colonize specific specie of insect larvae.

In addition to the works carried out in larval environment, other experiments were conducted in different microcosm condition. Furlaneto et al. (2000) evaluated the survival and conjugal transfer between *B. thuringiensis* strains in aquatic environment. The detected frequencies of conjugal transfer were low since this environment does not support vegetative cell multiplication allowing the spores formation rapidly. Similar results were also described by Thomas et al. (2001, 2002) in non susceptible insect larvae, where low plasmid transfer frequencies were detected as vegetative cell multiplication was also impaired under this condition.

B. thuringiensis is ubiquitous specie founded mainly in soil (Martin and Travers 1989) and this specie was considered a soil bacterium for a long time. Thus, several studies were made aiming to investigate the persistence of spores in soil samples (Saleh et al. 1970; Pruett et al. 1980; Pedersen et al. 1995; Guidi et al. 2011). Saleh et al. (1970) recovered spores of *B. thuringiensis* about 14–40 days after the soil inoculation, while Pruett et al. (1980) recovered spores 6 months after inoculation.

In addition, Guidi et al. (2011) measured the spatial distribution of *B. thuringiensis* subsp. *israelensis* spores in a wetland reserve where application was carried out with *B. thuringiensis* subsp. *israelensis* products since 1988. The authors detected a decrease in spores near to 96% and suggested that continuous accumulation due to regular treatments could be excluded. Therefore, natural soil generally does not support spores germination or even multiplication of vegetative cells, which can occurs in soil only in special conditions as nutrients supplementation, neutral values of pH, and sterilization (Saleh et al. 1970; Thomas et al. 2000; Vilas-Bôas et al. 2000).

The occurrence of the conjugation process in soil samples is also limited to special conditions, similarly to spores germination and multiplication of vegetative cells. Among the first studies aiming to detect conjugal transfer in soil samples using *Bacillus* strains, Van Elsas et al. (1987) detected that presence of nutrients in the soil stimulated plasmid transfer. Therefore, Vilas-Bôas et al. (1998) related the conjugal mating occurrence between two *B. thuringiensis* strains in sterile non supplemented soil samples with neutral values of pH. Subsequently, Vilas-Bôas et al. (2000) showed that the non-occurrence of spores germination and multiplication of vegetative cells in non-supplemented soil samples without pH correction (pH ~ 5.1), prevented the occurrence of conjugation in this environment, due to the lack of sufficient amounts of vegetative cells capable of exchange genetic material.

Several environmental conditions were used to evaluate the occurrence of conjugal transfer involving *B. thuringiensis* strains, the results of these studies demonstrated that the main limiting factor is the ability to allow spore germination, or simply the presence of viable vegetative cells, although the multiplication of these cells alone did not seem to influence dramatically in such event. Suzuki et al. (2004) reported that insect larvae feds with rearing diets containing *B. thuringiensis* strains with no specificity to kill the insect (toxin specificity) can multiply and exchange genetic information in insects killed mechanically, indicating that there was no correlation between the toxin specificity and the host specificity. Therefore, *B. thuringiensis* strains can colonize non-susceptible insects since these insects are killed mechanically.

Hu et al. (2004) used the pHT73-Em^R plasmid to evaluate the dynamics of the dispersion of the *cry1Ac* gene among species belonging to the *B. cereus* group in LB broth, demonstrating the transfer of this plasmid to several *B. thuringiensis* subsp. *kurstaki* strains, several *B. cereus* strains and *B. mycoides* and the absence of transfer to *B. weihenstephanensis*. All exconjugant strains produced bipyramidal crystalline inclusion bodies during sporulation and showed variable stability of the plasmid under non-selective conditions, which was dependent of the receptor strain. However, as all *B. thuringiensis* recipient strains used belongs to subspecie *kurstaki*, the same subspecies of the donor strain, it was suggested that the donor strain might prefer to establish conjugal mating with *kurstaki* strains. Thus, Hu et al. (2005) evaluated the transfer frequencies of the pBtoxis plasmid from *B. thuringiensis* subsp. *israelensis* to 15 potentially recipient strains, detecting only two recipient strains (one *B. thuringiensis* and one *B. cereus*), which were able to receive the plasmid. Additionally, Yuan et al. (2007) evaluated the conjugation kinetics of the pHT73-Em^R plasmid from *B. thuringiensis* subsp. *kurstaki* KT0 to six *B. cereus* group strains (including *B. thuringiensis*, *B. cereus,* and *B. mycoides*) in lepidopteran larvae, obtaining low transfer frequencies with the highest values reaching to 10^{-6} CFU/donor. However, some strains used in this study were the same strain utilized by Hu et al. (2004) and the results obtained by the two groups were conflicting, i.e., some recipient strains were able to conjugate in culture broth and unable to conjugate in lepidopteran larvae.

From the ecological view, a plausible explanation for the findings of Yuan et al. (2007) could be found in the differences in ecological niche between *B. thuringiensis*, *B. cereus* and *B. mycoides*. Thus, some *B. cereus* and *B. mycoides* strains did not show germination of spores and vegetative cells multiplication in larvae of *Spodoptera exigua*, *Plutella xyllostella* and *Helicoverpa armigera*, which prevent the occurrence of the conjugation process. Likewise, some *B. cereus* strains were able to conjugate in low frequencies, what may be a consequence of the presence of vegetative cells ingested by the larvae with the spores and crystals.

Assuming the existence of a complex environment composed by *B. thuringiensis* and *B. cereus* strains, one question arise: One specific strain has more preference for establish horizontal gene transfer with one strain than with the other, or the two-recipients gene transfer frequencies are similar to those in a mating system composed only by one donor and one recipient strain? Does the ecological niche have influence on the occurrence of the process? Thus, Santos et al. (2010) used one-recipient and two-recipients conjugal transfer systems to evaluated the conjugative efficiency of the pHT73-Em^R plasmid from *B. thuringiensis* KT0 to several *B. thuringiensis* and *B. cereus* recipient strains both *in vitro* and in *Bombyx mori* larvae, and also evaluated the multiplication of vegetative cells of the strains in dead larvae. The *B. thuringiensis* KT0 strain did not show preference for genetic exchange with the *B. thuringiensis* recipient strain over that with the *B. cereus* recipient strains. However, *B. thuringiensis* strains germinated and multiplied more efficiently than *B. cereus* strains in insect larvae and only *B. thuringiensis* maintained complete spore germination for at least 24 h in *B. mori* larvae. These findings showed that there is no positive association between bacterial multiplication efficiency and conjugation ability in infected insects in the tested conditions.

In addition to the studies conducted in insect larvae, water and soil, some studies were also developed in foodstuffs, since *B. cereus* is a species known to cause food contamination and therefore, foods might be evaluated as ecological niche for this species. Thus, some studies were conducted to characterize the behavior of *B. thuringiensis* and *B. cereus* strains in foodstuffs. Van der Auwera et al. (2007) evaluated the conjugative behavior of *B. thuringiensis* strains in LB broth, milk and rice pudding using two conjugative *B. thuringiensis* plasmid, pXO16 and pAW63, as well as the mobilizable plasmid pC194, in bi- and triparental matings. The highest conjugation frequencies were found in milk, with values approximately 10-fold higher as compared to liquid LB. Furthermore, when a strain of *B. cereus* was used as donor of the pXO16 (isolated from *B. thuringiensis*) plasmid to another *B. cereus* strain, the conjugation rates were 10^{-1} exconjugant/recipient and, therefore, similar to those values observed in conjugal mating between *B. thuringiensis* strains in insect larvae (Jarrett and Stephenson 1990; Vilas-Bôas et al. 1998). Subsequently, it was evaluated the dynamics of the transfer of the pAW63 plasmid in LB medium and in foodstuffs and showed higher transfer frequencies in foodstuffs when compared to those observed in LB, probably because of an earlier onset of conjugation in combination with a higher transfer rate and/or a longer mating period (Modrie et al. 2010).

Many evidences indicate that *B. thuringiensis* and *B. cereus* exhibit a variety of behaviors in different environments; these differences are related to the high rate of conjugation both within and between these species, germination of spores and multiplication of vegetative cells. However, many aspects of the biology of these bacteria remain unknown. Thus, Wilcks et al. (2008) shocked the scientific community by showing that strains of *B. thuringiensis* are able to establish its complete life cycle in the intestinal tract of gnotobiotic rats, including germination of spores, multiplication of vegetative cells and conjugal transfer of pXO16 plasmid. In the same year, Bizzarri and Bishop (2008), obtained the isolation of vegetative cells from the phylloplane, indicating once more the possibility of the life cycle of *B. thuringiensis* occurs in other environments and not only in the insect cadaver.

Therefore, as discussed above, *B. thuringiensis* is a cosmopolitan species found in soil, water, plants, insects and foodstuffs. The most of these environments might be considered as reservoirs of spores, but occasionally vegetative cells could occur in any of these environments, since provide ideal conditions such nutrients, pH, temperature, etc. However, the environment in which *B. thuringiensis* cells find the best conditions for multiplication is dead insect larvae, which might be considered as the ecological niche of this bacteria. Consequently, in this environment probably occurred the genome evolution of *B. thuringiensis* through different mechanisms such as genetic exchange. Thus, in the course of time, countless *B. thuringiensis* strains were submitted to the action of natural selection and genotypes more adapted to survive in this environment have been positively selected, rise two consequences: (i) most of *B. thuringiensis* strains could be more adapted to develop the complete life cycle in susceptible insect host than in other environments. (ii) *B. thuringiensis* shows greater competitive advantage that other bacterial species that occupy sporadically such an environment.

8.3 Conclusion and Perspectives

"*The surface of the Earth is far more beautiful and far more intricate than any lifeless world. Our planet is graced by life and one quality that sets life apart is its complexity slowly evolved through 4 billion years of natural selection*" (Carl Sagan, in the series Cosmos: A personal voyage). The natural selection acts on organisms, selecting positively more adapted phenotypes, which are the result of interaction between the genotypes and the environment in which the organisms lives.

In conclusion, the importance of conjugative transfer of plasmids carrying *cry* genes in *B. thuringiensis* strains, as well as of other large conjugative plasmids to other species belonging to the *B. cereus* group is widely recognized for the typical characteristics these species. However, plasmids are often seen as supporting genetic elements in the genome evolution of these bacteria and this is an unfair way of referring to a genetic group as important. Therefore, is must better understand these plasmids, carrying out the sequencing and annotation, know both their role in the bacteria and the favorable conditions to their dispersion, as well as how the vegetative cells containing different plasmids interact with different environments.

References

Aly C, Mulla MS, Federici BA (1985) Sporulation and toxin production by *Bacillus thuringiensis* var. *israelensis* in cadavers of mosquito larvae. J Invertebr Pathol 46:251–258

Amadio AF, Benintende GB, Zandomeni RO (2009) Complete sequence of three plasmids from *Bacillus thuringiensis* INTA-FR7-4 environmental isolate and comparison with related plasmids from the *Bacillus cereus* group. Plasmid 62:172–182

Andrup L (1998) Conjugation in Gram-positive bacteria and kinetics of plasmid transfer. APMIS Suppl 84:47–55

Andrup L, Damgaard J, Wassermann K (1993) Mobilization of small plasmids in *Bacillus thuringiensis* subsp. *israelensis* is accompanied by specific aggregation. J Bacteriol 175:6530–6536

Andrup L, Jorgensen O, Wilcks A, Smidt L, Jensen GB (1996) Mobilization of "non-mobilizable" plasmids by the aggregation-mediated conjugation system of *Bacillus thuringiensis*. Plasmid 36:75–85

Ankenbauer RG (1997) Reassessing forty years of genetic doctrine: retrotransfer and conjugation. Genetics 145:543–549

Bahl MI, Hansen LH, Sørensen SJ (2009) Persistence mechanisms of conjugative plasmids. Methods Mol Biol 532:73–102

Battisti L, Green B, Thorne C (1985) Mating system for transfer of plasmids among *Bacillus anthracis*, *Bacillus cereus*, and *Bacillus thuringiensis*. J Bacteriol 162:543–550

Bavykin SG, Lysov YP, Zakhariev V, Kelly JJ, Jackman J, Stahl DA, Cherni A (2004) Use of 16S rRNA, 23S rRNA, and *gyrB* gene sequence analysis to determine phylogenetic relationships of *Bacillus cereus* group micro-organisms. J Clin Microbiol 42:3711–3730

Beijersbergen A, Den Dulk-Ras A, Schilperoort RA, Hooykaas PJJ (1992) Conjugative transfer by the virulence system of *Agrobacterium tumefaciens*. Science 256:1324–1327

Ben-Dov E, Nissan G, Pelleg N, Manasherob R, Boussiba S, Zaritsky A (1999) Refined, circular restriction map of the *Bacillus thuringiensis* subsp. *israelensis* plasmid carrying the mosquito larvicidal genes. Plasmid 42:186–191

Berry C, O'Neil S, Ben-Dov E, Jones AF, Murphy L, Quail MA, Holden MTG, Harris D, Zaritsky A, Parkhill J (2002) Complete sequence and organization of pBtoxis, the toxin-coding plasmid of *Bacillus thuringiensis* subsp. *israelensis*. Appl Environ Microbiol 68:5082–5095

Bertolla F, Simonet P (1999) Horizontal gene transfers in the environment: natural transformation as a putative process for gene transfers between transgenic plants and microorganisms. Res Microbiol 150:375–384

Bizzarri MF, Bishop AH (2008) The ecology of *Bacillus thuringiensis* on the Phylloplane: colonization from soil, plasmid transfer, and interaction with larvae of *Pieris brassicae*. Microb Ecol 56:133–139

Broothaerts W, Mitchell HJ, Weir B, Kaines S, Smith LM, Yang W, Mayer JE, Roa-Rodriguez C, Jefferson RA (2005) Gene transfer to plants by diverse species of bacteria. Nature 433:629–633

Chang YH, Shangkuan YH, Lin HC, Liu HW (2003) PCR assay of the *groEL* gene for detection and differentiation of *Bacillus cereus* group cells. Appl Environ Microbiol 69:4502–4510

Chen L, Chen Y, Wood DW, Nester EW (2002) A new type IV secretion system promotes conjugal transfer in *Agrobacterium tumefaciens*. J Bacteriol 184:4838–4845

Chen I, Christie PJ, Dubnau D (2005) The Ins and Outs of DNA transfer in bacteria. Science 310:1456–1460

Cherif A, Brusetti L, Borin S, Rizzi A, Boudabous A, Khyami-Horani H, Daffonchio D (2003) Genetic relationship in the '*Bacillus cereus* group' by rep-PCR fingerprinting and sequencing of a *Bacillus anthracis*-specific rep-PCR fragment. J Appl Microbiol 94:1108–1119

Christie PJ (2004) Type IV secretion: the *Agrobacterium virB/D4* and related conjugation systems. Biochim Biophys Acta 1694:219–234

Christie PJ, Vogel JP (2000) Bacterial type IV secretion: conjugation systems adapted to deliver effector molecules to host cells. Trends Microbiol 8:354–360

Chumakov MI (2000) Transfer of genetic information from agrobacteria to bacterial and plant cells: membrane and supramembrane structures involved in transfer. Membr Cell Biol 14:309–331

Clark AJ, Adelberg EA (1962) Bacterial conjugation. Annu Rev Microbiol 16:289–319

Curtiss R III (1969) Bacterial conjugation. Annu Rev Microbiol 23:69–136

Daffonchio D, Cherif A, Borin S (2000) Homoduplex and heteroduplex polymorphisms of the amplified ribosomal 16S--23S internal transcribed spacers describe genetic relationships in the "*Bacillus cereus* group". Appl Environ Microbiol 66:5460–5468

Davison J (1999) Genetic exchange between bacteria in the environment. Mol Microbiol 42:73–91

Ferreira LHPL, Suzuki MT, Itano EM, Ono MA, Arantes OMN (2003) Ecological aspects of *Bacillus thuringiensis* in an oxisoil. Sci Agric Sin 60:19–20

Frost LS, Koraimann G (2010) Regulation of bacterial conjugation: balancing opportunity with adversity. Future Microbiol 5:1057–1071

Furlaneto L, Saridakis HO, Arantes OM (2000) Survival and conjugal transfer between *Bacillus thuringiensis* strains in aquatic environment. Braz J Microbiol 31:233–238

Garcillán-Barcia MP, Francia MV, De La Cruz F (2009) The diversity of conjugative relaxases and its application in plasmid classification. FEMS Microbiol Rev 33:657–687

González JM Jr, Carlton BC (1980) Patterns of plasmid DNA in crystalliferous and acrystalliferous strains of *Bacillus thuringiensis*. Plasmid 3:92–98

González JM, Carlton BC (1982) Plasmid transfer in *Bacillus thuringiensis*. In: Streips UN, Goodgal SH, Guild WR, Wilson GA (eds) Genetic exchange: a celebration and a new generation. Marcel Dekker, NY, pp 85–95

González JM Jr, Carlton BC (1984) A large transmissible plasmid is required for crystal toxin production in *Bacillus thuringiensis* variety *israelensis*. Plasmid 11:28–38

González JM Jr, Dulmage HT, Carlton BC (1981) Correlation between specific plasmids and delta-endotoxin production in *Bacillus thuringiensis*. Plasmid 5:352–365

González JM, Brown BJ, Carlton BC (1982) Transfer of *Bacillus thuringiensis* plasmids coding for delta-endotoxin among strains of *B. thuringiensis* and *B. cereus*. Proc Natl Acad Sci 79:6951–6955

Grohmann E, Muth G, Espinosa M (2003) Conjugative plasmid transfer in Gram-positive bacteria. Microbiol Mol Biol Rev 67:277–301

Guidi V, Patocchi N, Lüthy P, Tonolla M (2011) Distribution of *Bacillus thuringiensis* subsp. *israelensis* in soil of a Swiss Wetland reserve after 22 yrs of mosquito control. Appl Environ Microbiol 77:3663–3668

Han CS, Xie G, Challacombe JF, Altherr MR, Bhotika SS, Brown N, Bruce D, Campbell CS, Campbell ML, Chen J, Chertkov O, Cleland C, Dimitrijevic M, Doggett NA, Fawcett JJ, Glavina T, Goodwin LA, Green LD, Hill KK, Hitchcock P, Jackson PJ, Keim P, Kewalramani AR, Longmire J, Lucas S, Malfatti S, McMurry K, Meincke LJ, Misra M, Moseman BL, Mundt M, Munk AC, Okinaka RT, Parson-Quintana B, Reilly LP, Richardson P, Robinson DL, Rubin E, Saunders E, Tapia R, Tesmer JG, Thayer N, Thompson LS, Tice H, Ticknor LO, Wills PL, Brettin TS, Gilna P (2006) Pathogenomic sequence analysis of *Bacillus cereus* and *Bacillus thuringiensis* isolates closely related to *Bacillus anthracis*. J Bacteriol 188:3382–3390

Harrell LJ, Andersen GL, Wilson KH (1995) Genetic variability of *Bacillus anthracis* and related species. J Clin Microbiol 33:1847–1850

Hayman GT, Bolen PL (1993) Movement of shuttle plasmids from *Escherichia coli* into yeasts other than *Saccharomyces cerevisiae* using trans-kingdom conjugation. Plasmid 30:251–257

Heinemann JA (1999) Genetic evidence of protein transfer during bacterial conjugation. Plasmid 41:240–247

Heinemann JA, Ankenbauer RG (1993) Retrotransfer in *Escherichia coli* conjugation: bidirectional exchange or de novo mating? J Bacteriol 175:583–588

Heinemann JA, Scott HE, Williams M (1996) Doing the conjugative two-step: evidence of recipient autonomy in retrotransfer. Genetics 143:1425–1435

Helgason E, Caugant DA, Lecadet MM, Chen Y, Mahillon J, Lövgren A, Hegna I, Kvaløy K, Kolstø AB (1998) Genetic diversity of *Bacillus cereus/B. thuringiensis* isolates from natural sources. Curr Microbiol 37:80–87

Helgason E, Caugant DA, Olsen I, Kolstø A-B (2000a) Genetic structure of population of *Bacillus cereus* and *Bacillus thuringiensis* isolates associated with periodontitis and other human infections. J Clin Microbiol 38:1615–1622

Helgason E, Økstad OA, Caugant DA, Johansen HA, Fouet A, Mock M, Hegna I, Kolstø A-B (2000b). *Bacillus anthracis*, *Bacillus cereus*, and *Bacillus thuringiensis*—one species on the basis of genetic evidence. Appl Environ Microbiol 66:2627–2630

Helgason E, Tourasse NJ, Meisal R, Caugant DA, Kolstø A-B (2004) Multilocus sequence typing scheme for bacteria of the *Bacillus cereus* group. Appl Environ Microbiol 70:191–201

Hu X, Hansen BM, Eilenberg J, Hendriksen NB, Smidt L, Yuan Z, Jensen GB (2004) Conjugative transfer, stability and expression of a plasmid encoding a *cry1Ac* gene in *Bacillus cereus* group strains. FEMS Microbiol Lett 231:45–52

Hu X, Hansen BM, Yuan Z, Johansen JE, Eilenberg J, Hendriksen NB, Smidt L, Jensen GB (2005) Transfer and expression of the mosquitocidal plasmid pBtoxis in *Bacillus cereus* group strains. FEMS Microbiol Lett 245:239–247

Jarrett P, Stephenson M (1990) Plasmid transfer between strains of *Bacillus thuringiensis* infecting *Galleria mellonella* and *Spodoptera littoralis*. Appl Environ Microbiol 56:1608–1614

Jensen GB, Wilcks A, Petersen SS, Damgaard J, Baum JA, Andrup L (1995) The genetic basis of the aggregation system *in Bacillus thuringiensis* subsp. *israelensis* is located on the large conjugative plasmid pXO16. J Bacteriol 177:2914–2917

Jensen GB, Andrup L, Wilcks A, Smidt L, Poulsen OM (1996) The aggregation-mediated conjugation system of *Bacillus thuringiensis* subsp. *israelensis*: host range and kinetics. Curr Microbiol 33:228–236

Kashyap S, Amla DV (2007) Characterisation of *Bacillus thuringiensis kurstaki* strains by toxicity, plasmid profiles and numerical analysis of their *cry*IA genes. Afr J Biotechnol 6:1821–1827

Kaspar RL, Robertson DL (1987) Purification and physical analysis of *Bacillus anthracis* plasmids pXO1 and pXO2. Biochem Biophys Res Commun 149:362–368

Keim P, Kalif A, Schupp J, Hill K, Travis SE, Richmond K, Adair DM, Hugh-Jones M, Kuske CR, Jackson P (1997) Molecular evolution and diversity in *Bacillus anthracis* as detected by amplified fragment length polymorphism markers. J Bacteriol 179:818–824

Ko KS, Kim J-W, Kim J-M, Kim W, Chung S, Kim IJ, Kook Y-H (2004) Population structure of the *Bacillus cereus* group as determined by sequence analysis of six housekeeping genes and the *plcR* gene. Infect Immun 72:5253–5261

Koehler TM, Thorne CB (1987) *Bacillus subtilis* (natto) plasmid pLS20 mediates interspecies plasmid transfer. J Bacteriol 169:5271–5278

Kronstad JW, Schnepf HE, Whiteley HR (1983) Diversity of locations for *Bacillus thuringiensis* crystal protein genes. J Bacteriol 154:419–428

Lederberg J (1986) Forty years of genetic recombination in bacteria. Nature 324:627–628

Lederberg J, Tatum EL (1946) Gene recombination in *Escherichia coli*. Nature 158:558

Lereclus D, Lecadet MM, Ribier J, Dedonder R (1982) Molecular relationships among plasmids of *Bacillus thuringiensis*: conserved sequences through 11 crystalliferous strains. Mol Gen Genet 186:391–398

Lereclus D, Menou G, Lecadet MM (1983) Isolation of a DNA sequence related to several plasmids from Bacillus thuringiensis after a mating involving the Streptococcus faecalis plasmid pAM beta 1. Mol Gen Genet 191:307–313

Lereclus D, Lecadet MM, Klier A, Ribier J, Rapoport G, Dedonder R (1985) Recent aspects of genetic manipulation in *Bacillus thuringiensis*. Biochimie 67:91–99

Lereclus D, Mahillon J, Menou G, Lecadet MM (1986) Identification of Tn4430, a transposon of *Bacillus thuringiensis* functional in *Escherichia coli*. Mol Gen Genet 204:52–57

Llosa M, Gomis-Ruth FX, Coll M, De La Cruz F (2002) Bacterial conjugation: a two-step mechanism for DNA transport. Mol Microbiol 45:1–8

Mahillon J, Rezsöhazy R, Hallet B, Delcour J (1994) IS231 and other *Bacillus thuringiensis* transposable elements: a review. Genetica 93:13–26

Mahmood A, Kimura T, Takenaka M, Yoshida K (1996) The construction of novel mobilizable YAC plasmids and their behavior during trans-kingdom conjugation between bacteria and yeasts. Genet Anal 13:25–31

Martin PAW, Travers RS (1989) Worldwide abundance and distribution of *Bacillus thuringiensis* isolates. Appl Environ Microbiol 55:2437–2442

McDowell DG, Mann NH (1991) Characterization and sequence analysis of a small plasmid from *Bacillus thuringiensis* var. *kurstaki* strain HD1-DIPEL. Plasmid 25:113–120

Modrie P, Beuls E, Mahillon J (2010) Differential transfer dynamics of pAW63 plasmid among members of the *Bacillus cereus* group in food microcosms. J Appl Microbiol 108:888–897

Ohana B, Margalit J, Barak Z (1987) Fate of *Bacillus thuringiensis* subsp. i*sraelensis* under simulated field conditions. Appl Environ Microbiol 53:828–831

Oultram JD, Young M (1985) Conjugal transfer of plasmic pAMb1 from *Streptococcus lactis* and *Bacillus subtilis* to *Clostridium acetobutylicum*. FEMS Microbiol Lett 27:129–134

Pedersen JC, Damgaard PH, Eilenberg J, Hansen BM (1995) Dispersal of *Bacillus thuringiensis* var *kurstaki* in an experimental cabbage field. Can J Microbiol 41:118–125

Peruca APS, Vilas-Bôas GT, Arantes OMN (2008) Genetic relationships between sympatric populations of *Bacillus cereus* and *Bacillus thuringiensis*, as revealed by rep-PCR genomic fingerprinting. Mem Inst Oswaldo Cruz 103:497–500

Petras SF, Casida LE (1985) Survival of *Bacillus thuringiensis* spores in soil. Appl Environ Microbiol 50:1496–1501

Priest FG, Barker M, Baillie LWJ, Holmes EC, Maiden MCJ (2004) Population structure and evolution of the *Bacillus cereus* group. J Bacteriol 186:7959–7970

Prozorov AA (2003) Conjugation in bacilli. Microbiology 72:517–552

Pruett CJH, Burges HD, Wybom CH (1980) Effect of exposure to soil on potency and spores viability of *Bacillus thuringiensis*. J Invertebr Pathol 35:168–174

Radnedge L, Agron PG, Hill KK, Jackson PJ, Ticknor LO, Keim P, Andersen GL (2003) Genome differences that distinguish *Bacillus anthracis* from *Bacillus cereus* and *Bacillus thuringiensis*. Appl Environ Microbiol 69:2755–2764

Rasko DA, Rosovitz MJ, Økstad OA, Fouts DE, Jiang L, Cer RZ, Kolstø AB, Gill SR, Ravel J (2007) Complete sequence analysis of novel plasmids from emetic and periodontal *Bacillus cereus* isolates reveals a common evolutionary history among the *B. cereus*-group plasmids, including *Bacillus anthracis* pXO1. J Bacteriol 189:52–64

Raymond B, Johnston P, Nielsen-LeRoux C, Lereclus D, Crickmore N (2010a) *Bacillus thuringiensis*: an impotent pathogen? Trends Microbiol 18:189–194

Raymond B, Wyres KL, Sheppard SK, Ellis RJ, Bonsall MB (2010b) Environmental factors determining the epidemiology and population genetic structure of the *Bacillus cereus* group in the field. PLoS Pathog 6:e1000905

Reddy A, Battisti L, Thorne C (1987) Identification of self-transmissible plasmids in four *Bacillus thuringiensis* subspecies. J Bacteriol 169:5263–5270

Saleh SM, Harris RF, Allen ON (1970) Fate of *Bacillus thuringiensis* in soil: effect of soil pH and organic amendment. Can J Microbiol 16:667–680

Santos CA, Vilas-Bôas GT, Lereclus D, Suzuki MT, Angelo EA, Arantes OMN (2010) Conjugal transfer between *Bacillus thuringiensis* and *Bacillus cereus* strains is not directly correlated with growth of recipient strains. J Invertebr Pathol 105:171–175

Sia EA, Kuehner DM, Figurski DH (1996) Mechanism of retrotransfer in conjugation: prior transfer of the conjugative plasmid is required. *J Bacteriol* 178:1457–1464

Suzuki M, Lereclus D, Arantes OMN (2004) Fate of *Bacillus thuringiensis* in different insect larvae. Can J Microbiol 50:973–975

Takatsuka J, Kunimi Y (2000) Intestinal bacteria affect growth of *Bacillus thuringiensis* in larvae of the oriental tea tortrix, *Homona magnanima* Diakonoff (Lepidoptera: Tortricidae). J Invertebr Pathol 76:222–226

Thomas DJI, Morgan JAW, Whipps JM, Saunders JR (2000) Plasmid transfer betweem the *Bacillus thuringiensis* subspecies *kurstaki* and *tenebrionis* in laboratory culture and soil and in *Lepidopteran* and *Coleopteran larvae*. Appl Environ Microbiol 66:118–124

Thomas DJI, Morgan JAW, Whipps JM, Saunders JR (2001) Plasmid transfer between the *Bacillus thuringiensis* subspecies. *israelensis* strains in laboratory culture, river water, and dipteran larvae. Appl Environ Microbiol 67:330–338

Thomas DJI, Morgan JAW, Whipps JM, Saunders JR (2002) Transfer of plasmid pBC16 between *Bacillus thuringiensis* strains in non-susceptible larval. FEMS Microbiol Ecol 40:181–190

Timmery S, Modrie P, Minet O, Mahillon J (2009) Plasmid capture by the *Bacillus thuringiensis* conjugative plasmid pXO16. J Bacteriol 191:2197–2205

Top E, Vanrolleghem P, Mergeay M, Verstraete W (1992) Determination of the mechanism of retrotransfer by mechanistic mathematical modelling. *J Bacteriol* 174:5953–5960

Van der Auwera GA, Timmery S, Hoton F, Mahillon J (2007) Plasmid exchanges among members of the *Bacillus cereus* group in foodstuffs. Int J Food Microbiol 113:164–172

Van Elsas JD, Govaertj M, Van Veen JA (1987) Transfer of plasmid pFT30 between bacilli in soil as influenced by bacterial population dynamics and soil conditions. Soil Biol Biochem 19:639–647

Vilas-Bôas GFLT, Vilas-Bôas LA, Lereclus D, Arantes OMN (1998) *Bacillus thuringiensis* conjugation under environmental conditions. FEMS Microbiol Ecol 25:369–374

Vilas-Bôas LA, Vilas-Bôas GFLT, Saridakis HO, Lemos MVF, Lereclus D, Arantes OMN (2000) Survival and conjugation of *Bacillus thuringiensis* in a soil microcosm. FEMS Microbiol Ecol 31:255–259

Vilas-Bôas G, Sanchis V, Lereclus D, Lemos MV, Bourguet D (2002) Genetic differentiation between sympatric populations of *Bacillus cereus* and *Bacillus thuringiensis*. Appl Environ Microbiol 68:1414–1424

Vilas-Bôas GT, Peruca APS, Arantes OMN (2007) Biology and taxonomy of *Bacillus cereus*, *Bacillus anthracis* and *Bacillus thuringiensis*. Can J Microbiol 53:673–687

Waters VL, Guiney DG (1993) Processes at the nick region link conjugation, T-DNA transfer and rolling circle replication. Mol Microbiol 9:1123–1130

Wilcks A, Jayaswal N, Lereclus D, Andrup L (1998) Characterization of plasmid pAW63, a second self-transmissible plasmid in *Bacillus thuringiensis* subsp. *kurstaki* HD73. Microbiology 144:1263–1270

Wilcks A, Smidt L, Bahl MI, Hansen BM, Andrup L, Hendriksen NB, Licht TR (2008) Germination and conjugation of *Bacillus thuringiensis* subsp. *israelensis* in the intestine of gnotobiotic rats. J Appl Microbiol 104:1252–1259

Wozniak RA, Waldor MK (2010) Integrative and conjugative elements: mosaic mobile genetic elements enabling dynamic lateral gene flow. Nat Rev Microbiol 8:522–563

Yuan YM, Hu XM, Liu HZ, Hansen BM, Yan JP, Yuan ZM (2007) Kinetics of plasmid transfer among *Bacillus cereus* group strains within lepidopteran larvae. Arch Microbiol 187:425–431

Zhong C, Peng D, Ye W, Chai L, Qi J, Yu Z, Ruan L, Sun M (2011) Determination of plasmid copy number reveals the total plasmid DNA amount is greater than the chromosomal DNA amount in *Bacillus thuringiensis* YBT-1520. PLoS One 6(1):e16025

Chapter 9
Shuttle Vectors of *Bacillus thuringiensis*

Alejandra Ochoa-Zarzosa and Joel Edmundo López-Meza

Abstract *Bacillus thuringiensis* is characterized by the synthesis of parasporal crystals during sporulation; these are composed of one or more highly specific insecticidal or nematocidal endotoxin proteins. Genetic manipulation of *B. thuringiensis* has been achieved due to the development of stable shuttle vectors and the establishment of efficient transformation systems. Shuttle vectors of *B. thuringiensis* have been constructed using essentially replicons from resident plasmids from this bacterium that replicate by the theta mechanism. Also, these vectors have been developed using plasmid replicons from other Gram-positive bacteria or RCR plasmids. The transformation of *B. thuringiensis* with these vectors has been accomplished mainly through electroporation. The development of shuttle vectors with better characteristics and protocols with high transformation efficiency have greatly facilitated basic research and engineering of *B. thuringiensis*.

Keywords Shuttle vectors · Plasmids · Replication origins · Rolling circle · Transformation

9.1 Introduction

Bacillus thuringiensis is an aerobic, spore-forming, Gram-positive bacterium that synthesizes crystalline proteins during sporulation. These proteins have insecticidal properties by which *B. thuringiensis* has become an alternative of choice for control of insect pests. The genetic manipulation of *B. thuringiensis* has been difficult; however, the development of molecular tools began with the cloning of the first gene encoding for insecticidal proteins, which allowed the genetic manipulation of this bacterium.

Genetic characterization of plasmids from *B. thuringiensis* has been an invaluable tool in the development of stable cloning vectors as part of efforts to improve the properties of this bacterium. *B. thuringiensis* vectors essentially are shuttle vec-

J. E. López-Meza (✉) · A. Ochoa-Zarzosa
Centro Multidisciplinario de Estudios en Biotecnología-FMVZ,
Universidad Michoacana de San Nicolás de Hidalgo, km 9.5 Carretera Morelia-Zinapécuaro,
Posta Veterinaria, Morelia 58893, Michoacán, México
e-mail: elmeza@umich.mx

E. Sansinenea (ed.), *Bacillus thuringiensis Biotechnology,*

DOI 10.1007/978-94-007-3021-2_9,

tors that can be replicated in more than one host, and generally contain replication origins derived from resident plasmids of this bacterium. Basically, most of the *B. thuringiensis* shuttle vectors developed have *Escherichia coli* as an alternative host, since it is the most commonly used and best characterized organism for genetic manipulation. To date, shuttle vectors from *B. thuringiensis* are developed mainly from plasmids that replicate by tetha mechanism, although there have been attempts to construct such vectors from plasmids that replicate by rolling circle. In addition, the transformation of *B. thuringiensis* with shuttle vectors has been achieved essentially by electroporation.

In this chapter, we review some general aspects of *B. thuringiensis* plasmids and their use in the development of shuttle vectors for genetic manipulation of strains from this bacterium.

9.2 Plasmids in *Bacillus thuringiensis*

Plasmids are extrachromosomal DNA genetic elements that replicate autonomously; they have important roles in the evolution and adaptation of bacteria because they facilitate the genetic exchange in bacterial populations by a variety of ways, which include conjugation, mobilization, transformation and transduction (Del Solar et al. 1998; Gruss and Ehrlich 1989). In general, bacterial species simultaneously harbor multiple plasmids with a great diversity of valuable functional genes, such as genes related to drug resistance, pathogenicity, and others involved in different functions (Del Solar and Espinosa 2000). The replicons from plasmids have represented useful tools in the design of cloning vectors due to their ability to accept and maintain recombinant DNA (Khan 1997).

A common characteristic of many strains of *B. thuringiensis* is the presence of a complex arrangement of plasmid DNA. The number and size of these plasmids (2–250 kb) vary considerably among strains, which can constitute a substantial amount of the total genetic content of the bacterium, representing 10–20% of the genetic material of the cell (Gonzalez and Carlton 1980; Lereclus et al. 1982; McDowell and Mann 1991; Aronson 1993).

According to the replication mechanisms, plasmids from *B. thuringiensis* can be classified in two groups. (1) Plasmids that use a theta replication mechanism, which are normally large plasmids, such as p43, p44, p60, pAW63 and pBtoxis (Baum and Gilbert 1991; Wilcks et al. 1999; Tang et al. 2006); however, there are small plasmids such as pHT1030 that replicates by the theta mechanism (Lereclus and Arantes 1992). (2) Plasmids that usually replicate by a rolling-circle mechanism, which are mainly small plasmids, such as pTX14-1, pTX14-2, pTX14-3, pGI1, pGI2, pGI3, pUIBI1 and pBMBt1 (Andrup et al. 2003; López-Meza et al. 2003; Loeza-Lara et al. 2005).

The interest in the plasmids from *B. thuringiensis* has been focused on those of high molecular weight (>60 kb) because in them are located the insecticidal crystal protein genes (Höfte and Whiteley 1989). However, several reports indicate that some of *B. thuringiensis* plasmids have roles in conjugation (Battisti et al. 1985), contain transposons and insertion sequences (Lereclus et al. 1986; Mahillon et al. 1987) or temperate phages (Kanda et al. 1989), and have participation in the synthesis of β-exotoxin (Levinson et al. 1990). On the other hand, small plasmids from *B. thuringiensis* (< 15 kb) generally have been considered as "cryptic", because their functions are unknown. However, the functions attributed to these widely distributed plasmids essentially correspond to DNA replication and mobilization; but also have been reported other genes or elements, as the transposon Tn*4430*, a poison component of a poison-antidote system and insecticidal crystal protein-like gene (Madsen et al. 1993; Mahillon et al. 1988; McDowell and Mann 1991; Andrup et al. 2003; López-Meza et al. 2003; Loeza-Lara et al. 2005).

9.3 Shuttle Vectors of *Bacillus thuringiensis*

B. thuringiensis produces parasporal crystal proteins during sporulation whose genes are typically located on large plasmids (Schnepf et al. 1998). These genes have been cloned and genetically engineered to improve their insecticidal activity due that it has been developed efficient cloning vectors and transformation systems for *B. thuringiensis* (Bone and Ellar 1989; Lereclus et al. 1989; Mahillon et al. 1989; Masson et al. 1989; Schurter et al. 1989).

A key element in the genetic manipulation of *B. thuringiensis* has been the development of stable cloning vectors, especially the named shuttle vectors, which employ principally replication origins derived from resident *B. thuringiensis* plasmids. Shuttle vectors are cloning vectors that replicate in cells from more than one organism, so they have the origins of replication from various hosts, e.g. *E. coli* and *B. thuringiensis*.

9.3.1 Shuttle Vectors of B. thuringiensis Contain Replication Origins Derived from Resident Plasmids of this Bacterium

At the end of the 1980s and beginning of the 1990s of last century, were reported the first attempts to characterize origins of replication from *B. thuringiensis* resident plasmids, which lead to the development of shuttle vectors of this bacterium (Table 9.1). Lereclus et al. (1988) cloned two cryptic plasmids from *B. thuringiensis* subsp. *thuringiensis* LM2 (8.6 and 15 kb, designated as pHT1000 and pHT1030, respectively) in *E. coli* and analyzed their segregational stability in *B. subtilis*. Four recombinant plasmids were constructed for this purpose; pHT1000 and pHT1030 were ligated to vector pJH101 (a derivate of pBR322 that contains the cloramphenicol acetyl transferase (*cat*) gene of pC194) to obtain pHT1001, 1002, 1031 and 1032 recombinant plasmids. These vectors were introduced to *B. subtilis* and transformants were obtained, demonstrating their ability to replicate in this bacterium.

Table 9.1 Characteristics of some shuttle vectors of *Bacillus thuringiensis*

Vector	Features	Reference
pHT3101	*ori* pHT1030 (*B. thuringiensis*) *ori* pUC18 (*E. coli*) Low copy number (4 copies/chromosome) Ampicillin and erythromycin resistance markers	Lereclus et al. (1989)
pHT315	Similar to pHT3101, except that has high copy number (15 copies/chromosome)	Arantes and Lereclus (1991)
pEG597, pEG853 and pEG854	*ori* p43, p44, and p60 (*B. thuringiensis*) *ori* pTZ18u (*E. coli*) *cat* gene of pC194 to selection in *B. thuringiensis*	Baum et al. (1990)
pHBLBIV	*ori* pBLB (*B. thuringiensis*) *ori* pUC18 (*E. coli*) High copy number (16 copies/chromosome) Ampicillin and erythromycin resistance markers	Mesrati et al. (2005)
pEMB0557	*ori* p60 (*B. thuringiensis*) *ori* pBeloBAC11 (*E. coli*) Chloromycetin and erythromycin resistance markers Support large DNA fragments (at least of 70 kb)	Liu et al. (2009)
pEG146 and pEG147	*ori* pBC16 (*B. thuringiensis*) *ori* pUC18 (*E. coli*) Ampicillin and tetracycline markers	Von Tersch et al. (1991)
pBCX	*ori* pBC16 (*B. thuringiensis*) *ori* pBluescript II KS (*E. coli*) Ampicillin and tetracycline markers	Lertcanawanichakul and Wiwat (2000)

Further, the novel shuttle vector pHT3101 of *B. thuringiensis* was constructed using the plasmid pHT1030 (Lereclus et al. 1989). To achieve this, a 2.9 kb *BalI* fragment containing the replication region of pHT1030 was ligated to pUC18 vector together with a 1 kb *KpnI-BamHI* fragment from a derivative of pUC18 containing the erythromycin resistance gene (Em^r). pHT3101 (6.6 kb) shuttle vector allows selecting *E. coli* transformants with ampicillin and *B. thuringiensis* transformants with erythromycin. A disadvantage of pHT3101 vector is that the *Bam*HI and *Hind*III sites can not be used as unique cloning sites; however, pHT3101 and its derivates are among the most common vectors used for constructing recombinant *B. thuringiensis* strains (Park et al. 1998, 2003).

Shuttle vectors carrying the replication region of pHT1030 have a low copy number (about 4/chromosome), which could be a limiting. To solve this, Arantes and Lereclus (1991) reported the construction of novel shuttle vectors of *B. thuringiensis* with different copy numbers, which derive from pHT1035 (a plasmid of 6.3 kb bearing the replication region of pHT1030). Shuttle vectors pHT304 (about 4 copies/chromosome), pHT315 (about 15 copies/chromosome) and pHT370 (about 70 copies/chromosome) were constructed and used to clone the *cryIIIA* gene. When these vectors were introduced in *B. thuringiensis* strain *kurstaki* HD-1 Cry^{-}B by electroporation, only cells containing the pHT315 and pHT370 vectors showed detectable parasporal inclusions. In addition, there was no significant difference between the amounts of endotoxin produced by both cells, which indicates that above of 15 copies of plasmid per chromosome there are other factors that limit the synthesis of crystal protein.

In the early of 1990s of the last century, others research groups reported the development of novel shuttle vectors of *B. thuringiensis* containing replicons from resident plasmids of this bacterium. Baum et al. (1990) reported the cloning of seven replication origins from resident plasmids of *B. thuringiensis* HD263 and HD73 (subsp. *kurstaki*). Three of these replication origins, obtained from plasmids p43, p44, and p60, were used to construct the shuttle vectors pEG597, pEG853 and pEG854 that shown structural and segregational stability in *B. thuringiensis*. These vectors have the *cat* gene of pC194 inserted into the *Eco*RI site of the *E. coli* cloning vector pTZ18u to provide a selectable functional marker in *B. thuringiensis*. The functionality of these vectors was demonstrated cloning several insecticidal crystal protein genes, verifying the crystal protein production in *B. thuringiensis* and evaluating their insecticidal activity.

In the same way, Gamel and Piot (1992) reported the construction of pSB909 shuttle vector and its derivates, which contains the replicon of the 75-kb *B. thuringiensis* HD-73 plasmid and the erythromycin resistance gene. The vector pSB909.5 was stably maintained in transformed *B. thuringiensis* strains containing plasmids with similar or identical *ori* without a complete displacement of these native plasmids.

In order to improve the stability of shuttle vectors of *B. thuringiensis*, Mesrati et al. (2005) report the construction and characterization of a new cloning shuttle vector called pHBLBIV. This vector contains a 1.6 kb replicon of pBLB, a resident plasmid from *B. thuringiensis* BUPM101, which was ligated to *E. coli* pUC18 replicon containing the ampicillin and the erythromycin resistance genes (derivates of pHTBlue plasmid). The pHBLBIV vector has a relative high copy number (about 16 copies/chromosome), and was used successfully for the expression of vegetative insecticidal protein gene (*vip*) in *B. thuringiensis* BUPM106 Vip^{-} strain.

Until the early twenty-first century the shuttle vectors designed to *B. thuringiensis* showed limitations in their ability to maintain large DNA fragments, making difficult the cloning and expression of large DNA fragments, such as gene clusters. In an attempt to resolve these difficulties, Liu et al. (2009) provided a new tool for *B. thuringiensis* through the construction of the shuttle vector pEMB0557, that has the ability to support large DNA fragments. pEMB0557 contains the replicon of the plasmid p60 (*B. thuringiensis* subsp. *kurstaki* YBT-1520), and were incorporated the erythromycin and chloromycetin resistance genes to select transformants in *B. thuringiensis* or *E. coli*, respectively. This vector also contains the backbone of the BAC vector pBeloBAC11. To evaluate its functionality Liu et al. constructed a bacterial artificial chromosome library of *B. thuringiensis* strain CT-43 and showed that pEMB0557 was able to support large DNA fragments (at least of 70 kb).

9.3.2 Shuttle Vectors of B. thuringiensis Contain Replication Origins Derived from Plasmids of Other Gram-positive

Shuttle vectors for *B. thuringiensis* have also been developed using plasmid replicons from other Gram-positive bacteria. Bourgouin et al. (1990) report the construction of the bifunctional vector pBU4, which was obtained by ligation of EcoO109-restricted pUC19 and *Eco*RI-restricted pBC16-1, a miniplasmid of pBC16 (2.7 kb) from *B. cereus* that has lost an *Eco*RI fragment. The shuttle vector pBU4 was used to clone the binary toxin of *B. sphaericus* 1593 yielding the pGSP10 plasmid. This vector was transformed into the *B. thuringiensis* subsp. *israelensis* strain 4QS-72, which only contains the large plasmid encoding the Cyt and Cry endotoxins. Analysis of the recombinant *B. thuringiensis* strain showed that it produces the *B. thuringiensis* subsp. *israelensis* toxins in normal amounts along with the 51.4-kDa and 41.9-kDa proteins from *B. sphaericus*. However, according to results from toxicity tests against *Aedes aegypti*, *Culex pipiens* and *A. stephensi*, the toxic effects of the strain recombinant were not improved.

Similarly, pEG146 and pEG147 shuttle vectors were constructed incorporating the replicon of pBC16 (a 3 kb *Eco*RI fragment) into pUC18. Both shuttle vectors encode resistance to ampicillin and tetracycline but differ with respect to the orientation of the *Bacillus* replicon. Plasmids pEG146 and pEG147 and derivatives showed segregational stability in different *Bacillus* hosts and were used to express successfully insecticidal crystal protein genes (Von Tersch et al. 1991). Also, Lertcanawanichakul and Wiwat (2000) report the construction of a novel shuttle vector, pBCX (6.9 kb) that was constructed from the plasmid pBC16 (4.4 kb) and a 2.5 kb fragment of pBluescript II KS. This vector conferred ampicillin and tetracycline resistance in *E. coli* but only tetracycline resistance in *B. thuringiensis*. To evaluate the functionality of pBCX, a chitinase gene from *B. circulans* 4.1 was inserted into this vector to generate the pBX43 plasmid. The expression and stability of the recombinant plasmid were evaluated in *B. thuringiensis* subsp. *israelensis* strain c4Q272. The chitinase production was compared to pHYB43, a vector previously used to express chitinase genes in *B. thuringiensis* (Wiwat et al. 1996). pBX43 produces 3 times as much chitinase in *B. thuringiensis* subsp. *israelensis* in relation to the production reported for pHYB43.

9.3.3 Shuttle Vectors of B. thuringiensis Contain Replication Origins Derived from Plasmids that Replicate by the Rolling Circle Mechanism

Most of small plasmids of *B. thuringiensis* replicate by the rolling-circle mechanism, which groups them into the family of the highly interrelated RCR plasmids, so called because they use this replication mechanism and show a ssDNA intermediate during their replication process (Andrup et al. 2003). The RCR plasmids contain at least a double-strand replication origin (*dso*), which initiates the replication, and a single-strand origin (*sso*), which starts the generation of the new double strand from the ssDNA intermediate. The replication of RCR plasmids is mediated by a self-encoded Rep protein (Khan 2005). A distinctive characteristic of these plasmids is that they are structurally unstable when are used as cloning vectors; however, there are some efforts in order to test their usefulness in the design of shuttle vectors.

Zhang et al. (2007) cloned and characterized the RCR plasmid pBMB9741 from *B. thuringiensis* subsp. *kurstaki* YBT-1520. This plasmid was cloned into pBMB1105 (a derivate of pHT3101 without the replication region of *B. thuringiensis*) to generate pBMB1197, which was used to transform *B. thuringiensis* strain BMB171 (cured of plasmids) and other strains (including *B. cereus*) to evaluate its replication ability. pBMB1197 was stably maintained after 196–200 generations in both *B. thuringiensis* and *B. cereus* strains making it a suitable tool for *Bacillus* engineering.

Further, Li et al. (2009) reported the cloning and molecular characterization of pK1S-1 plasmid from *B. thuringiensis* subsp. *kurstaki* K1, which belongs to RCR group VII. To determine the replication properties, the 1.6 kb region containing the replicon of pKlS-1 (which contains the Rep protein and *dso*) was cloned into the *B. thuringiensis* pHTIK vector (ori-negative *B. thuringiensis*) and electroporated into a plasmid cured *B. thuringiensis* strain. Transformants were obtained demonstrating its replication ability.

9.4 Transformation of *Bacillus thuringiensis*

The first attempts to transform *B. thuringiensis* occurred before the cloning of the genes encoding the insecticidal crystal proteins; and were performed through conjugation experiments and protoplasts formed by the treatment of cells with lysozyme (Gonzalez and Carlton 1980; Alikhanian et al. 1981; Martin et al. 1981). Further, Heierson et al. (1987) reported the transformation of vegetative cells of *B. thuringiensis* after the induction of bacterial competence with Tris-sucrose. However, these strategies showed several problems such as low transformation efficiency and the difficulty to select transformants.

In the late 1980s of last century the first work where electroporation was used to transform *B. thuringiensis* appeared. The electroporation is a method that temporarily permeabilizes cell membranes to facilitate the entry of DNA. A brief electric pulse is given which causes a disruption of the lipid bilayer, resulting in the formation of aqueous pores in the membrane. Diverse protocols for transformation of *B. thuringiensis* using electroporation were published independently in the same year (Bone and Ellar 1989; Lereclus et al. 1989; Mahillon et al. 1989; Masson et al. 1989; Schurter et al. 1989). These studies showed that the transformation efficiency varies depending on the recipient strain, which may reflect differences in the restriction/modification systems of the strains. Efficiency varies between 10^2 and 10^5 transformants per μg DNA.

Further, Macaluso and Mettus (1991) showed that the transformation efficiency of *B. thuringiensis* depends on the source of plasmid DNA, being more efficient when non-methylated plasmid DNA is used. Recently, Peng et al. (2009) analyzed the effect of several conditions (DNA desalting, cell growth conditions, electroporation solutions, and electric fields) on transformation of *B. thuringiensis* in order to improve the efficiency. They reported transformation efficiency of 2×10^{10} per μg DNA with pHT304 vector, which is 10^4 times higher than previously reported. In conclusion, electroporation provided high transformation efficiency and made transformants easy to recognize and recover by using adequate selectable markers.

References

Alikhanian SI, Ryabchenko NF, Bukanov NO, Sakanyan VA (1981) Transformation of *Bacillus thuringiensis* subsp. *galleria* protoplasts by plasmid pBC16. J Bacteriol 146:7–9

Andrup L, Jensen GB, Wilcks A, Smidt L, Hoflack L, Mahillon J (2003) The patchwork nature of rolling-circle plasmids: comparison of six plasmid from two distinct *Bacillus thuringiensis* serotypes. Plasmid 49:205–232. doi:10.1016/S0147-619X(03)00015-5

Arantes O, Lereclus D (1991) Construction of cloning vectors for *Bacillus thuringiensis*. Gene 108:115–119. doi:10.1016/0378-1119(91)90495-W

Aronson AI (1993) The two faces of *Bacillus thuringiensis*: insecticidal proteins and post-exponential survival. Mol Microbiol 7:489–496. doi:10.1111/j.1365-2958.1993.tb01139.x

Battisti L, Green BD, Thorne CB (1985) Mating system for transfer of plasmids among *Bacillus anthracis*, *Bacillus cereus*, and *Bacillus thuringiensis*. J Bacteriol 162:543–550

Baum J, Gilbert MP (1991) Characterization and comparative sequence analysis of replication origins from three large *Bacillus thuringiensis* plasmids. J Bacteriol 173:5280–5289

Baum J, Coyle DM, Gilbert MP, Jany CS, Gawron-Burke C (1990) Novel cloning vectors for *Bacillus thuringiensis*. Appl Environ Microbiol 56:3420–3428

Bone EJ, Ellar DJ (1989) Transformation of *Bacillus thuringiensis* by electroporation. FEMS Microbiol Lett 58:171–178. doi:10.1111/j.1574-6968.1989.tb03039.x

Bourgouin C, Delécluse A, De La Torre F, Szulmajster J (1990) Transfer of the toxin protein genes of *Bacillus sphaericus* into *Bacillus thuringiensis* subsp. *israelensis* and their expression. Appl Environ Microbiol 56:340–344

del Solar G, Espinosa M (2000) Plasmid copy number control: an ever-growing story. Mol Microbiol 37:492–500. doi:10.1046/j.1365-2958.2000.02005.x

del Solar G, Giraldo R, Ruiz-Echevarria MJ, Espinosa M, Diaz-Orejas R (1998) Replication and control of circular bacterial plasmids. Microbiol Mol Biol Rev 62:434–464

Gamel PH, Piot JC (1992) Characterization and properties of a novel plasmid vector for *Bacillus thuringiensis* displaying compatibility with host plasmids. Gene 120:17–26. doi:10.1016/0378-1119(92)90004-9

Gonzalez JM, Carlton BC (1980) Patterns of plasmid DNA in crystalliferous and acrystalliferous strains of *Bacillus thuringiensis*. Plasmid 3:92–98. doi:10.1016/S0147-619X(80)90038-4

Gruss A, Ehrlich SD (1989) The family of highly interrelated single-stranded deoxyribonucleic acid plasmids. Microbiol Rev 53:231–241

Heierson A, Landén R, Lövgren A, Dalhammar G, Boman HG (1987) Transformation of vegetative cells of *Bacillus thuringiensis* by plasmid DNA. J Bacteriol 169:1147–1152

Höfte H, Whiteley HR (1989) Insecticidal crystal proteins of *Bacillus thuringiensis*. Microbiol Rev 53:242–255

Kanda K, Tan Y, Aizawa K (1989) A novel phage genome integrated into a plasmid in *Bacillus thuringiensis* strain AF101. J Gen Microbiol 135:3035–3041. doi:10.1099/00221287-135-11-3035

Khan SA (1997) Rolling-circle replication of bacterial plasmids. Microbiol Mol Biol Rev 61:442–455

Khan SA (2005) Plasmid rolling-circle replication: highlights of two decades of research. Plasmid 53:126–136. doi:10.1016/j.plasmid.2004.12.008

Lereclus D, Arantes O (1992) *spbA* locus ensures the segregational stability of pHT1030, a novel type of gram-positive replicon. Mol Microbiol 6:35–46. doi:10.1111/j.1365-2958.1992.tb00835.x

Lereclus D, Lecadet MM, Ribier J, Dedonder R (1982) Molecular relationships among plasmids of *Bacillus thuringiensis*: conserved sequences through 11 crystalliferous strains. Mol Gen Genet 186:391–398. doi:10.1007/BF00729459

Lereclus D, Mahillon J, Menou G, Lecadet MM (1986) Identification of Tn4430, a transposon of *Bacillus thuringiensis* functional in *Escherichia coli*. Mol Gen Genet 204:52–57. doi:10.1007/BF00330186

Lereclus D, Guo S, Sanchis V, Lecadet MM (1988) Characterization of two *Bacillus thuringiensis* plasmids whose replication is thermo sensitive in *Bacillus subtilis*. FEMS Microbiol Lett 49:417–422. doi:10.1111/j.1574-6968.1988.tb02768.x

Lereclus D, Arantes O, Chaufaux J, Lecadet MM (1989) Transformation and expression of a cloned endotoxin gene in *Bacillus thuringiensis*. FEMS Microbiol Lett 60:211–218. doi:10.1111/j.1574-6968.1989.tb03448.x

Lertcanawanichakul M, Wiwat C (2000) Improved shuttle vector for expression of chitinase gene in *Bacillus thuringiensis*. Lett Appl Microbiol 31:123–128. doi:10.1046/j.1365-2672.2000.00777.x

Levinson BL, Kasyan KJ, Chiu SS, Currier TS, Gonzalez JM (1990) Identification of β-exotoxin production, plasmids encoding β-exotoxin and a new exotoxin in *Bacillus thuringiensis* by using high-performance liquid chromatography. J Bacteriol 172:3172–3179

Li MS, Roh JY, Tao X, Yu ZN, Liu ZD, Liu Q, Xu HG, Shim HJ, Kim YS, Wang Y, Choi JY, Je YH (2009) Cloning and molecular characterization of a novel rolling-circle replicating plasmid, pK1S-1, from *Bacillus thuringiensis* subsp. *kurstaki* K1. J Microbiol 47:466–472. doi:10.1007/s12275-009-0020-2

Liu X, Peng D, Luo Y, Ruan L, Yu Z, Sun M (2009) Construction of an *Escherichia coli* to *Bacillus thuringiensis* shuttle vector for large DNA fragments. Appl Microbiol Biotechnol 82:765–772. doi:10.1007/s00253-008-1854-y

Loeza-Lara PD, Benintende G, Cozzi J, Ochoa-Zarzosa A, Baizabal-Aguirre VM, Valdez-Alarcón JJ, López-Meza JE (2005) The plasmid pBMBt1 from *Bacillus thuringiensis* subsp. *darmstadiensis* (INTA Mo14-4) replicates by the rolling circle mechanism and encodes a novel insecticidal crystal protein-like gene. Plasmid 54:229–240. doi:10.1016/j.plasmid.2005.04.003

López-Meza JE, Barboza-Corona JE, del Rincón-Castro C, Ibarra JE (2003) Sequencing and characterization of plasmid pUIBI-1 from *Bacillus thuringiensis* serovar *entomocidus* LBIT-113. Curr Microbiol 47:395–399. doi:10.1007/s00284-003-4041-5

Macaluso A, Mettus AM (1991) Efficient transformation of *Bacillus thuringiensis* requires nonmethylated plasmid DNA. J Bacteriol 173:1353–1356

Madsen S, Andrup L, Boe L (1993) Fine mapping and DNA sequence of replication functions of *Bacillus thuringiensis* plasmid pTX14-3. Plasmid 30:119–130. doi:10.1006/plas.1993.1039

Mahillon J, Seurinck J, Delcour J, Zabeau M (1987) Cloning and nucleotide sequence of different iso-IS231 elements and their structural association with the Tn4430 transposon in *Bacillus thuringiensis*. Gene 51:187–196. doi:10.1016/0378-1119(87)90307-6

Mahillon J, Hespel F, Pierssens AM, Delcour J (1988) Cloning and partial characterization of three small cryptic plasmids from *Bacillus thuringiensis*. Plasmid 19:169–173. doi:10.1016/0147-619X(88)90056-X

Mahillon J, Chungiatupornchai W, Decock J, Dierickx S, Michiels F, Peferoen M, Joos H (1989) Transformation of *Bacillus thuringiensis* by electroporation. FEMS Microbiol Lett 60:205–210. doi:10.1111/j.1574-6968.1989.tb03447.x

Martin PA, Lohr JR, Dean DH (1981) Transformation of *Bacillus thuringiensis* protoplasts by plasmid deoxyribonucleic acid. J Bacteriol 145:980–983

Masson L, Prefontaine G, Brousseau R (1989) Transformation of *Bacillus thuringiensis* vegetative cells by electroporation. FEMS Microbiol Lett 60:273–278. doi:10.1111/j.1574-6968.1989.tb03485.x

McDowell DG, Mann NH (1991) Characterization and sequence analysis of a small plasmid from *Bacillus thuringiensis* var. *kurstaki* HD1-DIPEL. Plasmid 25:113–120. doi:10.1016/0147-619X(91)90022-O

Mesrati LA, Karray MD, Tounsi S, Jaoua S (2005) Construction of a new high-copy number shuttle vector of *Bacillus thuringiensis*. Lett Appl Microbiol 41:361–366. doi:10.1111/j.1472-765X.2005.01733.x

Park HW, Ge B, Bauer LS, Federici BA (1998) Optimization of Cry3A yields in *Bacillus thuringiensis* by use of sporulation-dependent promoters in combination with the STAB-SD mRNA sequence. Appl Environ Microbiol 64:3932–3938

Park HW, Bideshi DK, Federici BA (2003) Recombinant strain of *Bacillus thuringiensis* producing Cyt1A, Cry11B, and the *Bacillus sphaericus* binary toxin. Appl Environ Microbiol 69:1331–1334

Peng D, Luo Y, Guo S, Zeng H, Ju S, Yu Z, Sun M (2009) Elaboration of an electroporation protocol for large plasmids and wild-type strains of *Bacillus thuringiensis*. J Appl Microbiol 106:1849–1858. doi:10.1111/j.1365-2672.2009.04151.x

Schnepf E, Crickmore N, Van Rie J, Lereclus D, Baum J, Feitelson J, Zeigler DR, Dean DH (1998) *Bacillus thuringiensis* and its pesticidal crystal proteins. Microbiol Mol Biol Rev 62:775–806

Schurter W, Geiser M, Mathe D (1989) Efficient transformation of *Bacillus thuringiensis* and *Bacillus cereus* via electroporation: transformation of acrystalliferous strains with a cloned delta-endotoxin gene. Mol Gen Genet 218:177–181

Tang M, Bideshi DK, Park HW, Federici BA (2006) Minireplicon from pBtoxis of *Bacillus thuringiensis* subsp. *israelensis*. Appl Environ Microbiol 72:6948–6954. doi:10.1128/AEM.00976-06

Von Tersch MA, Robbins HL, Jany CS, Johnson TB (1991) Insecticidal toxins from *Bacillus thuringiensis* subsp. *kenyae*: gene cloning and characterization and comparison with *B. thuringiensis* subsp. *kurstaki* CryIA(c) toxins. Appl Environ Microbiol 57:349–358

Wilcks A, Smidt L, Okstad OA, Kolsto AB, Mahillon J, Andrup L (1999) Replication mechanism and sequence analysis of the replicon of pAW63, a conjugative plasmid from *Bacillus thuringiensis*. J Bacteriol 181:3193–31200

Wiwat C, Lertcanawanichakul M, Siwayapram P, Pantuwatana S, Bhumiratana A (1996) Expression of chitinase-encoding genes from *Aeromonas hydrophila* and *Pseudomonas hydrophila* in *Bacillus thuringiensis* subsp. *israelensis*. Gene 179:119–126. doi:10.1016/S0378-1119(96)00575-6

Zhang Q, Sun M, Xu Z, Yu Z (2007) Cloning and characterization of pBMB9741, a native plasmid of *Bacillus thuringiensis* subsp. *kurstaki* strain YBT-1520. Curr Microbiol 55:302–307. doi:10.1007/s00284-006-0623-3

Chapter 10
Construction and Application in Plasmid Vectors of *Bacillus cereus* Group

Chengchen Xu, Yan Wang, Chan Yu, Lin Li, Minshun Li, Jin He, Ming Sun and Ziniu Yu

Abstract The *B. cereus* group (*B. cereus sense lato*) is a subgroup of related *Bacillus* containing six species: *B. thuringiensis*, *B. anthracis*, *B. cereus*, *B. weihenstephanensis*, *B. mycoides* and *B. pseudomycoides*. The first three species are opportunistic or pathogenic to insects or mammals, whereas the last three are generally regarded as nonpathogenic. Different types of plasmid vectors play an important role in studying gene function in genetically engineered bacteria or construction of mutant library. The researchers have constructed a variety of shuttle vectors in two or more different host strains. Expression vectors are widely used in construction of highly effective engineered strains. Resolution vectors can site-specifically transfer recombinant genes. Integration vectors could integrate foreign genes into the chromosome or plasmid of host strains and are free of antibiotic resistance genes, therefore eliminating security risks. In recent years, researchers have made great progress in the development of genetic manipulation vectors. Here we summarize latest progress in vector construction and their application in *Bacillus cereus* group.

Keywords *B. cereus* group · Shuttle vector · Integration vector · Resolution vector · Plasmid

10.1 Introduction

B. cereus group, a subgroup of *Bacillus*, includes *B. thuringiensis*, *B. anthracis*, *B. cereus*, *B. mycoides*, *B. pseudomycoides*, and *B. weihenstephanesis*. *B. cereus* groups are usually Gram-positive, rod-shaped, aerobic, and spore-forming (Kolsto et al. 2009). They have similar cell structure, physiological and biochemical characteristics, and genetic exchange systems (Koehler 2009).

B. cereus group is closely related to human activity. *B. anthracis* is a zoonotic pathogen and its pathogenic factors include toxins and capsules, encoded by the

Z. Yu (✉) · C. Xu · Y. Wang · C. Yu · L. Li · M. Li · J. He · M. Sun
State Key Laboratory of Agriculture Microbiology, National Engineering Research Center of Microbial Pesticides, College of Life Science and Technology, Huazhong Agricultural University, 430070 Wuhan, Hubei Province, P.R. China
e-mail: yz41@mail.hzau.edu.cn

E. Sansinenea (ed.), *Bacillus thuringiensis Biotechnology*,
DOI 10.1007/978-94-007-3021-2_10,

plasmids pXO1 and pXO2, respectively (Read et al. 2003). The toxins include the protective antigen (PA), lethal factor (LF), and edema factor (EF). The lethal toxin formed by the combination of PA and LF enters the cytoplasm through receptor-mediated endocytosis and causes toxicity (Barth et al. 2004). *B. anthracis* is a potent zoonotic pathogen. It can be transmitted to humans and animals in different ways, causing pneumonia, enteritis, and sepsis. It is a serious threat to both livestock production and human health (Koehler 2009). *B. anthracis* has also been used as a biological weapon in acts of bio-terrorism, which poses a great threat to mankind (Rao et al. 2010). Other PAs are non-toxic and can induce protective immunity. Many of these are the main components in vaccine preparations (Chitlaru and Shafferman 2009). All of these factors have made *B. anthracis* a hot topic in life science research.

B. thuringiensis is an insect pathogen. The microbial insecticide produced with *B. thuringiensis* is most widely used in the world. The insecticidal crystal proteins of *B. thuringiensis* have been used successfully in transgenic insect-resistant crops and play extremely important roles in pest control (Soberon et al. 2007). The main insecticidal component of *B. thuringiensis* is composed of insecticidal crystal proteins encoded by *cry* and *cyt*, VIP, and a nucleotide analogue thuringiensin (thu). Sixty-eight groups of ICP genes have been identified in 500 species. Of these, three groups in 35 species carry the *cyt* gene.

B. cereus is an opportunistic pathogen that contaminates food and can cause diarrhea-predominant and vomiting-predominant food poisoning. The thermostable toxins produced by *B. cereus* are cyclic dodecapeptide, which causes vomiting, while a variety of enterotoxins produced by *B. cereus* can cause diarrhea (Schoeni and Wong 2005).

B. weihenstephanesis is a psychrotolerant bacterium. It can grow at temperatures below 7°C (Soufiane and Cote 2010). *B. mycoides* typically forms non-pathogenic, rootlike colonies (Kolsto et al. 2009).

The vectors used in the *B. cereus* group research are usually various types of plasmid vectors, which are used to construct engineered *B. anthracis* bacteria and mutant libraries and study gene functions. *B. thuringiensis* plasmid vectors are used to expand insecticidal spectra and enhance virulence, while *B. anthracis* and *B. cereus* plasmid vectors are useful in the study of human pathogens.

B. thuringiensis, *B. anthracis*, and *B. cereus* in the *B. cereus* group are closely related evolutionarily and have been widely reported to be highly homogeneous at the genomic level (Slamti et al. 2004). Great advances have been made in the genetic manipulation of *B. cereus* group and the construction of applicable vectors. This review provides an overview of current research and applications of *B. cereus* group plasmid vectors and prospects for further development.

10.2 Shuttle Plasmid Vectors in the *B. cereus* Group

Manually constructed shuttle vectors have two different replication origins and selection markers and they can survive or replicate in two different types of host cells, which from origin cells and *E. coli*.

Several shuttle vectors can be used in the *B. cereus* group research. Some of these shuttle vectors come from *B. thuringiensis*. For example, shuttle vectors pEG597, pEG853, and pEG854 are constructed with replicons cloned from 65-, 67-, and 91-kb plasmids of *B. thuringiensis* subsp. *kurstaki* HD-263. A variety of insecticidal crystal proteins can be expressed in *B. thuringiensis* through the mediation of these shuttle vectors. The shuttle vector pHT3101 is constructed with a 2.9-kd DNA fragment carrying replication regions with plasmid pHT1030, which has been widely and successfully used in the cloning of *B. thuringiensis* and *B. subtilis*. Additional shuttle vectors pHT304, pHT315, and KHT370, which each has a different copy number, are constructed using a similar approach. They have not only been used in *B. thuringiensis* gene cloning and expression but also in the study of the effects of plasmid copy numbers on insecticidal crystal proteins expression.

Advances in *B. cereus* group study have increased the number of shuttle vectors in gene transfer and expression. To tackle the difficulties of stable expression of the chitinase gene in *B. thuringiensis*, the shuttle vector pBcX was constructed by recombining the 2.56-kb fragment of plasmid pBluescript II KS with plasmid pBc16. This is useful in cloning foreign DNA fragments (Lertcanawanichakul and Wiwat 2000). Sun et al. used a shuttle vector constructed using the small plasmid pBMB2062 cloned from *B. thuringiensis* subsp *kurstaki* strain YBT-1520 and was found capable of stabilizing the inheritance and expression of insecticidal crystal proteins (Sun et al. 2000).

Mesrati et al. (2005) connected the replicon of a high-copy, endogenous plasmid from *B. thuringiensis* to pUC18 of *E. coli* and constructed a short, stable, high-copy shuttle plasmid vector, pHBLBIV, which was shown the capable of transferring between *B. thuringiensis* strains and of expressing pesticidal crystal protein genes (vip) in trophophase. In 2009, Xia et al. (2009a) from the Seoul National University College of Agriculture and Life Sciences inserted *cry*1Ac into the *B. thuringiensis-E. coli* shuttle vector pHT1K, and then the *B. thuringiensis* shuttle vector pHT1K-1Ac. pHT1K-1Ac was successfully expressed in *Bacillus velezensis*, a species known with strong antifungal activity relevant to many plant diseases. This equipped recombinant strain with increasing toxicity, which made it possible to control both pests and fungal pathogens simultaneously in one plant.

The above mentioned vectors are mainly focus on shuttle function, gene expression, and high copy number. The shuttle vectors of *B. thuringiensis* and *B. cereus* can usually carry DNA fragments no larger than 20 kb, so it is necessary to construct shuttle vectors with higher carrying capacities. The *B. thuringiensis-E. coli* shuttle vector pEMB0557, constructed using the replicon *ori60* cloned from *B. thuringiensis* subsp. *kurstaki* YBT-1520, can carry at fragments of over 70 kb and can successfully express *cry*1B in the mutant strain *B. thuringiensis* BMB171. In addition, this vector can be used effectively in cloning *B. thuringiensis* silencing genes and expressing the gene clusters (Liu et al. 2009).

In recent years, a variety of shuttle vectors from *B. anthracis* and *B. cereus* strains have also been constructed for use in the study of pathogenesis and relevant gene functions. Based on vector pUB110, which has self-replication and shuttle capabilities, two *Bacillus-E. coli* shuttle vectors, pJB1 and pJB2a, were constructed. These vectors can be used in double-screening of the strong promoter of α-amylase and their shuttle function has been verified through electroporation in non-toxic strain of *B. anthracis* (Liu et al. 2006). Pomerantsev et al. (2009) constructed the shuttle plasmid vector pMR, which contained a small replicon, using the large plasmid pXO1 from *B. anthracis*. pMR can be transformed into *B. anthracis* and *B. cereus* respectively with good segregational stability.

Shuttle plasmid vectors are able to replicate in two or more hosts, which expands the range of target gene expression in *Bacillus cereus* flora. This is one reason why they are so widely used. However, there are still problems under actual use. For example, these plasmid vectors can be unstable, and may present safety concerns to the environment as they are likely to introduce resistance genes and other non-essential exogenous genes. In recent years, researchers have constructed many new vectors that are not subject to these problems.

10.3 Integration Vectors in the *B. cereus* Group

B. cereus group plasmids can take on unstable single-strand forms during replication, resulting in the loss of the plasmid vectors. One effective way is to construct integration vectors, which can be used to insert the target gene into the host chromosome or plasmid. This greatly improves the stability of target gene expression and these were turned into an important part of the genetic manipulation method. In *B. cereus* group studies, integration vectors are mainly used to construct knockout mutants and multi-functional engineered bacteria.

10.3.1 Homologous Recombination Integration Vectors

10.3.1.1 Constructed Gene Knock-out Mutants in the *B. cereus* Group

Homologous recombination is an effective tool used in the analysis of genes about unknown function. To construct mutants for such studies, the specific gene is site-specifically knocked out by homologous recombination so that the original gene function is lost. Several traditional methods can be used in gene replacement, but the frequency of transformation is low in *B. cereus* group and the probability of homologous recombination is small. In recent years, a variety of integration vectors have been constructed for generating mutants of *B. anthracis* and *B. cereus*. In particular, temperature-sensitive integration vectors can effectively knock out genes and then identify potential drug targets. For example, the integration vectors pKS1, pHY304, and pKSV7, constructed by inserting the upstream and downstream homologous arms of the target gene and the resistance gene into the shuttle vector, can knock out one or more genes in *B. anthracis* through homologous recombination. Such integration vectors can normally be replicated upon incubation at 30°C, but they are lost at 37–42°C. The strains that lose their plasmids and can still grow on plates are called positive clones (Shatalin and Neyfakh 2005; Jeng et al. 2003; Tang et al. 2009).

Arnaud et al. (2004) constructed the integration vector pMAD can generate mutants quickly. The positive clones that lose their plasmids at elevated temperature can facilitate the identification of mutants through blue/white screening on X-gal plates. The vector pMAD has been used successfully in the knockout of *tetB* in *B. cereus* strain ATCC 14579. Harvie and Ellar (2005) has used the integration vector pMUTIN4 to knock out the operon Fec in *B. cereus*, resulting in mutants unable to grow in environment where ferric citrate is the only source of iron. This operon can be used as a new target for anti-bacterial agents. Most vectors are designed to also leave antibiotic genes in the genomes of mutant strains, which limit the application of recombinant strains. In recent 2 years, a new integration vector, the Cre-LoxP system, which has high efficiency and no antibiotic residue, was constructed and successfully used to knock out *B. anthracis* genes. This integration vector is constructed using the host *B. anthracis* bacterial genome as a template to amplify the upstream and downstream homologous arms and by connecting the antibiotic resistance gene fragments with LoxP sites at both ends. The constructed vector is then transferred into the host and recombinant strains carrying resistance markers are harvested by screening. Plasmids with the Cre recombinase expression are then transferred into the recombinant strains and the resistance markers are removed before knockout mutants are harvested. This vector system has been used to remove a large fragment of *B. anthracis* pXO1 and to knock out the target gene in *B. anthracis* using pMAD as integration vector (Pomerantsev et al. 2009; Shatalin and Neyfakh 2005; Jeng et al. 2003; Tang et al. 2009; Arnaud et al. 2004; Harvie and Ellar 2005; Wang et al. 2009).

10.3.1.2 Constructed Genetically Engineered Bacteria in the *B. cereus* Group

Homologous recombination can be used not only to construct knockout mutants but also to insert new genes by integration, producing strains with new phenotypes. Integration vectors are widely used in the construction of genetically engineered bacteria and can expand insecticidal spectra and improve insecticidal activity by integrating new *insecticidal crystal proteins* into *B. thuringiensis* chromosomes and plasmids. The engineering strain produced by integrating *cry*1Ac into the chromosome of the pine symbiotic *B. cereus* strains 752 through the integration vector pEG601, was found to exert significant insecticidal activity against *Dendrolimus punctatus*. Similarly, integrating *cry*3A into the 75-kb plasmid of HD73 through homologous recombination using the temperature-sensitive vector pRN5101, produced a recombinant strain capable of killing both Lepidoptera and Coleoptera pests (Lereclus et al. 1992). By site-directed integration of *cry*1C into the chromosome using an integration vector carrying the *B. thuringiensis* chromosome fragment, the expression and stability of Cry1C in the recombinant strain can be significantly improved (Kalman et al. 1995). By integrating *cry*8Ca23A into the endogenous

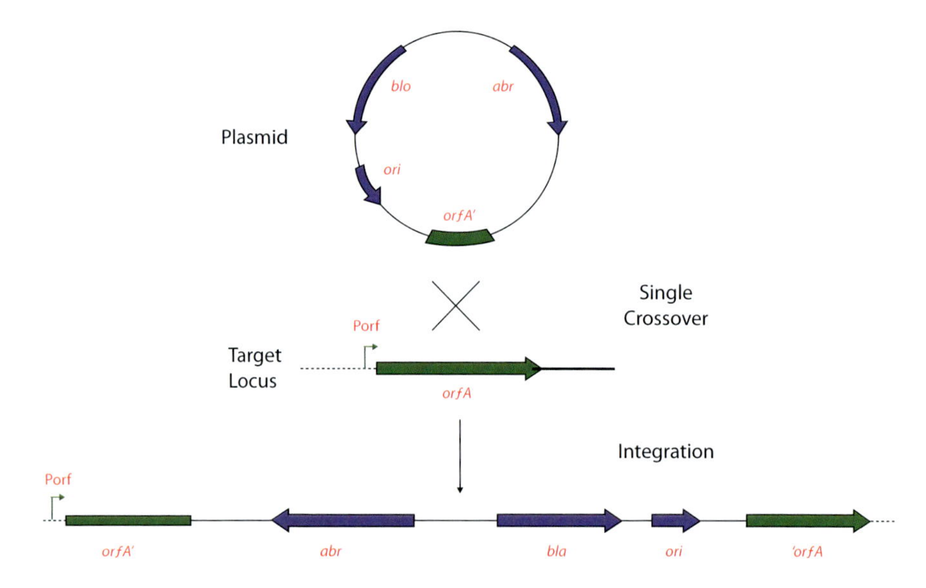

Fig. 10.1 Use of a basic integration vector to construct a knockout mutation in a hypothetical open reading frame, orfA. (Liang et al. 2007)

plasmid using the temperature-sensitive integration vector pMAD, the resulting engineering strain BIOT185 develops high efficiency against *Anomala corpulenta* larvae. In addition, the temperature-sensitive vector of the recombinant strain can be eliminated by raising the incubation temperature so that the new strain does not contain the antibiotic genes or other non-essential gene fragments (see Fig. 10.1). Such recombinant strains may be developed as environment-friendly microbial insecticides (Liu et al. 2010). In 2007, Fen et al. (Liu et al. 2010) by integrating VHb into the chromosomes of the *B. thuringiensis* mutant strain BMB171 using the integration vector pEG491, VHb expression can increase cell density and insecticidal crystal proteins production under low-ventilation in fermentation. The site-specific integration vector pKTF12, constructed using pKSV7, can site-specifically integrate cry1Ac into the *B. thuringiensis* XBU001 chromosome. The *cry*1Ac in the recombinant strain KCTF12 can be inherited and expressed stably, forming rhomboid crystals.

Target genes are imported into integration vectors by homologous recombination and directly integrated into the chromosomes or plasmids, overcoming the limitations of plasmid instability and providing new ideas and methods in the construction of multi-functional engineered bacteria of *B. cereus* group.

10.3.2 Transposon Integration Vectors

Transposon integration vectors use transposable elements to integrate the DNA fragment into the chromosomes or endogenous plasmids of host strain, forming insertional mutants or introducing new functional genes. Transposable elements can be classified into transposons and insertion sequences. Research performed by Qiu et al. has shown that the distribution of insertion sequences in the *B. cereus* group differs greatly between strains, indicating that the insertion and mobile abilities of various *B. cereus* group strains are different and that different strains have preferences for different types of insertion sequences (Qiu et al. 2010). Some insertion sequences, such as IS232, IS10, IS231A, etc., have been used to construct transposon vectors and successfully integrate *insecticidal crystal proteins* into the endogenous plasmids of host strains in *B. thuringiensis*.

The mechanism of transposon integration vectors involves transposase (TnpI) and resolvase (TnpA), which move DNA fragments between inverted repeat sequences at both ends of the transposon to other locations on the genome and integrate DNA fragments carried by the vector into the genome with the assistance of tnpA. Because the transposase gene is not transferred with the target gene, the target gene integrated into the host chromosome or plasmid can be inherited stably. Because of the instability of the vector, transposase is lost when the vector disappears, rendering the engineered bacteria less dangerous.

Many transposon integration vectors have been constructed and used in the study of genetic functions and in the construction of strains for genetic engineering. The transposable elements used in these vectors are usually transposons from host and other *Bacillus* bacteria. Sun et al. constructed the vectors pBMB-R14 and pBMB-F7 using the *B. thuringiensis* transposon Tn4430 (Sun et al. 2000), which integrated *cry*3A and *cry*1C into the *B. thuringiensis* chromosome and were found to be inherited stably and expressed effectively. These vectors could be eliminated by incubating the *B. thuringiensis* recombinant bacteria at high temperatures to avoid second integration. Similar vectors can be constructed using Tn5401. The integration vectors pTYP45 and pTYP26, constructed using the *Streptococcus* transposon Tn917, can integrate *insecticidal crystal proteins* into the *B. thuringiensis* chromosome.

In recent years, integration vectors constructed using transposons Tn10 and Mariner have been used to build *B. anthracis* and *B. thuringiensis* mutant libraries. Mariner is a good-quality transposon mutagen with no host preference. It can be randomly inserted into the coding and noncoding regions of the chromosome or plasmid, which is very beneficial to study of the function of different sites on the genome. Li et al. (2009) constructed the integration vectors pMarA333 and pMarB333 by the Mariner, that can easily detect the insertion site by adding the replication origin of *E. coli* between the two inverted repeats, with the transposition efficiency of 7.6% and 11.6%, respectively, in the mutant *B. thuringiensis* strain BMB171. These two vectors have significant potential application value in the construction of mutant libraries for *B. thuringiensis* and other Gram-positive bacteria. Wilson et al. (2007) constructed two integration vectors, pAW068 and pAW016 by the transposons Mariner and mini-Tn10, have been successfully used to create insertional mutants of *B. anthracis* and identify several unknown regulatory pathways affecting the gene expression relevant to *B. anthracis* virulence. They play important roles in the study of *B. anthracis* pathogenesis. These two vectors have also been widely used to create other Gram-positive insertional mutants.

10.4 Resolution Vectors in the *B. cereus* Group

In order to reduce the risks involved in the creation and use of engineering bacteria, resolution vectors utilizing the site-specific resolution characteristics of transposons are used to construct vectors that can site-specifically transfer recombinant genes. Resolution vectors can automatically resolve and eliminate resistance markers and other non-essential fragments, eliminating the security risks inherent in the release of genetically engineering bacteria. The name "resolution vector" comes from the fact that the key components in this vector are the resolution sites and resolvase. Two plasmids can be formed in this vector after recombination within the strain. Only plasmids containing the *B. thuringiensis* replication regions can survive, while other plasmids are automatically resolved. The *B. thuringiensis* genetically engineering strain WG001, constructed using self-constructed resolution vector, is the first genetically engineered insecticide in China. It has passed national security evaluations and been used as commercial production. The transposons used in the construction of resolution vectors in *B. thuringiensis* are Tn4430 and Tn5401. Wu et al. (2002) constructed the resolution shuttle vector pBMB1205, using the transposon Tn4430, can be transformed into the acrystalliferous mutant strain BMB171 after insertion of the spectinomycin resistance gene. The resistance gene can be resolved and eliminated under the action of the resolvase from the helper plasmid. The plasmid stability after resolution is 93% and the resolution rate is as high as 100%. This vector has several advantages over previous ones. The recombinant system comes from the highly virulent *B. thuringiensis* subsp. *kurstaki* strain and applies to the production of most *B. thuringiensis* preparations. The resolution vector constructed using the replication origin *ori*44 of *B. thuringiensis* subsp. *kurstaki* YBT-1520 plasmid retains its genetic stability after the vector is resolved and is not subject to plasmid incompatibility. However, Tn4430 is widespread in various *B. thuringiensis* subspecies (except *B. thuringiensis* subsp. *israelensis*) and it is difficult for the resolution vector to transform other subspecies of *B. thuringiensis*. There is a greater range of applications of the transposon Tn5401 from *B. thuringiensis* subsp. *morrisoni*.

Wu et al. (2007) constructed the resolution vector pBMB 5401 using transposon Tn5401 and then transformed it into the acrystalliferous mutant after inserting the target gene. The temperature-sensitive helper plasmid pEG922 is then imported into the mutant. The recombinant plasmid is integrated in vivo using integrase, eliminating the resistance genes and other non-essential fragments. The target gene is then expressed in the recombinant strain after resolution.

10.5 High-effective Expression Vectors in the *B. cereus* Group

In the study of target gene function and the construction of highly effective engineered strains, the target gene must be expressed effectively. This requires highly effective expression vectors. This can be done using fusion expression and co-expression, strong promoters, high copy plasmids, auxiliary protein and other methods.

10.5.1 High-effective Expression Vectors Constructed Based on Shuttle Vectors

In a large number of *B. cereus* group studies, a variety of high-expression shuttle vectors have been constructed for the cloning and expression of various functional genes. A variety of highly effective engineering strains have been successfully established.

10.5.1.1 Expression Vectors Constructed by Fusion Protein or Co-expression Methods

The expression vectors constructed using *B. thuringiensis* fusion protein or co-expression methods can enhance the activity and expression of insecticidal crystal proteins. This strategy is most commonly used in the construction of highly effective engineering strains and transgenic plants with highly insecticidal activity.

Some of the plasmid vectors with fusion expression or co-expression of insecticidal crystal proteins and other genes are constructed using the shuttle vector pHT315. These include the serine protease CDEP2 and *cry*1Ac fusion expression vector (Xia et al. 2009a), tobacoo chitinase *tchiB* and *cry*1Ac fusion expression vector (Ding et al. 2008), and *B. thuringiensis* chitinase gene and *cry*1Ac co-expression vector (Hu et al. 2009). All of these have increased expression of Cry1Ac, which increases insecticidal activity in the recombinant strains. The product of the expression of the myelin basic protease gene and *cyt*1Aa fusion expression vector is active in mouse myeloma cells, and it can be developed into effective anti-cancer drugs (Cohen et al. 2007).

Shuttle vector pHT304 has also been used to construct expression shuttle vectors, such as the expression vector pXL43, constructed using the fusion gene containing *cry*1Ac, and the neurotoxin gene, which is transferred into *B. thuringiensis*, leading to profound insecticidal activity in the recombinant strain (Xia et al. 2009b). Driss et al. (2011) constructed the fusion protein gene using the chitinase gene and *cry*1Ac was inserted downstream of the spore-dependent promoter BtI-BtII of *B. thuringiensis-E. coli* shuttle vector pHT*Blue*. The resulting recombinant expression plasmid pF is transferred into *B. thuringiensis* strain BNS3 so that the insecticidal activity of the recombinant strain against the *Ephestia kuehniella* (mediterranean flour moth) increases to 1.5 times than that of the wild strain.

10.5.1.2 Expression Vectors Constructed Using the Particular Promoters

Different gene expression effects may be caused by different expression vectors constructed with different promoters. The non-spore-dependent promoter *pro3a* of the *B. thuringiensis cry*3A gene can initiate transcription in the vegetative stage, causing insecticidal crystal proteins to be expressed early and extending the duration of expression. This has been widely used in the construction of *Bacillus* expression vectors. The prokaryotic expression vector pHT3AG, constructed by cloning the fusion gene containing green fluorescent protein gene *gfp* and the promoter *pro3a* into the shuttle vector pHT304 resulted in expression a large amount of green fluorescent protein and stable products in the vegetative stage.

Two resident strains, CQUBb and CQUBt, were isolated from the intestinal tract of *Apriona Germari Hope* larva. The recombinant bacteria were found capable of expressing GFP in large quantities during trophophase and the expression product showed good stability (Yang et al. 2010). Peng et al. (2010) constructed the recombinant expression vector pBMB1073, using the fusion of promoter *pro3a* and the toxin-binding region of the *Helicoverpa armigera* cadherin fragment (*HaCad1*) with *cry*1Ac, once inserted into the shuttle vector pHT304, is transferred into the *B. thuringiensis* mutant strain BMB171 (Fig. 10.2). The resulting recombinant strain promotes insecticidal activity against both *Helicoverpa armigera* (cotton bollworm) and *Spodoptera exigua* (beet army worm), whose virulence is increased by 5.1 and 6.5 times, respectively, over that of the recombinant strains expressing only the Cry1Ac protein.

Using the promoter from its own genome to construct expression vector can effectively prevent plasmid incompatibility and help vector transformation so that each vector is more stable (Zhang 2007). The α-amylase expression vector pHP14-amy 904, constructed via insertion of a promoter sequence from the thermostable α-amylase gene cloned from *B. cereus* 904 into the shuttle vector pHP14-promp, can be successfully expressed after it is transformed into *B. cereus* 905.

10.5.1.3 Expression Vectors Containing Helper Proteins

The *B. thuringiensis* helper proteins P20 and P19 can promote the expression of ICPs in *B. thuringiensis* (Shi et al. 2006a, b). A large number of studies have been performed on the construction of helper protein expression vectors. The expression vector P19, which contains helper protein, has been used to study the expression of ICPs. It is constructed by insertion of the *cry*11Aa promoter sequence and the ATG start codon from the full-length P19 gene and its coupled downstream gene *cry*11Aa into the shuttle vector pHT3101. The *B. thuringiensis* expression vector pHY2P is constructed by inserting a DNA fragment containing the p19 or p29 tandem genes into the shuttle vector pHT3101.

10.5.2 Other Expression Vectors

In addition, a variety of other expression vectors have also been constructed. *B. thuringiensis*, *B. cereus*, and *B. anthracis* are valuable in agriculture, medicine, and industry. They are easily distinguished from other aerobic *bacillus*, but they are difficult to distinguish from each other. The traditional classification system is also

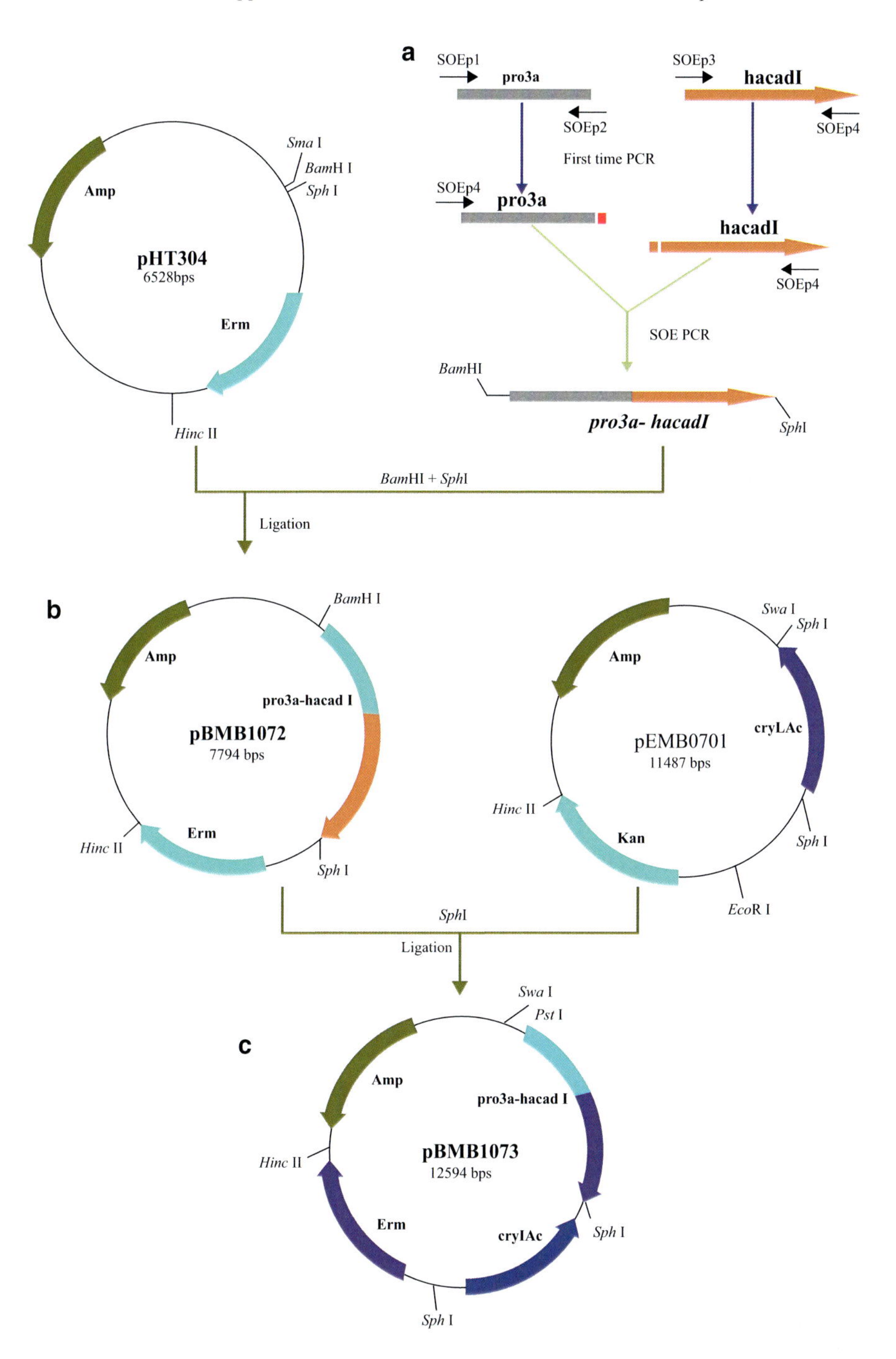

Fig. 10.2 Major steps in the construction of an *E. coli-B. thuringiensis* shuttle vector for co-expressing HaCad1 and Cry1Ac. (Peng et al. 2010)

in dispute. Although their phylogenetic relationship and pathogenicity have been analyzed extensively (see Chap. 6 of this part), the ecological characteristics of these *B. anthracis* bacteria in the natural environment are still poorly understood. As a result, the pAD43-25 vector containing *gfp* is constructed so that GFP can be expressed in *B. cereus* group with GFP acting as a marker in monitoring these strains in special environments and under various physiological conditions (Parente et al. 2008). The GFP-containing expression vectors pGFP5, pGFP8, and pGFP17 are constructed by inserting the *gftmut3a* from the GFP expression vector pGFP78 into shuttle vector pHY300PLY. Bright green fluorescence can be observed after these vectors are imported into the bio-control *B. cereus* B905, which has been used to detect the dynamics of the GFP-labeled recombinant in the environment and control objects (Zhao et al. 2010).

B. thuringiensis protein Cry10Aa has insecticidal activities against mosquitoes and blackflies, but its expression is low, and difficult to clone, express, even crystallize. A large amount of ICP was produced when *cry*10Aa was carried by the expression vector pSTAB, which has a great deal of activity against the *Aedes* (Hernandez-Soto et al. 2009).

The construction of high-expression *B. cereus* group vectors greatly improves the expression of target genes and has become an effective tool in the construction of engineered bacillus bacteria. These construction strategies also inspire the designs of vectors for use in other microorganisms.

10.6 Conclusion

B. cereus group includes microorganisms that have a major impact on human activity. Researchers have constructed many different types of vectors for the study of gene function in *B. thuringiensis*, *B. anthracis*, and *B. cereus* and for construction of a variety of engineering bacteria for various purposes. However, some vectors still have a limited scope of application and cannot be used widely. Future studies may find common promoters and construct vectors that can be cloned, expressed, and integrated into a variety of microorganisms and reduce the difficulty of genetic manipulation.

There are several methods that can potentially improve target gene expression through the construction of plasmid vectors with higher plasmid copy numbers and stronger promoters. However, there is still a dearth of highly expressed vectors in B. cereus group. It is therefore necessary to find new promoters.

We have sequenced the whole genomes of *B. thuringiensis* strains CT-43, YBT1520, YBT020, and BMB171 (He et al. 2010, 2011; Zhu et al. 2011). We anticipate that new promoters will be found among a large number of plasmid sequences here, and that these promoters will be used as construct efficient expression vectors. In the construction of genetically engineered *bacillus* bacteria, resolution vectors can be used to eliminate resistance markers and other non-essential fragments so as to relieve the security risks inherent in the environmental impact by those bacteria. This will play a major role in the development of genetic engineering research practices. The number of resolution vectors will increase, and they will become applicable in new ways. Study of the pathogenesis of *B. anthracis* and its relevant gene functions is of great significance, and it will require the development of integration vectors in the establishment of practical and effective approaches of gene knockout. A variety of vectors can be used to construct knockout mutants in *B. anthracis* and *B. cereus*, but vectors currently available in *B. thuringiensis* are limited. Efficient, site-specific *B. thuringiensis* mutants will bring new breakthroughs in the study of the functions of many regulatory proteins and new ways of expressing and synthesizing ICPs.

Acknowledgments This work received the financial support from the National Natural Science Foundation of China (Grant No. 30930004 and 30900016).

References

Arnaud M, Chastanet A, Debarbouille M (2004) New vector for efficient allelic replacement in naturally nontransformable, low-GC-content, gram-positive bacteria. Appl Environ Microbiol 70(11):6887–6891

Barth H, Aktories K, Popoff MR, Stiles BG (2004) Binary bacterial toxins: biochemistry, biology, and applications of common *Clostridium* and *Bacillus* proteins. Microbiol Mol Biol Rev 68(3):373–402

Chitlaru T, Shafferman A (2009) Proteomic studies of *Bacillus anthracis*. Future Microbiol 4:983–998

Cohen S, Cahan R, Ben-Dov E, Nisnevitch M, Zaritsky A, Firer MA (2007) Specific targeting to murine myeloma cells of Cyt1Aa toxin from *Bacillus thuringiensis* subsp *israelensis*. J Biol Chem 282(39):28301–28308

Ding XZ, Luo ZH, Xia LQ, Gao B, Sun YJ, Zhang YM (2008) Improving the insecticidal activity by expression of a recombinant *cry*1Ac gene with chitinase-encoding gene in acrystalliferous *Bacillus thuringiensis*. Curr Microbiol 56(5):442–446

Driss F, Rouis S, Azzouz H, Tounsi S, Zouari N, Jaoua S (2011) Integration of a recombinant chitinase into *Bacillus thuringiensis* parasporal insecticidal crystal. Curr Microbiol 62(1):281–288

Harvie DR, Ellar DJ (2005) A ferric dicitrate uptake system is required for the full virulence of *Bacillus cereus*. Curr Microbiol 50(5):246–250

He J, Shao X, Zheng H, Li M, Wang J, Zhang Q, Li L, Ziduo L, Sun M, Wang S, Yu Z (2010) The complete genome sequence of *Bacillus thuringiensis* mutant strain BMB171. J Bacteriol 192(15):4074–4075

He J, Wang J, Yin W, Shao X, Zheng H, Li M, Zhao Y, Sun M, Wang S, Yu Z (2011) The complete genome sequence of *Bacillus thuringiensis* subsp. *chinensis* strain CT-43. J Bacteriol 193(13):3407–3408

Hernandez-Soto A, Del Rincon-Castro MC, Espinoza AM, Ibarra JE (2009) Parasporal body formation via overexpression of the Cry10Aa toxin of *Bacillus thuringiensis* subsp. *israelensis*, and Cry10Aa-Cyt1Aa synergism. Appl Environ Microbiol 75(14):4661–4667

Hu SB, Liu P, Ding XZ, Yan L, Sun YJ, Zhang YM, Li WP, Xia LQ (2009) Efficient constitutive expression of chitinase in the mother cell of *Bacillus thuringiensis* and its potential to enhance the toxicity of Cry1Ac protoxin. Appl Microbiol Biotechnol 82(6):1157–1167

Jeng A, Sakota V, Li Z, Datta V, Beau B, Nizet V (2003) Molecular genetic analysis of a group A *Streptococcus* operon encoding serum opacity factor and a novel fibronectin-binding protein, SfbX. J Bacteriol 185(4):1208–1217

Kalman S, Kiehne KL, Cooper N, Reynoso MS, Yamamoto T (1995) Enhanced production of insecticidal proteins in *Bacillus thuringiensis* strains carrying an additional crystal protein gene in their chromosomes. Appl Environ Microbiol 61(8):3063–3068

Koehler TM (2009) *Bacillus anthracis* physiology and genetics. Mol Aspects Med 30(6):386–396

Kolsto AB, Tourasse NJ, Okstad OA (2009) What sets *Bacillus anthracis* apart from other *Bacillus* species? Annu Rev Microbiol 63:451–476

Lereclus D, Vallade M, Chaufaux J, Arantes O, Rambauds T (1992) Expansion of insecticidal host range of *Bacillus thuringiensis* by in vivo genetic recombination. Biotechnology 10(4):418–421

Lertcanawanichakul M, Wiwat C (2000) Improved shuttle vector for expression of chitinase gene in *Bacillus thuringiensis*. Lett Appl Microbiol 31(2):123–128

Li M, Li M, Yin W, He J, Yu Z (2009) Two novel transposon delivery vectors based on mariner transposon for random mutagenesis of *Bacillus thuringiensis*. J Microbiol Methods 78(2):242–244

Liang F, Shouwen C, Ming S, Yu Z (2007) Expression of vitreoscilla hemoglobin in *Bacillus thuringiensis* improve the cell density and insecticidal crystal proteins yield. Appl Microbiol Biotechnol 74(2):390–397

Liu RN, Huang HJ, Xu JH (2006) Construction of shuttle vector of anthrax protective antigen with upstream strong promoter. China Trop Med 9:1546–1548

Liu X, Peng D, Luo Y, Ruan LF, Yu ZN, Sun M (2009) Construction of an *Escherichia coli* to *Bacillus thuringiensis* shuttle vector for large DNA fragments. Appl Microbiol Biotechnol 82(4):765–772

Liu JJ, Yan G, Shu C, Zhao C, Liu CQ, Song FP, Zhou L, Ma JL, Zhang J, Huang DF (2010) Construction of a *Bacillus thuringiensis* engineered strain with high toxicity and broad pesticidal spectrum against coleopteran insects. Appl Microbiol Biotechnol 87(1):243–249

Mesrati LA, Karray MD, Tounsi S, Jaoua S (2005) Construction of a new high-copy number shuttle vector of *Bacillus thuringiensis*. Lett Appl Microbiol 41(4):361–366

Parente AF, Silva-Pereira I, Baldani JI, Tiburcio VH, Bao SN, De-Souza MT (2008) Construction of *Bacillus thuringiensis* wild-type S76 and Cry-derivatives expressing a green fluorescent protein: two potential marker organisms to study bacteria-plant interactions. Can J Microbiol 54(9):786–790

Peng D, Xu X, Ruan L, Yu ZN, Sun M (2010) Enhancing Cry1Ac toxicity by expression of the *Helicoverpa armigera* cadherin fragment in *Bacillus thuringiensis*. Res Microbiol 161(5):383–389

Pomerantsev AP, Camp A, Leppla SH (2009) A new minimal replicon of *Bacillus anthracis* plasmid pXO1. J Bacteriol 191(16):5134–5146

Qiu N, He J, Wang Y, Cheng G, Li MS, Sun M, Yu ZN (2010) Prevalence and diversity of insertion sequences in the genome of *Bacillus thuringiensis* YBT-1520 and comparison with other *Bacillus cereus* group members. FEMS Microbiol Lett 310(1):9–16

Rao SS, Mohan KV, Atreya CD (2010) Detection technologies for *Bacillus anthracis*: prospects and challenges. J Microbiol Methods 82(1):1–10

Read TD, Peterson SN, Tourasse N, Baillie LW, Paulsen IT, Nelson KE, Tettelin H, Fouts DE, Eisen JA, Gill SR, Holtzapple EK, Okstad OA, Helgason E, Rilstone J, Wu M, Kolonay JF, Beanan MJ, Dodson RJ, Brinkac LM, Gwinn M, Deboy RT, Madpu R, Daugherty SC, Durkin AS, Haft DH, Nelson WC, Peterson TD, Pop M, Khouri HM, Radune D, Benton JL, Mahamoud Y, Jiang L, Hance IR, Weidman TF, Berry KJ, Plaut RD, Wolf AM, Watkins KL, Nierman WC, Hazen A, Cline R, Redmond C, Thwaite JE, White O, Salzberg SL, Thomason B, Friedlander AM, Koehler TM, Hanna PC, Kolsto AB, Fraser CM (2003)The genome sequence of *Bacillus anthracis* Ames and comparison to closely related bacteria. Nature 423(6935):81–86

Schoeni JL, Wong AC (2005) *Bacillus cereus* food poisoning and its toxins. J Food Prot 68(3):636–648

Shatalin KY, Neyfakh AA (2005) Efficient gene inactivation in *Bacillus anthracis*. FEMS Microbiol Lett 245(2):315–319

Shi YX, Zeng SL, Yuan MJ, Sun F, Pang Y (2006a) Influence of accessory protein P19 from *Bacillus thuringiensis* on insecticidal crystal protein Cry11Aa. Acta Microbiol Sin 46(3):353–357

Shi YX, Yuan MJ, Chen JW, Sun F, Pang Y (2006b) Effects of helper protein P20 from *Bacillus thuringiensis* on vip3A expression. Acta Microbiol Sin 46(1):85–89

Slamti L, Perchat S, Gominet M, Boas GV, Fouet A, Mock M, Sanchis V, Chaufaux J, Gohar M, Lereclus D (2004) Distinct mutations in PlcR explain why some strains of the *Bacillus cereus* group are nonhemolytic. J Bacteriol 186(11):3531–3538

Soberon M, Pardo-Lopez L, Lopez I, Gomez I, Tabashnik BE, Bravo A (2007) Engineering modified *Bacillus thuringiensis* toxins to counter insect resistance. Science 318(5856):1640–1642

Soufiane B, Cote JC (2010) *Bacillus thuringiensis* serovars *bolivia*, *vazensis* and *navarrensis* meet the description of *Bacillus weihenstephanensis*. Curr Microbiol 60(5):343–349

Sun M, Wei F, Liu ZD, Yu ZN (2000) Cloning of plasmid pBMB2062 in *Bacillus thuringiensis* strain YBT-1520 and construction of plasmid vector with genetic stability. Acta Genat Sin 27(10):932–938

Tang HM, Liu XK, Gao MQ, Feng EL, Zhu L, Shi ZX, Liao XR, Wang HL (2009) Construction of putative S-layer protein deletion mutant in *Bacillus anthracis*. Lett Biotechnol 20(2):161–165

Wang YC, Jiang N, Zhang DW, Qiu Y, Yuan SL, Tao HX, Wang LC, Zhang ZS, Liu CJ (2009) Application of Cre-Loxp system in the gene knockout of *Bacillus anthracis* and the knockout of eag gene. Prog Biochem Biophys 36(7):934–940

Wilson AC, Perego M, Hoch JA (2007) New transposon delivery plasmids for insertional mutagenesis in *Bacillus anthracis*. J Microbiol Methods 71(3):332–335

Wu L, Sun M, Zhu CG, Zhang L, Yu ZN (2002) A novel resolution vector with *Bacillus thuringiensis* plasmid replicon *ori*44. Chin J Biotechnol 18(3):335–338

Wu H, Ye W, Yu Z, Sun M (2007) High efficient and stable expression of AiiA protein in *Bacillus thuringiensis*. Sci Agric Sin 40(10):2221–2226

Xia L, Zeng Z, Ding X, Huang F (2009a) The expression of a recombinant *cry*1Ac gene with subtilisin-like protease CDEP2 gene in acrystalliferous *Bacillus thuringiensis* by Red/ET homologous recombination. Curr Microbiol 59(4):386–392

Xia L, Long X, Ding X, Zhang YM (2009b) Increase in insecticidal toxicity by fusion of the cry1Ac gene from *Bacillus thuringiensis* with the neurotoxin gene *hwtx*-I. Curr Microbiol 58(1):52–57

Yang H, Rong R, Song FP, Sun CP, Wei J, Zhang J, Huang DF (2010) In vivo fluorescence observation of parasporal inclusion formation in *Bacillus thuringiensis*. Sci China Life Sci 53(9):1106–1111

Zhang LL (2007) Construction of engineering *Bacillus cereus* strain and expressing α-amylase gene. Master Degree Thesis. Inner Mongolia Agricultural University

Zhao JJ, Wang S, Wang Q, Mei R (2010) Cloning of promoter fragment from *Panebacillus polymyxa* M-1 and expression of its green fluorescent protein. J Agric Biotechnol 04:788–792

Zhu Y, Shang H, Zhu Q, Ji F, Wang P, Fu J, Deng Y, Xu C, Ye W, Zheng J, Zhu L, Ruan L, Peng D, Sun M (2011) Complete genome sequence of *Bacillus thuringiensis* serovar *finitimus* strain YBT-020. J Bacteriol 193(9):2379–2380

Chapter 11
Recombination in *Bacillus thuringiensis*

Lobna Abdelkefi-Mesrati and Slim Tounsi

Abstract Recombination plays a critical role in maintaining gene diversification and genome stability. There are two types of recombination: the site specific recombination and the homologous recombination. In this chapter we present the genetic recombination in *Bacillus thuringiensis*. The first part describes the site specific recombination, including transposition by transposons and transduction by phage, in this bacterium and its exploitation in the construction of recombinant strains of *B. thuringiensis* improving their production as bioinsecticides and their insecticidal activities and *B. thuringiensis* mutagenesis. In the second part we are interested by the homologous recombination and its role in the construction of improved *B. thuringiensis* strains and in gene disruption.

Keywords Site specific recombination · Transduction · Transposition · Homologous recombination · Transformation

11.1 Introduction

Recombination is a process or set of processes by which DNA molecules interact with one another to bring about a rearrangement of the genetic information or content in an organism. It plays a critical role in maintaining gene diversification and genome stability. In eukaryotic lineages, recombination involves meiosis and fertilization. It may seem somewhat surprising that bacteria can undergo recombination since the latter process requires two homologous DNA molecules; and bacteria have only one chromosome (and are therefore haploid). However, bacteria have mechanisms by which they can acquire extra DNA, which creates opportunities for recombination to occur. In fact, bacterial recombination is often classified according to the mechanism by which foreign DNA is introduced in the cell: transduction by acquisition of bacterial DNA via a bacteriophage, conjugation by acquisition of DNA directly from another bacterium or transformation by uptake of free DNA from the environment.

S. Tounsi (✉) · L. Abdelkefi-Mesrati
Biopesticides Team (LPIP), Centre of Biotechnology of Sfax, 3018 Sfax, Tunisia
e-mail: slim.tounsi@cbs.rnrt.tn

E. Sansinenea (ed.), *Bacillus thuringiensis Biotechnology*,
DOI 10.1007/978-94-007-3021-2_11, © Springer Science+Business Media B.V. 2012

Genetic recombination can be divided into two categories: general homologous recombination and site-specific recombination. These pathways are distinguished by their substrates DNAs, enzymes, and mechanisms. General recombination or general homologous recombination, which accounts for most recombination in the cell, is the exchange of genetic material between two molecules that have similar or identical sequences. However, site-specific recombination, a reaction in which DNA strands are broken and rejoined at determined positions of two target DNA sequences, does not depend on homology between the two DNA sequences involved in the recombination. It depends on enzymes that promote recombination between different regions in DNA, which may or may not have sequences in common. This type of recombination includes transposition by transposons, the integration and excision of prophages and other DNA elements, the inversion of invertible sequences, and the resolution of cointegrates by resolvases.

11.2 Site Specific Recombination in *B. thuringiensis*

11.2.1 Site Specific Recombination in B. thuringiensis via Transduction

11.2.1.1 Transduction: Definition

Transduction is a mechanism for the transfer of genetic material between cells by virus particles called phages or bacteriophages (in the case of bacteria). In fact, during transduction, genes from a host cell are incorporated into the genome of a bacteriophage and then carried to another host cell when the virus initiates another cycle of infection. In general transduction, any of the genes of the host cell may be involved in the process. However, only a few specific genes are transduced in special transduction. Transduction can not be accomplished by all bacteriophages. It is allowed only by phages that are classified as "temperate".

11.2.1.2 Phages of *B. thuringiensis*

About 83% of *B. thuringiensis* subspecies contain lysogenic phage (Ackermann et al. 1994). The first lysogenic phage was found in *B. thuringiensis berliner* (Yoder and Nelson 1960). The characterization (host range, plaque and particle morphology, serological specificities and one-step growth curves) of lysogenic and lytic phages of *B. thuringiensis* was reported by Chapman and Norris (1966) and Colasito and Rogoff (1969). By ultraviolet light and mitomycin C treatment, other phages were isolated and identified mainly from *B. thuringiensis galleria* (Zvenigorodskii et al. 1975; Rautenshtein et al. 1976). The decade 1980–1989 was

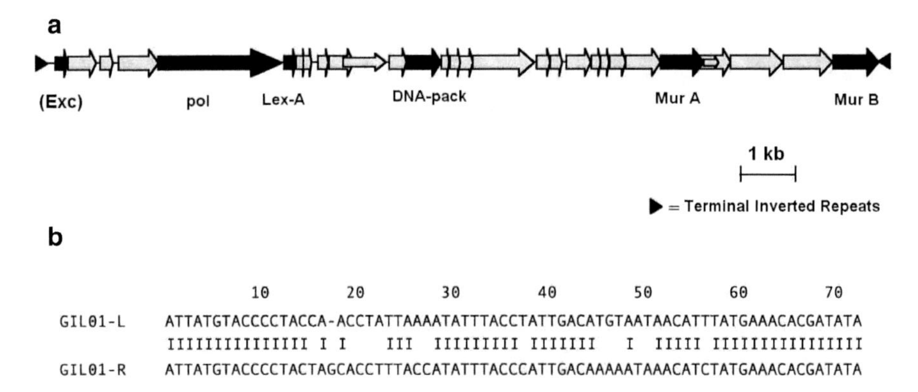

Fig. 11.1 Genetic organization of pGIL01/GIL01. **a** Thirty ORFs were reported, six of which showed similarity with known proteins (indicated by *black arrows*). (Exc), Excisionase; pol, DNA polymerase; Lex-A, Lex-A-like repressor; DNA-pack, DNA packaging protein; Mur A and B, muramidases. **b** The 73 bp Terminal Inverted Repeats present at both pGIL01 extremities shared more than 75% identity. (Verheust et al. 2003)

marked by research treating the biology, molecular weight, fingerprint, classification, prophage state of some lysogenic phages from *B. thuringiensis* strains. After that, all of the *B. thuringiensis* phages were classified into 25 types (Ackermann et al. 1995). In 1997, eight novel phages were assigned to family B1 according to their weight and enzymolysis (Azizbekian et al. 1997). After 2003, research shifted to sequencing and gene expression of *B. thuringiensis* phages. Some phages originating from *B. thuringiensis* including Bam35 (Stromsten et al. 2003), GIL01 (Fig. 11.1) (Verheust et al. 2003) and GIL16 (Verheust et al. 2005), have been completely sequenced. On the other hand, some genes have been cloned and expressed including *mur1* and *mur2* genes of GIL01, encoding two enzymes with peptidoglycan hydrolase activity (Verheust et al. 2004). Gaidelyte et al. (2006) showed that the penetration of Bam35 through the plasma membrane is a divalent-cation-dependent process, whereas adsorption and peptidoglycan digestion are not. In 2007, Thomas et al. sequenced the 218,948-bp genome of an atypical *Bacillus thuringiensis* phage 0305φ8-36 and identified the vibrion proteins by mass spectrometry. On the other hand, Hardies et al. (2007) showed that Phage 0305φ8-36 and BtI1 (a string of phage-like genes) are estimated to have diverged 2.0–2.5 billion years ago.

11.2.1.3 Construction of Improved Recombinant Strains and Chromosomal Mapping of *B. thuringiensis* Strains by Transduction

Thorne (1978) showed, for the first time, that phages CP-51 or CP-54 mediate generalized transduction in *B. thuringiensis*. CP-51 mediates also transduction of plasmid DNA, including pC194 of *B. thuringiensis*, into strains of *B. cereus*, *B. anthracis* and to strains of several *B. thuringiensis* subspecies. The frequency of transfer was as high as 10^{-5} transductants per PFU (Ruhfel et al. 1984). By a two step procedure (electroporation and transduction using phage CP-51 or CP-54Ber, respectively), Lecadet et al. (1992) and Kalman et al. (1995) constructed *B. thuringiensis* strains improved in their production of insecticidal proteins with new combinations of *cry* genes. In addition to strains improvement, various phage-based systems for mapping chromosomal genes have been developed by several groups. In 1984, Barsomian et al. used the temperate bacteriophages TP-13 and TP-18 to determine the order of markers in *B. thuringiensis* and concluded that TP-13 was useful for scanning large segments of the *B. thuringiensis* chromosome, and TP-18 was effective for ordering markers too closely linked for simple resolution with TP-13. In 1979, Perlak et al. showed that TP-13 was able to convert, with high frequency, an oligosporogenic mutant to spore positive and crystal positive one. Walter and Aronson (1991) showed that bacteriophage TP21 replicates as a plasmid and, under particular conditions, like phage lambda of *E. coli*, selectively transfers markers to *B. cereus*. This could be considered as a specialized transduction case in *B. thuringiensis*.

The construction of recombinant *B. thuringiensis* strains with new combinations of *cry* genes, by transduction, is of particular value for biotechnological applications. Indeed, this would expand the insecticidal spectrum of each strain, allowing native strains to have better insecticidal activity against more insect species. However, transduction efficiency depends on the ability of the host to phage transduction and on the stability of the introduced gene into the host.

11.2.2 Site Specific Recombination in B. thuringiensis via Transposition

11.2.2.1 Transposition of a Mobile Genetic Element: Definition

Transposition is the process whereby a mobile genetic element (transposon or insertion sequence) inserts itself into a new site on the same or another DNA molecule. Different transposons may transpose by different mechanisms. In many instances, transposition of the transposable genetic element results in removal of the element from the original site and insertion at a new site. However, in some cases the transposition event is accompanied by the duplication of the transposable genetic element. One copy remains at the original site and the other is transposed to the new site. The transposition event is mediated by a transposase coded by the transposable genetic element. Transposition does not require extensive DNA homology between the transposon and the target DNA. The phenomenon is therefore described as illegitimate or site-specific or non homologous recombination.

11.2.2.2 Mobile Genetic Elements of *B. thuringiensis*

B. thuringiensis is a rich source of mobile genetic elements. Several insertion sequences (IS231, IS232, IS240, ISBT1 and ISBT2), and Tn3-like transposons (Tn4430 and Tn5401) have been identified and found to be associated with *cry* genes (Mahillon et al. 1994). The IS231 and IS232 families, which display homology respectively with IS4 from *E. coli* and IS21 from *Pseudomonas aeruginosa*, are closely associated with *cry1A* genes on large plasmids in the *B. thuringiensis* subsp. *thuringiensis* strains Berliner 1715 and HD2 and *B. thuningiensis* subsp. *kurstaki* strains HD1 and HD73 (Kronstad and Whiteley 1984, 1986; Lereclus et al. 1984; Delécluse et al. 1989). The IS240 family, sharing homology with the IS26 of *Proteus vulgaris* and IS15 of *Salmonella panama*, has been shown to flank the *cry4A* gene of *B. thuringiensis israelensis* (Bourgouin et al. 1988). Although these associations, no evidence has been provided demonstrating a possible role for these elements in the evolution or mobility of the *cry* genes. Wang et al. (2008) showed that the transposon Tn4430 is directly associated with *B. thuringiensis* virulence. Usually, it is associated with *cry1A* genes in *B. thuringiensis* strains (Lereclus et al. 1984). It is delineated by 38-bp inverted repeats. This transposon harbors genes encoding a Tn3-like transposase and a resolvase or recombinase. The transposase of Tn4430 (a 113 kDa protein) shares homology with those of Tn3, Tn2l and Tn501. Through transpositional recombination, it generates the formation of co-integrates between both donor and target replicons leading to the duplication of Tn4430 molecules (Mahillon and Lereclus 1988). The Tn4430 resolvase (a 32 kDa protein) mediates the second step of the transposition, co-integrate resolution. It is unique among the resolvases of Tn3-like transposons. This resolvase displays some similarities with site-specific recombinases of the integrase family, such as Int of bacteriophage X, Cre of bacteriophage P1 or TnpA and TnpB of the Tn554 transposon (Mahillon and Lereclus 1988). Tn5401, a class II transposable element, unlike Tn4430, is not commonly found among different subspecies of *B. thuringiensis* and is not typically associated with known insecticidal crystal protein. This transposon shows distant homology to Tn4430, but shares a similar structural organization with it. The transposase (TnpA) and the integrase (TnpI) of this transposon are transcribed from a common promoter negatively regulated by TnpI. It is characterized by its unusually long 53-bp terminal inverted repeats (TIRs) (Fig. 11.2) (Baum 1994). Each of these TIRs contains a 38-bp sequence homologous to the 38- to 40-bp TIR sequence of Tn3-like transposons and an adjacent 12-bp sequence that binds TnpI (Baum et al. 1999). It transposed to both chromosomal and plasmid target sites but displayed an apparent preference for plasmid sites.

11.2.2.3 Construction of Improved Recombinant *B. thuringiensis* Strains

Biopesticides containing *B. thuringiensis* are environmentally friendly and effective in a variety of situations. However, their performance is often considered to be poorer than that of chemicals in terms of reliability, spectrum of activity,

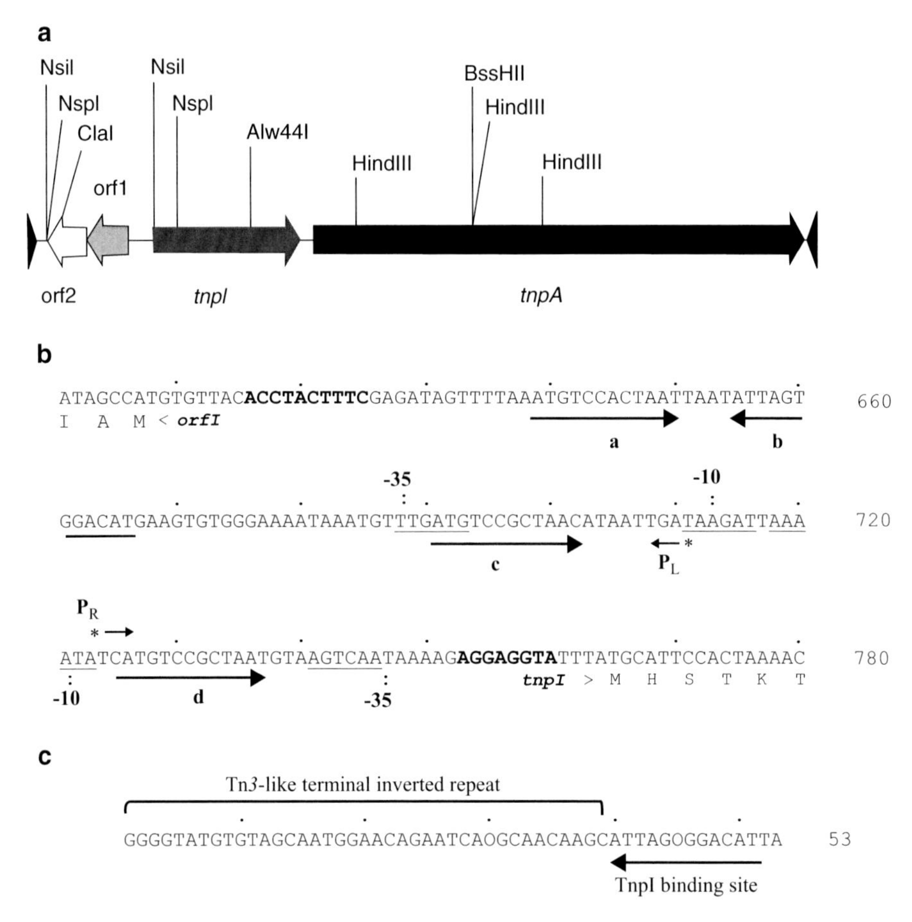

Fig. 11.2 Structural organization of Tn*5401* (**a**) and nucleotide sequences of the *tnpI-tnpA* promoter region (**b**) and Tn*5401* TIRs (**c**). Nucleotide positions in the promoter region are based on the published sequence of Tn*5401* (Baum 1994). The conserved 12-bp TnpI recognition sequence ATGTCCRCTAAY is indicated by the *arrows* in panels **b** and **c**. The terminal 38-bp sequence with homology to Tn*3*-like transposons is shown in panel C. The −35 and −10 regions of the *tnpI-tnpA* (PR) promoter (*single underline*) and the divergent *orf1* (PL) promoter (*double underline*) are shown, along with the corresponding transcriptional start sites (*). Ribosome binding sites are in boldface. (Baum et al. 1999)

speed of action, and cost effectiveness. Thus, genetic manipulation of *cry* genes in *B. thuringiensis* is desired to improve the potency and the spectrum of activity of their insecticidal toxins. For this purpose, site specific recombination has been used. Yue et al. (2005a, b) reported improved recombinant *B. thuringiensis* strains in their insecticidal activities, constructed by introducing additional *cry* genes (*cry1C* and *cry3A*) in their chromosomes, using an integrative and thermosensitive vector developed based on Tn4430. In order to avoid the dissemination of genetically engineered microorganisms, particularly those containing antibiotic resistance genes, Baum et al. (1996) developed a vector system that combines a *B. thuringiensis* plasmid replicon with an indigenous site-specific recombination system allowing the selective removal of ancillary or foreign DNA from the recombinant bacterium after introduction of the Cry-encoding plasmid vector. With the same goal, Sanchis et al. (1996, 1997) constructed a vector, based on the specific resolution site of the transposon Tn4430, useful for engineering strains free of antibiotic resistance genes. Based on the site-specific recombinase of Tn4430, Salamitou et al. (1997) constructed a genetic system, composed of two compatible plasmids, that reports the transient activation of a promoter by promoting the stable acquisition of an antibiotic resistance marker by the bacterium. To facilitate the study of the transfer of a self-transmissible plasmid in *B. thuringiensis* subsp. *kurstaki* strain HD73, pAW63, Wilcks et al. (1998) tagged the latter with the tetracycline resistance transposon Tn5401. Malvar and Baum (1994) constructed a recombinant *B. thuringiensis* overexpressing *cry3A* by Tn5401 disruption of *sop0F* gene.

11.2.2.4 *B. thuringiensis* Insertional Mutagenesis

Transposon mutagenesis is a powerful tool to unraveling the biology of *B. thuringiensis*. Many transposons, such as Tn917, Tn916, Tn10 and mariner, are able to effectively insert into the genome of *B. thuringiensis* and the insertion sites could be readily mapped. Tn917, a transposon from *Streptococcus faecalis* (Tomich et al. 1980), has been used to identify *B. thuringiensis* genes required for virulence and survival in a *Manduca sexta* (tobacco hornworm) septicaemia model (Steggles et al. 2006). This transposon has several insertion hot spots in chromosomes and preferentially insert into plasmids and non-coding regions of a genome. Tn916, a conjugation transposon, has a stronger bias for non-coding regions than Tn917 does (Garsin et al. 2004; Hoffmaster and Koehler 1997). When *Enterococcus faecalis* was the donor and *B. thuringiensis israelensis* the recipient, pregrowth in tetracycline increased the conjugative transposition frequency (Showsh and Andrews 1992). The mini-Tn10 tranposon has been widely applied in *B. thuringiensis* and some valuable genes involved in bacteriocin and beta-exotoxin I production, swarm cell differentiation, flagellin export, and secretion of virulence-associated proteins had been identified by its use (Kamoun et al. 2009; Salvetti et al. 2009; Ghelardi et al. 2002; Espinasse et al. 2002). Tn10 could not randomly insert into the chromosome, because it requires a symmetrical six-base-pair "GCTNAGC" as its target sequence (Pribil and Haniford 2003). In addition, the homology between Tn10 and its target could lead to the preference of this transposon (Monod et al. 1997). Moreover, Tn10-encoded transposase contacts a large stretch of the target DNA (~24 bp) with a symmetrical structure, which also may contribute to the selection of targets (Pribil and Haniford 2000). Li et al. (2009) constructed two mariner transposon delivery vector systems, pMarA333 and pMarB333, useful for construction of mutant libraries of *B. thuringiensis* and other *Bacillus* strains. The mariner tranposon could insert into a target DNA by a "cut and paste" reaction, which is catalyzed by mariner-encoded transposase and requires no obvious specific host factors (Vos et al. 1996).

11.3 Homologous Recombination in B. thuringiensis

11.3.1 B. thuringiensis Transformation, a Brief History

Transformation was first demonstrated in *Streptococcus pneumoniae* by Griffith (1928). Avery et al. (1944) proved that DNA was the transforming principle. Transformation refers to a form of genetic exchange in which recipient cell takes up DNA from the environment. This foreign DNA may be derived from unrelated species, such as bacteria, fungi, plants or animals, but most likely it is the remnants of DNA from dead bacterial cells. The ability of bacteria to maintain this DNA and replicate it during normal cell multiplication is the basis of cell transformation. In spite of *B. thuringiensis* strains have potential in bioinsecticides production, the lack of efficient techniques for genetic manipulation in this organism constituted an obstacle for genetic and physiological studies for many years. For *B. thuringiensis* transformation, several laboratories used the method of protoplast transformation developed by Chang and Cohen (1979) with some modifications. Due to the poor protoplastization process and the low regeneration frequency, efficiency of *B. thuringiensis* transformation was generally low and dependent on *B. thuringiensis* strains (Alikhanian et al. 1981; Martin et al. 1981; Fischer et al. 1984; Crawford et al. 1987; Rubinstein and Sanchez-Rivas 1988, 1989). In 1989, different efficient DNA transfer systems based on electroporation of vegetative *B. thuringiensis* cells (Bone and Ellar 1989; Lereclus et al. 1989; Mahillon et al. 1989; Masson et al. 1989; Schurter et al. 1989) were developed. These protocols for transformation of *B. thuringiensis* by electroporation, independently published in the same year, differed in the composition of buffers, the method for cells preparation and parameters of electric pulse (Zeigler 1999). They generally provided transformation efficiencies of 10^2–10^5 transformants per μg of plasmid DNA. The development of these effective DNA transfer systems by electroporation opened new avenues to improve the entomopathological potential of *B. thuringiensis*. In fact, transformation of this bacterium by electroporation allowed the expression of *cry* genes in their natural host and it became possible to change *B. thuringiensis* strains giving them new characteristics and thus make them more efficient in terms of pest management.

11.3.2 Construction of Improved B. thuringiensis Strains by Homologous Recombination

In 1986, Höfte et al. compared the amino acid sequence of a delta-endotoxin (named Bt2) from *B. thuringiensis Berliner* with those of *B. thuringiensis kurstaki strain* HD1 Dipel, *B. thuringiensis kurstaki* strain HD73 and *B. thuringiensis sotto* and suggested that homologous recombination between the different genes had occurred during evolution. In 1991, Delécluse et al. demonstrated that homologous recombination occurs in *B. thuringiensis* by disrupting the *cyt1A* gene, encoding the 28-kDa polypeptide of *Bacillus thuringiensis israelensis* crystals, from the 72-MDa resident plasmid. They deduced also that the absence of the Cyt1A protein in *B. thuringiensis israelensis* did not affect the crystallization of the other toxic components of the parasporal body but it abolished the hemolytic activity.

In order to enlarge the spectrum of the insecticidal activity of the wild strains of *B. thuringiensis*, and since it has been found that crystal genes introduced on plasmids are not stably maintained in *B. thuringiensis*, several teams used the homologous recombination to integrate additional *cry* gene into the chromosome. Thus, different improved *B. thuringiensis* strains were obtained using transformation and homologous recombination. After fusion of the P19 gene of *B. thuringiensis* subsp. *israelensis* with a chitinase gene of *B. licheniformis*, the construction was introduced in *B. thuringiensis* subsp. *aizawai* by transformation and integrated into the bacterium genome by homologous recombination. The resulting *B. thuringiensis* strain showed an improved insecticidal activity against *Spodoptera exigua* demonstrating that chromosomal integration might be used as a potential technique for strain improvement (Thamthiankul et al. 2004). After transformation of *B. thuringiensis* strain HD-73 with the recombinant plasmid pHT*cry1Ac-gfp*, the *gfp* gene was inserted into the large HD-73 endogenous plasmid pHT73 and fused with the 3′ terminal of the *cry1Ac* gene by homologous recombination. This Cry1Ac-GFP fusion protein is a useful tool for studying the function of the C-terminus of the Cry1A protoxin in insect midgets and to study the mechanisms of parasporal crystal formation and assembly (Yang et al. 2010). Homologous recombination was used to obtain a *B. thuringiensis* engineered strain with high toxicity and broad insecticidal spectrum against coleopteran insects. The construction was based on the integration of the *cry8Ca2* gene into the endogenous plasmid of BT185 which contains *cry8Ea1* (Liu et al. 2010). Liu et al. (2010) used the plasmid pMAD (Arnaud et al. 2004) for homologous recombination in *B. thuringiensis* and this after insertion of the gene of interest in the multiple cloning site. This procedure was based on the thermosensitive recombination system of the pMAD plasmid (Fig. 11.3) which contained a temperature-sensitive replication region named ori pE194ts, the *ermC* gene (erythromycin resistance gene), the *bgaB* gene (β-galactosidase gene) with a plcpB promoter (from *Staphylococcus aureus*), and the *bla* gene (β-lactamases gene) (Arnaud et al. 2004).

11.4 Conclusion

Genetic recombination is an excellent tool for generating *B. thuringiensis* strains with new and stable combination of *cry* genes and free of antibiotic resistance genes. This is of particular interest for biotechnological applications because leading to the construction of strains with better insecticidal activity against more insect species, and should facilitate regulatory approval for their development as a commercial

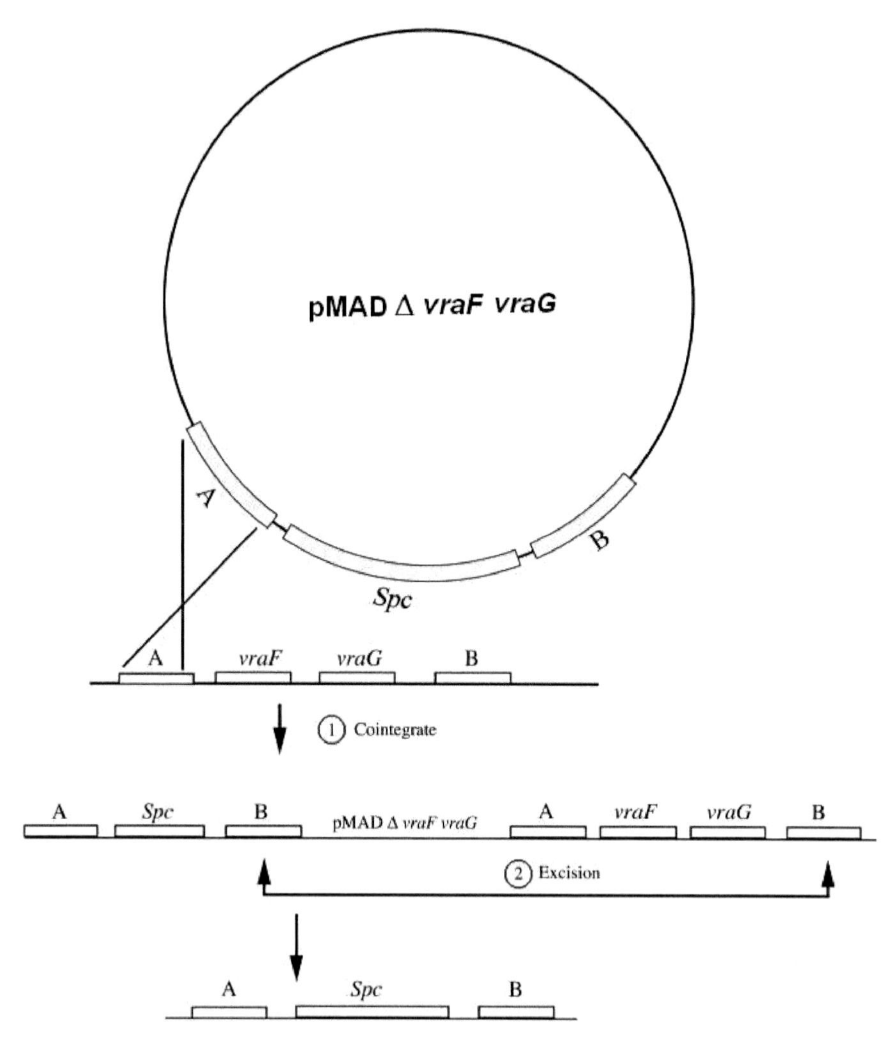

Fig. 11.3 Schematic representation of a two-step procedure used to obtain gene replacement by recombination. Areas labeled A and B represent DNA sequences located upstream and downstream from *vraF* and *vraG* genes. The *crossed lines* indicate crossover events. The integration of pMAD via homologous sequences can take place in area A or B. The cointegrate undergoes a second recombination event, regenerating the pMAD plasmid. Depending on whether the second recombination event occurs between the two homologous sequences in area A or B, the spectinomycin resistance marker will either remain in the chromosome (area B) or be excised along with the plasmid (area A). Gene replacement occurs only if the second recombination event occurs in area B, as shown. (Arnaud et al. 2004)

biopesticides. A guideline including well engineered *B. thuringiensis* strains will serve to specific and efficient biological control of pests.

References

Ackermann HW, Azizbekyan RR, Emadi Konjin HP (1994) New *Bacillus* bacteriophage species. Arch Virol 135:333–344

Ackermann HW, Azizbekyan RR, Bernier RL (1995) Phage typing of *Bacillus subtilis* and *Bacillus thuringiensis*. Res Microbiol 146:643–657

Alikhanian SL, Ryabchenko NF, Bukanov NO, Sakanyan VA (1981) Transformation of *Bacillus thuringiensis* subsp. *galleria* Protoplasts by Plasmid pBC16. J Bacteriol 146:7–9

Arnaud M, Chastanet A, Débarbouillé M (2004) New vector for efficient allelic replacement in naturally nontransformable, low-GC-content, gram-positive bacteria. Appl Environ Microbiol 70:6887–6891

Avery OT, MacLeod CM, McCarty M (1944) Studies on the chemical nature of the substance inducing transformation of pneumococcal types: induction of transformation by a desoxyribonucleic acid fraction isolated from *Pneumococcus* type III. J Exp Med 79:137–158

Azizbekian KR, Kuzin AI, Dobrzhanskaia EO (1997) Restriction analysis of DNA from phages isolated from type strains of *Bacillus thuringiensis*. Mikrobiologiia 66:247–253 (Russian)

Barsomian GD, Robillard NJ, Thorne CB (1984) Chromosomal mapping of *Bacillus thuringiensis* by transduction. J Bacteriol 157:746–750

Baum JA (1994) Tn5401, a new class II transposable element from *Bacillus thuringiensis*. J Bacteriol 176:2835–2845

Baum JA, Kakefuda M, Gawron-Burke C (1996) Engineering *Bacillus thuringiensis* bioinsecticides with an indigenous site-specific recombination system. Appl Environ Microbiol 62:4367–4373

Baum JA, Gilmer AJ, Light Mettus AM (1999) Multiple roles for TnpI recombinase in regulation of Tn5401 transposition in *Bacillus thuringiensis*. J Bacteriol 181:6271–6277

Bone EJ, Ellar DJ (1989) Transformation of *Bacillus thuringiensis* by electroporation. FEMS Microbiol Lett 58:171–178

Bourgouin C, Delecluse A, Ribier J, Klier A, Rapoport G (1988) A *Bacillus thuringiensis* subsp. *israelensis* gene encoding a 125-kilodalton larvicidal polypeptide is associated with inverted repeat sequences. J Bacteriol 170:3575–3583

Chang S, Cohen SN (1979) High frequency transformation of *Bacillus subtilis* protoplasts by plasmid DNA. Mol Gen Genet 168:111–115

Chapman HM, Norris JR (1966) Four new bacteriophages of *Bacillus thuringiensis*. J Appl Bacteriol 29:529–535

Colasito DJ, Rogoff MH (1969) Characterization of temperate bacteriophages of *Bacillus thuringiensis*. J Gen Virol 5:275–281

Crawford IT, Greis KD, Parks L, Streips UN (1987) Facile autoplast generation and transformation in *Bacillus thuringiensis* subsp. *kurstaki*. J Bacteriol 169:5423–5428

Delécluse A, Bourgouin C, Klier A, Rapoport G (1989) Nucleotide sequence and characterization of a new insertion element, IS240, from *Bacillus thuringiensis israelensis*. Plasmid 21:71–78

Delécluse A, Charles JF, Klier A, Rapoport G (1991) Deletion by in vivo recombination shows that the 28-kilodalton cytolytic polypeptide from *Bacillus thuringiensis* subsp. *israelensis* is not essential for mosquitocidal activity. J Bacteriol 173:3374–3381

Espinasse S, Gohar M, Lereclus D, Sanchis V (2002) An ABC transporter from *Bacillus thuringiensis* is essential for beta-exotoxin I production. J Bacteriol 184:5848–5854

Fischer HM, Lüthy P, Schweitzer S (1984) Introduction of plasmid pC194 into *Bacillus thuringiensis* by protoplast transformation and plasmid transfer. Arch Microbiol 139:213–217

Gaidelyte A, Cvirkaite-Krupovic V, Daugelavicius R, Bamford JK, Bamford DH (2006) The entry mechanism of membrane-containing phage Bam35 infecting *Bacillus thuringiensis*. J Bacteriol 188:5925–5934

Garsin DA, Urbach J, Huguet-Tapia JC, Peters JE, Ausubel FM (2004) Construction of an *Enterococcus faecalis* Tn917-mediated-gene-disruption library offers insight into Tn917 insertion patterns. J Bacteriol 186:7280–7289

Ghelardi E, Celandroni F, Salvetti S, Beecher DJ, Gominet M, Lereclus D, Wong AC, Senesi S (2002) Requirement of flhA for swarming differentiation, flagellin export, and secretion of virulence-associated proteins in *Bacillus thuringiensis*. J Bacteriol 184:6424–6433

Griffith F (1928) The significance of pneumococcal types. J Hyg 27:113–159

Hardies SC, Thomas JA, Serwer P (2007) Comparative genomics of *Bacillus thuringiensis* phage 0305φ8-36: defining patterns of descent in a novel ancient phage lineage. Virol J 4:97

Hoffmaster AR, Koehler TM (1997) The anthrax toxin activator gene *atxA* is associated with CO2-enhanced non-toxin gene expression in *Bacillus anthracis*. Infect Immun 65:3091–3099

Höfte H, de Greve H, Seurinck J, Jansens S, Mahillon J, Ampe C, Vandekerckhove J, Vanderbruggen H, van Montagu M, Zabeau M, Vaeck C (1986) Structural and functional analysis of a cloned delta endotoxin of *Bacillus thuringiensis Berliner* 1715. Eur J Biochem 161:273–280

Kalman S, Kiehne KL, Cooper N, Reynoso MS, Yamamoto T (1995) Enhanced production of insecticidal proteins in *Bacillus thuringiensis* strains carrying an additional crystal protein gene in their chromosomes. Appl Environ Microbiol 61:3063–3068

Kamoun F, Fguira IB, Tounsi S, Abdelkefi-Mesrati L, Sanchis V, Lereclus D, Jaoua S (2009) Generation of Mini-Tn10 transposon insertion mutant library of *Bacillus thuringiensis* for the investigation of genes required for its bacteriocin production. FEMS Microbiol Lett 294:141–149

Kronstad JW, Whiteley HR (1984) Inverted repeat sequences flank a *Bacillus thuringiensis* crystal protein gene. J Bacteriol 160:95–102

Kronstad JW, Whiteley HR (1986) Three classes of homologous *Bacillus thuringiensis* crystal-protein genes. Gene 43:29–40

Lecadet MM, Chaufaux J, Ribier J, Lereclus D (1992) Construction of novel *Bacillus thuringiensis* strains with different insecticidal activities by transduction and transformation. Appl Environ Microbiol 58:840–849

Lereclus D, Ribier J, Klier A, Menou G, Lecadet MM (1984) A transposon-like structure related to the delta-endotoxin gene of *Bacillus thuringiensis*. EMBO J 3:2561–2567

Lereclus D, Arantes O, Chaufaux J, Lecadet MM (1989) Transformation and expression of a cloned δ-endotoxin gene in *Bacillus thuringiensis*. FEMS Microbiol Lett 60:211–218

Li M, Li M, Yin W, He J, Yu Z (2009) Two novel transposon delivery vectors based on mariner transposon for random mutagenesis of *Bacillus thuringiensis*. J Microbiol Methods 78:242–244

Liu J, Yang G, Shu C, Zhao C, Liu C, Song F, Zhou L, Ma J, Zhang J, Huang D (2010) Construction of a *Bacillus thuringiensis* engineered strain with high toxicity and broad pesticidal spectrum against coleopteran insects. Appl Microbiol Biotechnol 87:243–249

Mahillon J, Lereclus D (1988) Structural and functional analysis of Tn4430: identification of an integrase-like protein involved in the co-integrate-resolution process. EMBO J 7:1515–1526

Mahillon J, Chungjatupornchai W, Decock J, Dierickx S, Michiels F, Peferoen M, Joos H (1989) Transformation of *Bacillus thuringiensis* by electroporation. FEMS Microbiol Lett 60:205–210

Mahillon J, Rezsöhazy R, Hallet B, Delcour J (1994) IS231 and other *Bacillus thuringiensis* transposable elements: a review. Genetica 93:13–26

Malvar T, Baum JA (1994) Tn5401 disruption of the *spo0F* gene, identified by direct chromosomal sequencing, results in CryIIIA overproduction in *Bacillus thuringiensis*. J Bacteriol 176:4750–4753

Martin PAW, Lahr JR, Dean DH (1981) Transformation of *Bacillus thuringiensis* protoplasts by plasmid deoxyribonucleic acid. J Bacteriol 145:980–983

Masson L, Préfontaine G, Brousseau R (1989) Transformation of *Bacillus thuringiensis* vegetative cells by electroporation. FEMS Microbiol Lett 60:273–278

Monod C, Repoila F, Kutateladze M, Tetart F, Krisch HM (1997) The genome of the pseudo T-even bacteriophages, a diverse group that resembles T4. J Mol Biol 267:237–249

Perlak FJ, Mendelsohn CL, Thorne CB (1979) Converting bacteriophage for sporulation and crystal formation in *Bacillus thuringiensis*. J Bacteriol 140:699–706

Pribil PA, Haniford DB (2000) Substrate recognition and induced DNA deformation by transposase at the target-capture stage of Tn10 transposition. J Mol Biol 303:145–159

Pribil PA, Haniford DB (2003) Target DNA bending is an important specificity determinant in target site selection in Tn10 transposition. J Mol Biol 330:247–259

Rautenshtein RI, Moskalenko LN, Bespalova IA (1976) Ultrastructure of bacteriophages specific for *Bacillus thuringiensis* var. *galleriae*. Mikrobiologiia 45:690–694 (Russian)

Rubinstein CP, Sanchez-Rivas C (1988) Production of protoplasts by autolytic induction in *Bacillus thuringiensis*: transformation and interspecific fusion. FEMS Microbiol Lett 52:67–72

Rubinstein CP, Sanchez-Rivas C (1989) Genetic manipulations in auto-induced protoplasts of *Bacillus thuringiensis*. Mem Inst Oswaldo Cruz, Rio de Janeiro 84:35–37

Ruhfel RE, Robillard NJ, Thorne CB (1984) Interspecies transduction of plasmids among *Bacillus anthracis*, *B. cereus*, and *B. thuringiensis*. J Bacteriol 157:708–711

Salamitou S, Agaisse H, Lereclus D (1997) A genetic system that reports transient activation of genes in *Bacillus*. Gene 20:121–126

Salvetti S, Celandroni F, Ceragioli M, Senesi S, Ghelardi E (2009) Identification of non-flagellar genes involved in swarm cell differentiation using a *Bacillus thuringiensis* mini-Tn10 mutant library. Microbiology 155:912–921

Sanchis V, Agaisse H, Chaufaux J, Lereclus D (1996) Construction of new insecticidal *Bacillus thuringiensis* recombinant strains by using the sporulation non-dependent expression system of *cryIIIA* and a site specific recombination vector. J Biotechnol 48:81–96

Sanchis V, Agaisse H, Chaufaux J, Lereclus D (1997) A recombinase-mediated system for elimination of antibiotic resistance gene markers from genetically engineered *Bacillus thuringiensis* strains. Appl Environ Microbiol 63:779–784

Schurter W, Geiser M, Mathé D (1989) Efficient transformation of *Bacillus thuringiensis* and *B. cereus* via electroporation: transformation of acrystalliferous strains with a cloned delta-endotoxin gene. Mol Gen Genet 218:177–181

Showsh SA, Andrews RE Jr (1992) Tetracycline enhances Tn916-mediated conjugal transfer. Plasmid 28:213–224

Steggles JR, Wang J, Ellar DJ (2006) Discovery of *Bacillus thuringiensis* virulence genes using signature-tagged mutagenesis in an insect model of septicaemia. Curr Microbiol 53:303–310

Stromsten NJ, Benson SD, Burnett RM (2003) The *Bacillus thuringiensis* linear double-stranded DNA phage Bam35, which is highly similar to the *Bacillus cereus* linear plasmid pBClin15, has a prophage state. J Bacteriol 185:6985–6989

Thamthiankul S, Moar WJ, Miller ME, Panbangred W (2004) Improving the insecticidal activity of *Bacillus thuringiensis* subsp. *aizawai* against *Spodoptera exigua* by chromosomal expression of a chitinase gene. Appl Microbiol Biotechnol 65:183–192

Thomas JA, Hardies SC, Rolando M, Hayes SJ, Lieman K, Carroll CA, Weintraub ST, Serwer P (2007) Complete genomic sequence and mass spectrometric analysis of highly diverse, atypical *Bacillus thuringiensis* phage 0305ϕ8-36. Virology 368:405–421

Thorne CB (1978) Transduction in *Bacillus thuringiensis*. Appl Environ Microbiol 35:1109–1115

Tomich PK, An FY, Clewell DB (1980) Properties of erythromycin-inducible transposon Tn917 in *Streptococcus faecalis*. J Bacteriol 141:1366–1374

Verheust C, Jensen G, Mahillon J (2003) pGIL01, a linear tectiviral plasmid prophage originating from *B. thuringiensis* serovar *israelensis*. Microbiology 149:2083–2092

Verheust C, Fornelos N, Mahillon J (2004) The *Bacillus thuringiensis* phage GIL01 encodes two enzymes with peptidoglycan hydrolase activity. FEMS Microbiol Lett 237:289–295

Verheust C, Fornelos N, Mahillon J (2005) GIL16, a new gram positive tectiviral phage related to the *Bacillus thuringiensis* GIL01 and the *Bacillus cereus* pBClin15 elements. J Bacteriol 187:1966–1973

Vos JC, De Baere I, Plasterk RH (1996) Transposase is the only nematode protein required for in vitro transposition of Tc1. Genes Dev 10:755–761

Walter TM, Aronson AI (1991) Transduction of certain genes by an autonomously replicating *Bacillus thuringiensis* phage. Appl Environ Microbiol 57:1000–1005

Wang J, Steggles JR, Ellar DJ (2008) Molecular characterization of virulence defects in *Bacillus thuringiensis* mutants. FEMS Microbiol Lett 280:127–134

Wilcks A, Jayaswal N, Lereclus D, Andrup L (1998) Characterization of plasmid pAW63, a second self-transmissible plasmid in *Bacillus thuringiensis* subsp. *kurstaki* HD73. Microbiology 144:1263–1270

Yang H, Rong R, Song F, Sun C, Wei J, Zhang J, Huang D (2010) In vivo fluorescence observation of parasporal inclusion formation in *Bacillus thuringiensis*. Sci China Life Sci 53:1106–1111

Yoder PE, Nelson EL (1960) Bacteriophage for *Bacillus thuringiensis* berliner and *Bacillus anthracis* cohn. J Insect Pathol 2:198–200

Yue C, Sun M, Yu Z (2005a) Improved production of insecticidal proteins in *Bacillus thuringiensis* strains carrying an additional *cry1C* gene in its chromosome. Biotechnol Bioeng 92:1–7

Yue C, Sun M, Yu Z (2005b) Broadening the insecticidal spectrum of Lepidoptera-specific *Bacillus thuringiensis* strains by chromosomal integration of *cry3A*. Biotechnol Bioeng 91:296–303

Zeigler DR (1999) *Bacillus* genetic stock center catalog of strains, Seventh Edition, Part 2: *Bacillus thuringiensis and Bacillus cereus*. 7th ed. *Bacillus* Genetic Stock Center, Columbus, Ohio, p 30. http://www.bgsc.org/Catalogs/Catpart2.pdf

Zvenigorodskii VI, Izakson IS, Kapitonova ON (1975) Identification of bacteriophages and study of the properties of phage-resistant mutants of *Bacillus thuringiensis* var. *galleriae*. Nauchnye Doki Vyss Shkoly Biol Nauki 5:92–98

Chapter 12
Genetic Improvement of Bt Strains and Development of Novel Biopesticides

Vincent Sanchis

Abstract This review describes how recombinant DNA technology has been used to improve *Bacillus thuringiensis* (Bt) products and overcome a number of the problems associated with Bt-based insect control measures. It will discuss how the knowledge of the genetics of Bt and of its insecticidal toxin genes, the understanding of their regulation and the development of cloning vectors has made possible the continuing improvement of first generation products. Several examples describing how biotechnology has been used to increase the production of insecticidal proteins in Bt, their persistence in the field by protecting them against UV degradation or to construct non-viable genetically modified strains, will be presented.

Keywords *Bacillus thuringiensis* · Biopesticides · ∂-endotoxins · Cry genes · Recombinant Bt strains · Asporogenic strains · Site-specific recombination

12.1 Introduction

B. thuringiensis or Bt is a spore-forming gram-positive pathogen that produces highly specific insecticidal proteins, called the δ-endotoxins, or Cry proteins, that accumulate as crystalline inclusions within the cell during sporulation. At the end of sporulation, the bacterial cells lyse and the spores and crystals are liberated. If ingested by susceptible insects (usually the larvae), the crystals are dissolved and the δ-endotoxins, which are protoxin molecules, are specifically cleaved by insect gut proteases. The resulting activated toxins recognize specific receptors on the surfaces of the midgut epithelium cells and cause cell lysis and the death of insect larvae. Commercial Bt products used as biopesticides generally consist of a mixture of spores and crystals, produced in large fermenters and applied as foliar sprays, much like synthetic insecticides. Numerous studies have shown that application of suspensions of crystal proteins and spores of *B. thuringiensis* are safe to mammals and birds and are safer for non-target insects than conventional insecticides (Siegel 2001). Given the undesirable effects of chemical insecticides, Bt-products—which

V. Sanchis (✉)
INRA, UMR1319-MICALIS, Equipe GME, La Minière
78285 Guyancourt Cedex, France
e-mail: vincent.sanchis@jouy.inra.fr

E. Sansinenea (ed.), *Bacillus thuringiensis Biotechnology*,
DOI 10.1007/978-94-007-3021-2_12,

present the advantage of having only a minor impact on the environment—have come to occupy a stable position of the insecticide market. Bt-formulations are used in agriculture, forestry and in human health for the elimination of vectors of diseases but they are often considered as less persistent in the field than other chemical products. The two main problems with Bt-products for pest control are their often narrow activity spectrum and high sensitivity to UV degradation (Pusztai et al. 1991; Behle et al. 1997) or other environmental factors. Therefore, the economic viability and acceptability of Bt biopesticides depends on the potency of the insecticidal toxins and the environmental stability of the crystals after spraying. An additional limitation of Bt products used as sprays is their inability to target insects feeding on internal tissues of the plants or sap sucking and soil dwelling pests that remain out of reach. Some of these problems have been circumvented by the development of genetically modified crops that constitutively express synthetic crystal proteins (Vaeck et al. 1987). Although the engineering of plants to express Bt *cry* genes has been especially helpful against pests that attack parts of the plant that are usually not well-protected by conventional insecticide application, this topic has been extensively reviewed elsewhere (Shelton et al. 2002) and is not within in the scope of this review. Apart from transgenic Bt-plants, the more traditional spray form of Bt is also very important in plant protection and is widely used (Liu and Tabashnik 1997; Sanchis and Bourguet 2008) by organic farmers. This review will describe the principal approaches and recombinant DNA techniques used to improve Bt strains and products used as sprays during the past two decades.

12.2 Discovery of New Bt Strains Harbouring Novel *Cry* Genes

The first step towards improving Bt strains naturally involved the isolation of new strains with new or higher insecticidal activity against a particular target pest as compared to previously known strains. A first important event for the development of Bt-products was the isolation by Dulmage, in 1970, of the isolate HD1 subsp. *kurstaki* from diseased *Pectinophora gossypiella* larvae (Dulmage 1970). This strain proved to be much more potent than the isolates in the existing commercial Bt products. A second milestone in the Bt development was the discovery by Goldberg and Margalit, in Israel, in 1976, of Bt subsp. *israelensis* (Goldberg and Margalit 1977). The new subspecies contained both Cry and cyt toxins in its inclusions and was highly active against biting dipteran mosquitoes and blackflies, the vectors of diseases such as dengue and malaria, thus extending the range of activity of Bt. Soon after, in 1983, another Bt subspecies was isolated that was active against plant feeding coleopteran larvae (Beetles), mainly chrysomelidae (Krieg et al. 1983). Intensive screening programs have followed and led to the isolation and characterization of new strains with different combinations of crystal proteins and to the cloning and DNA sequence determination of many new *cry* genes (Schnepf et al. 1985). Since the cloning of the first *cry* gene in 1981 (Schnepf and Whiteley 1981) the number of *cry* genes encoding Cry protein with novel target pests has expanded dramatically. To date, several thousand natural strains have been isolated from various geographical areas and from different sources, including grain dust, soil, insects and plants (Martin and Travers 1989; Smith and Couche 1991). These isolates produce well over 300 different crystal proteins classified into 60 major Cry protein classes on the basis of amino acid identity (Crickmore et al. 1998). Each individual Cry protein generally has a restricted spectrum of activity, but the entire Cry family is active against more than 500 insect species and against some other invertebrate pest species such as nematodes. A current list of Cry proteins can be found on the Internet at: http://www.lifesci.sussex.ac.uk/home/Neil_Crickmore/Bt/holo2.html.

12.3 Engineering of Novel Cry Proteins

An alternative to the isolation of novel insecticidal genes from naturally occurring Bt strains was to construct novel insecticidal genes for particular target insects. Indeed, since Cry proteins are homologous and have a similar mode of action, domain substitution and mutagenesis within various regions could yield new mutant toxins having new, broadened or improved toxicity (Pardo-Lopez et al. 2009). Mutagenesis of Cry toxins has first been used extensively to explore the participation of specific protein regions in the toxicity mechanism (Bravo et al. 2007) and some of these mutations have resulted in generation of Cry toxin molecules with improved toxicity (Rajamohan et al. 1996; Wu et al. 2000). This knowledge has been exploited to manipulate *cry* genes by creating chimeric toxins, using domain substitution, which have effectively resulted in production of novel toxins with a wider target spectrum or higher toxicity than the parental toxins from which were derived. For example, a Cry1Ac/1F chimeric toxin with superior toxicity to the fall armyworm, *S. frugiperda*, when compared to the parental CrylAc and CrylF toxins was engineered by Ecogen, a biotechnology company, and was introduced in Bt strain EG7826. This strain was registered as the active ingredient of the Lepinox™ bioinsecticide. A second lepidopteran-toxic EG7841 strain, harbouring a *cry1C* gene containing a mutation that resulted in improved toxicity towards the beet armyworm, *Spodoptera exigua*, was also registered as the active ingredient of the CryMax™ bioinsecticide. These two strains were readily approved and did not necessitated a registration package different than those required for naturally-occurring Bt strains because these genes, although modified by recombinant DNA techniques, were derived from Bt *cry* genes. Replacing domain III of a Cry1Ab toxin, which has low toxicity to *S. exigua*, by domain III of Cry1C also resulted in a hybrid toxin with a much higher toxicity against *S. exigua* than the parental toxin (de Maagd et al. 1996). A chimeric Cry1C/1Ab protein, carrying the C-terminal domain of Cry1Ab and N-terminal domains of cry1C was also constructed. This chimeric Cry1C/1Ab toxin was more toxic than the unmodified Cry1C toxin against various lepidopteran pests, including *Spodoptera littoralis* (Egyptian cotton leaf worm) and *Spodoptera exigua* (lesser [beet] armyworm) which were not readily controlled by other Cry toxins (Sanchis et al. 1999). A fusion toxin comprising a truncated version of Cry1Ba and domain II from Cry1Ia was also shown to be toxic to both coleopteran and lepidopteran pests (Naimov et al. 2001; Naimov et al. 2003). More recently, a hybrid Cry1Ab/3A toxin which was toxic to the western corn rootworm was also created (Walters et al. 2010). Finally, optimization of Bt crystal proteins through DNA shuffling has also been reported (Lassner and Bedbrook 2001; Craveiro et al. 2010). The screening of 4000 shuffled *cry1Ca* clones yielded some variants which were significantly more toxic against *Spodoptera exigua* larvae (Lepidoptera: Noctuidae) and toxic *Helicoverpa zea* (Lepidoptera: Noctuidae) a pest previously unaffected by the Cry1Ca toxin (Lassner and Bedbrook 2001). Directed evolution system based on phage display technology for producing toxins with improved binding to a receptor, and thus increased toxicity, has also been described (Ishikawa et al. 2007).

12.4 Construction of New Bt Strains by Conjugation

Amongst the critical milestones in the development of new Bt-products was the discovery that a natural conjugal plasmid exchange system exists in Bt and that the genes coding for many toxins were located on large plasmids which can be exchanged between different Bt strains during conjugation (González et al. 1982). This opened the way to developing strains with a pool of advantageous crystal proteins genes active against a given insect pest or creating strains containing optimized crystal gene combinations active against both lepidopteran and coleopteran larvae. For example, conjugation was used by Ecogen to construct strain EG2348, the active ingredient of the Condor® bioinsecticide product that contained a combination of *cry* genes encoding crystal proteins particularly active against specific lepidopteran pests of soybean crops. Another strain, EG2424, the active ingredient of Foil®, contained two plasmids, one carrying a *cry1Ac* gene whose product is active against the European corn borer, *Ostrinia nubilalis*, and a *cry3A* gene encoding a crystal protein with activity against the Colorado potato beetle, *Leptinotarsa decemlineata*, a coleopteran pest of potatoes (Baum 1998). This approach was successful for improving the insecticidal properties of *Bt* and its development as a biopesticide. Conjugal transfer of the *cry1Ac* gene into a *Bacillus megaterium* strain which persists in the cotton phyllosphere has also been reported (Bora et al. 1994). An additional advantage of the conjugational approach is that, as plasmid transfer is reported to also occur naturally among Bt strains in soil microcosms (Vilas-Boas et al. 1998, 2000), these genetically-modified Bt strains are not strictly transgenic organisms, an important consideration regarding approval and public acceptance. However, this approach was only applicable to *cry* genes carried by conjugative plasmids and could not be used to associate genes on plasmids from the same incompatibility group. Moreover, it could not be used to over express a given gene or to exploit cryptic *cry* genes or *cry* genes expressed only weakly in Bt.

12.5 Expression of Bt *Cry* Genes in Heterologous Microbial Hosts

A second strategy for improving the exploitation of Bt or increasing its entomopathogenic potential involved diversifying or improving the way the pesticidal Cry toxins were delivered, by using recombinant DNA technology. Several different delivery systems have been developed aiming to increase the persistence and efficiency of Bt Cry toxins in the field. The first of them was the production of Cry toxins active against lepidopteran and coleopteran pest larvae by the non-pathogenic bacterium *Pseudomonas fluorescens*. The bacteria were then killed by means of a physical chemical process after fermentation and the toxins remained enclosed in the cell wall of the dead microorganisms as crystalline inclusions. This process significantly increased their persistence in the environment by protecting them against degradation and inactivation by UV irradiation (Gaertner et al. 1993). This approach has been used by Mycogen, to produce two commercial products: MVP® for controlling lepidopterans and M-TRAK® for controlling coleopterans. In order to control root feeding insects inaccessible to conventional spray technology, the *cry1Ac* gene has also been transferred into the plant colonizers *Pseudomonas cepacia* (Stock et al. 1990) and *Azospirillum* spp. (Udayasuriyan et al. 1995), two bacteria that colonize the roots and leaves of many plants. Similarly, the *cry* genes from Bt were also introduced into a rhizosphere-inhabiting *Bacillus pumilus* isolate to create a delivery system control subterranean feeding insects such as *Agrotis orthogonia* (Selinger et al. 1998). However, only poor levels of expression were achieved probably due to instability of introduced plasmids. By contrast, *Rhizobium leguminosarum* a symbiotic nitrogen-fixing bacterium responsible for root nodule formation in many legume species was used successfully as a delivery vehicle for a *cry3A* gene (Skot et al. 1990) and control of various coleopteran soil-dwelling pests was described (Bezdicek et al. 1994). Using a different approach, Crop Genetics International (CGI) transferred a *cry* gene into *Clavibacter xyli* var. *cynodontis*, an endophytic bacterium that colonizes the internal tissues of various host plants without apparent harms to the hosts which makes endophytes a valuable tool for agricultural industries to protect plants from pests. The gene introduced into this bacterium encoded a CryIAc protein toxic to the larvae of *O. nubilalis* the European corn borer a pest of maize (Lampel et al. 1994). The authors found that this genetically modified endophyte was toxic in bioassays and had significant insecticidal activity *in planta*. This recombinant strain has been used to inoculate corn for the control of European corn borer infestation (Tomasino et al. 1995). Encouraging results have been obtained and a product called InCide® has been developed using this technology. Similarly, a *cry1Ac7* gene of *Bacillus thuringiensis* strain 234 was transferred into the endophytic bacterium *Herbaspirillum seropedica*, which was found in sugarcane, for controlling the sugarcane borer *Eldana saccharina* (Downing et al. 2000). It was shown that this genetically modified endophyte could inhibit the growth of the sugarcane borer larvae to some degree. A recombinant *B. sphaericus* strain, which had longer environmental persistence in the aquatic feeding zones of various *Aedes*, *Culex* and *Anopheles* mosquito species than Bt, was also constructed by integrating into its chromosome the dipteran-active Cry11A protein from *B. thuringiensis* subsp. *israelensis* (Poncet et al. 1997). Attempts have also been made to express Bt toxins used to control the mosquito vectors of diseases, in aquatic bacteria that inhabit the larval feeding zone (Federici et al. 2003). Various bacteria have been manipulated in this way including species of cyanobacteria belonging to the genus *Synechococcus* (Chungjatupornchai 1990; Soltes-Rak et al. 1993), *Caulobacter crescentus* (Thanabalu et al. 1992), *Anabaena* sp. (Khasdan et al. 2003) and *Asticcacaulis excentricus* (Armengol et al. 2005) but, to date, none of these recombinant organisms has been commercially developed.

12.6 Development of Recombinant *Bt* Strains

Conjugation was initially used to manipulate *Bt* strains genetically because no transformation procedure was available. In 1989, several laboratories reported that vegetative Bt cells were readily transformed with plasmid DNA by electroporation (Bone and Ellar 1989; Lereclus et al. 1989). This opened up new possibilities for improving the entomopathogenic potential of Bt making it possible to fully exploit the high natural diversity of Bt strains and toxins to create modified variants of these bacteria with greater efficacies. However, the construction of genetically engineered Bt strains first required the development of effective host/vector systems with vector plasmids able to replicate and persist in a stable manner in this bacterium. To achieve this purpose, plasmids carried by Bt were cloned and the regions required for their replication and stability identified (Arantes and Lereclus 1991; Baum and Gilbert 1991). A wide variety of shuttle vectors (*E. coli*/*Bt*) were constructed using the replication regions of these Bt resident plasmids (Baum et al. 1990; Arantes and Lereclus 1991; Chak et al. 1994) and used to introduce cloned *cry* genes into *Bt*. Heat-sensitive vectors capable of integration were also constructed and used to insert *cry* genes by homologous recombination into resident plasmids (Lereclus et al. 1992) or directly into the chromosome (Kalman et al. 1995). For example, Lereclus and collaborators, in 1992, introduced the coleopteran-specific *cry3A* gene into the resident plasmid of the strain *kurstaki* HD73 toxic to lepidopteran insects, using such a thermosensitive vector that was eliminated after recombination (Lereclus et al. 1992). The resulting strain had insecticidal activity against both lepidopterans and coleopterans insects. These thermosensitive plasmids were also used to disrupt genes involved in sporulation, such as *spo0F*, *spo0A*, *sigE* and *sigK* (see below), to produce asporogenic strains of Bt (Lereclus et al. 1995). For application in the field, recombinant strains of Bt from which antibiotic resistance genes and non-*Bt* DNA sequences were selectively eliminated, were also constructed using a second generation of vectors based on the specific resolution site of the Bt class II transposons, Tn*4430* and Tn*5401* (Baum et al. 1996; Sanchis et al. 1996). These new site-specific recombination vectors were designed specifically for selectively removing any unnecessary sequences such as antibiotic-resistant genes and non-Bt replication origins from *B. thuringiensis* after introduction of the plasmid into the host strain (Sanchis et al. 1997). Therefore, the recombinant Bt strains obtained by this procedure and vectors were free of non-Bt DNA and did not harbour any antibiotic resistance markers. In these vectors the DNA sequences, necessary for selection and replication in *E. coli* at the earliest stages of assembly, but undesirable in the final Bt recombinant strains, are flanked with two internal resolution sites (IRS) in the same orientation. In an appropriate host background (Bt strains containing Tn*4430* or Tn*5401*) site-specific recombination between the duplicate IRSs, catalyzed by the recombinases of Tn*4430* or Tn*5401* (Mahillon and Lereclus 1988; Baum 1995); this eliminates the resistance marker genes and other non-*Bt* DNA after introduction of the vector into the bacterium and selection of the transformants. The first live recombinant Bt product to be produced using this technology was Bt strain EG7673 that produces two types of Cry3 proteins active against beetles (Cry3A and Cry3Bb) and a protein active against caterpillars(Cry1Ac) and that was sold by Ecogen under the trade name of Raven®. Wang and collaborators also constructed a new strain by introducing the *cry3Aa7* gene into the UV17 strain, which produces Cry1Aa, Cry1Ac, Cry1Ca and Cry2Ab (Wang et al. 2006, 2008). The new strain was toxic to both lepidopteran and coleopteran insects. More recently, Bt strain engineering efforts have mainly focused in extending Bt host range and potency by creating strains harbouring unique combinations of crystal proteins not as yet found in nature and in increasing the duration of the Bt products applied in the field. To protect the crystal proteins from UV degradation Liu and collaborators initially used a *Streptomyces lividans* strain containing a recombinant plasmid allowing production of high amounts of melanin (Liu et al. 1993). The melanin obtained by the fermentation of *Streptomyces lividans* was added into a *Bacillus thuringiensis* var. *israelensis* formulation and it was seen that melanin was a photoprotective agent against UV light. A mutant of *B. thuringiensis* var. *kurstaki* producing melanin was also generated following UV irradiation of a wild-type strain culture. This mutant was more resistant to UV light than the parent strain and had greater insecticidal activity toward the diamondback moth, *Plutella xylostella* (Patel et al. 1996). In another example, a genetically engineered strain Bt TD841 that produced both melanin and a Cry1A protein was constructed, and its UV resistance was evaluated in the laboratory. Bioassays demonstrated that the UV resistance of this recombinant was enhanced 9.7-fold compared to its parental strain Bt HC42 after 4-h UV irradiation (Zhang et al. 2008). In another strategy aiming to increase the potency of Bt, a recombinant plasmid containing the *cry1Ac* gene from Bt and a chitinase gene from tobacco was also introduced into a Bt acrystalliferous strain by electroporation. The insecticidal activity of the transformant against *Helicoverpa armigera* was 11 and 18 times higher than a transformant expressing a single *cry1Ac* at 48 and 72 h respectively (Ding et al. 2008). More recently, Liu and collaborators also reported the construction of a new coleopteran-active strain BIOT185 that express both Cry-8ca2 and Cry8Ea1 toxic towards scarab insects (Liu et al. 2010).

12.7 Development of Asporogenic *Bt* Strains

The production of the toxin crystal is usually coupled to the sporulation phase of growth and is regulated and dependent on it (Agaisse and Lereclus 1995). Indeed most of the *cry* genes were found to be transcribed from promoters active only during sporulation. For the *cry1Aa* lepidopteran-active gene two overlapping transcription start sites have been mapped, defining two sporulation-specific promoters (BtI and BtII) used sequentially (Wong et al. 1983). Two genes encoding sigma factors, σ^{35} and σ^{28}, which specifically direct the transcription of the *cryl* genes were cloned and sequenced (Brown and Whiteley 1988, 1990; Adams et al. 1991). They were found to be homologous to the genes encoding the *B. subtilis* sporulation stage specific sigma factors σ^{E} and σ^{K}, respectively. Unlike to the other δ-endotoxin genes, the promoter that directs the transcription of the coleopteran *cry3A* gene is dissimilar to the σ^{35} and σ^{28} sporulation specific promoters and is activated at the onset of sporulation independently of the factors implicated in the initiation of sporulation (Agaisse and Lereclus 1994). Lecadet and collaborators, in 1992; reported that introduction of a *cry1Aa* gene into a strain harbouring a *cry3A* gene resulted in a two or threefold increase in the total amount of protein produced (Lecadet et al. 1992). This is probably because the expression systems of the *cry* genes are different, so they do not compete for rate-limiting elements of gene expression such as specific sigma factors (Agaisse and Lereclus 1995). An obvious implication of these findings was that it should also be possible to produce the non-sporulation dependent Cry3A toxin in a sporulation deficient genetic background. The use of molecular genetic techniques allowed the creation of defined mutations in a number of genes that controlled the sporulation process: *spo0F* (Malvar and Baum 1994), *spo0A* (Lereclus et al. 1995), *sigE* (Bravo et al. 1996) and *sigK* (Sanchis et al. 1999). The *cry3A* gene was the first expressed in a SpoOA^{-} mutant of Bt resulting in the production of large amounts of toxin (Lereclus et al. 1995). Soon after, a sporulation-dependent lepidopteran active *cry1C* gene was expressed in a sporulation-deficient Spo0A mutant of Bt by fusing the coding sequence of the *cry1C* gene to the non-sporulation dependent *cry3A* promoter (Sanchis et al. 1996). Another non-sporulating derivative of a wild-type Bt *kurstaki* strain was constructed by disrupting the chromosomal *sigK* gene, which encodes the σ^{28} late sporulation-specific sigma factor. This sporulation-deficient SigK^{-} Bt strain naturally carried a *cry1Ac* gene, highly active against *Ostrinia nubilalis* (European corn borer) and *Heliothis virescens* (tobacco budworm). In both cases, the toxins accumulated in the mother cell compartment forming crystal inclusions that remained encapsulated within the cell wall. A chimeric *cry1C/1Ab* gene was then introduced into this sporulation-deficient SigK^{-} strain via a site-specific recombination vector with high inheritable stability. The product of the *cry1C/1Ab* gene had a high activity against various lepidopteran pests, including *Spodoptera littoralis* and *Spodoptera exigua*. Therefore, the final recombinant strain, designated as AGRO2, produced large amounts of two different crystal proteins, which conferred it a broader activity spectrum than the parental strain, and the crystals produced remained encapsulated within the cells, which protected them from deactivation by ultraviolet radiation (Sanchis et al. 1999) as is the case for Bt toxins encapsulated in *P. fluorescens*. The encapsulation of the crystals in Bt ghost cells defective in sporulation is therefore an alternative to expressing Bt toxins in killed *Pseudomonas fluorescens* cells (see above) with the advantage that Bt is naturally capable of stably expressing *cry* genes to high levels. An additional advantage of developing asporogenic variants of Bt strains is that no viable spores would be present in the final product, thereby minimizing any environmental effects arising from the dissemination of large numbers of viable spores. These strains are thus a very safe biopesticides. In greenhouse and field trials, the AGRO2 strain applied at 80 g of ai/ha, showed good activity against a cabbage pest complex *Trichoplusia ni*, and *Plutella xylostella* (Sanchis et al. 1999).

12.8 Concluding Remarks

The increase of agricultural production and the development of microbial insecticides for effective insect control in the context of sustainable agriculture are two major challenges facing the world in the twenty-first century. Bt formulations used as sprays are among the safest microbial pesticides known but they only account for approximately 4% of the global insecticide spray market, estimated to be US$ 8 billion per annum in 2005. Its success in the past 20 years was essentially due to its ease of production, good specificity and the need to find alternatives to the widespread resistance to chemical insecticides, rather than to its safeness to humans and other non-target species. Since 1996, the most significant increase in the use of Bt and their insecticidal crystal proteins as a bio-control agent has been the wide adoption of transgenic crops expressing Bt Cry toxins (Sanchis 2011). Despite this large increase in usage of Bt crops, niche markets for Bt foliar sprays have remained, particularly for organic vegetable and fruit high value crops that have not been transformed with *cry* genes, as well as for bacterial insecticides to control vector mosquitoes. Indeed, Bt biological pesticides remain one of the few effective and non-synthetic insecticides that organic farmers can use in emergency to control insect infestations. They are also particularly well suited to Integrated Pest Management programs because they preserve natural predator or parasitoid populations much better than broad spectrum chemical pesticides. Another advantage of Bt sprays is that if they are used properly and over small areas the likelihood of insect resistance emerging under such usage is minimal. However, delivering a correct Bt dose to a susceptible stage of the pest before it has caused unacceptable damage is essential in developing a suitable pest managements strategies based on Bt. To achieve this goal, some of the limitations of Bt microbial preparations, such as potency, field instability and lack of capacity to reach cryptic pests, have been overcome in the past two decades by using molecular biology techniques. By cloning and expressing individual Cry proteins in acrystalliferous Bt strains it has been possible to identify those that were particularly active on a given target insect pest. Optimised combinations of *cry* genes have then been inserted into a Bt-host with well known safety and able to express these genes to high levels (25–30% of its total protein), using natural Bt plasmids as cloning vectors. The total amount of crystal protein produced by a recombinant strain has also been increased by combining various *cry* genes with different promoters into the same strain. Gene-disabling of the sporulation-specific neutral protease A in Bt has also led to yield enhancement of Cry proteins probably due to reduced crystal protein degradation (Donovan et al. 1997). Cry proteins have also been modified through protein engineering for increasing their toxicity and/or insecticidal spectrum. Diversifying the delivery of the toxin, by expressing genes encoding Cry proteins in diverse organisms occupying the same ecological environment as the targeted insects has also been explored, although until now these recombinant organisms have only been used for experimental purposes. The development of cloning vectors allowing elimination of unwanted DNA (e.g. antibiotic resistance markers, origins of replication that are functional in non-Gram-positive bacteria or more generally non-Bt DNA) when constructing Bt recombinant strains has also been useful in minimizing potential hazards resulting from the use genetically engineered organisms. Finally, genetically engineered sporulation-deficient Bt strains producing high amounts of Cry proteins have been made. These mutants that cannot survive in the environment are thus very safe recombinant biopesticides. All these elements have greatly facilitated regulatory approval and public acceptance for conducting large-scale field trials with recombinant bacteria. Although at present most of these genetically engineered Bt strains have not yet been developed commercially, there is much scope for development of these novel Bt strains into biopesticides. In the future, greater environmental concerns and tighter safety regulations may make these non-viable genetically modified bacteria an even much more attractive alternative to chemical insecticides.

References

Adams LF, Brown KL, Whiteley H (1991) Molecular cloning and characterization of two genes encoding sigma factors that direct transcription from a *Bacillus thuringiensis* crystal gene promoter. J Bacteriol 173:3846–3854

Agaisse H, Lereclus D (1994) Expression in *Bacillus subtilis* of the *Bacillus thuringiensis cryIIIA* toxin gene is not dependent on a sporulation-specific sigma factor and is increased in a *spo0A* mutant. J Bacteriol 176:4734–4741

Agaisse H, Lereclus D (1995) How does *Bacillus thuringiensis* produce so much insecticidal crystal protein? J Bacteriol 177:6027–6032

Arantes O, Lereclus D (1991) Construction of cloning vectors for *Bacillus thuringiensis*. Gene 108:115–119

Armengol G, Guevara OE, Orduz S, Crickmore N (2005) Expression of the *Bacillus thuringiensis* mosquitocidal toxin Cry11Aa in the aquatic bacterium *Asticcacaulis excentricus*. Curr Microbiol 51:430–433

Baum JA (1995) TnpI recombinase: identification of sites within Tn*5401* required for TnpI binding and site-specific recombination. J Bacteriol 177:4036–4042

Baum JA (1998) Transgenic *Bacillus thuringiensis*. Phytoprotection 79:127–130

Baum JA, Gilbert MP (1991) Characterization and comparative sequence analysis of replication origins from three large *Bacillus thuringiensis* plasmids. J Bacteriol 173:5280–5289

Baum JA, Coyle DM, Jany CS, Gilbert MP, Gawron-Burke C (1990) Novel cloning vectors for *Bacillus thuringiensis*. Appl Environ Microbiol 56:3420–3428

Baum JA, Kakefuda M, Gawron-Burke C (1996) Engineering *Bacillus thuringiensis* bioinsecticides with an indigenous site-specific recombination system. Appl Environ Microbiol 62:4367–4437

Behle RW, McGuire MR, Shasha BS (1997) Effects of sunlight and simulated rain on residual activity of *Bacillus thuringiensis* formulations. J Econ Entomol 90:1560–1566

Bezdicek DF, Quinn MA, Forse L, Heron D, Kahn ML (1994) Insecticidal activity and competitiveness of Rhizobium spp. containing the *Bacillus thuringiensis* subsp. *tenebrionis* endotoxin gene (*cry III*) in legume nodules. Soil Biol Biochem 26:1637–1646

Bone EJ, Ellar DJ (1989) Transformation of *Bacillus thuringiensis* by electroporation. FEMS Microbiol Lett 58:171–178

Bora RS, Murty MG, Shenbagarathi R, Sekar V (1994) Introduction of a lepidopteran-specifc insecticidal protein gene of *Bacillus thuringiensis* subsp. *kurstaki* by conjugal transfer into *Bacillus megaterium* strain that persists in cotton phyllosphere. Appl Environ Microbiol 60:214–222

Bravo A, Agaisse H, Salamitou S et al (1996) Analysis of *cryIAa* expression in *sigE* and *sigK* mutants of *Bacillus thuringiensis*. Mol Gen Genet 250:734–741

Bravo A, Gill SS, Soberon M (2007) Mode of action of *Bacillus thuringiensis* Cry and Cyt toxins and their potential for insect control. Toxicon 49:423–435

Brown KL, Whiteley HR (1988) Isolation of a *Bacillus thuringiensis* RNA polymerase capable of transcribing crystal protein genes. Proc Natl Acad Sci U S A 85:4166–4170

Brown KL, Whiteley HR (1990) Isolation of the second *Bacillus thuringiensis* RNA polymerase that transcribes from a crystal protein gene promoter. J Bacteriol 172:6682–6688

Chak KF, Tsen MY, Yamamoto T (1994) Expression of the crystal protein gene under the control of the a-amylase promoter in *Bacillus thuringiensis* strains. Appl Environ Microbiol 60:2304–2310

Chungjatupornchai W (1990) Expression of the mosquitocidal protein genes of *Bacillus thuringiensis* subsp. *israelensis* and the herbicideresistance gene *bar* in *Synechocystis* PCC 6803. Curr Microbiol 21:283–288

Craveiro KIC, Gomes JE Jr, Silva MCM et al (2010) Variant Cry1Ia toxins generated by DNA shuffling are active against sugarcane giant borer. J Biotechnol 145:215–221

Crickmore N, Zeigler D, Feitelson J et al (1998) Revision of the nomenclature for the *Bacillus thuringiensis* pesticidal crystal proteins. Microbiol Mol Biol Rev 62:807–813

de Maagd RA, Kwa MSG, van der Klei H et al (1996) Domain III substitution in *Bacillus thuringiensis* delta-endotoxin Cry1Ab results in superior toxicity for *Spodoptera exigua* and altered membrane protein recognition. Appl Environ Microbiol 62:1537–1543

Ding X, Luo Z, Xia L, Gao B, Sun Y, Zhang Y (2008) Improving the insecticidal activity by expression of a recombinant *cry1Ac* gene with chitinase-encoding gene in acrystalliferous *Bacillus thuringiensis*. Curr Microbiol 56:442–446

Donovan WP, Tan Y, Slaney AC (1997) Cloning of the *nprA* gene for neutral protease A of *Bacillus thuringiensis* and effect of *in vivo* deletion of *nprA* on insecticidal crystal protein. Appl Environ Microbiol 63:2311–2317

Downing KJ, Leslie G, Thomson JA (2000) Biocontrol of the sugarcane borer *Eldana saccharina* by expression of the *Bacillus thuringiensis cry1Ac7* and *Serratia marcescens chiA* genes in sugarcane-associated bacteria. Appl Environ Microbiol 66:2804–2810

Dulmage HD (1970) Insecticidal activity of HD1 a new isolate of *Bacillus thuringiensis* var. *alesti*. J Invertebr Pathol 15:232–239

Federici BA, Park H, Bideshi DK, Wirt MC, Johnson JJ (2003) Recombinant bacteria for mosquito control. J Exp Biol 206:3877–3885

Gaertner FH, Quick TC, Thompson MA (1993) CellCap: an encapsulation system for insecticidal biotoxin proteins. In: Kim L (ed) Advanced engineered pesticides. Marcel Dekker Inc, New York

Goldberg LJ, Margalit J (1977) A bacterial spore demonstrating rapid larvicidal activity against *Anopheles sergentii*, *Uranotaenia unguiculata*, *Culex univittatus*, *Aedes aegypti* and *Culex pipiens*. Mosq News 37:355–358

González JMJ, Brown BJ, Carlton BC (1982) Transfer of *Bacillus thuringiensis* plasmids coding for delta-endotoxin among strains of *B. thuringiensis* and *B. cereus*. Proc Natl Acad Sci U S A 79:6951–6955

Ishikawa H, Hoshino Y, Motoki Y et al (2007) A system for the directed evolution of the insecticidal protein from *Bacillus thuringiensis*. Mol Biotechnol 36:90–101

Kalman S, Kiehne KL, Cooper N et al (1995) Enhanced production of insecticidal proteins in *Bacillus thuringiensis* strains carrying an additional crystal protein gene in their chromosomes. Appl Environ Microbiol 61:3063–3068

Khasdan V, Ben-Dov E, Manasherob R, Boussiba S, Zaritsky A (2003) Mosquito larvicidal activity of transgenic Anabaena PCC 7120 expressing toxin genes from *Bacillus thuringiensis* subsp *israelensis*. FEMS Microbiol Lett 227:189–195

Krieg A, Huger AM, Langenbruch GA, Schnetter W (1983) *Bacillus thuringiensis* var. *tenebrionis*: a new pathotype effective against larvae of Coleoptera. Z Angew Entomol 96:500–508

Lampel JS, Canter GL, Dimock MB et al (1994) Integrative cloning, expression, and stability of the *cry1A(c)* gene from *Bacillus thuringiensis* subsp. *kurstaki* in a recombinant strain of *Clavibacter xyli* subsp. *cynodontis*. Appl Environ Microbiol 60:501–508

Lassner M, Bedbrook J (2001) Directed molecular evolution in plant improvement. Curr Opin Plant Biol 4:152–156

Lecadet M-M, Chaufaux J, Ribier J, Lereclus D (1992) Construction of novel *Bacillus thuringiensis* strains with different insecticidal specificities by transduction and by transformation. Appl Environ Microbiol 58:840–849

Lereclus D, Arantes O, Chaufaux J et al (1989) Transformation and expression of a cloned ∂-endotoxin gene in *Bacillus thuringiensis*. FEMS Microbiol Lett 60:211–218

Lereclus D, Vallade M, Chaufaux J, Arantes O, Rambaud S (1992) Expansion of the insecticidal hostrange of *Bacillus thuringiensis* by in vivo genetic recombination. Biotechnology 10:418–421

Lereclus D, Agaisse H, Gominet M et al (1995) Overproduction of encapsulated insecticidal crystal proteins in a *Bacillus thuringiensis spo0A* mutant. Biotechnology 13:67–71

Liu YB, Tabashnik BE (1997) Experimental evidence that refuges delay insect adaptation to *Bacillus thuringiensis*. Proc R Soc Lond B 264:605–610

Liu Y, Sui M, Ji D, Wu I, Chou C, Chen C (1993) Protection from ultraviolet irradiation by melanin of mosquitocidal activity of *Bacillus thuringiensis* var. *israelensis*. J Invertebr Pathol 62:131–136

Liu J, Yan G, Shu C et al (2010) Construction of a *Bacillus thuringiensis* engineered strain with high toxicity and broad pesticidal spectrum against coleopteran insects. Appl Microbiol Biotechnol 87:243–249

Mahillon J, Lereclus D (1988) Structural and functional analysis of Tn*4430*: identification of an integrase-like protein involved in the co-integrate-resolution process. EMBO J 7:1515–1526

Malvar T, Baum JA (1994) Tn*5401* disruption of the *spoOF* gene, identified by direct chromosomal sequencing, results in CryIIIA overproduction in *Bacillus thuringiensis*. J Bacteriol 176:4750–4753

Martin PA, Travers RS (1989) Worldwide abundance and distribution of *Bacillus thuringiensis* isolates. Appl Environ Microbiol 55:2437–2442

Naimov S, Weemen-Hendriks M, Dukiandjiev S, Maagd RA de (2001) *Bacillus thuringiensis* delta-endotoxin Cry1 hybrid proteins with increased activity against the Colorado potato beetle. Appl Environ Microbiol 67:5328–5330

Naimov S, Dukiandjiev S, Maagd RA de (2003) A hybrid *Bacillus thuringiensis* delta-endotoxin gives resistance against a coleopteran and a lepidopteran pest in transgenic potato. Plant Biotechnol J 1:51–57

Pardo-Lopez L, Munoz-Garay C, Porta H et al (2009) Strategies to improve the insecticidal activity of Cry toxins from *Bacillus thuringiensis*. Peptides 30:589–595

Patel KR, Wyman JA, Patel KA, Burden BJ (1996) A Mutant of *Bacillus thuringiensis* producing a dark-brown pigment with increased UV resistance and insecticidal activity. J Invertebr Pathol 67:120–124

Poncet S, Bernard C, Dervyn E, Cayley J, Klier A, Rapoport G (1997) Improvement of *Bacillus sphaericus* toxicity against dipteran larvae by integration, *via* homologous recombination, of the Cry11A toxin gene from *Bacillus thuringiensis* subsp. *israelensis*. Appl Environ Microbiol 63:4413–4420

Pusztai M, Fast M, Gringorten L et al (1991) The mechanism of sunlight-mediated inactivation of *Bacillus thuringiensis* crystals. Biochem J 273:43–47

Rajamohan F, Alzate O, Cotrill JA, Curtiss A, Dean DH (1996) Protein engineering of *Bacillus thuringiensis* delta-endotoxin: mutations at domain II of CryIAb enhance receptor affinity and toxicity toward gypsy moth larvae. Proc Natl Acad Sci U S A 93:14338–14343

Sanchis V (2011) From microbial sprays to insect-resistant transgenic plants: history of the biospesticide *Bacillus thuringiensis*. A review. Agron Sustain Dev 31:217–231. doi:10.1051/agro/2010027

Sanchis V, Bourguet D (2008) *Bacillus thuringiensis*: applications in agriculture and insect resistance management: a review. Agron Sustain Dev 28:11–20. doi:10.1051/agro:2007054

Sanchis V, Agaisse H, Chaufaux J et al (1996) Construction of new insecticidal *Bacillus thuringiensis* recombinant strains by using the sporulation non-dependent expression system of *cryIIIA* and a site specific recombination vector. J Biotechnol 48:81–96

Sanchis V, Agaisse H, Chaufaux J et al (1997) A recombinase-mediated system for elimination of antibiotic resistance gene markers from genetically engineered *Bacillus thuringiensis* strains. Appl Environ Microbiol 6:779–784

Sanchis V, Gohar M, Chaufaux J et al (1999) Development and field performance of a broad spectrum non-viable asporogenic recombinant strain of *Bacillus thuringiensis* with greater potency and UV resistance. Appl Environ Microbiol 69:4032–4039

Schnepf H, Whiteley HR (1981) Cloning and expression of the *Bacillus thuringiensis* crystal protein gene in *Escherichia coli*. Proc Natl Acad Sci U S A 78:2893–2897

Schnepf HE, Wong H, Whiteley HR (1985) The amino acid sequence of a crystal protein from *Bacillus thuringiensis* deduced from the DNA base sequence. J Biol Chem 260:6264–6272

Selinger LB, Khachatourians GG, Byers JR, Hynes MF (1998) Expression of a *Bacillus thuringiensis* ∂-endotoxin gene by *Bacillus pumilus*. Can J Microbiol 44:259–269

Shelton AM, Zhao JZ, Roush RT (2002) Economic, ecological, food safety, and social consequences of the deployment of Bt transgenic plants. Annu Rev Entomol 47:845–881

Siegel JP (2001) The mammalian safety of *Bacillus thuringiensis* based insecticides. J Invertebr Pathol 77:13–21

Skot L, Harrison SP, Nath A et al (1990) Expression of insecticidal activity in *Rhizobium* containing the ∂-endotoxin gene cloned from *Bacillus thuringiensis* subsp. *tenebrionis*. Plant Soil 127:285–295

Smith A, Couche GA (1991) The phylloplane as a source of *Bacillus thuringiensis* variants. Appl Environ Microbiol 57:311–315

Soltes-Rak E, Kushner DJ, Williams DD, Coleman JR (1993) Effect of promoter modification on mosquitocidal *cryIVB* gene expression in *Synechococcus* sp. strain PCC 7942. Appl Environ Microbiol 59:2404–2410

Stock CA, McLoughlin TJ, Klein JA et al (1990) Expression of a *Bacillus thuringiensis* crystal protein gene in *Pseudomonas cepecia*. Can J Microbiol 36:879–884

Thanabalu T, Hindley J, Brenner S, Oei C, Berry C (1992) Expression of the mosquitocidal toxins of *Bacillus sphaericus* and *Bacillus thuringiensis* subsp. *israelensis* by recombinant *Caulobacter crescentus*, a vehicle for biological control of aquatic insect larvae. Appl Environ Microbiol 58:905–910

Tomasino SF, Leister RT, Dimock MB, Beach RM, Kelly JL (1995) Field performance of *Clavibacter xyli* subsp. *cynodontis* expressing the insecticidal crystal protein genes cry1Ac of *Bacillus thuringiensis* against European corn borer in field corn. Biol Control 5:442–448

Udayasuriyan V, Nakamura A, Masaki H et al (1995) Transfer of an insecticidal protein gene of *Bacillus thuringiensis* into plant-colonizing *Azospirillum*. World J Microbiol Biotechnol 11:163–167

Vaeck M, Reynaerts A, Höfte H et al (1987) Transgenic plants protected from insect attack. Nature 328:33–37

Vilas-Boas GFL, Vilas-Boas LA, Lereclus D, Arantes OMN (1998) *Bacillus thuringiensis* conjugation under environmental conditions. FEMS Microbiol Ecol 25:369–374

Vilas-Boas LA, Vilas-Boas GFLT, Saridakis HO et al (2000) Survival and conjugation of *Bacillus thuringiensis* in a soil microcosm. FEMS Microbiol Ecol 31:255–255

Walters FS, Fontes CM de, Hart H, Warren GW, Chen JS (2010) Lepidopteran-active variable-region sequence imparts coleopteran activity in eCry3.1Ab, an engineered *Bacillus thuringiensis* hybrid insecticidal protein. Appl Environ Microbiol 76:3082–3088

Wang G, Zhang J, Song F, Wu J, Feng S, Huang D (2006) Engineered *Bacillus thuringiensis* GO33A with broad insecticidal activity against lepidopteran and coleopteran pests. Appl Microbiol Biotechnol 72:924–930

Wang G, Zhang J, Song F et al (2008) Recombinant *Bacillus thuringiensis* strain shows high insecticidal activity against *Plutella xylostella* and *Leptinotarsa decemlineata* without affecting nontarget species in the field. J Appl Microbiol 105:1536–1543

Wong HC, Schnepf HE, Whiteley HR (1983) Transcriptional and translational start sites for the *Bacillus thuringiensis* crystal protein gene. J Biol Chem 258:1960–1967

Wu SJ, Koller CN, Miller DL, Bauer LS, Dean DH (2000) Enhanced toxicity of *Bacillus thuringiensis* Cry3A delta-endotoxin in coleopterans by mutagenesis in a receptor binding loop. FEBS Lett 473:227–232

Zhang JT, Yan JP, Zheng DS, Sun YJ, Yuan ZM (2008) Expression of *mel* gene improves the UV resistance of *Bacillus thuringiensis*. J Appl Microbiol 10:5151–5157

Part III
Bt as Biopesticide: Applications in Biotechnology

Chapter 13
Genetically Modified *Bacillus thuringiensis* Biopesticides

Lin Li and Ziniu Yu

Abstract The reduction in crop production that results from damages of pests and pathogenic microbes has been the major prohibitive factor restricting the further development of agricultural production worldwide. In the long-term practice of controlling agricultural pests, multiple pest management, including biological approaches, are generally recognized as possible solutions to control pest infestation effectively. Bacterial pesticides, which constitute a series of leading biopesticidal products made from various naturally occurring or genetically modified insecticidal bacteria, have attracted increasing attention as a specific means of controlling agricultural and forestry pests.

Keywords Genetically modified Bt strains · Recombinant methods · Broader insecticidal spectrum · High insecticidal activity · Multifunctional strains

13.1 Introduction

The reduction in crop production that results from damages of pests and pathogenic microbes has been the major prohibitive factor restricting the further development of agricultural production worldwide. In the long-term practice of controlling agricultural pests, multiple pest management, including biological approaches, are generally recognized as possible solutions to control pest infestation effectively. Bacterial pesticides, which constitute a series of leading biopesticidal products made from various naturally occurring or genetically modified insecticidal bacteria, have attracted increasing attention as a specific means of controlling agricultural and forestry pests.

The spore-forming bacterium *Bacillus thuringiensis* (Bt) has the ability to synthesize a large amount of insecticidal crystal proteins (ICPs) during growth and metabolism. This organism has been exploited as a biopesticide for controlling agriculturally important pests for many years. In 1938, the world's first commercial Bt-based product, Sporeine, was launched in France (Sanahuja et al. 2011a). Sub-

L. Li (✉) · Z. Yu
State Key Laboratory of Agricultural Microbiology, Huazhong Agricultural University, Wuhan 430070, China
e-mail: lilin@mail.hzau.edu.cn

E. Sansinenea (ed.), *Bacillus thuringiensis Biotechnology*,
DOI 10.1007/978-94-007-3021-2_13,

sequently, with increasing understanding of its insecticidal properties, as well as its active ingredients, Bt-based pesticides have gradually developed strength, and ultimately have become a new industrial field at the global scale. Numerous investigations and applications of Bt-derived pesticides have proven that, in contrast to chemical and other types of pesticides, they are relatively highly insect-specific, remarkably safe for humans, other mammals, and non-target fauna, less or tardily resistant from target pests, maneuverable and cost-effective for production, and so on. Given these advantages, the industrialization of Bt pesticides has developed rapidly, making them the largest mass-produced and most widely used microbial pesticides in the 1980–1990s worldwide.

Like many other bacterial products, Bt pesticides are manufactured from the various strains that are initially screened from different natural habitats including soils, the bodies of dead pests, plant tissues, and water, among others. The insecticidal activity and nature of these strains directly affects the quality of Bt preparation. Thousands of naturally occurring Bt isolates have been identified since a Bt strain was first reported by the Japanese biologist Shigetane Ishiwatari during an investigation on wilt disease in silk worms in 1901. Some of these isolates have been used for mass production of Bt pesticides. However, although these naturally occurring strains have been proven successful in terms of facilitating fermentative culturing and specific insecticidal toxicity, several disadvantages were noted. The drawbacks of these strains in practical applications include excessive insecticidal specificity, low toxicity to aging or multiply assorted pests, relatively short persistence in the field, and evolving target pest resistance from the massive and repeated use of the agents. Apparently, overcoming these drawbacks by simply further screening new isolates from natural environments is infeasible. Therefore, with progress in molecular biology and genetic manipulations of various Bt strains, the development of genetically modified Bt derivatives as alternative strains for mass production of Bt pesticides has opened up new perspectives on Bt-based pesticides (Fig. 13.1). A variety of engineered Bt biopesticides has been approved for field release by the US EPA and the official regulatory agencies in other countries since the late 1980s (Table 13.1).

13.2 Strategies for Constructing Genetically Modified Bt Strains for Engineered Biopesticide Development

13.2.1 Essentials for Constructing Genetically Modified Bt Strains

13.2.1.1 ICP Genes

During its stationary phase, Bt produces several insecticidal proteins (known as δ-endotoxins, or Cry protein) that accumulate as parasporal crystalline inclusions. These endotoxin proteins are mainly encoded by the ICP genes normally located

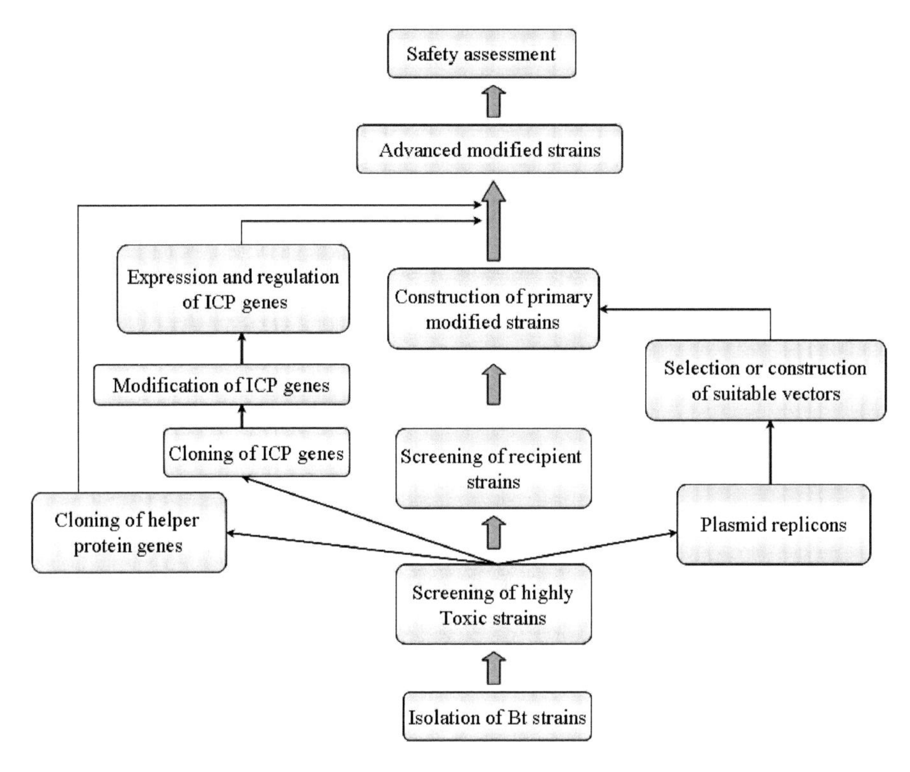

Fig. 13.1 Common strategy of constructing genetically modified strains of *Bacillus thuringiensis*

on the larger plasmids (a few of them probably on the chromosomal DNA) of the host strain (Hofte and Whiteley 1989). The ICP genes of various Bt subspecies/strains are different in their gene-type, existence states, expression level, and insecticidal specificity. For the Bt subsp. *tenebrionis* harboring the *cry3Aa* gene, the encoded Cry3Aa is coleoptera-active. However, for the Bt subsp. *kurstaki* (Btk) and subsp. *israelensis* (Bti) harboring *cry1A*, *cry2A*, and *cry11*, respectively, the corresponding proteins are less toxic to coleopteran pests (Keller and Langenbruch 1993). Cry1Ca has relatively high toxicity against *Spodoptera* pests, whereas other ICPs have remarkably low toxicity to such pests (Aronson et al. 1986; Hofte and Whiteley 1989). Even for those ICPs that are highly similar in protein structure, insecticidal activities against certain pests may also be significantly different. For example, among three main Cry1A proteins, the toxicity of Cry1Aa against *Bombyx mori* larvae is almost 400 times more than that of Cry1Ac. In contrast, Cry1Ac exhibits 10 times toxicity against the *Heliothis virescens* and *Trichoplusia ni* larvae over that of Cry1Aa, whereas Cry1Ab has the highest toxicity against *Spodoptera* larvae among these three Cry1A proteins (Ge et al. 1991). In addition, Cry1Ab is also the only Cry1A-type protein toxic to mosquito larvae (Yamamoto and Powell 1993). These differences constitute the varied insecticidal spectra and toxicity levels of different subspecies and strains, which provides a basis for combining insec-

Table 13.1 Partial registered genetically modified *B. thuringiensis* biopesticide products

Product	Producing bacteria or active ingredient	Main target pest	Manufacturer
Foil	Btk EG2424; conjugate strain	*Leptinotarsa decemlineata, Ostrinia nubilalis*	USA: Ecogen Inc.
Cutlass	Btk EG2371; conjugate strain	*Pseudaletia unipuncta, Spodoptera exigua, Trichoplusia ni, Hellula rogatalis, Plutella xylostella*	USA: Ecogen Inc.
Condor	Btk EG2348; conjugate strain	*Pseudoplusia includens, Anticarsia genimatalis, Flathypena scabra*	USA: Ecogen Inc.
MVP®	Btk δ-endotoxin encapsulated in killed *Pseudomonas fluorescens*, CellCap® product	Lepidopterans (esp *P. xylostella*)	USA: Mycogen Corporation
M-Trak®	Bt subsp. san diego δ-endotoxin encapsulated in killed *P. fluorescens*, CellCap® product	Coleopterans	USA: Mycogen Corporation
M-Peril®	Btk δ-endotoxin encapsulated in killed *P. fluorescens*, CellCap® product	*O. nubilalis*	USA: Mycogen Corporation
M-One®	Bt subsp. san diego δ-endotoxin encapsulated in killed *P. fluorescens*, CellCap® product	Coleopterans	USA: Mycogen Corporation
MYX1896®	Bt subsp. *san diego* δ-endotoxin encapsulated in killed *P. fluorescens*, CellCap® product	Coleopterans	USA: Mycogen Corporation
Mattch	Blend of Cry1Ac and Cry1C δ-endotoxin encapsulated in killed *P. fluorescens*, CellCap® product	Lepidopterans	USA: Mycogen Corporation
M/C	Cry1C δ-endotoxin encapsulated in killed *P. fluorescens*, CellCap® product	Lepidopterans	USA: Mycogen Corporation
SAN418	Btk conjugate strain	Lepidopterans, Coleopterans	USA: Sandoz
P. fluorescens P303	Transgenic *P. fluorescens* with Btk *cry* genes	Lepidopterans, Fungal pathogens	China: CAAS
Bt WG-001	Engineered Btk expressing Bti P20	Lepidopterans	China: Huazhong Agric. Univ.
BMB820-Bt	Engineered Btk expressing Bti P19	Lepidopterans	China: Huazhong Agric. Univ.
Bt-TnY	Engineered Btk expressing Bti Cyt1Aa and Cry11Aa	Lepidopterans, Dipterans	China: SUN YAT-SEN UNIV.

ticidal genes that do not occur naturally (Schnepf et al. 1998). With the regulatory effect of different ICP genes and/or the synergic toxicity of intracellularly expressed ICPs, the yielding and insecticidal activity of the engineered strain is thus probably increased.

The ICP genes harbored in various Bt subspecies and strains exhibit complex distribution patterns, which are remarkably variable in their types, copy numbers, localization, and existing states, which are referred to as "expressable" or "cryptic" genes. For instance, the HD-73 strain of Btk was found to harbor only *cry1Ac* located on a 50 MD plasmid (Gonzalez et al. 1981). However, the other strain of this subspecies, HD-1, harbors at least five different ICP genes. Among them, *cry1Aa*, *cry1Ac*, *cry2Aa*, and a "silent" *cry2Ab* were found to locate on a 110 MD plasmid, whereas *cry1Ab* separately located on a 44 MD plasmid (Baum and Malvar 1995). Thus, it is essential to investigate and identify the ICP gene resource of the target strains prior to cloning of a particular ICP gene, in which the possible omission of some "silent" ICP genes could be technically avoided.

The conventional methods for identifying Bt ICP gene types include molecular hybridization or expression product analysis. These methods are generally considered time consuming and have difficulty in obtaining clear results. Polymerase chain reaction (PCR)—based approaches, first introduced to identify Bt ICP genes by Carozzi et al. (1991), have been gradually attracted attention for ICP gene identification because of their ease and effectiveness, which allows the rapid identification of ICP genes of Bt wild-type strains. Gleave et al. (1993) investigated the distribution of *cryIV*-like insecticidal protein genes from 21 Bt serotypes using a similar method, and found that 7 Bt serotypes harbored *cryIV* genes among them. Subsequently, Kalman et al. simplified and accelerated the technique using a set of mixed oligonucleotide primers specific to the *cry1* gene (*cry1Aa–cry1F*). The sizes of PCR-amplified products were compared and the possible novel ICP gene was thus revealed (Kalman et al. 1993). Chak et al. used a new PCR-based technique to investigate the ICP gene types of 225 Bt strains in Taiwan (1994). They compared all known conserved *cryI*-type gene sequences, and then designed two pairs of universal oligonucleotide primers specific to these *cry1*-like ICP genes. Subsequent restriction fragment length polymorphism (RFLP) analysis of the PCR-amplified fragments revealed 14 distinct *cry*-type genes from 20 Bt wild-type strains (Kuo and Chak 1996). Considering its higher accuracy and rapidity, PCR-RFLP has been accepted as a universal procedure for identifying various ICP genes, as well as novel genes.

The first ICP gene, *cry1Aa*, was cloned from Btk HD-1-Dipel strain and was expressed in *Escherichia coli* in 1981 (Schnepf and Whiteley 1981). The localization of this gene was then verified via Tn5-mediated transposition, and the complete sequence of the *cry1Aa* gene was determined, which was deduced to encode an Mr 133.5 kDa peptide (Schnepf et al. 1985). As *cry1Aa* has been identified the functional hybridization probe, many ICP genes active to lepidopteran larvae were cloned and sequenced thereafter. The dipteran-active ICP genes were mainly originated from Bti and subsp. *morrisoni*. Ward et al. reported the cloning of the first mosquitocidal δ-endotoxin gene *cyt1A* from Bti, and characterized it via site-directed mutagenesis analysis and determined its sequence (Ward et al. 1988). Krieg et al. (1983) first isolated a coleopteran-active Bt subsp. *tenebrionis* strain. Several groups cloned and sequenced the corresponding gene *cry3A*, which encodes an Mr 73.1 kDa peptide that differed significantly from Cry1A-like ICPs (Herrnstadt et al. 1987; Hofte et al. 1987; Sekar et al. 1987). With the development of research, an increasing number of ICP genes were cloned and characterized, some of which exhibited novel insecticidal activities against homopteran, hymenopteran, and acarid pests, as well as the target pests from the phyla Platyhelminthes, Nemathelminthes, and Protozoa (Thompson and Gaertner 1990; O'Grady et al. 2007; Salehi Jouzani et al. 2008; Vazquez-Pineda et al. 2010).

13.2.1.2 Helper Protein Genes

Some Bt genes do not directly participate in the insecticidal process, but modulate the expression of other insecticidal proteins. These genes are designated as "helper protein genes." In Bti, the downstream DNA region of the gene *cry11A* flanks a gene that encodes an Mr 20 kDa peptide (Adams et al. 1989). This gene (*p20*) enhances the synthesis of Cyt1A in recombinant *E. coli* strains (Adams et al. 1989; Douek et al. 1992; Visick and Whiteley 1991). In fact, the *cyt1A* expression was found to be lethal to the *E. coli* host cells that lack *p20* (Douek et al. 1992). Moreover, the P20 protein also increased the production of other Bti toxins (Cry4A and Cry11A) and the truncated Cry1C (Rang et al. 1996; Visick and Whiteley 1991; Wu and Federici 1995; Yoshisue et al. 1992). Even Cry1Ac protoxin production is remarkably increased 2.5-fold and the size of the crystals formed is increased approximately threefold (Shao et al. 2001).

In the *cry2Aa* operon of the Btk HD-1 strain, the encoded ORF2 of *orf2* influences the crystallization of Cry2Aa (Crickmore and Ellar 1992). ORF2 is an Mr 29 kDa peptide structurally constituted with an unusual 15-amino acid motif that is repeated 11 times in tandem (Widner and Whiteley 1989). The *orf2* gene is not expressed under normal growth conditions (Widner and Whiteley 1989; Dankocsik et al. 1990; Hodgman et al. 1993) because of the absence of a functional promoter in the upstream DNA region of the *orf2* translational start codon (Hodgman et al. 1993). This gene was expressed when an additional promoter was added, but crystalline inclusions were not formed (Crickmore et al. 1994). Correspondingly, the *orf2* was identified as another helper gene. In Bt subsp. *thompsoni*, the two predominant toxin polypeptides are encoded by the genes *cry40* and *cry34*, which are located in an operon. The *cry40* gene also functions as a helper gene because its encoded polypeptide was found to be nontoxic (Brown 1993). Moreover, in Bti, the genes *p19*, *cry11A*, and *p20* are constituted in an operon (Dervyn et al. 1995). The gene *p19*, which encodes an Mr 19 kDa polypeptide (P19), is homologous to the *orf1* gene of the *cry2A* and *cry2C* operons, and enhances the initial expression of the *cyt1A* gene in *E. coli* (Liu et al. 1999). Therefore, these helper genes (*p19*, *p20*, *orf1*, and *orf2*) are useful for the genetically modified Bt strains in terms of improving the synthesis or stabilization of the crystal proteins (Agaisse and Lereclus 1995).

13.2.1.3 Vectors

Vectors are particularly important for constructing various Bt engineered strains because they effectively deliver the donor genes to the recipient strains. The commonly used vectors for constructing Bt engineered strains include the following:

Cloning and Expression Vectors The most developed cloning and expression vectors for Bt molecular biology and biotechnology are the shuttle plasmid vectors, which can replicate in both Bt and *E. coli* strains. Although most small plasmids that originate from other Gram-positive bacteria are frequently structurally unstable in Bt strains (Arantes and Lereclus 1991; Crickmore et al. 1990b), the approaches to construct the cloning and expression vectors using replicons originating from the stable resident plasmids of Bt strains, which do not replicate through an RC mechanism, has drawn increasing attention. In the Btk HD-263 and HD-73 strains, three replicons from 43, 44, and 60 MDa plasmids were cloned and further used to construct three compatible shuttle vectors, pEG597, pEG853, and pEG854 (Baum et al. 1990). The other plasmid replicon, *ori1030*, which originates from a small plasmid from Btk HD-73, was used to develop a set of cloning and expression vectors with favorable structural and segregational stability in many Bt recipient strains (Arantes and Lereclus 1991; Lereclus et al. 1989, 1992). These vectors have become the most widely used cloning and expression vectors for Bt ICP genes. Moreover, researchers at Sandoz cloned a 9.6 kb DNA fragment from a 75 kb plasmid of Btk HD-73 and deleted the unnecessary sequences for replication, thereby constructed a novel plasmid. The plasmid was found stable and compatible in Bt strains even in the absence of antibiotics (Gamel and Piot 1992).

Transposition Vectors The use of insertion sequences and transposons in constructing the transposition vectors are of particular interest as effective means for transferring or integrating heterologous genes into plasmids or chromosomes of the recipient strain. The insertion sequences *IS10*, *IS231A*, and *IS232*, were used to construct a set of transposition vectors including pGIC055, pGIC057, and pGI21 (Leonard et al. 1998; Mahillon and Kleckner 1992). Through transposition of insertion sequence or transposon-derived vectors, several ICP genes were integrated into the resident plasmid of the Bt recipient strain (Lereclus et al. 1992), and the chromosomal DNA of recipient strains such as *Bacillus subtilis* and *Pseudomonas fluorescens* (Calogero et al. 1989; Obukowicz et al. 1986b). Two transposons in Bt strains, namely, Tn4430 and Tn5401 were found. Both these transposons harbor a gene encoding transposase (*tnpA*) and another encoding resolvase (*tnpI*) between their two terminal inverted repeats (TIRs). Under the catalysis of transposase and resolvase, the DNA region between the two TIRs was shifted to other target positions (Mahillon et al. 1994). A series of transposition vectors using Tn4430 have also been developed (Sun et al. 2000). The structural genes of Tn4430 (*tnpI-tnpA*) were replaced with the multiple pUC18 cloning sites via splicing overlap extension PCR techniques. This modified structure and a fragment containing the *tnpI–tnpA* gene were inserted into a non-stable shuttle vector pBMB622N, resulting in the integrative vectors pBMB-F7 and pBMB-R14. These vectors transfer erythromycin-resistant genes into the chromosome of Bt BMB 171 strain, and could be used to deliver Bt ICP genes or other genes into the Bt chromosomes. Moreover, Yu et al. (1999) also described an approach to insert 8 kb target gene fragments (*cry1Aa* and *cry2Aa*) into the encoding frame of the *tnpA* gene of Tn917. Two transposition vectors, pTYP45 and pTYP26, were produced, which harbor *cry1Aa* and *cry2Aa*, respectively. These genes were further introduced into the chromosomes of the recipient strains using the additional plasmid pHtnpA (harboring the *tnpA*).

Integration Vectors Delecluse et al. reported an integrational plasmid vector, pHT633 (1991). They replaced a 0.55 kb *Bam*HI/*Hpa*I fragment (within the gene *cytA*) of pCB4 that harbors *cytA* with a 1.1 kb *Eco*RI/*Hpa*I fragment that harbors the entire *erm* gene. The pHT633 vector was produced by ligating the 3.1-kb *Eco*RV/*Eco*RI fragment into the integrational vector of pJH101 from *B. subtilis* (Ferrari et al. 1983). After introducing the pHT633 into the Bti recipient strain, homologous recombination through the Campbell mechanism occurred between the two truncated *cytA* fragments of pHT633 and the resident *cytA* gene in the 72 MD plasmid, allowing the entire *erm* gene to be inserted into the encoding frame of the *cytA* gene. In 1995, the researchers of Sandoz Agro Inc. constructed an integration vector, pSB210.2, which allowed the introduction of a foreign gene into the chromosome of a Bt recipient strain. This vector contained a Bt chromosomal DNA fragment that encodes a phosphatidylinositol-specific phospholipase C, which allows Bt subsp. *aizawai cry1C* to be targeted to the homologous region of the Btk chromosome. The recombinant strains show higher activity against *S. exigua* than the parent strains because of their newly acquired ability to produce the Cry1C protein (Kalman et al. 1995). Sansinenea et al. reported a one-step chromosomal integration approach (Sansinenea et al. 2010a, b). They constructed an integrative vector pES harboring Bti IPS82 chromosomal DNA fragments (4–10 kb long), and found one chromosomal sequence of 10 kb long that can functionally serve as a substrate for homologous recombination in the chromosomal DNA of Bti. The translocation features of Tn4430 and Tn5401 have also been used to construct a series of integration vectors, including pBMB-FLCE and pBMB-R14E, which can be used to introduce foreign genes into chromosomal DNA as efficient gene transferring tools (Yue et al. 2003, 2005a, b; Chaoyin et al. 2007; Sun and Yu 2000).

Resolution Vectors Given the transposase- and resolvase-mediated transposition and site-specific resolution nature of Tn4430 and Tn5401, resolution vector systems have been developed to combine a Bt plasmid replicon and an indigenous site-specific recombination system that allows for the selective removal of ancillary or foreign DNA from the recombinant bacterium after introduction of the Cry-encoding plasmid (Baum et al. 1996; Sanchis et al. 1997). This unique feature is of particular value in the development of environmentally safe Bt engineered strains. A similar approach has also been previously reported (Wu et al. 2000), in which a *cry1Ac10* gene transfer system based on the TnpI-mediated site-specific recombination mechanism of Bt transposon Tn4430 was developed. The *cry1Ac10* and the *ori1030* originating from a resident plasmid of a Bt wild-type strain were inserted

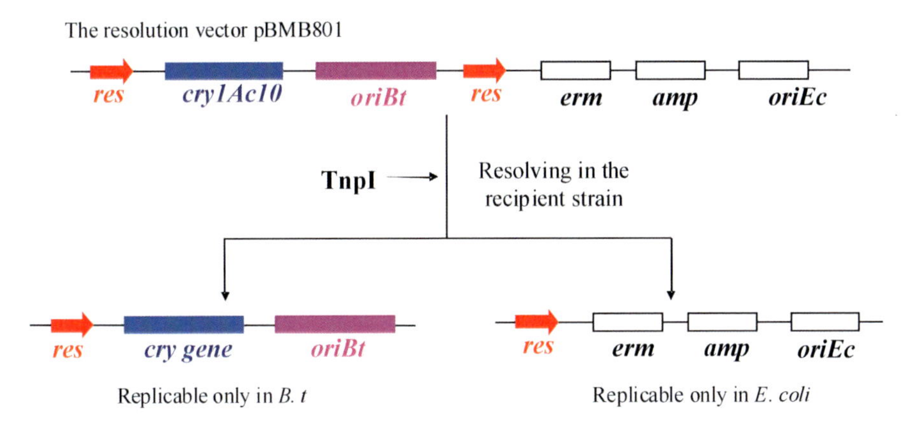

Fig. 13.2 Schematic representation of *B. thuringiensis* resolution vector pBMB801

into two copy sets of *res* sites, resulting in the recombinant resolution plasmid vector pBMB801. When pBMB801 was introduced into the plasmid-free recipient strain BMB171, its borne antibiotic resistance genes and other non-Bt DNAs were selectively eliminated (Fig. 13.2), resulting in a small recombinant plasmid containing only Bt-derived DNAs, which were very stable in the absence of antibiotics in the recipient strain. Thus, because the engineered Bt strain is very similar to naturally occurring Bt strains and it should trigger less environmental safety concerns. This strategy should facilitate regulatory approval for its development as a commercial biopesticide.

13.2.1.4 Recipient Strains

The common approach to constructing Bt engineered strains is the introduction of cloned ICP genes that have been genetically modified *in vitro* into the recipient strains through a vector-delivery system. Therefore, the features of the recipient strains, especially transformation and expression efficiency, have a significant effect on the capacity of the Bt engineered strains, such as insecticidal spectrum and toxicity.

***Bacillus* Strains** As the native hosts of ICP genes, various Bt wild-type strains and their derivatives are the most widely used recipient strains for constructing Bt engineered strains. Considering the insecticidal spectra and toxicity of different Bt subspecies and strains differ remarkably even within the same subspecies, they can be used as potential recipients for Bt engineered strains for different purposes. However, the presence of numerous resident plasmids and the incompatibility of foreign plasmids limit wild-type Bt strains for use as recipient strains. Alternatively, a series of plasmid-curing or acrystalliferous derivatives were found to be more suitable for use as recipient strains because they exhibit better transformation frequency and foreign gene-expression levels. These include Bti 4Q7 (Clark 1987), 4D10 (Yousten unpublished), Cry⁻B (Stahly et al. 1978), and IPS78/11 (Crickmore et al. 1990a). These acrystalliferous strains were mostly screened from Bti and they show uncured resident plasmids. A plasmid-free mutant strain BMB171, which originates from Btk, was obtained via step-up screening treatments such as increasing the growth temperature (to 42°C, 44°C, and 46°C) and adding 0.05% SDS as a plasmid-curing agent. The highest transformation frequency of BMB171 with several common plasmids was 10^7 transformants/μg DNA; its transformation efficiency and expression properties are better than that of the recipient strains discussed (Li et al. 2000; Li and Yu 1999). Moreover, in addition to Bt wild-type or acrystalliferous strains, various Bt ICP genes have also been successfully transferred into *B. subtilis* (Shivakumar et al. 1986, 1989; Calogero et al. 1989), *Bacillus sphaericus* (Poncet et al. 1994, 1997), *Bacillus megaterium* (Mettus and Macaluso 1990; Shivakumar et al. 1989), and *Bacillus pumilus* (Selinger et al. 1998), among others.

E. coli Expression of ICP genes with a long half-life and strong stabilization in the host strains are often required for high yielding Cry proteins. Although other strains have been fairly well exploited as specialized hosts, the many advantages of *E. coli* have ensured that remains a valuable host for the high rate production of various ICPs. As the common host of the cloned ICPs, *E. coli* strains have been used as the recipient strains for the expression of various ICP genes. Unlike the expression in Bt strains wherein the gene is primarily under transcriptional control during sporulation, the crystal protein is produced at all growth stages of the recombinant *E. coli* strain that expresses the cloned *cry1Aa* gene (Wong et al. 1983). Interestingly, Ge et al. (1990) reported that the maximal *cry1Ac* expression in *E. coli* JM103 reaches up to 48% cellular protein, forming crystals with the same lattice structure as the native crystals formed in Bt strains.

***Pseudomonas* Strain** Many *Pseudomonas* species are the dominant microorganisms in plant rhizospheres and leaves, and they are highly resistant to harsh environments. They are often considered as the ideal recipient strains for constructing ICP-producing engineered strains. Several CellCap products of Mycogen Corporation, such as MVP, M-Trak, M-One, and MYX1896 (Table 13.1), were manufactured from engineered *P. fluorescens* strains that express different ICP genes. Moreover, through the transposition and integration of vectors, several ICP genes have been integrated into the chromosome of root-colonizing pseudomonad strains to allow stable ICP expression (Obukowicz et al. 1986a, b, 1987; Waalwijk et al. 1991).

Other Recipient Strains Apart from the strains above, other recipients such as cyanobacteria (Chungjatupornchai 1990; Lluisma et al. 2001a; Lu et al. 1999b; Manasherob et al. 2002a, 2003b; Wu et al. 1997a), *Rhizobium* (Nambiar et al. 1990), and even the baculoviruses (Ribeiro and Crook 1993, 1998) have also been used as recipients for the expression of various ICP genes.

13.2.2 Genetically Recombinant Methods for Constructing Genetically Modified Bt Strains

In the natural state, the spontaneous transfer frequency of Bt ICP genes is often very low. Therefore, to construct genetically modified Bt strains, some specific methods are required to achieve the heterologous transfer and recombination between different host strains. These methods can be generalized as follows.

13.2.2.1 Plasmid Conjugation

Plasmid conjugation is an effective genetic exchange system for transferring plasmid-harbored ICP genes from one Bt host strain to another. Moreover, this method is considered more advantageous than other methods because of its technical feasibility for transferring multiple ICP genes in one-step experiment, as well as its inability to introduce ancillary or unnecessary DNA such as the *in vitro* artificial splicing fragments, antibiotic resistance genes, and so on, into the resultant conjugated strain. As a result, the process is easily approved for product registration because of reduced environmental risk. Carlton et al. (1990) described an engineered Bt strain EG2424 by the conjugation of a coleopteran-active Bt subsp. *tenebrionis* strain and a lepidopteran-active Btk one. In addition to direct conjugation, tri-parental conjugation is also a common method for transferring ICP genes. Lu et al. (1999a) constructed an engineered *Anabaena* PCC7120 strain containing a Bti *cry11A* gene through the helper plasmid pRL528 (carrying the *mob* gene and the genes encoding *Eco*47II and *Ava*I methylases). The construction was conducted via tri-parental conjugation with *E. coli* HB101 (bearing the recombinant plasmid and pRL528), *E. coli* HB101 (bearing the plasmid RP4), and *Anabaena* PCC7120. Both Southern and western blot analyses demonstrated the appearance and expression of Cry11A in the engineered *Anabaena* strain.

13.2.2.2 Transformation

Prior to ICP gene transfer through electroporation, Bt protoplast preparation is often needed to promote transformation efficiency (Fischer et al. 1984; Crawford et al. 1987; Martin et al. 1981). The technical development of electroporation has established the transformation of various expression vectors harboring different ICP genes into the recipient strains as the main method for constructing various engineered Bt strains. This method can be expediently used for transferring ICP genes that are artificially modified *in vitro* to increase insecticidal spectrum, improve ICP yield, and even induce unanticipated insecticidal activity. Crickmore et al. described an approach to introducing the cloned Bt entomocidal δ-endotoxin genes into several native Bt strains. In many cases, the resulting transformants expressed both their native toxins and the cloned toxin to produce strains with broader toxicity spectra. However, the introduction of the subsp. *tenebrionis* toxin gene into Bti produces a strain with activity against *Pieris brassicae* (cabbage white butterfly), an activity which neither parent strain possesses (Crickmore et al. 1990b).

13.2.2.3 Transduction

ICP gene transfer via transduction is also utilized for developing engineered Bt strains. Lecadet et al. (1992) described the methods for transferring the vector-harboring *cry1Aa* into several Bt subspecies through phage CP-54Ber-mediated transduction, with frequencies ranging from 5×10^{-8}–2×10^{-6} transductants per CFU. The introduction of the *cry1Aa* gene resulted in the formation of large bipyramidal crystals that were active against the insect *Plutella xylostella*. Kalman et al. also reported the genetic exchange of *cry1Ca* through the phage-mediated transduction of CP51 (Kalman et al. 1995).

13.2.2.4 Site-specific Recombination

Site-specific recombination is a genetic exchange system that allows ICP genes to shift secondarily within the recipient strain. Considering recombinant plasmids usually harbor antibiotic resistant genes and other heterologous DNA fragments aside from the ICP genes, which can be a significant environmental concern during the unlimited release of the plasmid, site-specific recombination can be used to develop environmentally friendlier Bt engineered strains. The shifting of ICP genes through site-specific recombination is often achieved through the following ways: (i) transfer of the introduced ICP genes into the resident plasmids. Lereclus et al. (1992) described the transfer of *cry3Aa* into a 50 MD plasmid of Btk HD-73 by cloning *cry3Aa* initially into a fragment of *IS232* and then inserting into a thermosensitive plasmid vector. Growth temperature is then elevated at non-permissive temperatures to eliminate the plasmid vector selectively. The resulting strain was found to harbor the newly introduced *cry3Aa* gene, which was integrated into a copy of *IS232* in the resident plasmid. Thus, this engineered Bt strain contained only a DNA of Bt origin, and displayed insecticidal activity against both lepidopteran and coleopteran pests; (ii) integration into the chromosomal DNA. Kalman et al. (1995) described a two-step site-specific recombination procedure for placing the *cry1Ca* from subsp. *aizawai* into the chromosomes of two Btk strains using an integration vector containing a Bt chromosomal fragment. The fragment encodes a phosphatidylinositol-specific phospholipase C, which then allows the targeting of *cry1Ca* into the homologous region of the Btk chromosome. Yue et al. (2005a, b) constructed two chromosome-integrated Bt engineered strains based on Bt Tn4430 harboring the genes *tnpI* and *tnpA*. With the mediation of TnpI-TnpA, the *cry1C* and *cry3Aa* genes were respectively integrated into the chromosome of the host strain. To prevent secondary integration, the integrative vector was eliminated by moving the recombinant cultures to 46°C for generations. Both resulting engineered strains exhibited additional insecticidal activities conferred by the introduced ICP gene; (iii) the transfer ICP genes through site-directed resolution recombination systems. This allows the cloning of the ICP genes into a resolution vector, followed by vector transfer into the recipient strain to allow the shifting of the ICP genes into a resolved plasmid containing the origin that can replicate in the recipient strain and stably retain the introduced ICP genes (Sanchis et al. 1997; Baum et al. 1996; Wu et al. 2000).

13.2.2.5 Protoplast Fusion

Protoplast fusion is a versatile technique for inducing genetic recombination of ICP genes between Bt subspecies and strains. Through this treatment, multiple ICP genes can be transferred via one-step process. It has been used to increase or induce additional insecticidal activity in the fusion between Bt strains (Fischer et al. 1984; Yari et al. 2002; Yu et al. 2001), or the novel strains between Bt and other *Bacillus* species (Fischer et al. 1984; Trisrisook et al. 1990).

13.3 Bt Engineered Strains with High Toxicity or Broader Insecticidal Spectrum

13.3.1 Construction of Bt Engineered Strains with Broader Insecticidal Spectrum

Most naturally occurring Bt strains are toxic to a limited variety of pests in their natural states. However, multiple mixed-species of pest populations often occur in crop fields. Therefore, biopesticides based on naturally occurring Bt strains are often unable to control these pest populations effectively. Since the 1980s, various ICP genes have been isolated and cloned continually, and the vectors and the expression systems have improved significantly. Engineered Bt strains can now be constructed via *in vitro* modification and recombination of various Bt ICP genes with different insecticidal activities to develop products with broader insecticidal spectra. Klier et al. (1983) reported that heterospecific mating between a *B. subtilis* strain and different Bt strains introduces the plasmid-coded crystal gene into both an acrystalliferous Bt strain and a Bti wild-type strain. The transcipient strains produced both types of δ-endotoxin, which are active on both lepidopteran (*Ephestia kuehniella*) and dipteran (*Anopheles stephensi*) larvae. Karamata and Luthy (1989) found that partial conjugants from Btk HD-73 and subsp. *tenebrionis* strain, which are toxic to *Trichoplusia ni* and *Phaedon cochleariae* larva, respectively, have broader toxicity to *Spodoptera lttoralis* larvae. Crickmore et al. (1990b) introduced the cloned Bt ICP gene into Bti wild-type strains, resulting in a strain active against *Pieris brassicae* (cabbage white butterfly), which neither parent strain possesses. Ecogen Inc. has also constructed several engineered Bt strains with broader pesticidal activity by introducing *cry3A* and *cry1C* into a Btk recipient strain. As a result, the engineered strain EG2424 has additional activity against coleopteran pests, and the engineered strain EG2371 has broader activity to the *Spodoptera* larvae. The commercial product, SAN418, manufactured from Sandoz, is also based on an engineered Bt strain simultaneously expressing *cry1* and *cry3* genes, resulting in combined pest-control activities to both lepidopteran and coleopteran larvae. Yue et al. (2005a, b) described methods for integrating *cry3Aa* and *cry1C* into the Btk YBT-1520 chromosome via an integrative and thermosensitive vector, resulting two engineered strains with broader pesticidal activities against *Rhyllodecta vulgatissima* and *S. exigua* larva, respectively. The degree of activity against *P. xylostella* third-instar later larvae was retained.

13.3.2 Construction of Bt Engineered Strains with High Insecticidal Activity

The ICP gene expression level in the engineered strains is a key factor that significantly affects the pesticidal activity of this strain. Various technical attempts have been performed to promote the toxicity of the engineered bacteria, which can be specialized as follows.

13.3.2.1 ICP Gene Expression in the Asporogenous (Spo$^-$) or Plasmid-free (or Partially Cured) Recipient Strains

Most Bt ICP genes express the corresponding proteins during sporulation of the host cells. Consequently, spore formation is correlated with ICP gene expression. Gene classes, including *cry1*, *cry2,* and *cry11*, are apparently sporulation-dependent ICP genes regulated by the sporulation, in which the expression levels are not remarkably increased in some Spo$^-$ mutant strains (Johnson 1981). However, for sporulation-independent ICP genes such as *cry3A*-like genes, 2.5-fold increase in the Cry3Aa protein in the asporogenous subsp. *morrisoni* strain EG1351 is observed, while the *hknA* gene is introduced into the EG1351 strain transcribed during vegetative growth from a sigma A-like promoter (Malvar et al. 1994). Lu et al. (2000) reported the construction of an engineered Bt strain with remarkably increased anti-lepidopteran activity. By transferring *cry1C* into the Bt wild-type strain YBT833, an engineered strain YBT833–2 without the resident *cry1Ab* gene is produced. However, it still produced the 130 and 65 kDa endotoxins. Bioassay results show that the toxicity of YBT833–2 to *P. xylostella* is significantly higher than that of its parental strain YBT833. Unfortunately, although the plasmid-free or partially plasmid-cured recipient strains are generally well suited for constructing the engineered Bt strains, in most cases, the resulting strains normally have lower toxicity than the wild-type strains.

13.3.2.2 Increasing ICP Gene Copy Numbers

Lereclus et al. (1989) found that the introduction of *cry1Aa* into an acrystalliferous strain (407cry^{-}) results in the increased expression of Cry1Aa, which could be ascribed to the increase in the copy number of the plasmid vector. This was further identified by ligating *cry3Aa* into plasmid vectors with different copy numbers. Ligation of this gene into the vector with a copy number of approximately 15 per equivalent chromosome, it conferred higher δ-endotoxin production compared with the vector with a lower copy number.

13.3.2.3 Increasing the Stability of ICP Genes

To overcome drawbacks such as the instability and negative effect of the introduced plasmids, Kalman et al. (1995) integrated the *cry1C* gene into the chromosomal DNA of the recipient strain through combined transformation and transduction procedures. Consequently, the engineered strain is very stable, with enhanced ICP production. Yu et al. (1999) reported that two cloned ICP genes were successfully integrated into Bt strains chromosomes using the transposon vector pTV1Ts (containing Tn917 from Streptococcus faecalis). Two recombinant transposon-derived plasmids containing the *cry11A* gene and *cyt1A* gene, respectively, were thus obtained and then transferred into Btk HD-1, resulting in recombinant strains HT26 and HT45, which contain the chromosome-integrated ICP genes. These strains express the corresponding ICP and formed normal parasposal crystals toxic to the third instars of the mosquito *Culex pipens fatigans*.

13.3.2.4 Utilizing the Synergistic Effect of Helper Proteins

Wu and Chang (1985) first revealed that the 26 kDa (now referred to as 27 kDa) protein from Bti has a synergistic mosquitocidal effect with both the 65 and 130 kDa protein. The subsequent study by Adams et al. (1989) also showed that the 20 kDa protein (P20) greatly enhances the expression of the 27 kDa protein (CytA) in *E. coli* and of the 130 kDa Cry protein. Wu and Federici (1993) found that P20 promotes crystal formation in the Bti species, which suggests that this protein chaperones CytA molecules during synthesis and crystallization, and concomitantly prevents CytA from interacting lethally with the Bt cells. Yoshisue et al. (1992) reported that P20 enhances the yield of two 130 kDa mosquitocidal proteins (CryIVA, CryIVB) in *E. coli*, which could be a posttranscriptional effect. Shao et al. (2001) investigated the effect of P20 on the Btk strain, and found that it significantly enhances Cry1Ac protoxin (133 kDa) production in a plasmid-free recipient strain. In the presence of P20, the yield of Cry1Ac protoxin increased 2.5-fold, and, on average, the resulting crystals were 3 times the size of those formed in the control strain without P20. Correspondingly, the recombinant strain that coexpressed P20 and Cry1Ac exhibited higher toxicity against *Heliothis armigera* and other lepidopteran

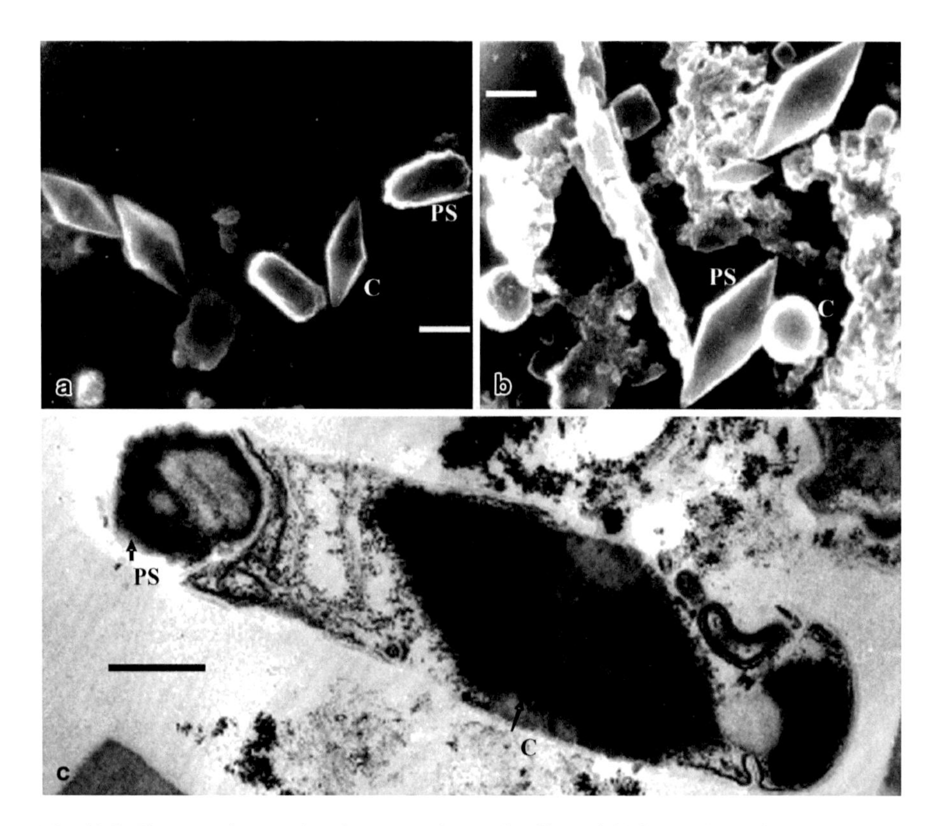

Fig. 13.3 Electron micrographs of spores and crystals of insecticidal crystal proteins produced by genetically modified *B. thuringiensis* strain WG-001. **a** preparation of YBT-1520; **b** preparation of WG-001; **c** one cell of WG-001. PS: parasporal body; C: crystal. Bars: 1 μm

larvae than the control. This distinct strain has been used to develop a genetically modified Bt biopesticide product called WG-001 (Fig. 13.3), which is the first officially approved engineered Bt biopesticide product in China. Moreover, the synergistic effect of P19 and other chaperon molecules have also been verified. These investigations provide strong evidence that the chaperons have great potential for constructing genetically modified Bt strains in terms of increasing the yielding of ICP and insecticidal activity.

To construct highly expressed engineered Bt strains using the chaperon molecules, Sirichotpakorn et al. (2001) used a cloned chitinase gene for co-expression with the regulatory gene *p19* and the toxin gene *cry11Aa1* in Bti strains 4Q2-72 and c4Q2-72. The engineered Bt strains expressing the chitinase exhibited enzyme activity approximately 2 times higher than the maximum activity of the parental strain. Meanwhile, the coapplication of crude chitinase from the cloned gene in the recombinant strain with cell suspensions of Bti 4Q2-72 and its transformants enhances larvicidal activity by 3-fold to 50-fold, improves sporulation, and cocrystallizes chitinase with crystal toxins. Chitinases have been proposed to elevate the larvicidal effect by perforating the peritrophic membrane (Pardo-Lopez et al. 2009), thereby increasing the accessibility of the Cry toxin to the epithelial membrane where the receptors are located. Therefore, the application of chitinase for enhancing the entomotoxicity of engineered Bt strains has received considerable attention since this century, and new strategies and potential have been continuously exploited (Driss et al. 2011; Vu et al. 2009; Hu et al. 2009; Barboza-Corona et al. 2009).

Shao and Yu (2004) constructed an engineered Btk strain that expresses the P20 protein, which increases the yield of crystal proteins and forms larger crystals compared with the wild control. P20 was effective in preventing the degradation of Cry1A *in vivo* in the wild-type strains. Thamthiankul et al. (2004) constructed a fused gene consisting of *p19* and a chitinase gene (*chiBlA*) from *B. licheniformis*, and then integrated it into the Bt subsp. *aizawai* BTA1 genome via homologous recombination. The resulting strain (INT1) showed growth and sporulation comparable with that of the wild-type strain. The bioassay results show that chitinase increases the activity of Bt subsp. *aizawai* against *S. exigua*, which suggests that chromosomal integration of chimeric *p19-chiBlA* can be used as a potential technique for strain improvement. In a recent study, the vegetative insecticidal protein (Vip) gene *vip3A* was redirected from the Bt subsp. *aizawai* strain SP41 to the sporulation stage by replacing its native promoter with the strong promoter *p19* of the *cry11Aa* operon. The resultant strain was found to express VIP during the spore stage at 20 and 48 h, which increases the toxicity of SP41 4.1- and 2.5-fold, respectively (Thamthiankul Chankhamhaengdecha et al. 2008). These results are corroborated by a previous investigation by Zhu et al. (2006). Recently, Xia et al. (2009a, b) reported that the expression of fusion genes with cDNA encoding the subtilisin-like serine protease gene CDEP2 or an enterokinase recognition site sequence, and the *cry1Ac* gene under the control of the native *cry1Ac* promoter in Btk Cry^-B recipient strain results in increased toxicity of the engineered strains against the third instar larvae of *H. armigera* or *Plutella xylostella*. Moreover, Peng et al. (2010) reported that the insect cadherin proteins function as synergists for Cry1A, Cry4Ba, and Cry3A toxicity against target insects. They studied the coexpression of the *H. armigera* cadherin gene (*hacad1*) and the *cry1Ac* gene in an acrystalliferous BMB171 strain. In their study, significant increases in insecticidal activity against *H. armigera* and *S. exigua* by 5.1-fold and 6.5-fold, respectively, were observed using the engineered Bt strain (BMB1073) compared with the strain that can only express the Cry1Ac protein. This finding can be a novel strategy for engineering strains and transgenic plants with higher insecticidal activity.

13.3.2.5 Modification of ICP Gene Promoters to Create Novel Activities

Promoter activity is particularly important in regulating the transcription of various Bt ICP genes. Considering most Bt ICP genes are normally expressed only during sporulation, producing extended crystal proteins during the vegetative growth phase by replacing the Bt ICP genes with a vegetative-active promoter. Mettus and Macaluso (1990) cloned the δ-endotoxin genes into the downstream region of a tetracycline resistance gene promoter, resulting in the expression of crystal proteins during vegetative growth, as well as sporulation in both Bt and *B. megaterium*. Since then, research focused on transcriptional regulation of ICP genes for developing valuable engineered Bt strains has been an important area. More recently, various studies related to the modification of Bt ICP gene promoters have been performed to increase the insecticidal activity of Bt strains. Hu et al. (2009) cloned a signal peptide-encoding sequence-deleted chitinase gene under the control of dual overlapping promoters with the Shine–Dalgarno sequence and the terminator sequence of the *cry1Ac3* gene in an acrystalliferous Cry^-B. The cloning resulted in chitinase overexpression in the form of inclusion bodies, which can potentiate the insecticidal effect of Cry1Ac because these inclusion bodies solubilize a wide range of alkaline pH values and exhibit chitinolytic activity. Perez-Garcia et al. (2010) investigated the regulatory role of a putative $sigma^H$-like promoter by forming various constructs consisting of the $sigma^H$-, $sigma^E$-, and $sigma^K$-like promoters, the 0A box, and the cry1Ac coding sequence (EK0AH), among others. Consequently, the construct EK0AH was found to have bigger crystal sizes, as well as higher *cry1Ac* gene expression levels than that of the other control constructs. Given that the expression of *vip3* genes occurs only during the vegetative growth phase of Bt, which is a limiting factor, Sellami et al. (2011) extended the synthesis of the Vip proteins to the sporulation phase by constructing an engineered Bt strain that expresses the *vip3LB* gene under the control of the sporulation dependent promoters BtI and BtII in the Bt BUPM 106 ($Vip3^-$) and BUPM 95 ($Vip3^+$) strains, respectively. This demonstrated the synthesis of Vip3LB and its toxicity against the second instar larvae of the lepidopteran insect *S. littoralis* for the recombinant BUPM 106.

13.4 Multifunctional Bt Engineered Strains

13.4.1 Constructing the Genetically-modified Bacteria with Dual Insecticidal and Antifungal Activity

Pseudomonads are a family of bacteria dominant in plant leaf- and root-colonizing microcosm. Constitutive synthesis of the Bt ICPs by the living plant-associated pseudomonads can potentially target pesticide delivery, minimize the need for repeated applications during the growing season, and even exhibit antifungal activity. For these purposes, Obukowicz et al. (1986a, b) pioneered the method of transferring the Bt ICP gene into the chromosome of *P. fluorescens* strains via Tn5-mediated transposition. Zhang et al. (1995) transferred the *cry1Ac* gene into the leaf-colonizing biocontrol bacterium *P. fluorescens* P303 for the expression of *B. thuringieneis* Cry protein. Consequently, the engineered *P. fluorescens* strain exhibits the dual activity of not only restraining the plant pathogen *Geumannomyces graminis* var. *tritici*, but also controlling corn borers and the diamond back moth. Interestingly, Raddadi et al. (2009) investigated the insecticidal and antifungal potential of 16 Bt strains. For fungal biocontrol, all the strains inhibited *Fusarium oxysporum* and *Aspergillus flavus* growth, and four strains had all or most of the antifungal determinants examined. The strain HD932 showed the widest spectrum of antifungal activity, which could be very promising in field application as polyvalent and safe Bt strains.

13.4.2 Constructing the Engineered Bt Strains with Extended Persistence Entomotoxicity

For many years, one of the disadvantages of Bt products is their short field persistence, and the products typically degraded before providing an acceptable level of control. In 1985, the Mycogen Corporation developed a novel bioencapsulation technology for Bt toxins through the invention of the CellCap encapsulation system, and has used the process since then to develop a series of advanced Bt formulations. MVP bioinsecticide was the first product based on this system, which contains a selected Btk endotoxin that is highly active against the diamondback moth (DBM) *P. xylostella*. Using the CellCap system, this toxin is encapsulated and stabilized within dead *P. fluorescens* cells that have been killed and fixed in the fermentor prior to harvest using proprietary physical and chemical processes. The effect of the CellCap bioencapsulation process in protecting the δ-endotoxin from environmental degradation is evaluated by applying MVP, and two other Bt products, javelin and Dipel 4L, to small plots of cabbage and broccoli. The results demonstrate that the MVP provided residual activity that was 2–3 times greater than the two other products containing unprotected toxin crystals. Moreover, laboratory tests evaluating the effects of ultraviolet radiation on the activity of MVP, and other commercial Bt products such as Dipel, Toarrow-CT, and Bacilex demonstrate that the persistence of MVP is 5–36 times greater than that of the other products (Soares and Quick 1990). Since 1991, six products containing selected Bt toxins produced using the CellCap system have been registered with the US EPA and other regulatory agencies. In 1995, the EPA approved Mycogen's Mattch bioinsecticide, which contains a blend of encapsulated Cry1Ac and Cry1C δ-endotoxins for the control of lepidopteran pests in vegetable crops (Panetta 2010).

13.4.3 Constructing the Engineered Bacteria with Entomotoxic and Nitrogen-fixing Activities

The pigeon pea grows well under nitrogen-limited conditions provided its roots are nodulated by the soil bacterium *Bradyrhizobium*. However, a dipteran pest, *Rivellia angulata*, significantly thwarts the nitrogen-fixing effect by feeding on the root nodule contents. To prevent the root nodules from damage by the larvae of this pest,

Nambiar et al. constructed a *Bradyrhizobium* engineered strain that expresses Bti Cry proteins via conjugative mobilization. Consequently, the construct was able to maintain stability in the absence of selection, which exhibited pest control and nitrogen fixation (Nambiar et al. 1990). With a similar purpose, Wu et al. introduced combinations of the Bti genes *cry4A*, *cry11A* and *p20* into the nitrogen-fixing cyanobacterium *Anabaena* sp. strain PCC 7120. The resultant constructs were found to be toxic to the third instar larvae of *Aedes aegypti* (Wu et al. 1997b). The introduced Bti genes were further demonstrated to be stably maintained in the transgenic *Anabaena* PCC7120, and the growth of the constructs was comparable to the wild-type strain under optimal growth conditions (Lluisma et al. 2001b). In addition, the expressed δ-endotoxin proteins were protected from UV-B damage (Manasherob et al. 2002b). This organism, which serves as a food source for mosquito larvae and multiplies in their breeding sites, may solve the environment-imposed limitations of Bti as a biological control agent for mosquitoes. Under simulated field conditions, the transgenic *Anabaena* PCC 7120 expressing the Bti genes had greater mosquito larvicidal activity than that of Bti primary powder (fun 89C06D) or wettable powder (WP) (Bactimos products) when either was mixed with silt or exposed to sunlight outdoors (Manasherob et al. 2003a).

13.5 Genetically Modified Bt Strains with Delayed Pest Resistance

Various native or transgenic Bt products have been overwhelmingly successful and beneficial, leading to effective pest control and reduction in the use of chemical pesticides. However, their deployment has also raised some negative effects particularly with the potential evolution of pest resistance (Sanahuja et al. 2011b). The first evidence of pest resistance to Bt spore-crystal protein complex was verified in 1985 (McGaughey 1985). Since then, numerous laboratory and field experiments have confirmed that the intensive use of a single Cry protein likely results in the evolution of pest resistance. Therefore, the development of genetically modified Bt strains with delayed or sustainable pest resistance has been one of the primary goals in the Bt industry.

To address the possibility of developing resistance in the target pests, various alternative or complementary approaches have been proposed so far. One robust approach is constructing an engineered strain by stacking multiple ICP genes that target the same pest through different mechanisms. The evolution of resistance to Bti is generally regarded as relatively slow despite their very large-scale application in some regions. Several previous investigations have also revealed that Bti Cyt1Aa acts as the principal factor responsible for delaying the evolution and expression of resistance to mosquitocidal Cry proteins (Wirth et al. 1997, 2000, 2005a, b). This raises the possibility of constructing an engineered Bt strain with delayed pest resistance by introducing the *cyt1Aa* gene. Park et al. has reported the construction of several recombinant larvicidal bacteria for reducing the cost of bacterial insecticides and combining their most potent mosquitocidal properties into individual strains (Park et al. 2003, 2005b). Aside from being at least 10-fold more toxic than the wild-type Bti or Bs, because this strain contains Cyt1Aa, it has the potential to prevent the development of resistance to the Bin protein (Park et al. 2005a). They further showed that combining Bs Bin with Bti proteins in a recombinant Bti strain significantly delays the evolution of resistance even under very heavy selection pressure. This implies that such a recombinant bacterial strain should exhibit longevity under the operational field conditions that aim to control nuisance and vector mosquitoes (Wirth et al. 2010).

Acknowledgments The authors are grateful to Dr. Chan Li for his practical helps. This work was supported by grants from the National Natural Science Foundation of China (item no. 31070111 and 30930004), and a grant from the Chinese National Program for High Technology Research and Development (item no. 2008AA02Z112).

References

Adams LF, Visick JE, Whiteley HR (1989) A 20-kilodalton protein is required for efficient production of the *Bacillus thuringiensis* subsp. *israelensis* 27-kilodalton crystal protein in *Escherichia coli*. J Bacteriol 171:521–530

Agaisse H, Lereclus D (1995) How does *Bacillus thuringiensis* produce so much insecticidal crystal protein? J Bacteriol 177:6027–6032

Arantes O, Lereclus D (1991) Construction of cloning vectors for *Bacillus thuringiensis*. Gene 108:115–119

Aronson AI, Beckman W, Dunn P (1986) *Bacillus thuringiensis* and related insect pathogens. Microbiol Rev 50(1):1–24

Barboza-Corona JE, Ortiz-Rodriguez T, de la Fuente-Salcido N, Bideshi DK, Ibarra JE, Salcedo-Hernandez R (2009) Hyperproduction of chitinase influences crystal toxin synthesis and sporulation of *Bacillus thuringiensis*. Antonie Van Leeuwenhoek 96:31–42

Baum JA, Malvar T (1995) Regulation of insecticidal crystal protein production in *Bacillus thuringiensis*. Mol Microbiol 18:1–12

Baum JA, Coyle DM, Gilbert MP, Jany CS, Gawron-Burke C (1990) Novel cloning vectors for *Bacillus thuringiensis*. Appl Environ Microbiol 56:3420–3428

Baum JA, Kakefuda M, Gawron-Burke C (1996) Engineering *Bacillus thuringiensis* bioinsecticides with an indigenous site-specific recombination system. Appl Environ Microbiol 62:4367–4373

Brown KL (1993) Transcriptional regulation of the *Bacillus thuringiensis* subsp. *thompsoni* crystal protein gene operon. J Bacteriol 175:7951–7957

Calogero S, Albertini AM, Fogher C, Marzari R, Galizzi A (1989) Expression of a cloned *Bacillus thuringiensis* delta-endotoxin gene in *Bacillus subtilis*. Appl Environ Microbiol 55:446–453

Carlton BC, Gawron-Burke MC, Johnson TB (1990) Exploiting the genetic diversity of *Bacillus thuringiensis* for the creation of new bioinsecticides. In: Proceedings and abstracts, the Vth International Colloquium on Invertebrate Pathology and Microbial Control, Adelaide, Australia, 20–24 Aug. 1990, pp 18–22

Carozzi NB, Kramer VC, Warren GW, Evola S, Koziel MG (1991) Prediction of insecticidal activity of *Bacillus thuringiensis* strains by polymerase chain reaction product profiles. Appl Environ Microbiol 57:3057–3061

Chak KF, Kao SS, Feng TY (1994) Characterization and cry gene typing of *Bacillus thuringiensis* isolates from Taiwan. In: Feng TY, Chak K-F, Smith R et al (eds) *Bacillus thuringiensis* biotechnology and environmental benefits, vol 1. Hua Shiang Yuan Publishing Co., Taipel, pp 105–123
Chaoyin Y, Wei S, Sun M, Lin L, Faju C, Zhengquan H, Ziniu Y (2007) Comparative study on effect of different promoters on expression of *cry1Ac* in *Bacillus thuringiensis* chromosome. J Appl Microbiol 103:454–461
Chungjatupornchai W (1990) Expression of the mosquitocidal-protein genes of *Bacillus thuringiensis* subsp. *israelensis* and the herbicide-resistance gene *bar* in *Synechocystis* PCC6803. Curr Microbiol 21:283–288
Clark BD (1987) Characterization of plasmids from *Bacillus thuringiensis* var. *israelensis*. Ph D thesis, The Ohio State University, USA
Crawford IT, Greis KD, Parks L, Streips UN (1987) Facile autoplast generation and transformation in *Bacillus thuringiensis* subsp. *kurstaki*. J Bacteriol 169:5423–5428
Crickmore N, Ellar DJ (1992) Involvement of a possible chaperonin in the efficient expression of a cloned CryIIA delta-endotoxin gene in *Bacillus thuringiensis*. Mol Microbiol 6:1533–1537
Crickmore N, Bone EJ, Ellar DJ (1990a) Genetic manipulation of *Bacillus thuringiensis*: towards an improved pesticide. Asp Appl Biol 24:17–24
Crickmore N, Nicholls C, Earp DJ, Hodgman TC, Ellar DJ (1990b) The construction of *Bacillus thuringiensis* strains expressing novel entomocidal delta-endotoxin combinations. Biochem J 270:133–136
Crickmore N, Wheeler VC, Ellar DJ (1994) Use of an operon fusion to induce expression and crystallisation of a *Bacillus thuringiensis* delta-endotoxin encoded by a cryptic gene. Mol Gen Genet 242:365–368
Dankocsik C, Donovan WP, Jany CS (1990) Activation of a cryptic crystal protein gene of *Bacillus thuringiensis* subspecies *kurstaki* by gene fusion and determination of the crystal protein insecticidal specificity. Mol Microbiol 4:2087–2094
Delecluse A, Charles JF, Klier A, Rapoport G (1991) Deletion by *in vivo* recombination shows that the 28-kilodalton cytolytic polypeptide from *Bacillus thuringiensis* subsp. *israelensis* is not essential for mosquitocidal activity. J Bacteriol 173:3374–3381
Dervyn E, Poncet S, Klier A, Rapoport G (1995) Transcriptional regulation of the *cryIVD* gene operon from *Bacillus thuringiensis* subsp. *israelensis*. J Bacteriol 177:2283–2291
Douek J, Einav M, Zaritsky A (1992) Sensitivity to plating of *Escherichia coli* cells expressing the *cryA* gene from *Bacillus thuringiensis* var. *israelensis*. Mol Gen Genet 232:162–165
Driss F, Rouis S, Azzouz H, Tounsi S, Zouari N, Jaoua S (2011) Integration of a recombinant chitinase into *Bacillus thuringiensis* parasporal insecticidal crystal. Curr Microbiol 62:281–288
Ferrari FA, Nguyen A, Lang D, Hoch JA (1983) Construction and properties of an integrable plasmid for *Bacillus subtilis*. J Bacteriol 154:1513–1515
Fischer HM, Luthy P, Schweitzer S (1984) Introduction of plasmid pC194 into *Bacillus thuringiensis* by protoplast transformation and plasmid transfer. Arch Microbiol 139:213–217
Gamel PH, Piot JC (1992) Characterization and properties of a novel plasmid vector for *Bacillus thuringiensis* displaying compatibility with host plasmids. Gene 120:17–26
Ge AZ, Pfister RM, Dean DH (1990) Hyperexpression of a *Bacillus thuringiensis* delta-endotoxin-encoding gene in *Escherichia coli*: properties of the product. Gene 93:49–54
Ge AZ, Rivers D, Milne R, Dean DH (1991) Functional domains of *Bacillus thuringiensis* insecticidal crystal proteins. Refinement of *Heliothis virescens* and *Trichoplusia ni* specificity domains on CryIA(c). J Biol Chem 266(27):17954–17958
Gleave AP, Williams R, Hedges RJ (1993) Screening by polymerase chain reaction of *Bacillus thuringiensis* serotypes for the presence of *cryV*-like insecticidal protein genes and characterization of a *cryV* gene cloned from *B. thuringiensis* subsp. *kurstaki*. Appl Environ Microbiol 59:1683–1687
Gonzalez JM Jr, Dulmage HT, Carlton BC (1981) Correlation between specific plasmids and delta-endotoxin production in *Bacillus thuringiensis*. Plasmid 5:352–365

Herrnstadt C, Gilroy TE, Sobieski DA, Bennett BD, Gaertner FH (1987) Nucleotide sequence and deduced amino acid sequence of a coleopteran-active delta-endotoxin gene from *Bacillus thuringiensis* subsp. *san diego*. Gene 57:37–46

Hodgman TC, Ziniu Y, Shen J, Ellar DJ (1993) Identification of a cryptic gene associated with an insertion sequence not previously identified in *Bacillus thuringiensis*. FEMS Microbiol Lett 114:23–29

Hofte H, Whiteley HR (1989) Insecticidal crystal proteins of *Bacillus thuringiensis*. Microbiol Rev 53(2):242–255

Hofte H, Seurinck J, Van Houtven A, Vaeck M (1987) Nucleotide sequence of a gene encoding an insecticidal protein of *Bacillus thuringiensis* var. *tenebrionis* toxic against Coleoptera. Nucleic Acids Res 15:7183

Hu SB, Liu P, Ding XZ, Yan L, Sun YJ, Zhang YM, Li WP, Xia LQ (2009) Efficient constitutive expression of chitinase in the mother cell of *Bacillus thuringiensis* and its potential to enhance the toxicity of Cry1Ac protoxin. Appl Microbiol Biotechnol 82:1157–1167

Johnson DE (1981) Preparing entomocidal products with oligosprogenic mutants of *Bacillus thuringiensis*. US Patent 4,277,546

Kalman S, Kiehne KL, Libs JL, Yamamoto T (1993) Cloning of a novel cryIC-type gene from a strain of *Bacillus thuringiensis* subsp. *galleriae*. Appl Environ Microbiol 59:1131–1137

Kalman S, Kiehne KL, Cooper N, Reynoso MS, Yamamoto T (1995) Enhanced production of insecticidal proteins in *Bacillus thuringiensis* strains carrying an additional crystal protein gene in their chromosomes. Appl Environ Microbiol 61:3063–3068

Karamata D, Luthy P (1989) Insecticidal hybrid bacteria from B.t. *kurstaki* and B.t. *tenebrionis*. US Patent 4,797,279

Keller B, Langenbruch GA (1993) Control of coleopteran pests by *Bacillus thuringiensis*. In: Entwist PF, Cory JS, Bailey MJ et al (eds) *Bacillus thuringiensis*, an environmental biopesticide: theory and practice. Wiley, Chichester, pp 171–191

Klier A, Bourgouin C, Rapoport G (1983) Mating between *Bacillus subtilis* and *Bacillus thuringiensis* and transfer of cloned crystal genes. Mol Gen Genet 191:257–262

Krieg A, Huger AM, Langenbruch GA, Schnetter W (1983) *Bacillus thuringiensis* var. *tenebrionis*, a new pathotype effective against larvae of *Coleoptera*. Z Angew Entomol 96:500–508

Kuo WS, Chak KF (1996) Identification of novel *cry*-type genes from *Bacillus thuringiensis* strains on the basis of restriction fragment length polymorphism of the PCR-amplified DNA. Appl Environ Microbiol 62:1369–1377

Lecadet MM, Chaufaux J, Ribier J, Lereclus D (1992) Construction of novel *Bacillus thuringiensis* strains with different insecticidal activities by transduction and transformation. Appl Environ Microbiol 58:840–849

Leonard C, Zekri O, Mahillon J (1998) Integrated physical and genetic mapping of *Bacillus cereus* and other gram-positive bacteria based on IS231A transposition vectors. Infect Immun 66:2163–2169

Lereclus D, Arantes O, Chaufaux J, Lecadet M (1989) Transformation and expression of a cloned delta-endotoxin gene in *Bacillus thuringiensis*. FEMS Microbiol Lett 51:211–217

Lereclus D, Vallade M, Chaufaux J, Arantes O, Rambaud S (1992) Expansion of insecticidal host range of *Bacillus thuringiensis* by *in vivo* genetic recombination. Biotechnology (NY) 10:418–421

Li L, Yu Z (1999) Transformation and expression properties of a *Bacillus thuringiensis* plasmid-free derivative strain BMB171. Chin J Appl Environ Biol 5:395–399

Li L, Yang C, Liu Z, Li F, Yu Z (2000) Screening of acrystalliferous mutants from *Bacillus thuringiensis* and their transformation properties. Acta Microbiol Sin 40:85–90

Liu Z, Sun M, Yu Z, Zaritsky A, Ben-Dov E, Manasherob R (1999) Preliminary study of *p19* gene from *Bacillus thuringiensis* subsp *israelensis*. Acta Microbiol Sin 39:114–119

Lluisma AO, Karmacharya N, Zarka A, Ben-Dov E, Zaritsky A, Boussiba S (2001a) Suitability of *Anabaena* PCC7120 expressing mosquitocidal toxin genes from *Bacillus thuringiensis* subsp. *israelensis* for biotechnological application. Appl Microbiol Biotechnol 57(1–2):161–166

Lluisma AO, Karmacharya N, Zarka A, Ben-Dov E, Zaritsky A, Boussiba S (2001b) Suitability of *Anabaena* PCC7120 expressing mosquitocidal toxin genes from *Bacillus thuringiensis* subsp. *israelensis* for biotechnological application. Appl Microbiol Biotechnol 57:161–166

Lu S, Liu Z, Dai J, Yu Z (1999a) Preliminary studies on cloning and expression of *Bacillus thuringiensis cry11A* gene in *Anabaena*. Acta Hydrobiol Sin 23:174–178

Lu S, Liu Z, Dai J, Yu Z (1999b) Preliminary studies on cloning and expression of *Bacillus thuringiensis cry11A* gene in *Anabaena*. Acta Hydrobiol Sin 23(2):174–178

Lu SQ, Liu ZD, Yu ZN (2000) The characterization of *Bacillus thuringiensis* strain YBT833 and its transformants that containing different ICP genes. Acta Genet Sin 27:839–844

Mahillon J, Kleckner N (1992) New IS10 transposition vectors based on a gram-positive replication origin. Gene 116:69–74

Mahillon J, Rezsohazy R, Hallet B, Delcour J (1994) *IS231* and other *Bacillus thuringiensis* transposable elements: a review. Genetica 93:13–26

Malvar T, Gawron-Burke C, Baum JA (1994) Overexpression of *Bacillus thuringiensis* HknA, a histidine protein kinase homology, bypasses early Spo mutations that result in CryIIIA overproduction. J Bacteriol 176:4742–4749

Manasherob R, Ben-Dov E, Xiaoqiang W, Boussiba S, Zaritsky A (2002a) Protection from UV-B damage of mosquito larvicidal toxins from *Bacillus thuringiensis* subsp. *israelensis* expressed in *Anabaena* PCC 7120. Curr Microbiol 45(3):217–220

Manasherob R, Ben-Dov E, Xiaoqiang W, Boussiba S, Zaritsky A (2002b) Protection from UV-B damage of mosquito larvicidal toxins from *Bacillus thuringiensis* subsp. *israelensis* expressed in *Anabaena* PCC 7120. Curr Microbiol 45:217–220

Manasherob R, Otieno-Ayayo ZN, Ben-Dov E, Miaskovsky R, Boussiba S, Zaritsky A (2003a) Enduring toxicity of transgenic *Anabaena* PCC 7120 expressing mosquito larvicidal genes from *Bacillus thuringiensis* ssp. *israelensis*. Environ Microbiol 5:997–1001

Manasherob R, Otieno-Ayayo ZN, Ben-Dov E, Miaskovsky R, Boussiba S, Zaritsky A (2003b) Enduring toxicity of transgenic *Anabaena* PCC 7120 expressing mosquito larvicidal genes from *Bacillus thuringiensis* ssp. *israelensis*. Environ Microbiol 5(10):997–1001

Martin PA, Lohr JR, Dean DH (1981) Transformation of *Bacillus thuringiensis* protoplasts by plasmid deoxyribonucleic acid. J Bacteriol 145:980–983

McGaughey WH (1985) Insect resistance to the biological insecticide *Bacillus thuringiensis*. Science 229(4709):193–195

Mettus AM, Macaluso A (1990) Expression of *Bacillus thuringiensis* delta-endotoxin genes during vegetative growth. Appl Environ Microbiol 56:1128–1134

Nambiar PT, Ma SW, Iyer VN (1990) Limiting an Insect Infestation of nitrogen-fixing root nodules of the pigeon pea (*Cajanus cajan*) by engineering the expression of an entomocidal gene in its root nodules. Appl Environ Microbiol 56:2866–2869

O'Grady J, Akhurst RJ, Kotze AC (2007) The requirement for early exposure of *Haemonchus contortus* larvae to *Bacillus thuringiensis* for effective inhibition of larval development. Vet Parasitol 150:97–103

Obukowicz MG, Perlak FJ, Kusano-Kretzmer K, Mayer EJ, Bolten SL, Watrud LS (1986a) Tn5-mediated integration of the delta-endotoxin gene from *Bacillus thuringiensis* into the chromosome of root-colonizing pseudomonads. J Bacteriol 168:982–989

Obukowicz MG, Perlak FJ, Kusano-Kretzmer K, Mayer EJ, Watrud LS (1986b) Integration of the delta-endotoxin gene of *Bacillus thuringiensis* into the chromosome of root-colonizing strains of pseudomonads using Tn5. Gene 45:327–331

Obukowicz MG, Perlak FJ, Bolten SL, Kusano-Kretzmer K, Mayer EJ, Watrud LS (1987) IS50L as a non-self transposable vector used to integrate the *Bacillus thuringiensis* delta-endotoxin gene into the chromosome of root-colonizing pseudomonads. Gene 51:91–96

Panetta JD (2010) Environmental and regulatory aspects: industry view and approach. In: Hall FR, Menn JJ (eds) Biopesticides: use and delivery. Humana Press Inc., USA, pp 473–484

Pardo-Lopez L, Munoz-Garay C, Porta H, Rodriguez-Almazan C, Soberon M, Bravo A (2009) Strategies to improve the insecticidal activity of Cry toxins from *Bacillus thuringiensis*. Peptides 30:589–595

Park HW, Bideshi DK, Federici BA (2003) Recombinant strain of *Bacillus thuringiensis* producing Cyt1A, Cry11B, and the *Bacillus sphaericus* binary toxin. Appl Environ Microbiol 69:1331–1334

Park HW, Bideshi DK, Federici BA (2005a) Synthesis of additional endotoxins in *Bacillus thuringiensis* subsp. *morrisoni* PG-14 and *Bacillus thuringiensis* subsp. *jegathesan* significantly improves their mosquitocidal efficacy. J Med Entomol 42:337–341

Park HW, Bideshi DK, Wirth MC, Johnson JJ, Walton WE, Federici BA (2005b) Recombinant larvicidal bacteria with markedly improved efficacy against culex vectors of west nile virus. Am J Trop Med Hyg 72:732–738

Peng D, Xu X, Ruan L, Yu Z, Sun M (2010) Enhancing Cry1Ac toxicity by expression of the *Helicoverpa armigera* cadherin fragment in *Bacillus thuringiensis*. Res Microbiol 161:383–389

Perez-Garcia G, Basurto-Rios R, Ibarra JE (2010) Potential effect of a putative sigma(H)-driven promoter on the over expression of the Cry1Ac toxin of *Bacillus thuringiensis*. J Invertebr Pathol 104:140–146

Poncet S, Delecluse A, Anello G, Klier A, Rapoport G (1994) Transfer and expression of the *cryIVB* and *cryIVD* genes of *Bacillus thuringiensis* subsp. *israelensis* in *Bacillus sphaericus* 2297. FEMS Microbiol Lett 117:91–95

Poncet S, Bernard C, Dervyn E, Cayley J, Klier A, Rapoport G (1997) Improvement of *Bacillus sphaericus* toxicity against dipteran larvae by integration, via homologous recombination, of the Cry11A toxin gene from *Bacillus thuringiensis* subsp. *israelensis*. Appl Environ Microbiol 63:4413–4420

Raddadi N, Belaouis A, Tamagnini I, Hansen BM, Hendriksen NB, Boudabous A, Cherif A, Daffonchio D (2009) Characterization of polyvalent and safe *Bacillus thuringiensis* strains with potential use for biocontrol. J Basic Microbiol 49:293–303

Rang C, Bes M, Lullien-Pellerin V, Wu D, Federici BA, Frutos R (1996) Influence of the 20-kDa protein from *Bacillus thuringiensis* ssp. *israelensis* on the rate of production of truncated Cry1C proteins. FEMS Microbiol Lett 141:261–264

Ribeiro BM, Crook NE (1993) Expression of full-length and truncated forms of crystal protein genes from *Bacillus thuringiensis* subsp. *kurstaki* in a baculovirus and pathogenicity of the recombinant viruses. J Invertebr Pathol 62:121–130

Ribeiro BM, Crook NE (1998) Construction of occluded recombinant baculoviruses containing the full-length *cry1Ab* and *cry1Ac* genes from *Bacillus thuringiensis*. Braz J Med Biol Res 31:763–769

Salehi Jouzani G, Seifinejad A, Saeedizadeh A, Nazarian A, Yousefloo M, Soheilivand S, Mousivand M, Jahangiri R, Yazdani M, Amiri RM, Akbari S (2008) Molecular detection of nematicidal crystalliferous *Bacillus thuringiensis* strains of Iran and evaluation of their toxicity on free-living and plant-parasitic nematodes. Can J Microbiol 54:812–822

Sanahuja G, Banakar R, Twyman RM, Capell T, Christou P (2011a) *Bacillus thuringiensis*: a century of research, development and commercial applications. Plant Biotechnol J 9(3):283–300

Sanahuja G, Banakar R, Twyman RM, Capell T, Christou P (2011b) *Bacillus thuringiensis*: a century of research, development and commercial applications. Plant Biotechnol J 9:283–300

Sanchis V, Agaisse H, Chaufaux J, Lereclus D (1997) A recombinase-mediated system for elimination of antibiotic resistance gene markers from genetically engineered *Bacillus thuringiensis* strains. Appl Environ Microbiol 63:779–784

Sansinenea E, Sanchez P, Anastacio E, Ibarra J, Olmedo G, Vazquez C (2010a) Homologous recombination to *Bacillus thuringiensis* chromosome in one step. Agrociencia 44:437–447

Sansinenea E, Vazquez C, Ortiz A (2010b) Genetic manipulation in *Bacillus thuringiensis* for strain improvement. Biotechnol Lett 32:1549–1557

Schnepf HE, Whiteley HR (1981) Cloning and expression of the *Bacillus thuringiensis* crystal protein gene in *Escherichia coli*. Proc Natl Acad Sci U S A 78:2893–2897

Schnepf HE, Wong HC, Whiteley HR (1985) The amino acid sequence of a crystal protein from *Bacillus thuringiensis* deduced from the DNA base sequence. J Biol Chem 260:6264–6272

Schnepf E, Crickmore N, Van Rie J, Lereclus D, Baum J, Feitelson J, Zeigler DR, Dean DH (1998) *Bacillus thuringiensis* and its pesticidal crystal proteins. Microbiol Mol Biol Rev 62:775–806

Sekar V, Thompson DV, Maroney MJ, Bookland RG, Adang MJ (1987) Molecular cloning and characterization of the insecticidal crystal protein gene of *Bacillus thuringiensis* var. *tenebrionis*. Proc Natl Acad Sci U S A 84:7036–7040

Selinger LB, Khachatourians GG, Byers JR, Hynes MF (1998) Expression of a *Bacillus thuringiensis* delta-endotoxin gene by *Bacillus pumilus*. Can J Microbiol 44:259–269

Sellami S, Jamoussi K, Dabbeche E, Jaoua S (2011) Increase of the *Bacillus thuringiensis* secreted toxicity against Lepidopteron larvae by homologous expression of the *vip3LB* gene during sporulation stage. Curr Microbiol 63:289–294

Shao Z, Yu Z (2004) Enhanced expression of insecticidal crystal proteins in wild *Bacillus thuringiensis* strains by a heterogeneous protein P20. Curr Microbiol 48:321–326

Shao Z, Liu Z, Yu Z (2001) Effects of the 20-kilodalton helper protein on Cry1Ac production and spore formation in *Bacillus thuringiensis*. Appl Environ Microbiol 67:5362–5369

Shivakumar AG, Gundling GJ, Benson TA, Casuto D, Miller MF, Spear BB (1986) Vegetative expression of the delta-endotoxin genes of *Bacillus thuringiensis* subsp. *kurstaki* in *Bacillus subtilis*. J Bacteriol 166:194–204

Shivakumar AG, Vanags RI, Wilcox DR, Katz L, Vary PS, Fox JL (1989) Gene dosage effect on the expression of the delta-endotoxin genes of *Bacillus thuringiensis* subsp. *kurstaki* in *Bacillus subtilis* and *Bacillus megaterium*. Gene 79:21–31

Sirichotpakorn N, Rongnoparut P, Choosang K, Panbangred W (2001) Coexpression of chitinase and the *cry11Aa1* toxin genes in *Bacillus thuringiensis* serovar *israelensis*. J Invertebr Pathol 78:160–169

Soares GG, Quick TC (1990) MVP, a novel bioinsecticide, for the control of diamondback moth. In Proceedings of the Second International Workshop, 10–14 Dec. 1990, Taiwan

Stahly DP, Dingman DW, Bulla LA Jr, Aronson AI (1978) Possible origin and function of the parasporal crystal in *Bacillus thuringiensis*. Biochem Biophys Res Commun 84:581–588

Sun M, Yu Z (2000) Recent developments in the biotechnology of *Bacillus thuringiensis*. Biotechnol Adv 18:143–145

Sun M, Yue C, Yu Z (2000) Construction of integrative vector with *Bacillus thuringiensis* transposon Tn4430. J Agric Biotechnol (Chn) 8:321–326

Thamthiankul S, Moar WJ, Miller ME, Panbangred W (2004) Improving the insecticidal activity of *Bacillus thuringiensis* subsp. *aizawai* against *Spodoptera exigua* by chromosomal expression of a chitinase gene. Appl Microbiol Biotechnol 65:183–192

Thamthiankul Chankhamhaengdecha S, Tantichodok A, Panbangred W (2008) Spore stage expression of a vegetative insecticidal gene increase toxicity of *Bacillus thuringiensis* subsp. *aizawai* SP41 against *Spodoptera exigua*. J Biotechnol 136:122–128

Thompson MA, Gaertner FH (1990) Novel *Bacillus thuringiensis* isolate having anti-protozoan activity. European Patent Application Number 0461799 A2

Trisrisook M, Pantuwatana S, Bhumiratana A, Panbangred W (1990) Molecular cloning of the 130-kilodalton mosquitocidal delta-endotoxin gene of *Bacillus thuringiensis* subsp. *israelensis* in *Bacillus sphaericus*. Appl Environ Microbiol 56:1710–1716

Vazquez-Pineda A, Yanez-Perez GN, Lopez-Arellano ME, Mendoza-de-Gives P, Liebano-Hernandez E, Bravo-de-la-Parra A (2010) Biochemical characterization of two purified proteins of the IB-16 *Bacillus thuringiensis* strains and their toxicity against the sheep nematode *Haemonchus contortus in vitro*. Transbound Emerg Dis 57:111–114

Visick JE, Whiteley HR (1991) Effect of a 20-kilodalton protein from *Bacillus thuringiensis* subsp. *israelensis* on production of the CytA protein by *Escherichia coli*. J Bacteriol 173:1748–1756

Vu KD, Yan S, Tyagi RD, Valero JR, Surampalli RY (2009) Induced production of chitinase to enhance entomotoxicity of *Bacillus thuringiensis* employing starch industry wastewater as a substrate. Bioresour Technol 100:5260–5269

Waalwijk C, Dullemans A, Maat C (1991) Construction of a bioinsecticidal rhizosphere isolate of *Pseudomonas fluorescens*. FEMS Microbiol Lett 77:257–263

Ward ES, Ellar DJ, Chilcott CN (1988) Single amino acid changes in the *Bacillus thuringiensis* var. *israelensis* delta-endotoxin affect the toxicity and expression of the protein. J Mol Biol 202:527–535

Widner WR, Whiteley HR (1989) Two highly related insecticidal crystal proteins of *Bacillus thuringiensis* subsp. *kurstaki* possess different host range specificities. J Bacteriol 171:965–974
Wirth MC, Georghiou GP, Federici BA (1997) CytA enables CryIV endotoxins of *Bacillus thuringiensis* to overcome high levels of CryIV resistance in the mosquito, *Culex quinquefasciatus*. Proc Natl Acad Sci U S A 94:10536–10540
Wirth MC, Walton WE, Federici BA (2000) Cyt1A from *Bacillus thuringiensis* restores toxicity of *Bacillus sphaericus* against resistant *Culex quinquefasciatus* (Diptera: Culicidae). J Med Entomol 37:401–407
Wirth MC, Jiannino JA, Federici BA, Walton WE (2005a) Evolution of resistance toward *Bacillus sphaericus* or a mixture of B. sphaericus +Cyt1A from *Bacillus thuringiensis*, in the mosquito, *Culex quinquefasciatus* (Diptera: Culicidae). J Invertebr Pathol 88:154–162
Wirth MC, Park HW, Walton WE, Federici BA (2005b) Cyt1A of *Bacillus thuringiensis* delays evolution of resistance to Cry11A in the mosquito *Culex quinquefasciatus*. Appl Environ Microbiol 71:185–189
Wirth MC, Walton WE, Federici BA (2010) Evolution of resistance to the *Bacillus sphaericus* Bin toxin is phenotypically masked by combination with the mosquitocidal proteins of *Bacillus thuringiensis* subspecies *israelensis*. Environ Microbiol 12:1154–1160
Wong HC, Schnepf HE, Whiteley HR (1983) Transcriptional and translational start sites for the *Bacillus thuringiensis* crystal protein gene. J Biol Chem 258:1960–1967
Wu D, Chang FN (1985) Synergism in mosquitocidal activity of 26 and 65 kDa proteins from *Bacillus thuringiensis* subsp. *israelensis* crystal. FEBS Lett 190:232–236
Wu D, Federici BA (1993) A 20-kilodalton protein preserves cell viability and promotes CytA crystal formation during sporulation in *Bacillus thuringiensis*. J Bacteriol 175:5276–5280
Wu D, Federici BA (1995) Improved production of the insecticidal CryIVD protein in *Bacillus thuringiensis* using *cryIA(c)* promoters to express the gene for an associated 20-kDa protein. Appl Microbiol Biotechnol 42:697–702
Wu X, Vennison SJ, Huirong L, Ben-Dov E, Zaritsky A, Boussiba S (1997a) Mosquito larvicidal activity of transgenic *Anabaena* strain PCC 7120 expressing combinations of genes from *Bacillus thuringiensis* subsp. *israelensis*. Appl Environ Microbiol 63(12):4971–4974
Wu X, Vennison SJ, Huirong L, Ben-Dov E, Zaritsky A, Boussiba S (1997b) Mosquito larvicidal activity of transgenic *Anabaena* strain PCC 7120 expressing combinations of genes from *Bacillus thuringiensis* subsp. *israelensis*. Appl Environ Microbiol 63:4971–4974
Wu L, Sun M, Yu Z (2000) A new resolution vector with *cry1Ac10* gene based on *Bacillus thuringiensis* transposon Tn4430. Acta Microbiol Sin 40:264–269
Xia L, Long X, Ding X, Zhang Y (2009a) Increase in insecticidal toxicity by fusion of the *cry1Ac* gene from *Bacillus thuringiensis* with the neurotoxin gene *hwtx-I*. Curr Microbiol 58:52–57
Xia L, Zeng Z, Ding X, Huang F (2009b) The expression of a recombinant *cry1Ac* gene with subtilisin-like protease CDEP2 gene in acrystalliferous *Bacillus thuringiensis* by Red/ET homologous recombination. Curr Microbiol 59:386–392
Yamamoto T, Powell GK (1993) *Bacillus thuringiensis* crystal proteins: recent advances in understanding its insecticidal activity. In: Leo K (ed) Advanced engineered pesticides. Marcel Dekker, Inc., New York, pp 3–42
Yari S, Inanlou DN, Yari F, Saleh M, Behrokh F, Akbarzadeh A (2002) Effects of protoplast fusion on delta-endotoxin production in *Bacillus thuringiensis* spp. (H14). Iran Biomed J 6:25–29
Yoshisue H, Yoshida K, Sen K, Sakai H, Komano T (1992) Effects of *Bacillus thuringiensis* var. *israelensis* 20-kDa protein on production of the Bti 130-kDa crystal protein in *Escherichia coli*. Biosci Biotechnol Biochem 56:1429–1433
Yu J, Pang Y, Li J, Yu R, Tang M (1999) Construction of insecticidal engineered *Bacillus thuringiensis* using the transposon Tn917. Acta Sci Nat Uni Sunyasen 38:52–57
Yu J, Pang Y, Tang M, Xie R, Tan L, Zeng S, Yuan M, Liu J (2001) Highly toxic and broad-spectrum insecticidal *Bacillus thuringiensis* engineered by using the transposon Tn917 and protoplast fusion. Curr Microbiol 43:112–119
Yue CY, Sun M, Chen SW, Yu ZN (2003) Construction of insecticidal recombinant *Bacillus thuringiensis* using an integrative vector. J Genet Genomics 30:737–742

Yue C, Sun M, Yu Z (2005a) Broadening the insecticidal spectrum of Lepidoptera-specific *Bacillus thuringiensis* strains by chromosomal integration of *cry3A*. Biotechnol Bioeng 91:296–303
Yue C, Sun M, Yu Z (2005b) Improved production of insecticidal proteins in *Bacillus thuringiensis* strains carrying an additional *cry1C* gene in its chromosome. Biotechnol Bioeng 92:1–7
Zhang GY, Zhang J, Peng Y, Zhao J, Chen C, Xu Y, Huang D (1995) Construction of genetically engineered strains of *Pseudomonas fluorescens* against plant pathogen and insect pests by electrophorating transformation. Sci Agric Sin 28:8–13
Zhu C, Ruan L, Peng D, Yu Z, Sun M (2006) Vegetative insecticidal protein enhancing the toxicity of *Bacillus thuringiensis* subsp *kurstaki* against *Spodoptera exigua*. Lett Appl Microbiol 42:109–114

Chapter 14
Bacillus thuringiensis Recombinant Insecticidal Protein Production

H. Ernest Schnepf

Abstract The initial excitement of applying recombinant DNA technology to insecticidal proteins of *Bacillus thuringiensis* was subdued by regulatory caution about the technology and the genetic complexity of the proteins themselves. While seven biopesticide products containing recombinant proteins were eventually manufactured, expression of the proteins in Gram negative and Gram positive bacteria is predominantly for discovery and mode of action work. Regulatory studies for the registration of transgenic plants requires microbially produced insecticidal proteins in the tens of grams. Transgenic plants now dominate the production of *B. thuringiensis* recombinant insecticidal proteins, with expression technology yielding more than 10-fold higher levels than the earliest registered plants and worldwide use on nearly 60 million hectares.

Keywords Recombinant insecticides · Protein production · Protein expression · Bacterial expression · Plant expression

14.1 Introduction

Bacillus thuringiensis (Bt) was discovered somewhat over 100 years ago (Beegle and Yamamoto 1992; Yamamoto 2001). Since then, scientific work on Bt has moved from basic descriptive microbiology and insect pathology to the gamut of modern biology. Commercially, Bt products have ranged from essentially sporulated cultures to breeding traits of commercial crops that provide in-plant protection from insects. Studies involving Bt range from molecular biology and protein crystallography to population genetics and food product safety (Shelton et al. 2002; Tabashnik et al. 2009; US Environmental Protection Agency 2011a; Yamamoto 2001). While initially known for its lepidopteran activity, the Bt pesticidal range has expanded to a large number of proteins spanning a wide range of activities and several distinct

Dedicated to the memory of Lee F. Adams III, 1958–2010

H. E. Schnepf (✉)
7954 Handel Ct. San Diego, San Diego, CA 92126-3010, USA
e-mail: heschnepf@gmail.com

E. Sansinenea (ed.), *Bacillus thuringiensis Biotechnology*,

DOI 10.1007/978-94-007-3021-2_14,

protein structural classes (Crickmore et al. 1998; de Maagd et al. 2003; Schnepf et al. 1998).

Prior to the development of recombinant DNA methods, Bt provided a striking example of natural high-level protein expression and accumulation, with 20% of the dry weight of sporulated cultures being insecticidal crystals (Lilley et al. 1980). Improved production of insecticidal material was balanced between cost of production and efficacy. There initially was controversy as to how potency should be defined, resulting in the activity-based international unit standard, which did not necessarily correlate with the amount of crystal protein, or the spore count (Beegle and Yamamoto 1992). We now better understand this situation as a combination of: crystal proteins that control a target insect and spore-supplied factors such as zwittermycin (reviewed in (Schnepf et al. 1998)) or insecticidal Vip proteins following germination in the insect gut (Donovan et al. 2001).

As recombinant DNA methods were applied to Bt, a number of crystal toxin genes were cloned and expressed in *Escherichia coli*, *Bacillus subtilis*, *Pseudomonas fluorescens*, and a number of other hosts, some with the intent of high-level fermentation and others for environmental colonization (reviewed in (Schnepf et al. 1998)). The advent of electroporation allowed the otherwise minimally transformable Bt to also be used as an expression host, and opened an avenue for genetic modification of Bt strains and improved sprayable Bt products. By this time, plants had also been successfully engineered to express Bt toxin genes for insect control (see below). Oddly, the regulatory path for transgenic plants seemed more lenient in many ways than that for genetically modified microorganisms. For example, approval of the Mycogen killed *P. fluorescens* product required documentation of a 14 log kill, which was nearly as heroic as it sounds (Panetta 1993). As transgenic insect-protected plants were commercialized, major higher value markets for sprayable Bts disappeared, reducing the commercial interest in recombinant Bt strain improvement (Yamamoto 2001).

Initially, recombinant Bt protein expression was primarily for analytical use to determine the target range of the individual gene products of a Bt strain, with amounts of protein required in the multi-milligram range needed for lab bioassays (reviewed in (Yamamoto 2001)), because of the regulatory restrictions on environmental release of recombinant bacteria. (An exception to this was the killed *P. fluorescens* products from Mycogen, which were being scaled up to production runs of >10,000 l over this same period (Panetta 1993)). Somewhat larger amounts of highly purified protein were needed for crystallographic studies (Li et al. 1991, Table 14.1). In the late 1980s and early 1990s regulatory discussions were progressing on the registration of insect-protected transgenic plants expressing Bt toxin proteins. While the plants themselves were producing the Bt proteins in the range of micrograms toxin per gram plant material, toxicology, environmental fate, and resistance management studies to support initial and continuing product registration, required tens of grams of protein, sometimes at a very high level of purity. Since it would require vast amounts of plant material to yield gram levels of Bt proteins (and even then possibly not at the required level of purity per gram test substance for toxicology), these materials were produced in bacteria, with so-called

Table 14.1 Publicly available Bt crystal protein structure coordinates and their expression sources

PDB ID	Protein	Expression	References
2ZTB	Cry46Aa/parasporin-2	*E. coli*	Akiba et al. (2009)
3EB7	Cry8Ea1	*B. thuringiensis*	Guo et al. (2009)
2RCI	Cyt2Ba	*E. coli*	Cohen et al. (2008)
2C9K	Cry4Aa	*E. coli*	Boonserm et al. (2006)
2D42	Non-toxic crystal protein	Native *B. thuringiensis*	Akiba et al. (2006)
1W99	Cry4Ba	*B. thuringiensis serovar israelensis*	Boonserm et al. (2005)
1JI6	Cry3Bb1	*E. coli*	Galitsky et al. (2001)
I5P	Cry2Aa	*B. thuringiensis*	Morse et al. (2001)
1CIY	Cry1Aa	*E. coli*	Grochulski et al. (1995)
1CBY	Cyt2Aa	*E. coli*	Li et al. (1996)
1DLC	Cry3Aa	Native *B. thuringiensis*	Li et al. (1991)

bridging studies performed to demonstrate equivalence of the plant- and bacterially-produced proteins (reviewed in (Evans 2004)). As noted below, only a few of these studies used test material from a recombinant Bt host.

For the purpose of this chapter, the term recombinant protein will mean reintroduction of a cloned DNA into a host as a means of producing protein. While there have been many hosts for Bt protein expression, the focus will be on *E. coli* (and to a lesser extent *P. fluorescens*), *B. thuringiensis*, and plants. Notable successes will be mentioned, however, more time will be spent on modifications that incrementally improve high production levels, unresolved issues with low-expressing proteins, or maintaining functionality of the expression host.

14.2 Pesticidal Protein Expression in Gram Negative Hosts

Historically, *E. coli* was the first recombinant expression host for any cloned gene. Expression of Bt crystal proteins in *E. coli* was first used to document the isolation of cloned genes. While native Bt expression signals often function in *E. coli*, and in some cases, can result in production of crystals (Whiteley et al. 1987), their primary use has been basic lab-scale detection of crystal proteins for further study. Bt proteins of most interest were usually Cry1 or Cry3 proteins that are typically well expressed and well tolerated by *E. coli* (reviewed in (Gawron-Burke and Baum 1991)). Ge et al. (1990) presented a systematic study of shake flask production of Cry1Ac, demonstrating the utility of the Ptac/pKK223-3 promoter and host JM103 to produce up to about 0.3 mg/ml, albeit with evidence of derepression for all strains. Boonserm et al. (2004) expressed P_{tac}-P_{cry4A}-driven *cry4Aa* in *E. coli* to produce purified protein for crystallography. Peer-reviewed publications on the production of toxicology material for product registration are: Gustafson et al. (1997)

yielding over 50 g of purified Cry3Aa from a 1000 l fermentation of MB101, and Huang et al. (2007) reporting the purification to 98% purity of 31 g of Cry34Ab1 and 37 g of Cry35Ab1 from 75 l high-density fermentations of an IPTG-inducible *P. fluorescens* expression system. The productivity of Cry34Ab1 and Cry35Ab1 in the latter study was estimated at 3–4 g/l, while they report expression in *E.coli* at well under 1 g/l.

For high-level crystal protein expression in Gram negative hosts, it is advisable to configure the gene for such a host, and regulate expression so that no inclusions form without some form of induction to maintain stability of the host and the gene. Sequences associated with the sporulation-inducible Bt promoters can be reasonably active promoters in *E. coli*, which may be problematic for expression and stability (Whiteley et al. 1987; Schnepf et al. 1987). *Bacillus* ribosome binding sites typically work well in *E. coli*, however, *Bacillus* effectively uses the UUG initiation codon, which is not well utilized in *E. coli*; for example, a UUG to AUG initiation codon change in *cry1Ba* markedly improves expression (Brizzard et al. 1991). An unusual observation in Schnepf et al. (1987) was that the native *cry1Aa* promoter/translation start fused in frame to *E. coli* beta-galactosidase resulted in production of LacZ protein to a significant percentage of cell protein, substantially more protein than equivalently configured Cry1Aa, and enough to induce likely LacY toxicity. This implies that even the relatively well expressed and tolerated Cry1Aa has significant barriers to expression in *E.coli* subsequent to translation initiation.

Inefficiencies related to translational or transcriptional elongation and coupling, as well as mRNA and protein stability, are more difficult to analyze and correct in a targeted manner. With transgenic plants, synonymous codon gene reconstruction was a solution to expression problems thought mostly to be related to adventitious plant RNA processing signals in the *Bacillus* coding sequences (Perlak et al. 1991). The utility of a similar approach in *E. coli* was shown when the toxin-encoding portion of the *cry4Aa* gene was re-synthesized with host codon usage and expressed as an N-terminal GST fusion peptide (Hayakawa et al. 2008) with a resulting fivefold increase in expression and threefold increase in the recovery of fully active protein. An alternative approach, using an *E. coli* expression host over-producing rare tRNAs, was successfully used for a *cry1Ac* gene (Hire et al. 2008), although the unsupplemented low expression was a bit puzzling. A recent study has indicated that bacterially "optimized" synthetic human genes generally outperform tRNA supplementation (Maertens et al. 2010). While the cost of gene synthesis has decreased, screening expression host strains may be adequate and much less expensive.

Post-translational problems could be a factor in these and other cases. Expression of the original *cyt1Aa* clones in *E. coli* without co-expression of a 20 kDa protein (P20) from a closely linked gene was virtually absent (Adams et al. 1989), and then only in small amounts. An additional study showed that P20 (which was itself expressed well enough to make antibodies) interacted directly with Cyt1Aa and could stabilize a heat shock sensitive LacZX90 protein at 42 °C. That study also showed that temperature sensitive defects in the chaperonins GroEL and DnaK would allow synthesis of Cyt1Aa at 42 °C (Visick and Whiteley 1991). Apparently co-expression of GroEL or GroES with Cyt1Aa (Manasherob et al. 2001) cannot substitute for P20. Thus, the P20 appears to act as a chaperone to partially protect the host from the deleterious effects of Cyt1Aa, and although Cyt1Aa interacts with elements of the heat shock regulon, over-expression of *groEL* or *groES* does not substitute. A recent report has indicated that insecticidally active Cyt1Aa alone can be produced in *E. coli* using a tightly regulated promoter; however, there is still a dramatic reduction of cell viability upon induction (Sazhenskiy et al. 2010). Interestingly, other structurally similar proteins can be well expressed in *E. coli* (Cohen et al. 2008; Li et al. 1996), suggesting, as expected, that sequence-specific protein interactions account for the Cyt1Aa expression defect.

While the importance of Cyt1Aa in mosquito vector control drives continued study of its expression, problems with other genes and proteins often are not published, an exception being *cry13A* and *cry14A* (Wei et al. 2003). Finally, Thompson and Schwab (1996) showed that substituting the Cry1Ab or Cry1Ac protoxin segment for that of Cry1Fa improves expression of the resultant chimeric protein with respect to native Cry1Fa in *P. fluorescens*, however a mechanism was not presented (see below for a similar result in Bt). This modification was later used in transgenic plants (see below).

The continued work on expression of a wide variety of proteins in *E. coli* for a number of uses, often for detailed study of isolated human gene products, has shown that expression of any particular protein is far from routine. Significant progress has been made in expression vectors, host strains and potential co-expression partners that could be useful for either expression of a difficult to express protein or production optimization of a protein needed in very large amounts (Sørensen and Mortensen 2005; Terpe 2006; Structural Genomics Consortium 2008; Makino et al. 2011).

14.3 Pesticidal Protein Expression in *B. thuringiensis*

When molecular genetic methods began to be applied to Bt toxins, there was considerable interest in translating the newly acquired information into Bt sprayable biopesticides. However, the lack of a practical means of introducing cloned DNA into Bt, coupled with concerns regarding the release of genetically engineered organisms, meant that the majority of the molecular analysis went into the study of toxins rather than product development (Gawron-Burke and Baum 1991). In the late 1980s, several labs reported transformation of Bt using electroporation, opening the way for targeted genetic modification. There ensued a proliferation of cloning vectors, many of which were more stably maintained than the available vectors from *Staphylococcus aureus*, *B. subtilis* and *Bacillus cereus* (for example (Gawron et al. 1991; Baum et al. 1990; Arantes and Lereclus 1991)). Additional methods were developed or adapted from other organisms for genetic analysis and manipulation of Bt to create Bts with new insecticidal gene combinations (Crickmore et al. 1990; Gawron-Burke and Baum 1991; Lecadet et al. 1992).

The understanding of Bt toxin gene regulation progressed dramatically, with significant levels of expression control involving promoters and RNA polymerase sigma factors, sequences that can stabilize the 3' and 5' ends of mRNA, and apparent chaperone-like proteins that can improve crystal formation (Agaisse and Lereclus 1995; Baum et al. 1999; Baum and Malvar 1995; Schnepf et al. 1998; Yamamoto 2001). Additionally, Gilmer and Baum (1999) showed that Cry1Fa was accumulated in larger amounts when a Cry1Ac protoxin segment was exchanged for its own, the type of modification later used in recombinant Bt strains and transgenic plants (see below). A number of regulatory sequences and potential crystallization partners were then available for expressing Bt toxin genes. While crystal proteins accumulate post-exponentially, there are two different promoter classes: one recognized by the major vegetative sigma factor that is not under sporulation control, and the other recognized by two successive mother cell sporulation sigma factors. The *cry3A* and *cry3B* promoters are in the former class, and together with their 5' RNA stabilizing sequences, can provide high-level protein production, for example a Cry1C/Cry1Ab protein using the *cry3A* promoter (Sanchis et al. 1999), or the *cry2A* operon using the *cry3B* promoter (Baum and Malvar 1995). The class under sporulation control includes *cry1Aa*, *cry1Ba*,(and probably most *cry1* genes), *cry4A*, *cry4B*, *cyt1Aa* and *cry11A*/P19 (Baum and Malvar 1995; Dervyn et al. 1995). An interesting observation of Baum et al. (1996) was that addition of a *cry3Bb* transcription unit to a *cry1Ac*-containing Bt kurstaki strain resulted in additive protein production of Cry3Bb and Cry1Ac, while the addition of *cry1* genes results in sub-additive protein production, possibly due to regulatory factor competition. Aronson (1994) also noted the compensatory relationship between Cry1 production and gene content. A more recent comparison of *cry3A*-promoted and *cry1Ac*-promoted *cry-1Ac* genes also concluded that the *cry3A* system resulted in more protein production (Chaoyin et al. 2007). There have also been reports using other promoters to express Bt crystal proteins during vegetative growth, for example the *tetR* promoter from the plasmid pBC16 (Mettus and Macaluso 1990), and the alpha amylase promoter of *B. subtilis* (Chak et al. 1994). Laboratory-scale production of crystal proteins by Bts could reach about 0.5 g/l (Agaisse and Lereclus 1995; Baum et al. 1990) with recombinant production of single proteins at about half that amount.

Several groups developed recombination systems to place cloned genes stably into Bt cells without the use of antibiotics. In one case (Kalman et al. 1995; Yamamoto 2001) a homologous recombination system was used and in the others, a site specific recombination system was employed (Baum et al. 1996; Sanchis et al. 1997). In all three cases, the strains were successfully tested in the field (Sanchis et al. 1999; Yamamoto 2001), and the Ecogen strains (Baum et al. 1999) became registered products: Raven®, CRYMAX® and Lepinox®; the latter two of which are available for sale as of 2011 (Table 14.2). The other two had at least entered the registration process for Sandoz and AgrEvo, respectively. The Pesticide Action Network Database (Kegley et al. 2011) also indicates some registration related test activity with recombinant Bts for Abbott Labs and one other unidentified organization.

Baum et al. (1999) mentioned the possibility of improving the insecticidal capabilities of Bt host strains, for example by enhancing expression Vip3 proteins or altering the regulation of zwittermycin. More recent reports have shown that the *vip3A* gene can be reprogrammed to express during sporulation with either the

Table 14.2 Applied biopesticides registered with the U.S. EPA containing recombinant *B. thuringiensis* crystal proteins

Product	Company	Crystal protein (copies of gene)	Form	Years registered with U.S. EPA[a]
M-Trak®	Mycogen	Cry3Aa[R]	Killed *P. fluorescens*	1991–2003
MVP®, MVPII®	Mycogen	Cry1Ac[R]	Killed *P. fluorescens*	1991–2003
Raven®	Ecogen	Cry1Ac(2), Cry3A, Cry3Bb[R]	*B. thuringiensis*	1995–2005
MATTCH®	Mycogen	Cry1Ac[R] and Cry1Ca/Cry1Ab protoxin[R]	Each in killed *P. fluorescens*	1995–2004
M/C®	Mycogen	Cry1Ca/Cry1Ab protoxin[R]	Killed *P. fluorescens*	1996–2003
CRYMAX®	Ecogen	Cry1Ac(3), Cry2A, Cry1Ca[R]	*B. thuringiensis*	1996–2011+
Lepinox®	Ecogen	Cry1Aa, Cry1Ac(2), Cry2A, Cry1Fa/1Ac protoxin[R]	*B. thuringiensis*	1996–2011+

[R] Recombinant protein product
+ Still available as of 2011
[a] Kegley et al. (2011)

cry1C promoter (Song et al. 2008) or the *cry11A*/P19 (Thamthiankul Chankhamhaengdecha et al. 2008) promoter. In the former case, Vip3A accumulated during sporulation and inclusion formation could be induced using a fusion to the Cry1C protoxin segment, however, activity of the resulting protein was very low; in the latter, whole culture insecticidal activity was clearly improved. Since the zwittermycin biosynthetic gene cluster has been identified by Kevany et al. (2009) and others, and methods for manipulating large DNAs in Bt are being developed (Liu et al. 2009), contemplating alteration of its synthesis is possible. Genome alterations to introduce crystal protein genes were noted above, and related methods have been used to eliminate enterotoxin genes (Klimowicz et al. 2010). More drastic genome reduction in *B. subtilis* (Morimoto et al. 2008) led to elongation of the transition phase to sporulation and enhancement of recombinant enzyme synthesis, which appears to be the same period over which Cry3A and Vip3 proteins are produced (see above).

14.4 Bacterial Expression Applications Summary

The preceding sections have shown that there are generally good methods for producing Bt insecticidal proteins in Gram negative organisms and in Bt, with probably fewer difficulties in obtaining expression in Bt, but certainly with more facility for gene manipulation in *E. coli*. Bioassays and mode-of-action studies employ either host, with some labs having a preference for particular purposes. Larger amounts of protein were required for crystallographic work, especially for the earlier structures. As shown in Table 14.1, of the 11 structures currently in the Protein Data Bank (PDB), two were from native Bt strains, three from Bt expression sources and six from *E. coli* expression sources. At the sprayable product level, seven recombinant Bt crystal protein products received EPA registration (Table 14.2) and were produced in batches that were some multiple of 10,000 l. Four of these were killed bacterial CellCap® products from Mycogen Corp., were registered between 1991 and 1996 and were phased out in 2003 and 2004 at least in part because their new owner wanted to consolidate around a Bt production process. The *P. fluorescens*-based CellCap® system can produce some proteins at over 25 g/l under optimized conditions (Squires et al. 2004), but would have a different production process from Bt, particularly post fermentation. The recombinant Bt products were from Ecogen and were registered in 1995 and 1996. Two of the three are still in available, and the third, Raven®, was abandoned in 2005, possibly due to lack of a market. In an assessment of insecticidal activity of an enterotoxin deleted Bt *kurstaki* strain (Klimowicz et al. 2010), crystal protein production of the deleted and parent strains was estimated at about 10 g/l in 7.5 l fermenters.

Evans (2004) reviewed the production of toxicology materials for transgenic plant registrations. Table 14.3 shows the source and highest concentration of the active ingredient in toxicology test substances for a number of commercially approved transgenic plants (typically administered at about 5000 mg substance per gram body weight for oral tests). The concentration of the active ingredient in the test substance depends on its overall purity, not just percent of total protein. It also needs to be fully insecticidally active, another prerequisite of the test (Evans 2004). As shown in Table 14.3, a native Bt strain was used for part of the Event 176 maize registration, recombinant Bt strains were used for Cry2Ab cotton and some tests for Cry3Bb corn, *P. fluorescens* was used for event 281/3006 cotton and cry1Fa, Cry34Ab, and Cry35Ab corn. The remainder were produced in *E. coli* with little to no description of the expression system, production process, or protein productivity (except as noted above and the pET BL21(DE3) system for vip3A (US Environmental Protection Agency 2008)).

14.5 Expression in Transgenic Plants

Streatfield (2007) presented a thorough review of methods to achieve high-level protein expression in plants as a means of recombinant protein production, and Liu (2009) reviewed gene expression in transgenic maize. As noted in those reviews, expression of the same construct from different chromosomal positions can vary dramatically, can itself affect plant phenotype, and requires the examination of tens to thousands of independent transformation events to find one that is optimal. For Bt proteins, the plant- pest interaction and agronomic performance constrain the various factors affecting protein expression. On the one hand, resistance management

Table 14.3 Microbial expression of Bt insecticidal proteins for toxicology and environmental fate studies

Event	Oral toxicity test dose[a]	Year registered	Reference
Event 176 Cry1Ab maize	3280 mg/kg *Bt kurstaki* HD1-9	1995	US EPA (2011c)
MON531 Cry1Ac cotton	4200 mg/kg *E.coli*	1995	US Environmental Protection Agency (2001)
MON810 Cry1Ab maize	4000 mg/kg *E. coli*	1996	US Environmental Protection Agency (2010a)
Bt11 Cry1Ab maize	3535 mg/kg *E. coli*	1996	US Environmental Protection Agency (2010a)
TC1507 Cry1F maize	576 mg/kg *P. fluorescens*	2001	US Environmental Protection Agency (2010a)
MON 15985 Cry2Ab cotton	1450 mg/kg Bt EG7699	2002	US Environmental Protection Agency (2002)
MON88017 Cry3Bb maize	3780 mg/kg Bt, 2700 mg/kg *E.coli*	2003[b]	US Environmental Protection Agency (2010b)
281/3006 Cry1Fa cotton	375 mg/kg *P. fluorescens*	2004	US Environmental Protection Agency (2005)
281/3006 Cry1Ac cotton	700 mg/kg *P. fluorescens*	2004	US Environmental Protection Agency (2005)
59122-7 Cry34Ab1 maize	2700 mg/kg *P. fluorescens*	2005	US Environmental Protection Agency (2010c)
59122-7 Cry35Ab1 maize	1850 mg/kg *P. fluorescens*	2005	US Environmental Protection Agency (2010c)
MIR604 mCry3A maize	2377 mg/kg *E. coli*	2006	US Environmental Protection Agency (2007)
MON 89034 cry1A. 105 maize	2072 mg/kg *E.coli*	2008	US Environmental Protection Agency (2010d)
MON 89034 CTP-Cry2Ab2 maize	1000 mg/kg *E.coli*	2008	US Environmental Protection Agency (2010d)
COT102 Vip3Aa19 cotton	3675 mg/kg *E. coli*	2008	US Environmental Protection Agency (2008)
COT67B Cry1Ab cotton	1830 mg/kg *E.coli*	2008	US Environmental Protection Agency (2008)
MIR162 Vip3Aa20 maize	3675 mg/kg *E. coli*	2008	US Environmental Protection Agency (2009)
MON 87701 CTP-Cry1Ac Soybean	1460 mg/kg *E.coli*	2010	US Environmental Protection Agency (2010e)

[a] Highest dose of Bt protein in ~5000 mg/kg test substance with identified source organism
[b] For a different event, MON863, of the same Bt protein

considerations demand the pest organism be exposed to a high dose everywhere on the plant it is present (Bravo et al. 2011; Shelton et al. 2002; Tabashnik et al. 2009). On the other, the plants expressing the Bt protein must have top tier levels of crop productivity for grower acceptance. The use of regulatory information to assess expression levels is for two reasons. First, regulatory agencies carefully scrutinize the analytical methods for expression determination for the purpose of insect resistance management, toxicology and environmental fate, and second, the expression levels are likely to be the best effort of the respective research organizations to balance high levels of insect control and good agronomic performance over a wide range of growing conditions. Additionally, the source of the transgenic crop needs to be "traceable", which requires making DNA-based tests to identify specific transformation events, and this often means a patent has been filed describing the transgene insertion and flanking plant DNA (often to the full base sequence).

The expression of Bt insecticidal proteins in plants is by now a well-known story of very poor expression of the bacterial sequences due to a number of deleterious processes during gene expression in plants, including mis-splicing (van Aarssen et al. 1995), rapid degradation (De Rocher et al. 1998) and premature polyadenylation (Diehn et al. 1998) of the transgene RNA. Repairing even a few of the offending sites in the bacterial gene can greatly improve message formation (Perlak et al. 1991; van Aarssen et al. 1995), however, the preferred solution has been to rebuild the bacterial genes with synonymous codons so that they are more plant-like. There are a number of approaches to designing the transgenes: in one, the most preferred codon is favored for each amino acid unless deleterious sequences need to be avoided (Koziel et al. 1993), while most of the other methods focus on removing deleterious sequences while avoiding rarely used codons for a particular plant species (reviewed in (Liu 2009)). The corn optimized synonymous Bt gene synthesized by Koziel et al. (1993) was reportedly 64% identical to the native Bt gene, and it is possible to have the same protein encoded by plant-like sequences from different groups that are only 80–90% identical.

Other expression control elements are needed as well. Promoter/enhancer combinations seem to be evolving from viral cauliflower (CAMV 35S, e35S) and figwort (FMV) mosaic viruses to a variety of genomic promoters (Table 14.4). The CAMV 35S promoter has been used in both corn and cotton as has the maize ubiquitin (Zm Ubi) promoter. Placing a strongly expressed 5' untranslated region and an intron before the transgene can substantially improve expression, particularly in monocots (Liu 2009), but also in dicots (Parra et al. 2011). This intron mediated enhancement works through little understood mechanisms thought to range from transcription initiation through RNA processing and transport, or translation; however increased transcription elongation may mediate a major portion of the effect (Parra et al. 2011). Nearly all of the recent transformation events include an explicit 5' untranslated region (UTR) and intron (Table 14.4). Sequences mediating transcription termination and polyadenylation at the 3' end of the transgene transcript may be important for efficient gene expression (Liu 2009), but also help convince regulators that no unintended gene products are being made (Rosati et al. 2008). The most popular 3' elements are from nopaline synthase (*nos*) and ORF25 from the Agrobacterium T-DNA, with CAMV and plant 3' UTR's also in use (Table 14.4). The MIR162 maize event also contains an intron following the vip3Aa20 ORF (Long et al. 2009, Table 14.4). There is a general trend of using monocot elements in corn, such as the rice actin intron, wheat peroxidase promoter and wheat Hsp 17 3' element, and using dicot elements in cotton, such as the *Arabidopsis* actin promoter,

Table 14.4 Plant expression context and integration

Event	Promoter	Leader	Bt-derived ORF	Terminator	Trasformation method/ insertion	References
Event 176 Cry1Ab maize	Zm* PEPC	–	cry1Ab trunc	CAMV 35S	Biolistic	CERA (2010a)
Event 176 Cry1Ab maize	Zm Ca dep PK	–	cry1Ab trunc	CAMV 35S	Biolistic	CERA (2010a)
MON531 Cry1Ac cotton	CaMV e35S	–	cry1Ac	Soybean 7S	Agro/1 intact cassette with partial 3'1Ac+3'7s placed 5' and second partial with 7s	CERA (2010b)
MON810 Cry1Ab maize	5'trunc 307 bp CaMV e35S	Zm Hsp70 intron	cry1Ab (5' 2448 bp; 3' trunc)	Zm HECT 3' antisense DNA. (Δnos 3')	Biolistic/5' maize discontinuous from 3'	Rosati et al. (2008)
Bt11 Cry1Ab maize	CaMV 35S	Zm Adh1S IVS 6 intron	cry1Ab trunc	Agro Nos3'	Direct protoplast transformation/1copy	US Environmental Protection Agency(2010a)
TC1507 Cry1F maize	Zm Ubi-1	Zm Ubi-1 5'utr intron	cry1Fa2L604F trunc	3'Agro orf25	Biolistic/Intact cry1F CAMV 35s pat cassette+scrambled fragments and maize DNA 3' and 5'	Barbour et al. (2007)
MON 15985 Cry2Ab cotton	CaMV e35S	Petunia 5'UTR Hsp70+Ath CPT2 transit peptide	CTP-cry2Ab2	Agro nos3'utr	Biolistic/cassette intact except very 3' of nos+scrambled DNA each end	Huber et al. (2007)
MON88017 Cry3Bb maize	CaMV e35S	Wheat CAB 5' leader+rice actin 1 intron	cry3Bb1	Wheat Hsp17 3'	Agro/1 insert	Beazley et al. (2008)
281/3006 Cry1Fa cotton	Agro (4OCS) DeltaMas 2'	–	cry1Fa	3' Agro ORF25	Agro/1 insert+contiguous partial pat cassette	Song et al. (2011)

Table 14.4 (continued)

Event	Promoter	Leader	Bt-derived ORF	Terminator	Trasformation method/ insertion	References
281/3006 Cry1Ac cotton	Zm Ubi-1	Zm Ubi-1 exon1, intron1	cry1Ac	3' Agro ORF25	Agro/1 insert	Song et al. (2011)
59122-7 Cry34Ab1 maize	Zm Ubi-1	Zm Ubi-1 exon1, intron1	cry34Ab1	Potato PinII	Agro/1 insert, ~20 bases missing each end.	Bing et al. (2008)
59122-7 Cry35Ab1 maize	Wheat peroxidase	–	cry35Ab1	Potato PinII	Cry34Ab1 and Cry35Ab1 on the same cassette	Bing et al. (2008)
MIR604 mCry3A maize	Zm MTL metallothionein like	–	mcry3A	Agro nos3'utr	Agro/1 insert ~40 bp missing each end	US Environmental Protection Agency (2007)
MON 89034 cry1A.105 maize	CAMV e35S	Wheat CAB 5' leader+rice actin 1 intron	cry1A.105	Wheat Hsp17 3'	Agro/1 insert with cry1A.105 and CTP-cry2Ab cassette	Anderson et al. (2008)
MON 89034 CTP-Cry2Ab2 maize	FMV	Zm Hsp70 intron, Zm transit peptide (TS-SSU-CTP)	CTP-cry2Ab2	Agro nos3'utr	Intact except 5' partial enhancer deletion to e35S; copy of right border on left end	Anderson et al. (2008)
Cot102 Vip3A cotton	Ath Act2	Ath Act2 intron	vip3Aa19	Agro nos3'utr	Agro/1 insert; intact	Ellis et al. (2006)
Cot67B Cry1Ab cotton	Ath Act2	Ath Act2 intron	cry1Ab[b]	Agro nos3'utr	Agro/1 insert; intact	US Environmental Protection Agency (2008)
MIR162 Vip3A maize	Zm Ubi	Zm Ubi first intron	vip3Aa20	Zm PEPC Intron#9 CaMV 35S 3'	Agro/1 intact insert, Maize chrom 5 w 58 bp displacement	Long et al. (2009)
MON 87701 Cry1Ac Soybean	Ath RbcS4	Ath RbcS leader and CTP1 transit peptide	CTP-cry1Ac	Soybean 7S	Agro/1 insert; intact	Gao et al. (2009)

[a] Source organism abbreviations: Zm—*Zea mays*; Ath—*Arabidopsis thaliana*; Agro—*Agrobacterium tumefaciens* T-DNA

[b] Restores a protoxin sequence missing in Cry1Ab vs. Cry1Aa or Cry1Ac

first intron (Ath Act2) and transit peptides (CTP1 and CPT2), petunia Hsp70 5'UTR and soybean 3' elements.

The plant transformation methods have evolved as well, with essentially all of the transformations done with Agrobacterium. Methods for obtaining low copy events with fewer rearrangements, efficient transformation of corn, and the use of vectors with a second T-DNA for a selectable marker that can subsequently be segregated from the insect resistance locus (MON 89034 and MON 87701) are now common (Table 14.4). The integration site complexity of biolistic transformation and its consequences are illustrated by the early corn event MON810, which was a full length *cry1Ab* construct with an Agrobacterium *nos* terminator for the polyadenylation site. More recent analysis of the expressed cry1Ab gene in this event (Rosati et al. 2008) showed that the *cry1Ab* gene was truncated and the *nos* terminator missing. Additionally, the 3' integration site was in a putative maize HECT E3 ubiquitin ligase oriented opposed to *cry1Ab*, and the alternative splicing at the 3' end of the *cry1Ab* ORF resulted in an 18 amino acid extension to the Cry1Ab protein in addition to the two amino acid extension predicted from the genomic sequence. Scrambling at the transformed DNA ends is also evident with TC1507 maize and MON 15985 cotton (Barbour et al. 2007; Huber et al. 2007, Table 14.4).

Another potential issue with expressing Bt insecticidal genes in plants is brought up in two patent publications from Monsanto (Bogdanova et al. 2009; Corbin and Romano 2006). They state that many of the Cry1 and Cry2 Bt insecticidal proteins are not expressible in the cytoplasm above roughly 10 ppm (i.e. 10 μg/g fresh weight) without some degree of deleterious consequences on the plant (Bogdanova et al. 2009). In certain cases, such as Cry2Ab, the effects of cytoplasmic expression can make obtaining efficacious plants difficult; however targeting the protein to the chloroplast through N-terminal fusion of a targeting peptide can sometimes alleviate the problem in both corn and cotton, and results in much higher levels of expression (Corbin and Romano 2006). This modification has been present in commercial cotton since 2002 and corn since 2008 (Table 14.4). Bogdanova et al. (2009) present genes of a chimeric crystal protein Cry1A.105, essentially Cry1Ac with domain III from Cry1Fa, one optimized for corn with no targeting peptide and one optimized for dicots with a targeting peptide. The non-targeted protein has been in commercial corn since 2008 (Table 14.4), however the targeted version is not in commercial cotton despite apparently being in field tests since at least 2008 (Monsanto Company Press Release 2008). The decreased expression and deleterious effects of chloroplast targeted Cry2Aa (Corbin and Romano 2006) reinforces the empirical nature of expression improvement by chloroplast targeting. The complexity of the chloroplast import process (Li and Chiu 2010), and lack of gene expression analysis suggest it would be premature to conclude that improved protein accumulation by chloroplast targeting is solely at a post-translational protein interaction level.

The data in Table 14.5 are an attempt to interpret expression levels for the particular registered transformation events from various data sources, reducing tables of data to rows of numbers of 1–2 significant digits. Despite an EPA Scientific Advisory Board request to have data as protein per unit dry weight (the current standard) (US Environmental Protection Agency 2000), the fresh weight basis is generally

Table 14.5 Insecticidal protein expression levels in transgenic Bt plants

Event	Pollen	Leaf	Whole plant	Seed/kernel	Root	References
Event 176 Cry1Ab maize	1–3[a]	2[a]	< 1[a]	<< 1[a]	<< 1[a]	CERA (2010a); US EPA (2011c)
MON531 Cry1Ac cotton	< 1[a]	2[a]	< 1[a]	3[a]	–	CERA (2010b); Risk Assessment and Risk, Management Plan (2002)
MON810 Cry1Ab maize	<< 1	50–120	20–120	< 1	20–40	US Environmental Protection Agency (2010a)
Bt11 Cry1Ab maize	< 1	25–100	20	5	5	US Environmental Protection Agency (2010a)
TC1507 Cry1F maize	20	20	5	2	6	US Environmental Protection Agency (2010a)
MON 15985 Cry2Ab cotton	< 1[a]	5–25[a]	10–20[a]	50[a]	–	CERA (2010b); Risk Assessment and Risk, Management Plan (2002)
MON88017 Cry3Bb maize	15	400	–	9	200	Nguyen and Jehle (2009); US Environmental Protection Agency (2010b)
281/3006 Cry1Fa cotton	< 1[a]	8[a]	–	4[a]	< 1[a]	US Environmental Protection Agency (2005)
281/3006 Cry1Ac cotton	1[a]	1.5[a]	–	< 1[a]	< 1[a]	US Environmental Protection Agency (2005)
59122-7 Cry34Ab1 maize	70	50–200	30–80	50	45	US Environmental Protection Agency (2010c)
59122-7 Cry35Ab1 maize	<< 1	40–90	5–15	1	5–10	US Environmental Protection Agency (2010c)
MIR604 mCry3A maize	–	20	7–25	1	5–30	US Environmental Protection Agency (2007)
MON 89034 cry1A.105 maize	10	70–500	100–400	6	10–80	US Environmental Protection Agency (2010d)
MON 89034 CTP-Cry2Ab2 maize	< 1	160	40–130	1	20–60	US Environmental Protection Agency (2010d)
COT102 Vip3Aa19 cotton	4	40–300	25	7	15	US Environmental Protection Agency (2008)
COT67B Cry1Ab cotton	10	60–160	25	30	15	US Environmental Protection Agency (2008)
MIR162 Vip3Aa20 maize	100	180	75	100	40	US Environmental Protection Agency (2009)
MON 87701 CTP-Cry1Ac soybean	2	200–350	–	5	<< 1	US Environmental Protection Agency (2010e)

[a] µg/g tissue or fresh weight

what is available for earlier transgenic plants and especially cotton. The original data cover a more than 1000-fold range of values with insecticidal relevance usually starting at about 1 μg/g dry weight. Assume that numbers have a standard error of a good 50%, and ranges of values for leaf, root, and plant developmental stages are only shown if they are twofold or more. Where the same or similar data are available on a fresh weight and dry weight basis, the dry weight data tend to be roughly 6–10 times higher than the fresh weight data for leaves, and lower or nonexistent in less succulent tissue (not shown).

There are several points to make about the expression data in Table 14.5. First, Event 176 appears to have been a low-expressing event, 8–20 times lower than MON810 on a fresh weight basis (not shown) and clearly outside the fresh-to-dry weight conversion mentioned above. Additionally, Event 176 had very low expression in the pith of the stalk, leading to susceptibility to stalk tunneling of second generation European corn borer (Walker et al. 2000), which in turn led to resistance management concerns and the withdrawal of its registration. Second, expression of Cry1Fa in TC1507 appears somewhat lower than Cry1Ac in MON810 and Bt11, which in turn are roughly at the low end of CTP-Cry2Ab and generally lower than Cry1A.105 in the stacked event MON 89034. Continuing in corn, expression of all rootworm control proteins except MIR604 mCry3A (i.e. Cry3Bb, Cry34, Cry35) appears higher in leaves than in roots, with leaf expression of Cry3Bb1 in the range of lepidoptran-active proteins. Expression of the more recent vip3Aa20 event is as high as or higher than the other Lepidoptera control proteins in corn, except for the upper range of Cry1A.105, and is the highest of those events in the kernel, where insect control is becoming more of a concern.

In cotton, leaf expression of the Cry1Ac proteins in Widestrike® (281/3006) and Bollgard® I & II (MON531) appear roughly similar, and the respective Cry1Fa and CTP-Cry2Ab proteins in the stacked events are expressed at up to roughly an order of magnitude higher. Interestingly, expression of Cry1Ab in COT67B cotton overlaps the range of maize CTP-Cry2Ab expression. This would seem higher than the expected expression per dry weight of Cry1Ac in the cotton events above, probably higher than Cry1Fa, and possibly in the same range as CTP-Cry2Ab. Expression of Vip3Aa19 is over a similar and possibly higher range to that of COT67B Cry1Ab, and a similar, but broader range to Vip3Aa20 in maize. CTP-Cry1Ac expression in soybean leaf tissue appears consistently higher than that of the other Cry1Ac-expresssing events.

It was natural to combine, or stack, traits, including insect control proteins, in a single plant. An important aspect of getting regulatory approval for such stacked traits is to show that expression of each of the traits in the stack is at the same level as the trait expressed by itself. For insect control, which requires the highest practical dose for insect resistance management, a fall off in expression would clearly present a problem. The gene combinations can be with other types of traits, like herbicide resistance, or with other insect control traits. So far, the expression information comparing stacked traits vs. single trait near isogenic corn and cotton has shown a good correspondence (available at (US Environmental Protection Agency 2011b)). Where there have been doubled traits, such as the same selectable marker with two different stacked insect control genes, the level of marker protein has roughly doubled. While the stacked Bt cotton, and apparently several of the stacked corn products have been successful, it may be too soon to comment on the newer, more complex stacks, since their introgression may be behind the most advanced seed production lines. The well-established modeling of insect resistance management of stacked traits was recently reviewed (Tabashnik et al. 2009).

14.6 Plant Expression Summary

Overall, the expression of Bt insecticidal proteins seems to have been increasing with more recent constructs. Given the overall complexity of the expression process, it is difficult to ascribe improvements to anything in particular, except when there is some simple change like adding a transit peptide and seeing a statistical improvement over a number of transformation events when such a change works. Starting with the synthetic Bt gene, there has clearly been an increase in the understanding of RNA processing in various plants that may have influenced the design process. Some of the newer promoter, leader and intron sequences may have played a role as well, such as the *Arabidopsis* actin elements. Shifting from biolistic to *Agrobacterium* transformation in corn could also have played a significant role by providing more expressible integration sites. There seems to be an observational basis for an upper level of expression of some genes without phenotypic effects in certain plants, as noted above. The ability to stack traits in an additive manner, even with genes like *cry1Fa* and *cry1A.105* in corn, or *cry1Fa/cry1Ab*-protoxin and *cry1Ac* in cotton, that are of the same class, suggests that if some process is limiting productive expression, it is quite specific to individual genes and not to the class as a whole. Another issue with the improved levels of expression and stacking is the total amount of recombinant protein produced. Nguyen and Jehle (2009) estimated that MON 88017 produced 905 g/ha Cry3Bb1 with lower productivity numbers than Monsanto's registration document. If one was to add Cry2Ab, Cry1A.105, Cry1Fa, Cry34Ab, Cry35Ab, and a glyphosphate tolerance protein, all found in a recent stacked corn line, the total is obviously more than the few micrograms per gram plant tissue found in the initial corn and cotton events. While this level of overall expression is impressive, as new traits are added, the regulatory equivalence between genetically modified and non-modified plants may become an issue.

14.7 Conclusion

It has been 31 years since the biotechnology of Bt was a spot on an X-ray film. In 2010 there were 59 million hectares planted in Bt crops world-wide (James 2010). In the mean time, thousands of Bt strains have been isolated and screened in many ways, hundreds of Bt insecticidal genes have been cloned and sequenced, and at least tens of genes have been expressed in plants. Going back to the production of Bt proteins, *E. coli* is still the workhorse organism for cloning, characterization, many mode of action studies and material for regulatory studies. The use of *E. coli* to characterize individual human gene products is driving expression technology in ways that should continue to facilitate its use with Bt proteins. Using Bt as a recombinant expression host has some advantages for expressing native cloned genes and for avoiding certain types of contaminants for mode of action studies. It remains to be seen whether more recombinant Bt agricultural pesticides will be produced in light of an expanding stable of Bt plants; improved recombinant Bt *isrealensis* for public health use may be feasible, but would undoubtedly require government or non-profit support for registration. For transgenic plants, integration site complexity has been vastly improved, marker free events will aid registration and stacking issues, and the number of useful expression elements is expanding. Strong expression elements that discriminate better between above ground and below ground expression would be very useful. The upper end of current expression levels may be adequate for reasonably active insecticidal proteins. Having a system to prototype, analyze and improve expression of proteins problematic for plants seems like the holy grail of expression analysis in light of the pseudo-random nature of T-DNA integration, position effects on expression, and the expense of regenerating plants for evaluation. Barring that, the current evolution of Bt protein expression improvement, which has already done pretty well, will continue.

References

Adams LF, Visick JE, Whiteley HR (1989) A 20-kilodalton protein is required for efficient production of the *Bacillus thuringiensis* subsp. *israelensis* 27-kilodalton crystal protein in *Escherichia coli*. J Bacteriol 171:521–530

Agaisse H, Lereclus D (1995) How does *Bacillus thuringiensis* produce so much insecticidal crystal protein? J Bacteriol 177:6027–6032

Akiba T, Higuchi K, Mizuki E, Ekino K, Shin T, Ohba M, Kanai R, Harata K (2006) Nontoxic crystal protein from *Bacillus thuringiensis* demonstrates a remarkable structural similarity to beta-pore-forming toxins. Proteins 63:243–248

Akiba T, Abe Y, Kitada S, Kusaka Y, Ito A, Ichimatsu T, Katayama H, Akao T, Higuchi K, Mizuki E, Ohba M, Kanai R, Harata K (2009) Crystal structure of the parasporin-2 *Bacillus thuringiensis* toxin that recognizes cancer cells. J Mol Biol 386:121–133

Anderson HM, Allen JR, Groat JR, Johnson SC, Kelly RA, Korte J, Rice JF (2008) Corn plant and seed corresponding to transgenic event MON89034 and methods for detection and use thereof. US Patent Application 20080260932 A1

Arantes O, Lereclus D (1991) Construction of cloning vectors for *Bacillus thuringiensis*. Gene 108:115–119

Aronson AI (1994) Flexibility in the protoxin composition of *Bacillus thuringiensis*. FEMS Microbiol Lett 117:21–27

Barbour E, Bing JW, Cardineau GA, Cressman RF Jr, Gupta M, Locke MEH, Hondred D, Keaschall JW, Koziel MG, Meyer TE, Moellenbeck D, Narva KE, Nirunsuksiri W, Ritchie SW, Rudert ML, Sanders CD, Shao A, Stelman SJ, Stucker DS, Tagliani LA, Van Zante WM (2007) Corn event TC1507 and methods for detection thereof. US Patent 7,288,643

Baum JA, Malvar T (1995) Regulation of insecticidal crystal protein production in *Bacillus thuringiensis*. Mol Microbiol 18:1–12

Baum JA, Coyle DM, Gilbert MP, Jany CS, Gawron-Burke C (1990) Novel cloning vectors for *Bacillus thuringiensis*. Appl Environ Microbiol 56:3420–3428

Baum JA, Kakefuda M, Gawron-Burke C (1996) Engineering *Bacillus thuringiensis* bioinsecticides with an indigenous site-specific recombination system. Appl Environ Microbiol 62:4367–4373

Baum JA, Johnson TB, Carlton BC (1999) *Bacillus thuringiensis* natural and recombinant bioinsecticide products. In: Hall FR, Menn JJ (eds) Methods in biotechnology, vol 5: biopesticides: use and delivery. Humana Press, Totowa, pp 189–210

Beazley KA, Coombe TR, Groth ME, Hinchey TB, Pershing JC, Vaughn TT, Zhang B (2008) Corn plant Mon88017 and compositions and methods for detection thereof. US Patent Application 20080028482 A1

Beegle CC, Yamamoto T (1992) History of *Bacillus thuringiensis* Berliner research and development. Can Entomol 124:587–616

Bing JW, Cressman RF Jr, Gupta M, Hakimi SM, Hondred D, Krone TL, Hartnett Locke ME, Luckring AK, Meyer SE, Moellenbeck D, Narva KE, Olson PD, Sanders CD, Wang J, Zhang J, Zhong G-Y (2008) Corn event DAS-59122-7 and methods for detection thereof. US Patent 7,323,556

Bogdanova NN, Corbin DR, Malvar TM, Perlak FJ, Roberts JK, Romano CP (2009) Nucleotide sequences encoding insecticidal proteins. US Patent Application 20090238798 A1

Boonserm P, Pornwiroon W, Katzenmeier G, Panyim S, Angsuthanasombat C (2004) Optimised expression in Escherichia coli and purification of the functional form of the *Bacillus thuringiensis* Cry4Aa delta-endotoxin. Protein Expr Purif 35:397–403

Boonserm P, Davis P, Ellar DJ, Li J (2005) Crystal structure of the mosquito-larvicidal toxin Cry-4Ba and its biological implications. J Mol Biol 348:363–382

Boonserm P, Mo M, Angsuthanasombat C, Lescar J (2006) Structure of the functional form of the mosquito larvicidal Cry4Aa toxin from *Bacillus thuringiensis* at a 2.8-angstrom resolution. J Bacteriol 188:3391–4401

Bravo A, Likitvivatanavong S, Gill SS, Soberón M (2011) *Bacillus thuringiensis*: a story of a successful. Insect Biochem Mol Biol 41:423–431

Brizzard BL, Schnepf HE, Kronstad JW (1991) Expression of the *cryIB* crystal protein gene of *Bacillus thuringiensis*. Mol Gen Genet 231:59–64

CERA (2010a) GM Crop Database. Center for Environmental Risk Assessment (CERA), ILSI Research Foundation, Washington DC. SYN-EV176-9 (176). http://cera-gmc.org/index.php?evidcode%5B%5D=176&auDate1=&auDate2=&action=gm_crop_database&mode=Submit

CERA (2010b) GM Crop Database. Center for Environmental Risk Assessment (CERA), ILSI Research Foundation, Washington D.C. MON-ØØ531-6, MON-ØØ757-7. http://cera-gmc.org/index.php?evidcode%5B%5D=MON531%2F757%2F1076&auDate1=&auDate2=&action=gm_crop_database&mode=Submit

Chak KF, Tseng MY, Yamamoto T (1994) Expression of the crystal protein gene under the control of the alpha-amylase promoter in *Bacillus thuringiensis* strains. Appl Environ Microbiol 60:2304–2310

Chaoyin Y, Wei S, Sun M, Lin L, Faju C, Zhengquan H, Ziniu Y (2007) Comparative study on effect of different promoters on expression of cry1Ac in *Bacillus thuringiensis* chromosome. J Appl Microbiol 103:454–461

Cohen S, Dym O, Albeck S, Ben-Dov E, Cahan R, Firer M, Zaritsky A (2008) High-resolution crystal structure of activated Cyt2Ba monomer from *Bacillus thuringiensis* subsp. *israelensis*. J Mol Biol 380:820–827

Corbin DR, Romano CP (2006) Plants transformed to express Cry2A delta-endotoxins. US Patent 7,064,249

Crickmore N, Nicholls C, Earp DJ, Hodgman TC, Ellar DJ (1990) The construction of *Bacillus thuringiensis* strains expressing novel entomocidal delta-endotoxin combinations. Biochem J 270:133–136

Crickmore N, Zeigler DR, Feitelson J, Schnepf E, Van Rie J, Lereclus D, Baum J, Dean DH (1998) Revision of the nomenclature for the *Bacillus thuringiensis* pesticidal crystal proteins. Microbiol Mol Biol Rev 62:807–813

de Maagd RA, Bravo A, Berry C, Crickmore N, Schnepf HE (2003) Structure, diversity, and evolution of protein toxins from spore-forming entomopathogenic bacteria. Annu Rev Genet 37:409–433

De Rocher EJ, Vargo-Gogola TC, Diehn SH, Green PJ (1998) Direct evidence for rapid degradation of *Bacillus thuringiensis* toxin mRNA as a cause of poor expression in plants. Plant Physiol 117:1445–1461

Dervyn E, Poncet S, Klier A, Rapoport G (1995) Transcriptional regulation of the cryIVD gene operon from *Bacillus thuringiensis* subsp. *israelensis*. J Bacteriol 177:2283–2291

Diehn SH, Chiu WL, De Rocher EJ, Green PJ (1998) Premature polyadenylation at multiple sites within a *Bacillus thuringiensis* toxin gene-coding region. Plant Physiol 117:1433–1443

Donovan WP, Donovan JC, Engleman JT (2001) Gene knockout demonstrates that vip3A contributes to the pathogenesis of *Bacillus thuringiensis* toward Agrotis ipsilon and Spodoptera exigua. J Invertebr Pathol 78:45–51

Ellis DM, Negrotto DV, Shi L, Shotkoski FA, Thomas CR (2006) Cot102 insecticidal cotton. US Patent Application 20060130175 A1

Evans SL (2004) Producing proteins derived from genetically modified organisms for toxicology and environmental fate assessment of biopesticides. In: Parekh SR (ed) The GMO handbook: genetically modified animals, microbes and plants in biotechnology. Humana Press, Totowa, pp 53–83

Galitsky N, Cody V, Wojtczak A, Ghosh D, Luft JR, Pangborn W, English L (2001) Structure of the insecticidal bacterial delta-endotoxin Cry3Bb1 of *Bacillus thuringiensis*. Acta Crystallogr D Biol Crystallogr 57:1101–1109

Gao A-G, Kolacz KH, Macrae TC, Miklos JA, Paradise MS, Perlak FJ, Dressel Toedebusch AS (2009) Soybean plant and seed corresponding to transgenic event MON87701 and methods for detection thereof. US Patent Application 20090130071 A1

Gawron-Burke C, Baum JA (1991) Genetic manipulation of *Bacillus thuringiensis* insecticidal crystal protein genes in bacteria. Genet Eng (N Y) 13:237–263

Ge AZ, Pfister RM, Dean DH (1990) Hyperexpression of a *Bacillus thuringiensis* delta-endotoxin-encoding gene in *Escherichia coli:* properties of the product. Gene 93:49–54

Gilmer AJ, Baum JA (1999) Chimeric lepidopteran-toxic crystal proteins. US Patent 5,965,428

Grochulski P, Masson L, Borisova S, Pusztai-Carey M, Schwartz JL, Brousseau R, Cygler M (1995) *Bacillus thuringiensis* CryIA(a) insecticidal toxin: crystal structure and channel formation. J Mol Biol 254:447–464

Guo S, Ye S, Liu Y, Wei L, Xue J, Wu H, Song F, Zhang J, Wu X, Huang D, Rao Z (2009) Crystal structure of *Bacillus thuringiensis* Cry8Ea1: an insecticidal toxin toxic to underground pests, the larvae of *Holotrichia parallela*. J Struct Biol 168:259–266

Gustafson ME, Clayton RA, Lavrik PB, Johnson GV, Leimgruber RM, Sims SR, Bartnicki DE (1997) Large-scale production and characterization of *Bacillus thuringiensis* subsp. *tenebrionis* insecticidal protein from *Escherichia coli*. Appl Microbiol Biotechnol 47:255–261

Hayakawa T, Howlader MT, Yamagiwa M, Sakai H (2008) Design and construction of a synthetic *Bacillus thuringiensis* Cry4Aa gene: hyperexpression in *Escherichia coli*. Appl Microbiol Biotechnol 80:1033–1037

Hire RS, Makde RD, Dongre TK, D'souza SF (2008) Characterization of the cry1Ac17 gene from an indigenous strain of *Bacillus thuringiensis* subsp. *kenyae*. Curr Microbiol 57:570–557

Huang KX, Badger M, Haney K, Evans SL (2007) Large scale production of *Bacillus thuringiensis* PS149B1 insecticidal proteins Cry34Ab1 and Cry35Ab1 from *Pseudomonas fluorescens*. Protein Expr Purif 53:325–330

Huber SA, Roberts JK, Shappley ZW, Doherty S (2007) Cotton event MON15985 and compositions and methods for detection thereof. US Patent 7,223,907

James C (2010) Global Status of Commercialized Biotech/GM Crops: 2010. ISAAA Brief No. 42. ISAAA, Ithaca, NY

Kalman S, Kiehne KL, Cooper N, Reynoso MS, Yamamoto T (1995) Enhanced production of insecticidal proteins in *Bacillus thuringiensis* strains carrying an additional crystal protein gene in their chromosomes. Appl Environ Microbiol 61:3063–3068

Kegley SE, Hill BR, Orme S, Choi AH (2011) PAN Pesticide Database, Pesticide Action Network, North America (San Francisco, CA). http://www.pesticideinfo.org

Kevany BM, Rasko DA, Thomas MG (2009) Characterization of the complete zwittermicin A biosynthesis gene cluster from *Bacillus cereus*. Appl Environ Microbiol 75:1144–1155

Klimowicz AK, Benson TA, Handelsman J (2010) A quadruple-enterotoxin-deficient mutant of *Bacillus thuringiensis* remains insecticidal. Microbiology 156:3575–3583

Koziel MG, Beland GL, Bowman C, Carozzi NB, Crenshaw R, Crossland L, Dawson J, Desai N, Hill M, Kadwell S, Launis K, Lewis K, Maddox D, McPherson K, Meghji MR, Merlin E, Rgodes R, Warren GW, Wright M, Evola SV (1993) Field performance of elite transgenic maize plants expressing an insecticidal protein derived from *Bacillus thuringiensis*. Biotechnology 11:194–200

Lecadet MM, Chaufaux J, Ribier J, Lereclus D (1992) Construction of novel *Bacillus thuringiensis* strains with different insecticidal activities by transduction and transformation. Appl Environ Microbiol 58:840–849

Li HM, Chiu CC (2010) Protein transport into chloroplasts. Annu Rev Plant Biol 61:157–180

Li JD, Carroll J, Ellar DJ (1991) Crystal structure of insecticidal delta-endotoxin from *Bacillus thuringiensis* at 2.5 A resolution. Nature 353:815–821

Li J, Koni PA, Ellar DJ (1996) Structure of the mosquitocidal delta-endotoxin CytB from *Bacillus thuringiensis* sp. *kyushuensis* and implications for membrane pore formation. J Mol Biol 257:129–152

Lilley M, Ruffell RN, Somerville HJ (1980) Purification of the insecticidal toxin in crystals of *Bacillus thuringiensis*. J Gen Microbiol 118:1–11

Liu D (2009) Design of gene constructs for transgenic maize. In: Scott MP (ed) Methods in molecular biology: transgenic maize, vol 526. Humana Press, a part of Springer Science + Business Media, USA

Liu X, Peng D, Luo Y, Ruan L, Yu Z, Sun M (2009) Construction of an *Escherichia coli* to *Bacillus thuringiensis* shuttle vector for large DNA fragments. Appl Microbiol Biotechnol 82:765–772

Long N, Bottoms J, Meghji M, Hart H, Que Q, Pulliam D (2009) Corn Event MIR162. US Patent Application 20090300784 A1

Maertens B, Spriestersbach A, von Groll U, Roth U, Kubicek J, Gerrits M, Graf M, Liss M, Daubert D, Wagner R, Schäfer F (2010) Gene optimization mechanisms: a multi-gene study reveals a high success rate of full-length human proteins expressed in *Escherichia coli*. Protein Sci 19:1312–1326

Makino T, Skretas G, Georgiou G (2011) Strain engineering for improved expression of recombinant proteins in bacteria. Microb Cell Factor 10:32

Manasherob R, Zaritsky A, Ben-Dov E, Saxena D, Barak Z, Einav M (2001) Effect of accessory proteins P19 and P20 on cytolytic activity of Cyt1Aa from *Bacillus thuringiensis* subsp. *israelensis* in *Escherichia coli*. Curr Microbiol 43:355–364

Mettus AM, Macaluso A (1990) Expression of *Bacillus thuringiensis* delta-endotoxin genes during vegetative growth. Appl Environ Microbiol 56:1128–1134

Monsanto Company Press Release (2008) Monsanto confirms safety of research cotton in Texas. http://monsanto.mediaroom.com/index.php?s=43&item=666

Morimoto T, Kadoya AR, Endo K, Tohat M, Sawda K, Liu S, Ozawa T, Kodama T, Kakeshita H, Kageyama Y, Manabe K, Kanaya S, Ara K, Ozaki K, Ogasawara N (2008) Enhanced recombinant protein productivity by genome reduction in *Bacillus subtilis*. DNA Res 15:73–81

Morse RJ, Yamamoto T, Stroud RM (2001) Structure of Cry2Aa suggests an unexpected receptor binding epitope. Structure 9:409–417

Nguyen HT, Jehle JA (2009) Expression of Cry3Bb1 in transgenic corn MON88017. J Agric Food Chem 57:9990–9996

Panetta JD (1993) Engineered microbes: the CellCap® system. In: Kim L (ed) Advanced engineered pesticides. Marcel Dekker, New York, pp 379–392

Parra G, Bradnam K, Rose AB, Korf I (2011) Comparative and functional analysis of intron-mediated enhancement signals reveals conserved features among plants. Nucleic Acids Res 39:5328–5337

Perlak FJ, Fuchs RL, Dean DA, McPherson SL, Fischhoff DA (1991) Modification of the coding sequence enhances plant expression of insect control protein genes. Proc Natl Acad Sci USA 88:3324–3328

Risk Assessment and Risk, Management Plan (2002) Application for licence for dealings involving an intentional release into the environment, DIR 012/2002. Commercial release of Bollgard II® cotton. Monsanto Australia Ltd. http://cera-gmc.org/docs/decdocs/06-300-001.pdf. Accessed Sept 2002

Rosati A, Bogani P, Santarlasci A, Buiatti M (2008) Characterisation of 3¢ transgene insertion site and derived mRNAs in MON810 YieldGard maize. Plant Mol Biol 67:271–281

Sanchis V, Agaisse H, Chaufaux J, Lereclus D (1997) A recombinase-mediated system for elimination of antibiotic resistance gene markers from genetically engineered *Bacillus thuringiensis* strains. Appl Environ Microbiol 63:779–784

Sanchis V, Gohar M, Chaufaux J, Arantes O, Meier A, Agaisse H, Cayley J, Lereclus D (1999) Development and field performance of a broad-spectrum nonviable asporogenic recombinant strain of *Bacillus thuringiensis* with greater potency and UV resistance. Appl Environ Microbiol 65:4032–4039

Sazhenskiy V, Zaritsky A, Itsko M (2010) Expression in Escherichia coli of the native *cyt1Aa* from *Bacillus thuringiensis* subsp. *israelensis*. Appl Environ Microbiol 76:3409–3411

Schnepf HE, Wong HC, Whiteley HR (1987) Expression of a cloned *Bacillus thuringiensis* crystal protein gene in *Escherichia coli*. J Bacteriol 169:4110–4118

Schnepf E, Crickmore N, Van Rie J, Lereclus D, Baum J, Feitelson J, Zeigler DR, Dean DH (1998) *Bacillus thuringiensis* and its pesticidal crystal proteins. Microbiol Mol Biol Rev 62:775–806

Shelton AM, Zhao JZ, Roush RT (2002) Economic, ecological, food safety, and social consequences of the deployment of bt transgenic plants. Annu Rev Entomol 47:845–881

Song R, Peng D, Yu Z, Sun M (2008) Carboxy-terminal half of Cry1C can help vegetative insecticidal protein to form inclusion bodies in the mother cell of *Bacillus thuringiensis*. Appl Microbiol Biotechnol 80:647–654

Song P, Tagliani LA, Pellow JW (2011) Cry1F and Cry1Ac transgenic cotton lines and event-specific identification thereof. US Patent 7,883,850

Sørensen HP, Mortensen KK (2005) Advanced genetic strategies for recombinant protein expression in *Escherichia coli*. J Biotechnol 115:113–128

Squires CH, Retallack DM, Chew LC, Ramseier TM, Schneider JC, Talbot HW (2004) Heterologous protein production in *P. fluorescens*. Bioprocess Int 2:54–59

Streatfield SJ (2007) Heterologous protein production in plants. Plant Biotechnol J 5:2–15

Structural Genomics Consortium (2008) Protein production and purification. Nat Methods 5:135–146

Tabashnik BE, Van Rensburg JB, Carrière Y (2009) Field-evolved insect resistance to Bt crops: definition, theory, and data. J Econ Entomol 102:2011–2025

Terpe K (2006) Overview of bacterial expression systems for heterologous protein production: from molecular and biochemical fundamentals to commercial systems. Appl Microbiol Biotechnol 72:211–222

Thamthiankul Chankhamhaengdecha S, Tantichodok A, Panbangred W (2008) Spore stage expression of a vegetative insecticidal gene increase toxicity of *Bacillus thuringiensis* subsp. *aizawai* SP41 against *Spodoptera exigua*. J Biotechnol 136:122–128

Thompson M, Schwab GE (1996) Delta-endotoxin expression in *pseudomonas fluorescens*. US Patent 5,527,883

US Environmental Protection Agency (2000) SAP Report No. 99-06A, 4 Feb 2000. Characterization and non-target organism data requirements for protein plant-pesticides. http://www.epa.gov/scipoly/sap/meetings/1999/december/report.pdf

US Environmental Protection Agency (2001) Registration action document—*Bacillus thuringiensis* plant-incorporated protectants. http://www.epa.gov/oppbppd1/biopesticides/pips/bt_brad.htm
US Environmental Protection Agency (2002) *Bacillus thuringiensis* Cry2Ab2 protein and its genetic material necessary for its production in cotton (Chemical PC Code: 006487) AMENDED. http://www.epa.gov/oppbppd1/biopesticides/ingredients///tech_docs/brad_006487.pdf
US Environmental Protection Agency (2005) *Bacillus thuringiensis* Cry1F (synpro) and Cry1Ac (synpro)Construct 281/3006 Insecticidal Crystal Proteins as expressed in cotton (Chemical PC Codes: 006512 and 006513, respectively). http://www.epa.gov/oppbppd1/biopesticides/ingredients/tech_docs/brad_006512.pdf
US Environmental Protection Agency (2007) Modified Cry3A protein and the genetic material necessary for its production (via elements of pZM26) in event MIR604 corn SYN-IR604-8. http://www.epa.gov/oppbppd1/biopesticides/pips/mcry3a-brad.pdf
US Environmental Protection Agency (2008) *Bacillus thuringiensis* modified Cry1Ab (SYN-IR67B-1) and Vip3Aa19 (SYN-IR102-7) insecticidal proteins and the genetic material necessary for their production in COT102 X COT67B cotton. http://www.epa.gov/oppbppd1/biopesticides/ingredients/tech_docs/brad_006529.pdf
US Environmental Protection Agency (2009) *Bacillus thuringiensis* Vip3Aa20 insecticidal protein and the genetic material necessary for its production (via elements of vector pNOV1300) in event MIR162 maize (OECD Unique Identifier: SYN-IR162-4) (PC Code: 006599). http://www.epa.gov/oppbppd1/biopesticides/ingredients/tech_docs/brad_006599.pdf
US Environmental Protection Agency (2010a) Cry1Ab and Cry1F *Bacillus thuringiensis* (Bt) corn plant-incorporated protectants. http://www.epa.gov/oppbppd1/biopesticides/pips/cry1f-cry1ab-brad.pdf
US Environmental Protection Agency (2010b) *Bacillus thuringiensis* Cry3Bb1 protein and the genetic material necessary for its production (Vector PV-ZMIR13L) in MON 863 corn (OECD Unique Identifier: MON-ØØ863-5) (PC Code: 006484), *Bacillus thuringiensis* Cry3Bb1 protein and the genetic material necessary for its production (Vector PV-ZMIR39) in MON 88017 corn (OECD Unique Identifier: MON-88Ø17-3) (PC Code: 006498). http://www.epa.gov/oppbppd1/biopesticides/pips/cry3bb1-brad.pdf
US Environmental Protection Agency (2010c) *Bacillus thuringiensis* Cry34Ab1 and Cry35Ab1 proteins and the genetic material necessary for their production (PHP17662T-DNA) in event DAS-59122-7 corn (OECD Unique Identifier: DAS-59122-7) (PC Code: 006490). http://www.epa.gov/oppbppd1/biopesticides/pips/cry3435ab1-brad.pdf
US Environmental Protection Agency (2010d) *Bacillus thuringiensis* Cry1A.105 and Cry2Ab2 insecticidal proteins and the genetic material, necessary for their production in corn (PC Codes: 006515 (Cry2Ab2), 006514 (Cry1A.105)). http://www.epa.gov/oppbppd1/biopesticides/pips/mon-89034-brad.pdf
US Environmental Protection Agency (2010e) *Bacillus thuringiensis* Cry1Ac protein and the genetic material (Vector PV-GMIR9) necessary for its production in MON 87701 (OECD Unique Identifier: MON 877Ø1-2) soybean (PC Code: 006532). http://www.epa.gov/oppbppd1/biopesticides/pips/bt-cry1ac-protien.pdf
US Environmental Protection Agency (2011) Introduction to biotechnology regulation for pesticides. http://www.epa.gov/pesticides/biopesticides/regtools/biotech-reg-prod.htm#overview
US Environmental Protection Agency (2011) Current & previously registered section 3 PIP registrations. http://www.epa.gov/oppbppd1/biopesticides/pips/pip_list.htm. Accessed 15 Feb 2011
US EPA (2011c) Biopesticide fact sheet *Bacillus thuringiensis* cry1Ab delta-endotoxin and the genetic material necessary for its production (Plasmid Vector pCIB4431) in corn (Event 176) (006458). http://cera-gmc.org/docs/decdocs/01-290-041.pdf. Accessed 30 July 2011
van Aarssen R, Soetaert P, Stam M, Dockx J, Gosselé V, Seurinck J, Reynaerts A, Cornelissen M (1995) Cry IA(b) transcript formation in tobacco is inefficient. Plant Mol Biol 28:513–524
Visick JE, Whiteley HR (1991) Effect of a 20-kilodalton protein from *Bacillus thuringiensis* subsp. *israelensis* on production of the CytA protein by *Escherichia coli*. J Bacteriol 173:1748–1756

Walker KA, Hellmich RL, Lewis LC (2000) Late-instar European corn borer (Lepidoptera: Crambidae) tunneling and survival in transgenic corn hybrids. J Econ Entomol 93:1276–1285
Wei JZ, Hale K, Carta L, Platzer E, Wong C, Fang SC, Aroian RV (2003) *Bacillus thuringiensis* crystal proteins that target nematodes. Proc Natl Acad Sci USA 100:2760–2765
Whiteley HR, Schnepf HE, Tomczak K, Lara JC (1987) Structure and regulation of the crystal protein gene of *Bacillus thuringiensis*. In: Maramorosch K (ed) Biotechnology advances in invertebrate pathology and cell culture. Academic Press, San Diego, pp 13–27
Yamamoto T (2001) One hundred years of *Bacillus thuringiensis* research and development: discovery to transgenic crops. J Insect Biotechnol Sericol 70:1–23

Chapter 15
Bt Crops: Past and Future

Anais S. Castagnola and Juan Luis Jurat-Fuentes

Abstract The development and commercialization of transgenic plants expressing insecticidal toxin genes from the bacterium *Bacillus thuringiensis* (Bt) has revolutionized agriculture in the past two decades. Development of this revolutionary insect pest control technology was facilitated by the identification and characterization of insecticidal Bt proteins and advancements in plant transformation and genetic engineering. While commercialization of this technology is currently limited to a number of countries, these transgenic "Bt crops" are replacing in most cases conventional crop varieties due to their insect resistance, lower spraying requirements, and higher yields. However, concerns related to the increasing adoption of this technology include gene flow to wild relatives, evolution of resistance in target pests, and unintended effects on the environment. In this chapter, we discuss key events in the history of Bt crop development and summarize current regulations aimed at reducing the risks associated with increased adoption of this technology. By analyzing the history of Bt transgenic crops and the current marketplace trends and issues, we aim to examine the outlook of current and impending Bt crops as well as potential issues that may emerge during their future use.

Keywords Bt crops · Toxin genes · Risk of Bt crops · Effects of Bt crops · Resistance

15.1 Introduction

Insecticidal products based on the bacterium *Bacillus thuringiensis* (Bt) have been used for decades to control lepidopteran (caterpillars), dipteran (mosquitoes and black flies), and coleopteran (beetle larvae) pests (Sanchis 2010). It was estimated that Bt products represent about 80% of all biopesticides sold worldwide (Whalon and Wingerd 2003). A major driver in this adoption of Bt products has been the global proliferation of certified organic production, which relies heavily on Bt for insect control. The specificity and high relative toxicity of the insecticidal proteins

J. L. Jurat-Fuentes (✉) · A. S. Castagnola
Department of Entomology and Plant Pathology, University of Tennessee, 2431 Joe Johnson Drive, 205 Ellington Plant Sciences Building, Knoxville, TN 37996, USA
e-mail: jurat@utk.edu

E. Sansinenea (ed.), *Bacillus thuringiensis Biotechnology*,

DOI 10.1007/978-94-007-3021-2_15,

produced by Bt results in high efficacy and environmental safety when compared to available synthetic pesticides. However, several factors have limited higher adoption of Bt products, including short persistence and low residual activity due to environmental degradation, and poor control of tunneling or root-feeding pests (Sanahuja et al. 2011). These limitations directed the interest in developing alternative systems for more persistent and direct delivery of Bt toxins to control agricultural pests. For example, encapsulation of Bt toxins in non-pathogenic *Pseudomonas fluorescens* cells that were killed before release resulted in increased resistance to environmental degradation and toxicity (Gaertner et al. 1993). Undoubtedly, the most efficient delivery system developed to control lepidopteran and coleopteran pests is the transformation of plants with insecticidal Bt genes. These transgenic "Bt crops" are protected from insect attack by expression of the transformed Bt toxin genes and accumulation of Bt toxins in the plant tissues. Direct delivery to insects feeding on the plant minimizes exposure to non-target fauna and allows management of otherwise difficult to control tunneling and root-feeding pests.

15.2 Developments Conducive to Bt Crops

15.2.1 Research on Bt Toxin Genes

Early research on the identification of Bt proteins responsible for insecticidal activity and progress on methods to transform plants were vital to the development of Bt crops. Discovery, characterization, and classification of Bt isolates using flagellar H antigen typing of vegetative cells (de Barjac and Bonnefoi 1968) greatly facilitated the identification of strains producing toxins with potential to control specific pests. More recently, determination of insecticidal activity has been mostly focused on bioassays with purified Bt toxins (van Frankenhuyzen 2009), because strains within a single Bt serotype can produce multiple and diverse insecticidal components.

Among several virulence factors, Bt cells produce toxins that are responsible for insecticidal activity, and include vegetative insecticidal proteins (Vip), crystal (Cry), and cytolytic (Cyt) toxins. While Vip toxins are produced during the vegetative growth phase (Estruch et al. 1996), Cry and Cyt toxins are synthesized during sporulation (Hannay and Fitz-James 1955) and late exponential growth phase (Salamitou et al. 1996). Both Cry and Cyt toxins are stored as parasporal crystalline bodies (Bulla et al. 1977), which may be composed of single or multiple toxins contributing to insecticidal activity (Crickmore et al. 1995). Determination and compilation of the activity range for individual Bt toxins (van Frankenhuyzen 2009) allows for the identification of high potency toxins as optimal candidates for expression in transgenic crops to control key insect pests. Due to their earlier discovery and characterization, *cry* toxin genes have been predominantly used for plant transformation, although finding and characterization of *vip* genes has allowed their use to produce transgenic Bt crops (Table 15.1).

Table 15.1 Commercially available and projected Bt crops. (Center for Environmental Risk Assessment 2009)

Event(s)	Trade name(s)	Toxin(s)	M[a]	Targeted pest(s)[b]
Syngenta Maize				
Bt11	Agrisure CB/LL, Agrisure GT/CB/LL	Cry1Ab	D	ECB
MIR604	Agrisure RW, Agrisure RW/GT	Cry3Aa	A	WCR
Bt11, MIR604	Agrisure 3000GT, Agrisure CB/LL/RW	Cry1Ab, Cry3Aa	D, A	ECB, WCR
Bt 11, MIR162	Agrisure Viptera 3110	Cry1Ab, Vip3Aa20	D, A	ECB, FAW, CEW, BCW, WBC
Bt11, MIR604, MIR162	Agrisure Viptera 3111	Cry1Ab, Cry3Aa Vip3Aa20	D, A, A	ECB, WCR, FAW, CEW, BCW, WBC,
Syngenta Cotton				
COT102, COT67B	VipCot	Vip3Aa19, Cry1Ab	A	CBW, TBW, PBW, FAW, BAW, SBL, CL, CLP
Monsanto Maize				
MON810	YieldGard Corn Borer	Cry1Ab	B	ECB
MON863	YieldGard RW	Cry3Bb1	B	CRW
MON810, MON863	YieldGard VT Triple, Yield-Gard Plus	Cry1Ab, Cry3Bb1	B, B	CRW, ECB
MON89034	Genuity VT Double PRO	Cry1A.105/ Cry2Ab2	A	CEW, ECB, FAW
MON89034, MON88017	Genuity VT Triple PRO	Cry1A.105/ Cry2Ab2, Cry3Bb1	A, A	CEW, CRW, ECB, FAW
MON89034, TC1507, MON88017, DAS-59122-7	Genuity SmartStax	Cry1A.105/ Cry2Ab2, Cry1Fa2, Cry3Bb1, Cry34/35Ab1	A, B, B, A	BCW, CEW, CRW, ECB, FAW, WBC, SCB, SWCB, SCSB, CEW, SCB, WBC, WCR
Monsanto Cotton				
MON531	Genuity Bollgard, Ingard	Cry1Ac	A	PBW, TBW
MON15985	Genuity Boll-guard II	Cry1Ac/ Cry2Ab2	A/B	CBW, PBW, TBW
Pioneer (DuPont) and Dow Agrosciences Maize				
DAS-06275-8		Cry1F	A	BCW, ECB, FAW, WBC, SWCB, CEW

Table 15.1 (continued)

Event(s)	Trade name(s)	Toxin(s)	M[a]	Targeted pest(s)[b]
TC1507	Herculex I	Cry1F	B	BCW, ECB, FAW, WBC, SWCB, CEW
MON810,	Optimum Intrasect	Cry1Ab	B	ECB
TC1507, MON810		Cry1F, Cry1Ab	B, B	ECB, WBC, BCW, FAW
DAS-59122-7	Herculex RW, Optimum AcreMax RW	Cry34/35Ab1	A	WCR
DAS-59122-7, TC1507	Optimum AcreMax 1, Herculex Xtra	Cry34/35Ab1, Cry1F	A, B	WCR, BEC, ECB, FAW, WBC
Pioneer (DuPont) and Dow Agrosciences Cotton				
3006-210-24, 281-24-236	WideStrike	Cry1Ac, Cry1F	A, A	CBW, PBW, TBW, ECB, SBL, BAW, FAW

[a] Method of transgene insertion. Abbreviations are D- Direct DNA transfer, B- Biolistics, A- Agro-mediated transformation.

[b] Common insect name abbreviations are: ECB- European corn borer; WCR- Western corn rootworm; FAW- Fall armyworm; CEW- Corn earworm; BCW- Black cutworm; WBC- Western bean cutworm; BEC- Bean cutworm; CBW- Cotton bollworm; TBW- Tobacco budworm; PBW- Pink bollworm; BAW- Beet armyworm; SBL- Soybean looper; CL- Cabbage looper; Cotton leaf perforator; SCB- Sugar cane borer; SWCB- Southwestern corn borer; SCSB- Southern cornstalk borer; SPB- Spotted bollworm

The genes encoding Cry (Kronstad et al. 1983) and Vip (Wu et al. 2004; Franco-Rivera et al. 2004) toxins are located in plasmids or the bacterial chromosome. Cloning of the first Cry toxin gene and its expression as an active insecticidal toxin in *Escherichia coli* (Schnepf and Whiteley 1981) suggested the potential for transformation of Cry toxin genes in diverse organisms to enhance efficacy or increase activity range. Currently, more than 400 Bt toxin genes have been cloned and sequenced, including 218 Cry and 28 Vip toxin holotypes (Crickmore et al. 2011). Thorough characterization of the mechanisms directing Bt toxin gene expression (Agaisse and Lereclus 1995), and the small number of genetic loci implicated in controlling this process, greatly facilitated genetic manipulations for efficient expression in heterologous systems (Andrews et al. 1987). A diverse range of microorganisms were initially transformed with Bt toxin genes including alternative *Bacillus* spp. (Shivakumar et al. 1989), *P. fluorescens* (Gaertner et al. 1993; Obukowicz et al. 1986), *Rhyzobium* spp. (Skøt et al. 1990), and *Clavibacter xyli* (Lampel et al. 1994). Expression of Bt toxin genes in plants required the development of efficient plant transformation and selection methods and the identification of efficient promoter sequences to direct Bt toxin gene expression.

15.2.2 Plant Expression Systems

The identification of the tandem duplication 35S promoter from the Cauliflower Mosaic Virus (CaMV35S) (Kay et al. 1987) and the ubiquitin (ubi) maize (*Zea mays*) promoter (Christensen et al. 1992), were crucial to achieve enhanced expression of non-plant genes *in planta*, including Bt toxin genes. Although promoters driving expression systemically were initially used for Bt crops, there have also been later examples of the use of tissue-preferred promoters for targeted expression. This strategy is especially desirable when the target pest specializes on feeding on a particular plant tissue. For example, expression of the *cry34Ab1/cry35Ab1* toxin genes to control the root-feeding larvae of the Western corn rootworm (*Diabrotica virgifera*) has been recently achieved using the *Triticum aestivum* (wheat) peroxidase gene promoter driving root transgene expression (Gao et al. 2004). Expression of the transgene can also be targeted to a particular organelle by using a transit peptide gene, a proposed strategy to help prevent escape of the Bt transgene to wild relatives. For example, the chloroplast transit peptide gene (*cab22L*) from *Petunia hybrida* was used to drive expression of the *cry9c* gene to maize chloroplast in StarLink corn (Jansens et al. 1997). After this product was withdrawn from the market, alternative promoters such as the rice *rbcS* and its transit peptide sequence (*tp*) have been used to target and enhance expression of *cry* genes in chloroplasts (Kim et al. 2009). Alternatively, a chloroplast expression vector containing sites for homologous recombination and the chloroplast promoter Prrn was used for directing chroroplast expression of *cry1Ac* (McBride et al. 1995) or *cry2Aa* in tobacco (Kota et al. 1999), and *cry1Ab* in cabbage (Liu et al. 2008).

15.2.3 Bt Toxin Gene Transformation and Selection of Transformants

The discovery of effective promoters and the design of *in planta* expression cassettes also advanced the development of effective plant transformation methods. The development of various *Agrobacterium tumefaciens* protocols (Zambryski et al. 1983) allowed for stable transformation of Bt genes into dicotyledoneous plants, such as tomato or tobacco. In this system, the *Agrobacterium* Ti plasmid vector is modified by removing tumor-generating portions and inserting the Bt toxin gene of interest and selectable markers so that plant infection resulted in systemic gene integration and subsequent regeneration of transgenic plants. Alternative transformation methods such as electroporation and biolistic bombardment (Gordon-Kamm et al. 1990), which could also be used to transform monocotyledoneous crops (maize, rice, or wheat), were also used for development of Bt crops (Koziel et al. 1993). In these methods, the plasmid DNA containing a Bt toxin gene of interest and selectable markers are delivered into plant cells using pores induced by electric shock (electroporation) or by coating the plasmid with a heavy metal particle and then using it to bombard plant embryos with a gene gun.

After transformation, successful transformants are usually selected using diverse antibiotic resistance genes. The most commonly used selection gene for Bt crops has been the neomycin phosphotransferase II (NPTII) gene (*neo*) from *E. coli* (Betz et al. 2000), which inactivates antibiotics like kanamycin and neomycin. An alternative selection approach that has been used for Bt crops is the growth of transformants containing the mannose-6-phosphate isomerase (*Pmi*) gene from *E. coli* on media containing mannose as the only carbon source (Long et al. 2007). Other examples of markers used for selection of plants transformed with Bt genes include the aminoglycoside 39-adenyltransferase (*aad*A) gene that confers resistance to spectinomycin-streptomycin (Kota et al. 1999), and the hygromycin-B phosphotransferase (*aph4*) gene, which allows for transformed cell selection on culture medium containing hygromycin (Llewellyn et al. 2007).

15.2.4 Early Development and Challenges for Bt Crops

Once Bt toxin genes of interest were identified and transformation and selection methods were optimized public and private research groups quickly began to experiment with the goal to develop transgenic Bt plants. Initial attempts to transform full-length crystal toxin genes resulted in plant toxicity (Barton et al. 1987), shifting interest to the expression of truncated crystal toxin genes encoding the insecticidal N-terminal toxin half, which were found not to be toxic to the plant. Truncated *cry1A* toxin genes were used to develop transgenic tomato (Fischhoff et al. 1987), tobacco (Vaeck et al. 1987; Barton et al. 1987), and cotton (Perlak et al. 1990). Although some of these initial transgenic Bt plants presented resistance to feeding by selected lepidopteran larvae, it was recognized that the levels of toxin gene expression were still low and performance considered insufficient for commercialization. Low levels of toxin mRNA in the transformed plants suggested that the toxin transcripts were unstable, possibly due to inefficient posttranscriptional processing or rapid turnover (Barton et al. 1987; Murray et al. 1991). This phenomenon was originally attributed to the AT-rich nature of Bt toxin genes, which resulted in recognition by the plant cell regulatory mechanisms as foreign, triggering polyadenylation and mRNA instability. Translational efficiency was greatly improved by modification of the Bt toxin gene to match plant-preferred coding sequences, resulting in up to 100-fold higher levels of Cry1Ab toxin in transgenic tobacco and tomato (Perlak et al. 1991). Laboratory and field tests using transgenic tobacco lines containing this truncated *cry1Ab* toxin gene under the control of the CaMV 35S promoter (Carozzi et al. 1992) provided evidence of effective protection against tobacco hornworm (*Manduca sexta*) and tobacco budworm (*Heliothis virescens*) (Warren et al. 1992). Maize plants transformed with an optimized synthetic *cry1Ab* gene were reported to be resistant to European corn borer (*Ostrinia nubilalis*) under field conditions (Koziel et al. 1993). Further increase in the efficiency of Cry1Ab toxin production in tobacco was achieved using point mutations to avoid inefficient cryptic splice sites that inhibited nuclear transcript processing and transport to the cytoplasm (Aarssen et al. 1995). Even higher levels of Cry toxin accumulation were reported for expression of the *cry1Ac* toxin gene in tobacco chloroplasts, with accumulation of Cry1Ac protoxin to 5% of the total soluble protein in tobacco leaves (McBride et al. 1995).

Prior to the 1960s, intellectual property rights were not readily enforced and the value of scientifically formulated technologies was not realized. A group of biotech companies acknowledged the significance of the above biotechnological developments and shifted the majority of its funding initiatives from chemical manufacturing to basic biotechnology. Companies like Monsanto or Mycogen began experimenting with a variety of promoters, antibiotic resistance genes, transformation-tissue culture systems, and investigating novel Bt insecticidal proteins. These important investments in new technologies compelled the protection of intellectual property rights to secure economic returns. By implementing measures for agricultural companies to protect their investments, the commercial viability of insecticidal transgenic technology was further explored (Horsch 1993). The patentable portions of the transgenic Bt technologies included promoters, expression strategies, selection markers and techniques, as well as transgenic plants and traits expressed. The strengthening of intellectual property rights clearly indicated the commercial success and marketability of Bt transgenic crops as pest control products, as exemplified in a 14-fold increase in expenditure on research and development of plant breeding-related science and technology derived from private funding sources. Public expenditure patterns, which also contribute to investment trends, changed little during this time, further solidifying the causal agent of progress to intellectual property law enforcement (Fernandez-Cornejo 2004). The actions of early agriculture biotechnology companies permanently restructured their commercial identity and led the field of agricultural biotechnology from predominantly academic to the most competitive pest management marketplace in history.

Transgenic crops fell under the jurisdiction of the Federal Insecticide, Fungicide, and Rodenticide Act (FIFRA) and were therefore considered by regulatory agencies as pesticides for registration purposes (Earl 1983). Concerns related to changes in Bt toxin behavior after expression in a plant host prevented the use of established data advocating safety of Bt sprays to support safety of Bt crops. Thus, companies soliciting registration were requested to present data detailing toxicological tests with a wide range of organisms (including vertebrates, nematodes, and non-target insects) to support the safety of Bt crops at the toxin amounts produced by the plants. These tests had to be presented for each of the individual insecticidal transgenic traits contained in a plant, so that registration of Bt crops expressing multiple Bt toxin genes would require toxicological assays for each of the toxins produced by the plant. Concerns over the development of resistance to Bt crops, which would also affect alternative Bt-based products, resulted in an additional requisite during registration for a resistance management strategy. These resistance management programs included recommendations for growers, development of resistance monitoring protocols, and resistance control procedures (Matten et al. 1996).

15.3 Commercialization and Performance of "First Generation" Bt Crops

The first commercially available Bt crop was potato expressing the *cry3A* toxin gene (NewLeaf), a product manufactured by an affiliate of Monsanto (NatureMark) that was commercialized in 1995. The expressed *cry3A* gene derived from Bt subsp. *tenebrionis* was selected due to its high activity against larvae of the Colorado potato beetle (*Leptinotarsa decemlineata*), one of the most economically relevant pests of potato. The amount of Cry3Aa protein expressed within the foliage tissue of the NewLeaf Russet Burbank potatoes was 0.1–0.2% of total leaf protein, representing about a 100-fold higher concentration than the dose needed to kill 95% of neonate *L. decemlineata* larvae, thus fulfilling a high-dose requirement (Perlak et al. 1993). Unlike any pest management tool beforehand, NewLeaf potatoes were protected from *L. decemlineata* during all of its life stages and throughout the entire growing season, resulting in significant reductions in insecticide use. Sequent NewLeaf varieties commercialized in the late 1990s also provided resistance to aphid-transmitted potato viruses (Lawson et al. 2001). Despite remarkable product performance (Stark 1997), sales and marketing of NewLeaf potato were suspended in 2001 due to issues related to public concerns over transgenic potatoes being used for human consumption.

Alternative transgenic Bt crops were commercialized shortly after NewLeaf potatoes, including maize and cotton, which have since remained the most relevant markets for transgenic seed. Transgenic cotton varieties expressing the *cry1Ac* gene (Bollgard I in the U.S. and Ingard in Australia) were commercialized to control the cotton budworm (*H. virescens*) and the pink bollworm (*Pectinophora gosypiella*) in the U.S. and *Helicoverpa* spp. in Australia. In the first year after its introduction, Bollgard I technology was adopted by more than 5700 growers in the United States, and contributed to reduction of more than a quarter of a million gallons in chemical insecticide use (Fraley 1996). Similarly, the average number of sprays to control *Helicoverpa* spp. in Australia was reduced by 56% for Ingard cotton during the first 6 years after commercialization (Pitt 2003). Transgenic Bt cotton 'stacked' varieties containing genes for herbicide resistance were introduced a year later. This technology was rapidly accepted by growers due to its cost-effectiveness under pest pressure (Martin and Hyde 2001), with an estimated 37% adoption rate in the U.S. by 2001 (Fernandez-Cornejo and McBride 2002). Concerns with development of resistance led to capping of commercial deployment of Ingard cotton at 30% of the cotton area in Australia until 2004, when two-toxin gene varieties became available (Pitt 2003). The growing popularity of transgenic Bt cotton resulted in further reductions in the number of required insecticidal applications in the U.S. by 1999 (Carriere et al. 2001). Similarly, adoption of Bt cotton in China and India has been reported to result in increased production yields and reduced insecticidal applications (Wu et al. 2008; Pray et al. 2002; Qaim and Zilberman 2003). However, efficient control of targeted pests has resulted in increased populations of secondary pests not controlled by Bt toxins (Lu et al. 2010).

Another early success story of the commercialization of transgenic Bt crops was the effective control of the European corn borer (*O. nubilalis*) by Bt maize (Koziel et al. 1996). Prior to the advent of Bt maize, *O. nubilalis* pest populations were especially devastating due to the burrowing feeding behavior of the larvae, which feed on the whorl, leaf axils, and sheath before boring into the stalk and becoming protected to chemical or foliar insecticidal sprays. Larvae of *O. nubilalis* are highly susceptible to Cry1Ab, which was the toxin gene selected for production of the first round of registered Bt maize varieties, which were based on events 176 (KnockOut from Syngenta and NatureGard from Mycogen), Bt11 (Agrisure from Northup King), or MON810 (Yieldgard from Monsanto) (Sanahuja et al. 2011). Ensuing Bt maize products expressing the *cry1Fa* (event TC1507 in Herculex from Dow AgroSciences) or *cry9C* (event CBH-351 in StarLink from Aventis CropScience) toxin genes (Table 15.1) were commercialized to target *O. nubilalis* and additional selected species of armyworm (*Spodoptera* spp.). By the year 2000, transgenic Bt maize represented over 85% of all corn grown worldwide (Shelton et al. 2002), and it currently represents 65% of the corn grown in the U.S. (Economic Research Service-USDA 2011). These high levels of adoption have resulted in area-wide elimination of *O. nubilalis* along the U.S. corn belt, benefiting both farmers growing transgenic and non-Bt maize (Hutchison et al. 2010). The use of transgenic Bt maize has also been reported to significantly reduce accumulation of ear molds and associated mycotoxins (Hammond et al. 2004; Wu 2006; Dowd 2000), contributing to food safety (Kershen 2006). On the other hand, Cry9C from StarLink maize represented the first case of unintended entry of transgenic grain in the human food supply. This maize event was approved for domestic animal feed but not for human consumption, yet it was detected in taco shells. This detection resulted in Aventis requesting cancellation of the StarLink registration and the U.S. Food and Drug Administration recommending testing of more than 4 million bushels of corn for the presence of Cry9C until 2007, when monitoring efforts ceased due to lack of detection of significant levels of toxin residue. While there were no documented human allergy cases related to Cry9C, media exposure of the unintended introduction of transgenic Bt maize in the human food supply resulted in public objections to commercialization of transgenic crops.

15.4 Risks Associated with the Use of Bt Crops

Regulation of transgenic Bt crops in the U.S. is currently directed by the Coordinated Framework for the Regulation of Biotechnology, including branches of the Department of Agriculture Animal and Plant Health Inspection Service (USDA-APHIS), the Environmental Protection Agency (EPA) and the Food and Drug Administration (FDA). While USDA-APHIS regulates issuance of permits for field release, the EPA assesses the human, environmental and non-target safety of the transgene itself, while FDA assesses the food quality of transgenic crops. Although the commercialization of Bt crops is recognized as one of the most relevant events in the history of agricultural pest control, a number of potential risks associated to the environmental safety and future utility of this technology have been considered. Most relevant identified risks include potential toxicity to non-targets, escape of transgenes in the environment, and development of resistance in targeted insect populations. Although these issues are still a matter of extensive research, strategies to minimize these risks have been proposed and implemented in some cases.

15.4.1 Effects of Bt Crops on Non-target Organisms

The effect of Bt crops on non-target organisms is one of the mandatory components of product registration, although there are also available studies testing safety of Bt crops to non-targets under field conditions. Using NewLeaf potatoes as an initial case study, Stark (1997) reported no detrimental effects on non-target, beneficial and predatory insect populations, probably due to reductions in insecticidal applications. Specifically, lady beetles (Coccinellidae) and their important feeding behavior on aphids were unaffected by transgenic potatoes producing Cry3Aa toxin which are insecticidal to closely related *L. decemlineata* larvae (Dogan et al. 1996). Results from meta-analysis studies suggest that Bt crops in general support lower numbers of beneficial insects compared to conventional crops when no insecticides are used (Marvier et al. 2007), which may be due to prey number reduction in Bt plants. In contrast, these studies also found that when insecticidal sprays were used, Bt crops supported higher levels of beneficial insects compared to non-Bt crops, due to more intensive applications needed for non-Bt crops. A compilation of laboratory studies on the influence of Bt crops on 48 species of beneficial predatory insects supports a generally negative impact of Bt crops on predaceous and parasitoid insects (Lovei et al. 2009), but whether this is due to the transgene presence or reduction in insect prey populations is unclear. Green lacewings (*Chrysoperla carnea*) have been extensively used as a non-target model organism. Adult lacewings feeding on Bt maize pollen expressing Cry1Ab or Cry3Bb1 did not show significant impact on survival rate, pre-oviposition period, fecundity, fertility or dry weight (Li et al. 2008). Lacewing larvae are not directly affected by Bt toxins (Rodrigo-Simon et al. 2006), although detrimental effects resulting from low quality or reduced availability of prey in Bt crop fields have been reported (Romeis et al. 2004). Aphids, which are another non-target insect model, do not accumulate Bt toxin after feeding on transgenic Bt plants, which was shown to prevent unintended exposure of predators (Lawo et al. 2009). A meta-analysis of independent studies supported lack of negative effects for Bt crops on honey bees (*Apis mellifera*). Research on potential effects on non-target Lepidopteran populations by Bt crops has been mostly limited to the Monarch butterfly (*Danaus plexippus*) (Lang and Otto 2010). This focus is probably attributed to the wide attention (Shelton and Sears 2001) surrounding a single publication (Losey et al. 1999) presenting conclusions that were later demonstrated to be unfounded (Gatehouse et al. 2002; Sears et al. 2001).

The possible leaching of Bt toxins from Bt crop residues into nearby water bodies and the potential effect on aquatic fauna has also been met with controversy (Waltz 2009). While deposition of Bt plant tissue in the proximity of streams by wind and surface runoff can result in leaching of Bt toxins in the water (Tank et al. 2010; Viktorov 2011), their insecticidal properties on aquatic insects have not been demonstrated. The effect of transgenic Bt crop detritus on aquatic ecosystems should be comprehensively explored. In soil ecosystems, Bt crops have generally been found not to adversely affect symbiotic arbuscular mycorrhizal fungi (Liu 2010). While short-term activity shifts in bacterial communities were reported in the presence of residue from Bt corn expressing Cry1Ab toxin (Mulder et al. 2006), no changes in soil microbial population composition or activity were detected in a 2-year field study with the same Bt corn varieties (Oliveira et al. 2008). Overall, it is considered that the ecological benefits of reduced synthetic pesticide usage greatly outweigh the minimal consequences to beneficial insect populations (Gatehouse et al. 2011).

15.4.2 Bt Transgene Escape

The potential escape of Bt genes from transgenic crops into the environment has also been an issue of concern. Some reports suggested introgression into wild maize relatives (Quist and Chapela 2001), although this conclusion was quickly refuted with issues of contamination and methodology (Christou 2002). Two main categories of strategies, biological and non-biological, have been considered to prevent transgene escape from transgenic plants. Non-biological methods revolve around mechanical control where pollinating flowers are removed or transgenic crops are secluded from non-transgenic varieties (Rong et al. 2007; Kausch et al. 2010). Various molecular strategies for biotechnology-based gene containment have been proposed, including complete transgene excision from seed and pollen (Luo et al. 2007), and seed (Daniell 2002) and male sterility (He et al. 1996).

15.4.3 Insect Resistance to Bt Crops

Resistance to Bt crops has arguably been the main concern related to increased adoption of this technology, mostly because resistance to a specific Bt crop may result in cross-resistance to multiple Bt-based products, including microbial pesticides. Resistance to Bt toxins has been described in a number of laboratory-selected insect strains, evidencing the genetic potential for evolution of resistance to Bt toxins in the field. In the majority of laboratory cases, resistance is associated with alterations in toxin binding to midgut receptors, which is generally transmitted as a single autosomal recessive gene (Ferré and Van Rie 2002). In agreement with these laboratory reports, resistance to commercial Bt sprays was reported to result from reduced Bt toxin binding in strains of *Plutella xylostella* (Ferré et al. 1991) and *Plodia interpunctella* (Van Rie et al. 1990). Based on these data and results from predictive models, three main strategies to delay resistance to Bt crops were outlined (Gould 1988; Tabashnik 1989): use of refugia in conjunction with expression of high levels of the Bt toxin gene in the plant, rotations between plants expressing Bt toxins with diverse mode of action, and pyramiding of multiple toxin genes within a plant. The initial high dose/refugia strategy used to delay evolution of resistance to single-toxin Bt crops combined the use of mandated 20% refugia and the expression of high Bt toxin doses (25-fold the dose killing 99% of the pest population) in the plant. Exceptions and differences in this regulation occur when there are pests for which Bt crops do not fulfill the high dose requirement, in which case manufacturer recommended refuge amounts can be raised to as much as 50% of the crop. Refugia are typically areas contiguous to Bt crop fields, and this proximity greatly increases the probability of mating between resistant insects emerging from Bt crops and non-selected insects growing in the refuge. Mating events generate heterozygotes, which combined with a lethal toxin dose decreases the frequency of resistant individuals, assuming that resistance is transmitted as a recessive trait (Gould 1998). The success of this high dose/refuge strategy is greatly dependent on constant expression of Bt toxin in the plant throughout the season and grower compliance with refuge planting regulations. Both tenets are difficult to achieve, as expression of Bt toxin genes varies depending on diverse factors (Adamczyk et al. 2001, 2009; Adamczyk and Sumerford 2001), and refuge compliance is known to vary among growers and years (Gray 2010; Bourguet et al. 2005). In addition, a crucial condition for the effectiveness of refugia is the pest movement throughout a particular crop. In cases of pests with limited movement, such as Western corn rootworm larvae, modeling studies suggest that as grower compliance to refugia decreases resistance emergence sharply increases (Pan et al. 2011). In Bt cotton, resistance models predict faster evolution of resistance populations when larvae move between plants and do not discriminate based on plant genotype for host selection (Heuberger et al. 2011). As an alternative to mandated refugia, non-transgenic crops and wild plant relatives have been used as viable refugia for Bt crops in China (Qiao et al. 2010). However, recent reports suggest emergence of resistance to Bt cotton in *Helicoverpa armigera* populations in a region of Northern China with a history of high rate of Bt cotton adoption (Zhang et al. 2011). In India, difficulties in monitoring compliance and the lack of control over illegal transgenic varieties have been suggested as important issues facing resistance management programs for Bt crops (Jayaraman 2001, 2002). Despite these issues and after more than a decade of use, field evolved resistance to Bt crops can be considered a rare event keeping in mind the level of adoption. Recent reports of field evolved resistance to Bt crops (Tabashnik et al. 2009; Storer et al. 2010; van Rensburg 2007; Dhurua and Gujar 2011; Zhang et al. 2011) are usually correlated with sub-optimal crop growth conditions and most have not been reported to result in crop losses. However, these reports of field-evolved resistance to crops expressing a single Bt toxin have further incentivized the development of alternative approaches to delay resistance in second and third generation Bt crops.

15.5 "Second" and "Third" Generation Bt Crops

High adoption of single transgene Bt crop technology and concerns on potential evolution of insect resistance to these crops promoted the development of "second generation" Bt crops (cotton and maize), which we define in this chapter as those varieties expressing multiple Bt genes with diverse (pyramided) mode of action. Based on the importance of toxin binding to specific receptors in the insect midgut for toxicity and resistance, Bt toxins recognizing alternative midgut receptors are considered as having diverse mode of action by regulatory agencies and therefore are optimum candidates for pyramiding. The use of these pyramided Bt toxin genes greatly reduces the probability of resistance evolution, as target insects would need to develop simultaneous mutations in diverse toxin receptors to acquire resistance (Roush 1998; Zhao et al. 2003). Pyramiding of *cry1Ac* and *cry2Ab* (Chitkowski et al. 2003) or *cry* and *vip* (Estruch et al. 1997) toxin genes are examples of this strategy to reduce the rate of resistance evolution in second generation Bt crops. However, cross-resistance observed in laboratory-selected insect strains suggests that mechanisms such as altered toxin proteolysis (Oppert et al. 1997) or enhanced midgut healing (Martinez-Ramirez et al. 1999) could potentially result in resistance to multiple Bt toxins. In addition, coexistence of single and two-toxin Bt crops may result in faster evolution of resistance to pyramided traits (Zhao et al. 2005).

Expression of multiple toxins in second generation Bt crops (Table 15.1) achieves increased control of target pests with low susceptibility to single-toxin Bt crops. For instance, pyramiding of the *cry2Ab2* gene with a chimeric toxin gene (*cry1A.105*) composed of portions of the *cry1Ab*, *cry1Ac*, and *cry1Fa* genes, into maize plants (event MON89034, Yieldgard line of products) expanded the range of control to armyworms (*Spodoptera* spp.) and the black cutworm (*Agrotis ipsilon*). Similarly, Agrisure Viptera maize (events MIR162 and Bt11) producing Vip3Aa20 and Cry1Ab toxins can control corn earworm (*Helicoverpa zea*) and the fall armyworm (*Spodoptera frugiperda*), which display low susceptibility to Cry1A toxins. Control of these pest larvae is also accomplished in cotton varieties producing Cry1Ac and Cry1Fa (events DAS 21023-5 and DAS 24236-5, respectively, WideStrike line of products) or Cry1Ac and Cry2Ab (event 15985, Bollgard II) toxin pyramids in cotton (Stewart et al. 2001; Jackson et al. 2004). In addition to expanding range of activity to pest species within the same taxonomic order, second generation Bt maize varieties have also been developed to control both lepidopteran and coleopteran pests. For example, the Genuity Vt triple line of maize products produces Cry1Ab and Cry2Ab toxins targeting lepidopteran larvae (event MON89034-3) and Cry3Bb toxin to control root-feeding Western corn rootworm (*D. virgifera*) larvae (event MON88017-3).

Reductions in mandated refuge size from 20–5% have been proposed for pyramided Bt crops based on the increased toxicity against targeted pests and expression of multiple Bt toxins with diverse mode of action. However, it is important to consider that despite the predicted slower evolution of resistance, the evolutionary processes involved in resistance to pyramided Bt crops will be the same as those driving resistance to single toxin varieties (Ives et al. 2011). An alternative to the inherent compliance problems observed with the use of spatial refugia is the use of seed mixtures (refuge in a bag), which is currently being adopted by industry in Bt crops expressing multiple toxins. While this strategy would guarantee the existence of non-Bt refuge plants in fields planted with Bt crops, some models suggest that insect movement between plants may accelerate evolution of resistance (Heuberger et al. 2011). Pests continuously moving and feeding on Bt and non-Bt plants would undergo increased selection pressure in mixtures versus spatial refugia (Ives et al. 2011), which may allow rare broad-spectrum resistance mechanisms to emerge.

The "third generation" of Bt crops, which we define in this chapter as transgenic plants currently in the pipeline for commercialization, are expected to address activity and resistance management issues related to previous Bt crops. These crops are transformed to express pyramids of Bt toxins targeting lepidopteran and coleopteran pests, or combinations of Bt toxins with alternative insecticidal technologies, such as RNA interference (RNAi). Apart from the effects of toxin pyramiding on delaying resistance evolution, expression of multiple toxins also addresses cases of previous Bt crop varieties not fulfilling the high dose tenet for a particular pest. For example, SmartStax maize from Monsanto and Dow AgroSciences (events MON89034, TC1507, MON88017, and DAS-59122-7) will produce Cry1A.105, Cry2Ab2, and Cry1Fa2 toxins to effectively control a wide range of lepidopteran larvae in addition to producing Cry3Bb1 and Cry34Ab1/Cry35Ab1 toxins to achieve increased efficacy against *Diabrotica* ssp. larvae. Combining Bt toxins and insecticidal components with alternative mode of action, such as RNA interference (RNAi) (Baum et al. 2007) or protease inhibitors (Cui et al. 2011), in third generation Bt crops, should address current issues regarding the definition of diverse mode of action used for Bt toxins. Potential challenges with these new Bt crops would include the design and implementation of effective resistance management practices, demonstration that the effects of combining multiple Bt toxins can be predicted from integration of individual toxin assessments, and that novel insecticidal components do not have unforeseen negative impacts on plant innate immune responses to herbivory.

15.6 Conclusions and Future Prospects

So far, introduction of transgenic Bt crops has had major positive ecological and agricultural consequences (Betz et al. 2000). Despite the highly publicized interchange between non-peer reviewed scientific opinions over yield comparisons (Sheridan 2009), overall the use of Bt cotton or Bt maize has been reported to result in increased farm income benefits (Brookes and Barfoot 2009). While assessment of the environmental impact of Bt crops is highly dependent on variables considered, unintended effects pale in comparison to those of alternative pest suppression methods (Naranjo 2009). In the U.S. alone, the use of Bt maize and cotton has resulted in consistent reductions in chemical insecticide use from 1996–2009, totaling 64.2 million pounds (Benbrook 2009). In addition to these economic and environmental benefits, Bt crops can also enhance food safety, as in the case of mycotoxin level reductions reported for Bt maize compared to non-Bt varieties (Hammond et al. 2004).

Current trends in the advancement of Bt crops include the integration of transgenes into other essential crops worldwide, enhancing the quality and efficacy of the transgenes through genetic engineering, discovering new insecticidal Bt toxins (Kaur 2006), and coupling Bt toxins with non-Bt insecticidal components such as protease inhibitors (Vaughan 2003) or RNAi (Baum et al. 2007). These initiatives can broaden activity range, increase insecticidal potency, and further delay the evolution of insect resistance. For example, the development of Bt rice (Yang et al. 2011) has extraordinary commercial appeal for agricultural biotechnology (Xia et al. 2011), and has recently been suggested to be effective in preventing non-targeted pest outbreaks (Chen et al. 2011). The approval of transgenic Bt brinjal as the first genetically modified crop for human consumption in India is expected to result in benefits observed for other Bt crops, although approval has been initially met with controversy (Seetharam 2010). Other non-essential crops, like sunflower, are being also explored for commercial viability (Cantamutto and Poverene 2007).

Advances on the characterization of the Bt toxin mode of action have allowed for design of improved Bt toxins amenable to expression in Bt crops. Genetically engineered Cry3A toxins (mCry3A and eCry3.1Ab) displaying increased activity against *D. virgifera* larvae (Walters et al. 2008, 2010) are currently used in some Bt maize varieties (Table 15.1). Modified (Mod) Cry1A toxins targeting lepidopteran larvae and effective against larvae lacking the primary Cry1A toxin receptor (Soberón et al. 2007) are a clear alternative to currently used lepidopteran-specific traits. The use of enhancers of Bt toxicity has also been proposed as a strategy to attain high dose mortality levels in pests with low susceptibility to Bt toxins (Gao et al. 2011; Chen et al. 2007), although their efficacy in transgenic plants has not been reported.

We are at an important moment in the Bt crop marketplace, with many issues simultaneously impacting the future of this technology. Past controversy surrounding some of the publications addressing these issues has hindered advancement of Bt crop research to secure long-term utility of Bt-based biotechnology. New regulatory principles that deviate from the longstanding idea of how selection pressure is established within a Bt crop environment are being developed to address risks related to introduction of second and third generation Bt crops, especially in environments dominated by single Bt gene crop varieties. Commercialization of Bt crops in alternative developing and industrialized countries will result in regulatory challenges due to agricultural infrastructure distinctions, including enforcement, reporting and policing. These issues will have to be addressed using scientifically sound and peer-reviewed discourse exchange to advance dependable knowledge regarding environmental safety of Bt crops and to maintain utility of Bt-based pesticidal products for future generations.

References

Aarssen R, Soetaert P, Stam M, Dockx J, Gosselé V, Seurinck J, Reynaerts A, Cornelissen M (1995) *cryIA(b)* transcript formation in tobacco is inefficient. Plant Mol Biol 28:513–524

Adamczyk JJ Jr, Sumerford DV (2001) Potential factors impacting season-long expression of Cry1Ac in 13 commercial varieties of Bollgard cotton. J Insect Sci 1:13

Adamczyk JJ Jr, Adams LC, Hardee DD (2001) Field efficacy and seasonal expression profiles for terminal leaves of single and double *Bacillus thuringiensis* toxin cotton genotypes. J Econ Entomol 94:1589–1593

Adamczyk JJ Jr, Perera O, Meredith WR (2009) Production of mRNA from the *cry1Ac* transgene differs among Bollgard lines which correlates to the level of subsequent protein. Transgenic Res 18:143–149

Agaisse H, Lereclus D (1995) How does *Bacillus thuringiensis* produce so much insecticidal crystal protein? J Bacteriol 177:6027–6032

Andrews RE, Faust RM, Wabiko H, Raymond KC, Bulla LA (1987) The biotechnology of *Bacillus thuringiensis*. Crit Rev Biotechnol 6:163–232

Barton KA, Whiteley HR, Yang NS (1987) *Bacillus thuringienis* delta-endotoxin expressed in transgenic *Nicotiana tabacum* provides resistance to lepidopteran insects. Plant Physiol 85:1103–1109

Baum JA, Bogaert T, Clinton W, Heck GR, Feldmann P, Ilagan O, Johnson S, Plaetinck G, Munyikwa T, Pleau M, Vaughn T, Roberts J (2007) Control of coleopteran insect pests through RNA interference. Nat Biotechnol 25:1322–1326

Benbrook C (2009) Impacts of genetically engineered crops on pesticide use: the first thirteen years. The Organic Center (http://www.organic-center.org), Critical Issue Report: The First Thirteen years, Boulder, CO

Betz FS, Hammond BG, Fuchs RL (2000) Safety and advantages of *Bacillus thuringiensis*-protected plants to control insect pests. Regul Toxicol Pharmacol 32:156–173

Bourguet D, Desquilbet M, Lemarie S (2005) Regulating insect resistance management: the case of non-Bt corn refuges in the US. J Environ Manag 76:210–220

Brookes G, Barfoot P (2009) Global impact of biotech crops: Income and production effects 1996–2007. AgBioForum 12:184–208

Bulla LA, Kramer KJ, Davidson LI (1977) Characterization of entomocidal parasporal crystal of *Bacillus thuringiensis*. J Bacteriol 130:375–383

Cantamutto M, Poverene M (2007) Genetically modified sunflower release: opportunities and risks. Field Crops Res 101:133–144

Carozzi NB, Warren GW, Desai N, Jayne SM, Lotstein R, Rice DA, Evola S, Koziel MG (1992) Expression of a chimeric CaMV 35S *Bacillus thuringiensis* insecticidal protein gene in transgenic tobacco. Plant Mol Biol 20:539–548

Carriere Y, Dennehy TJ, Pedersen B, Haller S, Ellers-Kirk C, Antilla L, Liu YB, Willott E, Tabashnik BE (2001) Large-scale management of insect resistance to transgenic cotton in Arizona: can transgenic insecticidal crops be sustained? J Econ Entomol 94:315–325

Center for Environmental Risk Assessment (2009) GM Crop Database, International Life Sciences Institute Research Foundation. http://cera-gmc.org/index.php?action=gm_crop_database. Accessed 7 Sept 2011

Chen J, Hua G, Jurat-Fuentes JL, Abdullah MA, Adang MJ (2007) Synergism of *Bacillus thuringiensis* toxins by a fragment of a toxin-binding cadherin. Proc Natl Acad Sci USA 104:13901–13906

Chen Y, Tian JC, Wang W, Fang Q, Akhtar ZR, Peng YF, Cui H, Guo YY, Song QS, Ye GY (2011) Bt rice expressing Cry1Ab does not stimulate an outbreak of its non-target herbivore, *Nilaparvata lugens*. Transgenic Res. doi:10.1007/s11248-011-9530-x

Chitkowski RL, Turnipseed SG, Sullivan MJ, Bridges WC (2003) Field and laboratory evaluations of transgenic cottons expressing one or two *Bacillus thuringiensis* var. *kurstaki* Berliner proteins for management of noctuid (Lepidoptera) pests. J Econ Entomol 96:755–762

Christensen AH, Sharrock RA, Quail PH (1992) Maize polyubiquitin genes: structure, thermal perturbation of expression and transcript splicing, and promoter activity following transfer to protoplasts by electroporation. Plant Mol Biol 18:675–689

Christou P (2002) No credible scientific evidence is presented to support claims that transgenic DNA was introgressed into traditional maize landraces in Oaxaca, Mexico. Transgenic Res 11:III–V

Crickmore N, Bone EJ, Williams JA, Ellar DJ (1995) Contribution of the individual components of the δ-endotoxin crystal to the mosquitocidal activity of *Bacillus thuringiensis* subsp. *israelensis*. FEMS Microbiol Lett 131:249–254

Crickmore N, Zeigler DR, Schnepf E, Van Rie J, Lereclus D, Baum J, Bravo A, Dean DH (2011) *Bacillus thuringiensis* toxin nomenclature. http://www.lifesci.sussex.ac.uk/Home/Neil_Crickmore/Bt/. Accessed 23 Sept 2011

Cui J, Luo J, Werf WVD, Ma Y, Xia J (2011) Effect of Pyramiding Bt and CpTI Genes on resistance of cotton to *Helicoverpa armigera* (Lepidoptera: Noctuidae) under laboratory and field conditions. J Econ Entomol 104:673–684

Daniell H (2002) Molecular strategies for gene containment in transgenic crops. Nat Biotechnol 20:581–586

de Barjac H, Bonnefoi A (1968) A classification of strains of *Bacillus thuringiensis* Berliner with a key to their differentiation. J Invertebr Pathol 11:335–347

Dhurua S, Gujar GT (2011) Field-evolved resistance to Bt toxin Cry1Ac in the pink bollworm, *Pectinophora gossypiella* (Saunders) (Lepidoptera: Gelechiidae), from India. Pest Manag Sci 67:898–903

Dogan EB, Berry RE, Reed GL, Rossignol PA (1996) Biological parameters of convergent lady beetle (Coleoptera: Coccinellidae) feeding on aphids (Homoptera: Aphididae) on transgenic potato. J Econ Entomol 89:1105–1108

Dowd PF (2000) Indirect reduction of ear molds and associated mycotoxins in *Bacillus thuringiensis* corn under controlled and open field conditions: utility and limitations. J Econ Entomol 93:1669–1679

Earl C (1983) Biotechnology regulation- rules for freed organisms planned. Nature 306:5

Economic Research Service-USDA (2011) Adoption of genetically engineered crops in the U.S. http://www.ers.usda.gov/Data/BiotechCrops/. Accessed 23 Sept 2011

Estruch JJ, Warren GW, Mullins MA, Nye GJ, Craig JA, Koziel MG (1996) Vip3A, a novel *Bacillus thuringiensis* vegetative insecticidal protein with a wide spectrum of activities against lepidopteran insects. Proc Natl Acad Sci USA 93:5389–5394

Estruch JJ, Carozzi NB, Desai N, Duck NB, Warren GW, Koziel MG (1997) Transgenic plants: an emerging approach to pest control. Nat Biotechnol 15:137–141

Fernandez-Cornejo J (2004) The seed industry in U.S. agriculture—an exploration of data and information on crop seed markets, regulation, industry structure, and research and development. Agriculture Information Bulletin, U.S. Department of Agriculture, Economic Research Service, Washington

Fernandez-Cornejo J, McBride WD (2002) Adoption of bioengineered crops. U.S. Department of Agriculture, Economic Research Service, Washington

Ferré J, Van Rie J (2002) Biochemistry and genetics of insect resistance to *Bacillus thuringiensis*. Annu Rev Entomol 47:501–533

Ferré J, Real MD, Van Rie J, Jansens S, Peferoen M (1991) Resistance to the *Bacillus thuringiensis* bioinsecticide in a field population of *Plutella xylostella* is due to a change in a midgut membrane receptor. Proc Natl Acad Sci USA 88:5119–5123

Fischhoff DA, Bowdish KS, Perlak FJ, Marrone PG, McCormick SM, Niedermeyer JG, Dean DA, Kusano-Kretzmer K, Mayer EJ, Rochester DE, Rogers SG, Fraley RT (1987) Insect tolerant transgenic tomato plants. Nat Biotechnol 5:807–813

Fraley RT (1996) Bollgard cotton performance. Science 274:1994

Franco-Rivera A, Benintende G, Cozzi J, Baizabal-Aguirre VM, Valdez-Alarcón JJ, López-Meza JE (2004) Molecular characterization of *Bacillus thuringiensis* strains from Argentina. Antonie Van Leeuwenhoek 86:87–92

Gaertner FH, Quick TC, Thompson MA (1993) CellCap: an encapsulation system for insecticidal biotoxin proteins. In: Kim L (ed) Advanced engineered pesticides. Marcel Dekker, Inc., NY

Gao Y, Schafer BW, Collins RA, Herman RA, Xu XP, Gilbert JR, Ni WT, Langer VL, Tagliani LA (2004) Characterization of Cry34Ab1 and Cry35Ab1 insecticidal crystal proteins expressed in transgenic corn plants and *Pseudomonas fluorescens*. J Agric Food Chem 52:8057–8065

Gao Y, Jurat-Fuentes JL, Oppert B, Fabrick JA, Liu C, Gao J, Lei Z (2011) Increased toxicity of *Bacillus thuringiensis* Cry3Aa against *Crioceris quatuordecimpunctata*, *Phaedon brassicae* and *Colaphellus bowringi* by a *Tenebrio molitor* cadherin fragment. Pest Manag Sci 67:1076–1081

Gatehouse AMR, Ferry N, Raemaekers RJM (2002) The case of the monarch butterfly: a verdict is returned. Trends Genet 18:249–251

Gatehouse AMR, Ferry N, Edwards MG, Bell HA (2011) Insect-resistant biotech crops and their impacts on beneficial arthropods. Philos Trans R Soc Lond B Biol Sci 366:1438–1452

Gordon-Kamm WJ, Spencer TM, Mangano ML, Adams TR, Daines RJ, Start WG, O'Brien JV, Chambers SA, Adams WR Jr, Willetts NG, Rice TB, Mackey CJ, Krueger RW, Kausch AP, Lemaux PG (1990) Transformation of maize cells and regeneration of fertile transgenic plants. Plant Cell Online 2:603–618

Gould F (1988) Evolutionary biology and genetically engineered crops: consideration of evolutionary theory can aid in crop design. Bioscience 38:26–33

Gould F (1998) Sustainability of transgenic insecticidal cultivars: integrating pest genetics and ecology. Annu Rev Entomol 43:701–726

Gray ME (2010) Relevance of traditional integrated pest management (IPM) strategies for commercial corn producers in a transgenic agroecosystem: a bygone era? J Agric Food Chem 59:5852–5858

Hammond BG, Campbell KW, Pilcher CD, Degooyer TA, Robinson AE, McMillen BL, Spangler SM, Riordan SG, Rice LG, Richard JL (2004) Lower fumonisin mycotoxin levels in the grain of Bt corn grown in the United States in 2000–2002. J Agric Food Chem 52:1390–1397

Hannay CL, Fitz-James P (1955) The protein crystals of *Bacillus thuringiensis* Berliner. Can J Microbiol 1:694–710

He SC, Abad AR, Gelvin SB, Mackenzie SA (1996) A cytoplasmic male sterility-associated mitochondrial protein causes pollen disruption in transgenic tobacco. Proc Natl Acad Sci USA 93:11763–11768

Heuberger S, Crowder DW, Brevault T, Tabashnik BE, Carriere Y (2011) Modeling the effects of plant-to-plant gene flow, larval behavior, and refuge size on pest resistance to Bt cotton. Environ Entomol 40:484–495

Horsch RB (1993) Commercialization of genetically engineered crops. Philos Trans R Soc Lond B Biol Sci 342:287–291

Hutchison WD, Burkness EC, Mitchell PD, Moon RD, Leslie TW, Fleischer SJ, Abrahamson M, Hamilton KL, Steffey KL, Gray ME, Hellmich RL, Kaster LV, Hunt TE, Wright RJ, Pecinovsky K, Rabaey TL, Flood BR, Raun ES (2010) Areawide suppression of European corn borer with Bt maize reaps savings to non-Bt maize growers. Science 330:222–225

Ives AR, Glaum PR, Ziebarth NL, Andow DA (2011) The evolution of resistance to two-toxin pyramid transgenic crops. Ecol Appl 21:503–515

Jackson RE, Bradley JR Jr, Van Duyn JW, Gould F (2004) Comparative production of *Helicoverpa zea* (Lepidoptera: Noctuidae) from transgenic cotton expressing either one or two *Bacillus thuringiensis* proteins with and without insecticide oversprays. J Econ Entomol 97:1719–1725

Jansens S, Vliet A van, Dickburt C, Buysse L, Piens C, Saey B, DeWulf A, Gossele V, Paez A, Gobel E, Peferoen M (1997) Transgenic corn expressing a Cry9C insecticidal protein from *Bacillus thuringiensis* protected from European corn borer damage. Crop Sci 37:1616–1624

Jayaraman KS (2001) Illegal Bt cotton in India haunts regulators. Nat Biotechnol 19:1090

Jayaraman KS (2002) Poor crop management plagues Bt cotton experiment in India. Nat Biotechnol 20:1069

Kaur S (2006) Molecular approaches for identification and construction of novel insecticidal genes for crop protection. World J Microbiol Biotechnol 22:233–253

Kausch AP, Hague J, Oliver M, Li Y, Daniell H, Mascia P, Watrud LS, Stewart CN (2010) Transgenic perennial biofuel feedstocks and strategies for bioconfinement. Biofuels 1:163–176
Kay R, Chan A, Daly M, McPherson J (1987) Duplication of CaMV 35S promoter sequences creates a strong enhancer for plant genes. Science 236:1299–1302
Kershen DL (2006) Health and food safety: the benefits of Bt-corn. Food Drug Law J 61:197–235
Kim E, Suh S, Park B, Shin K, Kweon S, Han E, Park S-H, Kim Y, Kim J-K (2009) Chloroplast-targeted expression of synthetic *cry1Ac* in transgenic rice as an alternative strategy for increased pest protection. Planta 230:397–405
Kota M, Daniell H, Varma S, Garczynski SF, Gould F, Moar WJ (1999) Overexpression of the *Bacillus thuringiensis* (Bt) Cry2Aa2 protein in chloroplasts confers resistance to plants against susceptible and Bt-resistant insects. Proc Natl Acad Sci USA 96:1840–1845
Koziel MG, Beland GL, Bowman C, Carozzi NB, Crenshaw R, Crossland L, Dawson J, Desai N, Hill M, Kadwell S, Launis K, Lewis K, Maddox D, McPherson K, Meghji MR, Merlin E, Rhodes R, Warren GW, Wright M, Evola SV (1993) Field performance of elite transgenic maize plants expressing an insecticidal protein derived from *Bacillus thuringiensis*. Nat Biotechnol 11:194–200
Koziel MG, Carozzi NB, Desai N, Warren GW, Dawson J, Dunder E, Launis K, Evola SV (1996) Transgenic maize for the control of european corn corer and other maize insect pests. Ann N Y Acad Sci 792:164–171
Kronstad JW, Schnepf HE, Whiteley HR (1983) Diversity of locations for *Bacillus thuringiensis* crystal protein genes. J Bacteriol 154:419–428
Lampel JS, Canter GL, Dimock MB, Kelly JL, Anderson JJ, Uratani BB, Foulke JS, Turner JT (1994) Integrative cloning, expression, and stability of the *cryIA(c)* gene from *Bacillus thuringiensis* subsp. *kurstaki* in a recombinant strain of *Clavibacter xyli* subsp. *cynodontis*. Appl Environ Microbiol 60:501–508
Lang A, Otto M (2010) A synthesis of laboratory and field studies on the effects of transgenic *Bacillus thuringiensis* (Bt) maize on non-target Lepidoptera. Entomol Exp Appl 135:121–134
Lawo NC, Wackers FL, Romeis J (2009) Indian Bt cotton varieties do not affect the performance of cotton aphids. PLoS One 4:e4804
Lawson EC, Weiss JD, Thomas PE, Kaniewski WK (2001) NewLeaf Plus (R) Russet Burbank potatoes: replicase-mediated resistance to potato leafroll virus. Mol Breed 7:1–12
Li Y, Meissle M, Romeis J (2008) Consumption of Bt maize pollen expressing Cry1Ab or Cry3Bb1 does not harm adult green lacewings, *Chrysoperla carnea* (Neuroptera: Chrysopidae). PLoS One 3:e2909
Liu W (2010) Do genetically modified plants impact arbuscular mycorrhizal fungi? Ecotoxicology 19:229–238
Liu CW, Lin CC, Yiu JC, Chen JJW, Tseng MJ (2008) Expression of a *Bacillus thuringiensis* toxin (*cry1Ab*) gene in cabbage (*Brassica oleracea L.* var. *capitata L.*) chloroplasts confers high insecticidal efficacy against *Plutella xylostella*. Theor Appl Genet 117:75–88
Llewellyn DJ, Mares CL, Fitt GP (2007) Field performance and seasonal changes in the efficacy against *Helicoverpa armigera* (Hubner) of transgenic cotton expressing the insecticidal protein Vip3A. Agric For Entomol 9:93–101
Long N, Bottoms J, Meghji M, Hart H, Que Q, Pulliam D (2007) Corn Event MIR162. United States Patent US 20090300784 A1, 15 July 2009
Losey JE, Rayor LS, Carter ME (1999) Transgenic pollen harms monarch larvae. Nature 399:214
Lovei GL, Andow DA, Arpaia S (2009) Transgenic insecticidal crops and natural enemies: a detailed review of laboratory studies. Environ Entomol 38:293–306
Lu Y, Wu K, Jiang Y, Xia B, Li P, Feng H, Wyckhuys KA, Guo Y (2010) Mirid bug outbreaks in multiple crops correlated with wide-scale adoption of Bt cotton in China. Science 328:1151–1154
Luo KM, Duan H, Zhao DG, Zheng XL, Deng W, Chen YQ, Stewart CN, McAvoy R, Jiang XN, Wu YH, He AG, Pei Y, Li Y (2007) 'GM-gene-deletor': fused loxP-FRT recognition sequences

dramatically improve the efficiency of FLP or CRE recombinase on transgene excision from pollen and seed of tobacco plants. Plant Biotechnol J 5:263–274
Martin MA, Hyde J (2001) Economic considerations for the adoption of transgenic crops: the case of Bt corn. J Nematol 33:173–177
Martinez-Ramirez AC, Gould F, Ferre J (1999) Histopathological effects and growth reduction in a susceptible and a resistant strain of *Heliothis virescens* (Lepidoptera: Noctuidae) caused by sublethal doses of pure Cry1A crystal proteins from *Bacillus thuringiensis*. Biocontrol Sci Technol 9:239–246
Marvier M, McCreedy C, Regetz J, Kareiva P (2007) A meta-analysis of effects of Bt cotton and maize on nontarget invertebrates. Science 316:1475–1477
Matten SR, Lewis PI, Tomimatsu G, Sutherland DWS, Anderson N, ColvinSnyder TL (1996) The US Environmental Protection Agency's role in pesticide resistance management. In: Brown TM (ed) Molecular genetics and evolution of pesticide resistance. Amer Chemical Soc, Washington
McBride KE, Svab Z, Schaaf DJ, Hogan PS, Stalker DM, Maliga P (1995) Amplification of a Chimeric *Bacillus* gene in chloroplasts leads to an extraordinary level of an insecticidal protein in tobacco. Nat Biotechnol 13:362–365
Mulder C, Wouterse M, Raubuch M, Roelofs W, Rutgers M (2006) Can transgenic maize affect soil microbial communities? PLoS Comput Biol 2:e128
Murray EE, Stock C, Eberle M, Sekar V, Rocheleau TA, Adang MJ (1991) Analysis of unstable RNA transcripts of insecticidal crystal protein genes of *Bacilus thuringiensis* in transgenic plants and electroporated protoplasts. Plant Mol Biol 16:1035–1050
Naranjo SE (2009) Impacts of Bt crops on non-target invertebrates and insecticide use patterns. In: CAB Reviews: Perspectives in Agriculture, Veterinary Science, Nutrition and Natural Resources, CABI, Wallingford, UK.
Obukowicz MG, Perlak FJ, Kusano-Kretzmer K, Mayer EJ, Watrud LS (1986) Integration of the delta-endotoxin gene of *Bacillus thuringiensis* into the chromosome of root-colonizing strains of pseudomonads using Tn5. Gene 45:327–331
Oliveira AP, Pampulha ME, Bennett JP (2008) A two-year field study with transgenic *Bacillus thuringiensis* maize: effects on soil microorganisms. Sci Total Environ 405:351–357
Oppert B, Kramer KJ, Beeman RW, Johnson D, McGaughey WH (1997) Proteinase-mediated insect resistance to *Bacillus thuringiensis* toxins. J Biol Chem 272:23473–23476
Pan ZQ, Onstad DW, Nowatzki TM, Stanley BH, Meinke LJ, Flexner JL (2011) Western corn rootworm (Coleoptera: Chrysomelidae) dispersal and adaptation to single-toxin transgenic corn deployed with block or blended refuge. Environ Entomol 40:964–978
Perlak FJ, Deaton RW, Armstrong TA, Fuchs RL, Sims SR, Greenplate JT, Fischhoff DA (1990) Insect resistant cotton plants. Nat Biotechnol 8:939–943
Perlak FJ, Fuchs RL, Dean DA, McPherson SL, Fischhoff DA (1991) Modification of the coding sequence enhances plant expression of insect control protein genes. Proc Natl Acad Sci USA 88:3324–3328
Perlak FJ, Stone TB, Muskopf YM, Petersen LJ, Parker GB, McPherson SA, Wyman J, Love S, Reed G, Biever D et al (1993) Genetically improved potatoes: protection from damage by Colorado potato beetles. Plant Mol Biol 22:313–321
Pitt G (2003) Implementation and impact of transgenic Bt cottons in Australia. ICAC Recorder 21:14–19
Pray CE, Huang J, Hu R, Rozelle S (2002) Five years of Bt cotton in China—the benefits continue. Plant J 31:423–430
Qaim M, Zilberman D (2003) Yield effects of genetically modified crops in developing countries. Science 299:900–902
Qiao F, Huang J, Rozelle S, Wilen J (2010) Natural refuge crops, buildup of resistance, and zero-refuge strategy for Bt cotton in China. Sci China Life Sci 53:1227–1238
Quist D, Chapela IH (2001) Transgenic DNA introgressed into traditional maize landraces in Oaxaca, Mexico. Nature 414:541–543
Rodrigo-Simon A, Maagd RA de, Avilla C, Bakker PL, Molthoff J, Gonzalez-Zamora JE, Ferre J (2006) Lack of detrimental effects of *Bacillus thuringiensis* Cry toxins on the insect predator

Chrysoperla carnea: a toxicological, histopathological, and biochemical analysis. Appl Environ Microbiol 72:1595–1603
Romeis J, Dutton A, Bigler F (2004) *Bacillus thuringiensis* toxin (Cry1Ab) has no direct effect on larvae of the green lacewing *Chrysoperla carnea* (Stephens) (Neuroptera: Chrysopidae). J Insect Physiol 50:175–183
Rong J, Lu BR, Song Z, Su J, Snow AA, Zhang X, Sun S, Chen R, Wang F (2007) Dramatic reduction of crop-to-crop gene flow within a short distance from transgenic rice fields. New Phytol 173:346–353
Roush RT (1998) Two-toxins strategies for management of insecticidal transgenic crops: can pyramiding succeed where pesticide mixtures have not? Philos Trans R Soc Lond B Biol Sci 353:1777–1786
Salamitou S, Agaisse H, Bravo A, Lereclus D (1996) Genetic analysis of *cryIIIA* gene expression in *Bacillus thuringiensis*. Microbiology 142:2049–2055
Sanahuja G, Banakar R, Twyman RM, Capell T, Christou P (2011) *Bacillus thuringiensis*: a century of research, development and commercial applications. Plant Biotechnol J 9:283–300
Sanchis V (2010) From microbial sprays to insect-resistant transgenic plants: history of the biospesticide *Bacillus thuringiensis*. A review. Agron Sustain Dev 31:217–231
Schnepf HE, Whiteley HR (1981) Cloning and expression of the *Bacillus thuringiensis* crystal protein gene in *Escherichia coli*. Proc Natl Acad Sci USA 78:2893–2897
Sears MK, Hellmich RL, Stanley-Horn DE, Oberhauser KS, Pleasants JM, Mattila HR, Siegfried BD, Dively GP (2001) Impact of Bt corn pollen on monarch butterfly populations: a risk assessment. Proc Natl Acad Sci USA 98:11937–11942
Seetharam S (2010) Should the Bt brinjal controversy concern healthcare professionals and bioethicists? Indian J Med Ethics 7:9–12
Shelton A, Sears M (2001) The monarch butterfly controversy: scientific interpretations of a phenomenon. Plant J 27:483–488
Shelton AM, Zhao J-Z, Roush RT (2002) Economic, ecological, food safety, and social consequences of the deployment of Bt transgenic plants. Annu Rev Entomol 47:845–881
Sheridan C (2009) Report claims no yield advantage for Bt crops. Nat Biotechnol 27:588–589
Shivakumar AG, Vanags RI, Wilcox DR, Katz L, Vary PS, Fox JL (1989) Gene dosage effect on the expression of the delta-endotoxin genes of *Bacillus thuringiensis* subsp. *kurstaki* in *Bacillus subtilis* and *Bacillus megaterium*. Gene 79:21–31
Skøt L, Harrison SP, Nath A, Mytton LR, Clifford BC (1990) Expression of insecticidal activity in *Rhizobium* containing the δ-endotoxin gene cloned from *Bacillus thuringiensis* subsp. *tenebrionis*. Plant Soil 127:285–295
Soberón M, Pardo-López L, López I, Gómez I, Tabashnik BE, Bravo A (2007) Engineering modified Bt toxins to counter insect resistance. Science 318:1640–1642
Stark DM (1997) Risk assessment and criteria for commercial launch of transgenic plants. Proceedings of the 3rd Ifgene Workshop, Dornach, Switzerland
Stewart SD, Adamczyk JJ Jr, Knighten KS, Davis FM (2001) Impact of Bt cottons expressing one or two insecticidal proteins of *Bacillus thuringiensis* Berliner on growth and survival of noctuid (Lepidoptera) larvae. J Econ Entomol 94:752–760
Storer NP, Babcock JM, Schlenz M, Meade T, Thompson GD, Bing JW, Huckaba RM (2010) Discovery and characterization of field resistance to Bt maize: *Spodoptera frugiperda* (Lepidoptera: Noctuidae) in Puerto Rico. J Econ Entomol 103:1031–1038
Tabashnik BE (1989) Managing resistance with multiple pesticide tactics: theory, evidence, and recommendations. J Econ Entomol 82:1263–1269
Tabashnik BE, Van Rensburg JB, Carrière Y (2009) Field-evolved insect resistance to Bt crops: definition, theory, and data. J Econ Entomol 102:2011–2025
Tank JL, Rosi-Marshall EJ, Royer TV, Whiles MR, Griffiths NA, Frauendorf TC, Treering DJ (2010) Occurrence of maize detritus and a transgenic insecticidal protein (Cry1Ab) within the stream network of an agricultural landscape. Proc Natl Acad Sci USA 107:17645–17650
Vaeck M, Reynaerts A, Höfte H, Jansens S, De Beuckeleer M, Dean C, Zabeau M, Van Montagu M, Leemans J (1987) Transgenic plants protected from insect attack. Nature 328:33–37

van Frankenhuyzen K (2009) Insecticidal activity of *Bacillus thuringiensis* crystal proteins. J Invertebr Pathol 101:1–16
van Rensburg JBJ (2007) First report of field resistance by stem borer, *Busseola fusca* (Fuller) to Bt-transgenic maize. S Afr J Plant Soil 24:147–151
Van Rie J, McGaughey WH, Johnson DE, Barnett BD, Van Mellaert H (1990) Mechanism of insect resistance to the microbial insecticide *Bacillus thuringiensis*. Science 247:72–74
Vaughan H (2003) GM plants and protection against insects—Alternative strategies based on gene technology. Acta Agric Scand B 53:34–40
Viktorov AG (2011) Transfer of Bt corn byproducts from terrestrial to stream ecosystems. Russ J Plant Physiol 58:543–548
Walters FS, Stacy CM, Lee MK, Palekar N, Chen JS (2008) An engineered chymotrypsin/cathepsin G site in domain I renders *Bacillus thuringiensis* Cry3A active against Western corn rootworm larvae. Appl Environ Microbiol 74:367–374
Walters FS, de Fontes CM, Hart H, Warren GW, Chen JS (2010) Lepidopteran-active variable-region sequence imparts coleopteran activity in eCry3.1Ab, an engineered *Bacillus thuringiensis* hybrid insecticidal protein. Appl Environ Microbiol 76:3082–3088
Waltz E (2009) Battlefield. Nature 461:27–32
Warren GW, Carozzi NB, Desai N, Koziel MG (1992) Field evaluation of transgenic tobacco containing a *Bacillus thuringiensis* insecticidal protein gene. J Econ Entomol 85:1651–1659
Whalon ME, Wingerd BA (2003) Bt: mode of action and use. Arch Insect Biochem Physiol 54:200–211
Wu F (2006) Mycotoxin reduction in Bt corn: potential economic, health, and regulatory impacts. Transgenic Res 15:277–289
Wu ZL, Guo WY, Qiu JZ, Huang TP, Li XB, Guan X (2004) Cloning and localization of *vip3A* gene of *Bacillus thuringiensis*. Biotechnol Lett 26:1425–1428
Wu KM, Lu YH, Feng HQ, Jiang YY, Zhao JZ (2008) Suppression of cotton bollworm in multiple crops in China in areas with Bt toxin-containing cotton. Science 321:1676–1678
Xia H, Lu BR, Xu K, Wang W, Yang X, Yang C, Luo J, Lai FX, Ye WL, Fu Q (2011) Enhanced yield performance of Bt rice under target-insect attacks: implications for field insect management. Transgenic Res 20:655–664
Yang Z, Chen H, Tang W, Hua H, Lin Y (2011) Development and characterisation of transgenic rice expressing two *Bacillus thuringiensis* genes. Pest Manag Sci 67:414–422
Zambryski P, Joos H, Genetello C, Leemans J, Montagu MV, Schell J (1983) Ti plasmid vector for the introduction of DNA into plant cells without alteration of their normal regeneration capacity. EMBO J 2:2143–2150
Zhang H, Yin W, Zhao J, Jin L, Yang Y, Wu S, Tabashnik BE, Wu Y (2011) Early warning of cotton bollworm resistance associated with intensive planting of Bt cotton in China. PLoS One 6:e22874
Zhao JZ, Cao J, Li Y, Collins HL, Roush RT, Earle ED, Shelton AM (2003) Transgenic plants expressing two *Bacillus thuringiensis* toxins delay insect resistance evolution. Nat Biotechnol 21:1493–1497
Zhao JZ, Cao J, Collins HL, Bates SL, Roush RT, Earle ED, Shelton AM (2005) Concurrent use of transgenic plants expressing a single and two *Bacillus thuringiensis* genes speeds insect adaptation to pyramided plants. Proc Natl Acad Sci USA 102:8426–8430

Chapter 16
A Review of the Food Safety of *Bt* Crops

Bruce G. Hammond and Michael S. Koch

Abstract There is a 50-year history of safe use and consumption of agricultural food crops sprayed with commercial *Bt* (*Bacillus thuringiensis*) microbial pesticides and a 14 year history of safe consumption of food and feed derived from *Bt* crops. This review summarizes the published literature addressing the safety of Cry insect control proteins found in both *Bt* microbial pesticides and those introduced into *Bt* agricultural crops. A discussion on the species-specific mode of action of Cry proteins to control target insect pests is presented. This information provides the scientific basis for the absence of toxicity of Cry proteins towards non-target organisms that has been confirmed in numerous mammalian toxicology studies. A human dietary exposure assessment for Cry proteins has also been provided which includes information that food processing of *Bt* crops such as maize leads to loss of functionally active Cry proteins in processed food products. Lastly the food and feed safety benefits of *Bt* crops are briefly summarized including lower insecticide use and reduction in fumonisin mycotoxin contamination of grain.

Keywords Food safety · Cry proteins · Toxicity · Bt crops · Consumption of Cry proteins

16.1 Background

Bt is a common Gram positive, spore-forming aerobic bacterium that is found in a variety of environmental sources such as soil, water, plant surfaces, grain dust, dead insects etc. (Federici and Siegel 2008). As part of its normal life cycle, the bacteria produce one or more insecticidal proteins in parasporal bodies when nutrients become insufficient to support bacterial growth. *Bt* microorganisms have been used for many years as a tool to control larval insect pests that feed on agricultural crops. They are also widely used in certified organic agricultural food production in the United States, Europe, and other countries. When the *Bt* microbial formulations are applied to the leaves of agricultural crops, the vegetative cells, spores,

B. G. Hammond (✉) · M. S. Koch
Monsanto company, Product safety center, 800 North Lindbergh Blvd,
Bldg C1N, 63167 St Louis, MO, USA
e-mail: bruce.g.hammond@monsanto.com

E. Sansinenea (ed.), *Bacillus thuringiensis Biotechnology*,
DOI 10.1007/978-94-007-3021-2_16, © Springer Science+Business Media B.V. 2012

and insecticidal proteins in the formulation are consumed by the larval insect. The insecticidal proteins exist as protoxins and are converted to active insect toxins by proteases in the alkaline environment of the lepidopteran insect gastrointestinal tract. The activated toxins bind to specific receptors on the membranes of target insect mid-gastrointestinal tract epithelial cells and form pores in the membranes allowing water and electrolytes from the gastrointestinal tract juices to enter the cell. The epithelial cells swell and lyse, leading to electrolyte imbalance in the insect hemolymph causing paralysis so that the insect stops eating and dies. The *Bt* spores can also germinate and colonize the insect body, allowing the bacteria to reproduce (WHO/IPCS 1999; Betz et al. 2000; OECD 2007; Federici and Siegel 2008).

Bt insecticidal crystal proteins, Cry (for crystal) and Cyt (for cytolytic) proteins, as well as VIPs (vegetative insecticidal proteins produced during the vegetative phase) are the major insecticidal proteins. There can be considerable variations in the amino acid content and structure of each of these proteins as they exist in different strains of *Bt*.

As discussed previously, *Bt* Cry proteins are one of the insecticidal components in *Bt* microbial commercial products widely used as biological insecticides for over 50 years. *Bt* was first discovered in Japan in 1901 and later rediscovered in Germany (Sanchis 2010). Field trials were subsequently carried out in Europe and the United States with *Bt* microbes to investigate their insecticidal properties; the first commercial *Bt* microbial formulation was launched in France in 1938 (Sanchis 2010). *Bt* microbial pesticides are highly regarded as environmentally-friendly due to their species-specificity (controlling only target insect pest species) and their lack of environmental persistence (WHO/IPCS 1999; Betz et al. 2000; OECD 2007; Federici and Siegel 2008). China has been probably the biggest user of *Bt* microbial pesticides where, over the last few decades, tens of thousands of tons of various *Bt* microbial formulations have been topically applied on agricultural food crops (rice, vegetables, maize), in forests and to potable water to control mosquitoes and other larval insects that are vectors of human disease (WHO/ICPS 1999; Ziwen 2010). According to recent data, there were at least 180 registered *Bt* microbial products in the United States (EPA 1998) and over 120 microbial products in the European Union. There are reported to be approximately 276 *Bt* microbials registered in China (Huang et al. 2007). *Bt* microbial pesticides were first registered in the US in 1961 (Betz et al. 2000).

The efficacy of *Bt* microbials applied to the surface of leaves is limited by the fact that the formulation can be washed off by rain and the Cry proteins are inactivated by sunlight within a few days of application (Federici and Siegel 2008). With the development of biotechnology, it has been possible to introduce the genes coding for Cry proteins into plants so that Cry proteins are expressed in the plant and are produced throughout the growing season to provide protection against insect pests. At present, most commercial *Bt* crops are based on Cry proteins, although VIPs are now being introduced into agricultural crops (EPA 2004). To date, Cyt proteins have not been introduced into commercial *Bt* crops.

16.2 Regulatory Guidance for the Safety Assessment of Cry Proteins

According to guidance from the US Environmental Protection Agency (Mendelsohn et al. 2003), the European Food Safety Agency (2011), and Codex (2009), the food safety assessment of any insecticidal protein introduced into food/feed crops through genetic engineering should include:

1. information on the biochemical characterization of the introduced protein including the amino acid sequence, molecular weight, post-translation modifications (if any), and a description of the function.
2. Assessment of amino acid sequence similarity between the protein and any known protein mammalian toxins using bioinformatics tools to search curated data bases of amino acid sequences of proteins (eg. NCBI Entrez Protein, PIR, UniProt-Swiss-Prot etc.).
3. Assessment of stability of the protein to heat or food processing conditions.
4. Assessment of potential degradation in appropriate and validated *in vitro* gastric and intestinal model systems.
5. High dose acute toxicology testing to confirm the absence of toxicity to mammals (EPA requirement only). The US EPA requires high dose acute toxicology testing in rodents with either *Bt* microbial pesticides or Cry proteins that end up in food and feed crops (McClintock et al. 1995; Betz et al. 2000). The rationale for requiring high dose acute testing is that Cry proteins act through an acute mode of action to kill insect pests.

EFSA (2011) considers that acute toxicology testing provides little value, but may require a repeat dose 28 day toxicology study where there is considered to be insufficient safety information on the introduced protein.

16.3 The Species-specific Acute Mode of Action of Cry Proteins

The general mode of action of Cry proteins has been studied extensively and reviewed in a number of publications (WHO/IPCS 1999; Betz et al. 2000; Siegel 2001; OECD 2007; Bravo et al. 2007; Federici and Siegel 2008; Soberón et al. 2010). Cry proteins are not contact insecticides like chemical pesticides, but must be ingested and activated by proteases in the gastrointestinal tract of target insect pests. The activated Cry toxins bind to specific receptors on mid-gastrointestinal tract epithelial cells of target larval insects and this, results in oligomerization of the Cry toxin monomers. The toxin complex translocates into the cellular membrane of the gastrointestinal tract epithelial cells and forms pores that cause osmotic shock and cell lysis leading to death of the insect (Federici and Siegel 2008; Soberón et al. 2010). Some Cry protein binding receptors have been identified such as cadherin-like glycoproteins, and glycosylphosphatidyl-inositol (GPI) membrane anchored receptors such as aminopeptidase N or alkaline phosphatase (Soberón et al. 2010). Binding of Cry proteins to the aforementioned receptors is not sufficient of itself to cause toxicity to the insect as oligomerization must also occur to form the pores in the membrane. This may explain the observation that some Cry proteins can bind to insect mid-gastrointestinal tract epithelial cells, but since oligomerization does not follow, no toxicity to the insect occurs (Federici and Siegel 2008; Soberón et al. 2010).

Non-target organisms such as humans, rhesus monkeys, cattle, mice, rats, rabbits, other non-target insects etc., lack high affinity Cry protein binding-receptors (Sacchi et al. 1986; Hofmann et al. 1988a, b; Wolfersberger et al. 1986; Van Rie et al. 1989, 1990; Lambert et al. 1996; Mendelsohn et al. 2003; Griffiths et al. 2005; Shimada et al. 2006; OECD 2007). In contrast to these studies in mammals demonstrating an absence of specific high affinity receptors to bind Cry proteins, one study reported binding of Cry1Ac protein to the mouse jejunum (Vazquez-Padron et al. 2000). However, this binding appeared to be non-specific because extremely high (non-physiological) concentrations of Cry1Ac protein were incubated *in vitro* with mouse intestinal brush border membrane vesicles (BBMV) at 1 μg Cry protein/1 μg BBMV. In target insects, Cry proteins bind avidly to receptors on insect BBMV at much lower concentrations (0.00001–0.001 μg Cry protein/μg BBMV) and surface plasmon resonance experiments indicate that binding to the cadherin receptor occurs at nM concentrations (Hofmann et al. 1988a; Sacchi et al. 1986; Soberón et al. 2010). This is supported by the observation that the LD50 dose of Cry proteins is in the low ng/larval insect range (Federici and Siegel 2008). Low affinity binding of Cry proteins to rat BBMV has been reported but was considered to be non-specific as it was not displaceable by non-iodinated Cry protein, whereas Cry protein binding to target insect BBMV was readily displaced (Hofmann et al. 1988b). In bovine epithelial cells (Shimada et al. 2006), low level binding of Cry1Ab protein to the cytoskeletal protein actin was detected. However, no binding to extracellular proteins such as aminopeptidase N, cadherins and alkaline phosphatase was detected on bovine epithelial cells; these aforementioned proteins have been identified as receptors for Cry protein binding on target insect mid-gastrointestinal tract epithelia. The absence of high affinity binding of Cry proteins on mammalian gastrointestinal tract epithelial cells may be due in part to the absence of a glycosylating enzyme, BL2, in mammalian gastrointestinal tract cells. This enzyme, which is present in target insect gastrointestinal tract cells, produces the specific sugar residues that facilitates recognition and binding by Cry proteins to the aforementioned aminopeptidase N and alkaline phosphatase receptors (Federici and Siegel 2008; Soberón et al. 2010). The absence of specific Cry binding receptors on the mammalian gastrointestinal tract epithelial cells can explain, in part, the absence of cellular toxicity when high, non-physiological concentrations of Cry1Ab protein were incubated with sheep rumen epithelial cells, whereas incubation with a positive control toxin, valinomycin, caused apoptosis and reduced cell viability (Bondzio et al. 2008).

Furthermore, the *in vitro* concentration of Cry protein used by Vazquez-Padron (2000) was many orders of magnitude higher than the potential dietary exposures the mammalian digestive tract might encounter from consumption of food derived from *Bt* maize (Hammond and Cockburn 2008). As will be discussed later in section 16.6, processing (*e.g.* cooking, etc.) of *Bt* maize into human food has been reported to denature and inactivate Cry proteins further reducing residual functionally active Cry protein residues in food (Hammond and Jez 2011). Any residual (~ppb) levels of functionally active Cry protein that survived food processing would also be expected to be digested in the gastrointestinal tract (see Sect. 16.4).

16.4 Potential Digestibility of Cry Proteins

The normal fate of most ingested dietary proteins is hydrolytic digestion and/or degradation to either individual amino acids or small peptides that are subsequently absorbed to provide amino acids for protein synthesis in the body (Delaney et al. 2008). A validated assay to assess the potential digestibility of proteins has been developed *in vitro* using a fixed ratio of pepsin to protein and at pH 1.2 and 2.0 that is designed to simulate conditions in the stomach (Thomas et al. 2004). A similar *in vitro* test to simulate intestinal digestion using pancreatin has also been developed. Cry proteins are readily degraded by pepsin when tested *in vitro* using the aforementioned pepsin assay (Okunuki et al. 2001; Herman et al. 2003; EPA 2001; Thomas et al. 2004; Cao et al. 2010; Guimaraes, et al. 2010). An alternative *in vitro* digestive model has been recently proposed using higher pH and a lower pepsin/Cry protein ratio where Cry1Ab protein is more slowly degraded (Guimaraes et al. 2010). Others have suggested that differences in pH and pepsin concentration using *in vitro* digestibility assays had only small effects on digestion of proteins of intermediate stability to pepsin and no effects on proteins that were either stable or resistant to pepsin digestion (Ofori-Anti et al. 2008)

In pigs and calves, Cry protein fragments were observed in the GI tract, but none were detected in the liver, spleen and lymph nodes (Chowdhury et al. 2003a, b) indicating they were too large to be systemically absorbed intact from the gastrointestinal tract. Farm animals are generally fed much higher levels of maize in the diet than humans, and the maize is generally not processed resulting in higher dietary exposure to Cry proteins. Human dietary exposure, in contrast, would be much lower due to the lower consumption of maize and the fact that human food derived from maize is processed. Maize is subjected to a variety of processing conditions such as cooking that denatures Cry proteins causing them to lose insecticidal activity (Sect. 16.6). Denaturation also makes proteins more susceptible to degradation by proteases (Herman et al. 2006) including Cry1Ab protein (Okunuki et al. 2001) so that potential dietary exposure to functionally active Cry proteins in food derived from maize is very low.

16.5 Toxicology Testing of Cry Proteins

Genes coding for Cry proteins that were similar to those derived from *Bt* microorganisms have been introduced into a variety of different agricultural crops. These Cry proteins are expressed at low (ppm) levels in plants since those levels are sufficient to control targeted insect pests (Hammond and Cockburn 2008). As described earlier, only Cry and VIP proteins are currently used in registered *Bt* crops, making the insecticidal complexity of the crop much simpler than that of the components in *Bt* microbial pesticide formulations. As a group, the Cry protein family contains considerable diversity, enabling *Bt* strains to kill different kinds of larval insect pests. The currently commercialized *Bt* crops used for food and feed are mainly *Bt* maize and cotton. Cry proteins produced by *Bt* plants registered in the US and other countries include Cry1Ab, Cry1Ac, Cry1F, Cry1A.105, Cry1Ac, Cry2Ab2, Cry3Bb and Cry34Ab1, and Cry35Ab1. VIP proteins have also been registered in the US for use in agricultural crops (EPA 2004). New *Bt* soybean varieties (US) that have the *cry1Ac* gene and *Bt* rice varieties (China) with the *cry1C* gene are currently going through regulatory review. The food safety of introduced Cry1Ac and Cry1C proteins has already been demonstrated (Betz et al. 2000; Cao et al. 2010).

16.5.1 Acute Toxicity Testing

As shown in Table 16.1, mice are not adversely affected even when fed acute, high dosages of Cry proteins that are thousands to millions of times higher than doses acutely toxic to target insect pests (Hammond and Cockburn 2008). The mouse is a relevant model for such testing because it is known to be susceptible to the toxicity of known mammalian protein toxins (Delaney et al. 2008). In such testing the Cry1Ab protein was administered to mice at a dose level of 4000 mg/kg/day and produced no adverse effects. An adult human would have to consume approximately 900,000 kg of uncooked *Bt* maize grain in 1 day to attain a similar acute dose of Cry1Ab protein administered to mice (Hammond and Cockburn 2008).

Numerous reviews summarizing results of animal toxicology studies with *Bt* microbial pesticides and individual Cry proteins and the long history of safe use of *Bt* microbial products and *Bt* crops support the safety of Cry proteins (Fisher and Rosner 1959; Siegel and Shadduck 1989; McClintock et al. 1995; WHO/ICPS 1999; Betz et al. 2000; Siegel 2001; OECD 2007; Federici and Siegel 2008). In contrast to certain chemical insecticides, no significant human illnesses have been attributed to the use of *Bt* microbial pesticides in agriculture (WHO/ICPS 1999; Federici and Siegel 2008). This correlates well with the results in human safety testing conducted on volunteers in the early days of safety assessment of *Bt* microbials. These individuals were fed 10^{10} *Bt* spores for 5 days or inhaled 10^{9} *Bt* spores with no reported adverse effects (Siegel and Shadduck 1989).

Table 16.1 Acute toxicity studies in mice with Cry proteins

Cry protein	NOAEL (mg/kg)	Reference
Cry1Ab	4000	Betz et al. (2000)
Cry1Ab/Cry1Ac fusion protein	5000	Xu et al. (2010)
Cry1A.105	2072	EPA (2008a)
Cry 1Ac	4200	Betz et al. (2000)
Cry1C	5000	Cao et al. (2010)
Cry2Aa	4011	Betz et al. (2000)
Cry2Ab	1450	Betz et al. (2000)
Cry2Ab2	2198	EPA (2008b)
Cry3A	5220	Betz et al. (2000)
Cry3Bb	3780	Betz et al. (2000)
Cry1F	576	EPA (2001)
Cry34Ab1	2700	Juberg et al. (2009)
Cry35Ab1	1850	Juberg et al. (2009)
VIP3A	3675	EPA (2004)

16.5.2 Allergenicity and Immunogenicity Assessment

According to the aforementioned Codex guidelines (2009), the assessment of potential allergenicity of introduced proteins is carried out by comparing the biochemical characteristics of the introduced protein to characteristics of known allergens. A protein is not likely to be allergenic if: (1) the protein is from a non-allergenic source; (2) the protein represents only a very small portion of the total plant protein; (3) the protein does not share structural similarities to known allergens based on amino acid sequence homology comparisons to known allergens using bioinformatics search tools, and (4) the protein has the potential to be digested as confirmed when incubated *in vitro* with simulated digestion fluids. All Cry proteins expressed in *Bt* crops have been assessed for potential allergenicity according to the recommendations of the aforementioned Codex guidelines. Those that are used in commercial *Bt* crops do not fit the profile of known protein allergens, *ie.* they are digested in simulated gastric fluid, are generally present at low (ppm) levels in grain and much lower levels in food, and are not structurally related to known allergens based on bioinformatics searches. Following commercial production and use of thousands of tons of *Bt* microbial formulations over the last few decades, there is no evidence of allergic reactions in workers who manufacture or apply *Bt* microbials to agricultural crops and forests (Siegel 2001; Federici and Siegel 2008). Similarly, for the millions of tons of *Bt* crops produced since the 1990s, there have been no reports of allergenic reactions in those that handle or consume the grain/seed. There has been a report of immunologic responses in workers who apply *Bt* microbial formulations to agricultural crops, but this reaction is classically observed when humans are exposed to "foreign" or non-human proteins. However, the immunologic responses were attributed to other bacterial proteins present in the *Bt* microbial formulation, and not the Cry proteins (Siegel 2001; Federici and Siegel 2008).

Cry1Ac protein that is present in *Bt* microbial formulations has been shown to be immunogenic in mice following intraperitoneal (IP), intragastric (IG), intranasal (IN) or intra-rectal (IR) administration (Moreno-Fierros et al. 2000; Vazquez-Padron et al. 2000). Systemic and mucosal immune responses with production of specific IgG, IgM and IgA antibodies were reported. In a separate study, administration of GM rice containing Cry1Ab protein (88% amino acid sequence homology to Cry1Ac protein), or spiked with Cry1Ab protein, was also reported to have induced an immune response in rats (Kroghsbo et al. 2008). Another study reported that feeding mice either 1 month or 18 months of age with MON810 maize resulted in some alterations in the intestinal and peripheral immune cell populations (Finamore et al. 2008).

The biological relevance of these studies to assessing potential health risks from human consumption of foods derived from *Bt* crops can be questioned on several fronts. Administration of Cry proteins by IN, IP and IR routes of exposure do not necessarily predict risks from IG or dietary intake because in some cases, the researchers bypassed the protective barrier of the gastrointestinal tract by injecting or administering proteins by other routes. When they did employ the IG route in mice, they gave doses of Cry proteins far in excess of potential human intakes and included Maalox® to neutralize the pH of the gastrointestinal tract, thereby compromising normal physiological conditions for digestion of Cry1Ac protein by pepsin. Mice were often dosed with 100 µg Cry1Ac protein which exceeds potential human intake by approximately 5000-fold.[1] This discrepancy may be greater still, since maize is normally processed into human foods and not consumed raw. Processing (e.g., cooking) denatures Cry1Ab protein (see Sect. 16.6), so the actual dietary intakes of intact Cry1Ab protein from consumption of processed foods is anticipated to be far less. It has been estimated that grain processing could reduce levels of functionally active protein by approximately two orders of magnitude (Hammond and Jez 2011). Based on the calculations in footnote 1, this would result in an actual human intact Cry1Ab intake of approximately 0.008 µg/kg body weight; a level 500,000-fold lower than the levels used on the study demonstrating immunogenic effects. In addition to human dietary irrelevance of the doses tested in mice, an attempt to reproduce this work in mice given Cry1Ab protein, failed to detect anti-Cry IgG antibodies at the 100 µg/mouse dose (these authors did not include Maalox along with the oral dosed Cry1Ab protein) (Adel-Patient et al. 2010). These authors attributed the discrepancy between their study and earlier studies (Moreno-Fierros et al. 2000; Vazquez-Padron et al. 2000) to the possible presence of *E. coli* endotoxin in the Cry1Ac preparations used on the earlier studies. The previous authors used engineered *E. coli* to produce Cry1Ac but did not apparently check the preparations for endotoxin contamination. Adel-Patient et al. (2010) reported that

[1] 100 µg Cry1Ac/25 gm mouse ~4000 µg/kg body weight; human intake of Cry1Ab protein from consumption of MON 810 maize (YIELDGARD Corn Borer®) was estimated to be 0.008 µg/kg body weight (Hammond and Jez 2011).

® Maalox—registered trademark of Novartis.

® YIELDGARD Corn Borer—registered trademark of Monsanto Technology, LLC.

IG administration of purified Cry1Ab protein had no impact on immune response in mice and confirmed the earlier reports of the immunogenicity of Cry1Ac administered by IP injection to mice, without any evidence of allergenicity. The other study (Finamore et al. 2008) reporting possible alterations of intestinal and peripheral immune cell populations in young and old mice fed large amounts of MON 810 maize in the diet can be questioned since the alterations were often of small magnitude, and sometimes in opposite direction when measured at different study intervals. It is difficult to interpret the biological relevance of the changes that were observed since no historical information was provided on normal variation of the measured parameters in control mice. A much bigger question is the relevance of findings in mice to predicting possible immune effects in humans.

While the mouse model has some similarities to human immunological mechanisms (Adel-Patient et al. 2010), in general, animal models have not been considered to have been sufficiently validated to be able to accurately predict potential allergenic or immunologic effects in humans from dietary exposure to proteins (Goodman et al. 2008; Thomas et al. 2009; Codex 2009). As a practical matter, there has been widespread dietary exposure to Cry proteins from application of *Bt* microbial formulations applied to vegetables and other crops for many decades (see Sect. 16.7) and there have been no reports that the immune system of humans is at risk from dietary exposures. Given the comparatively low dietary exposures to Cry proteins from consumption of foods derived from *Bt* crops (or for that matter, application of *Bt* microbials to food crops) the potential to induce an immune response in humans was considered to be unlikely (Guimaraes et al. 2010).

16.6 Assessment of Food Processing on Cry Protein Biological Activity

The functional activity of proteins including Cry proteins is dependent on their three-dimensional structure and the combination of various environmental forces (electrostatic forces, van der Waals interactions, hydrogen bonds, and hydrophobic interactions) that help to maintain that structure in the cell (Branden and Tooze 1991; Creighton 1993). In general, protein structures are only marginally stable under a limited range of physiological conditions and are easily disrupted by changes such as increases in temperature, variation of pH, or physical disruption that overcome the forces keeping them folded properly (Creighton 1993). These changes typically result in denaturation of proteins which leads to a drastic change in protein structure. Typically, denaturation does not involve changes in the primary structure (amino acid sequence) of a protein (i.e., degradation of the polypeptide chain), but disrupts the secondary, tertiary, and quaternary (if applicable) structure of denatured proteins. Consequently, there is a complete loss of biological function, as the denatured polypeptide is more like a random coil than a folded protein.

Changes such as increased temperature, altered pH, physical disruption occur routinely during processing of the seed/grain into human food. For example, during the processing of maize and soybeans into food fractions, heating, high pressure extrusion, mechanical shearing, changes in pH, and the use of reducing agents are all employed and will unfold a native protein structure and/or alter the primary structure of a protein by hydrolysis of peptide bonds (Kilara and Sharkasi 1986; Meade et al. 2005). In typical processing of maize and soybeans into food fractions, temperatures of 95–100°C are commonly encountered (Berk 1992; Duensing et al. 2003; Rooney and Serna-Saldivar 2003). These elevated temperatures can lead to irreversible denaturation and loss of protein function (de Luis et al. 2009; Thomas et al. 2007), although this does not alter the nutritional value of the denatured protein as a source of dietary amino acids. Cooking proteins aids their digestion in the gastrointestinal tract as proteases (e.g., pepsin and trypsin), are able to cleave the random coil of a denatured protein more quickly and efficiently compared with the same protein in its native three dimensional conformation (Herman et al. 2006).

Precipitation is often used to remove proteins from other food fractions like lipids, as changes in the physical properties of a protein can reduce its solubility leading to aggregation or precipitation. Multimeric proteins can also dissociate into monomers resulting in loss of function (Schultz and Liebman 2002; Meade et al. 2005). It may still be possible to detect epitopes on denatured, aggregated or precipitated proteins using immunologic detection methods. This has been observed for some introduced proteins but not others (Terry et al. 2002; Grothaus et al. 2006; Margarit et al. 2006; Thomas et al. 2007; de Luis et al. 2009; Codex 2010). In general, immunologic methods for detecting introduced proteins is not often done because of the effects of denaturation of proteins in processed food; immunologic detection methods are mostly reserved for testing raw agricultural commodities (Margarit et al. 2006; Bogani et al. 2008; de Luis et al. 2009). Even if you can still detect proteins through antibody recognition of sequence-specific epitopes, the introduced proteins most likely have lost their functional activity as discussed below.

Cry proteins have been subjected to *in vitro* heat stability studies to determine if they maintain their insecticidal activity after cooking. Cry1Ab, Cry1F, Cry34Ab1 and Cry35Ab1 proteins were cooked to temperatures ranging from 60–90°C for periods ranging from 10–30 min (EFSA 2005a, 2007; de Luis et al. 2009). *Bt* soybean seeds containing Cry1Ac protein were also subjected to cooking that approximated temperatures used during processing of soybeans into meal (EPA 2010). All Cry proteins lost insecticidal activity after cooking when tested in insect bioassays. These Cry proteins were subjected to similar cooking conditions used when maize grain and soybeans are processed into human foods. The impact of food processing on the functional activity of introduced proteins may be relevant for other processed crops such as rice that are cooked before consumption. It is likely that there is minimal, if any dietary exposure to functionally active Cry proteins when food products derived from processed food crops are consumed.

Overall, the impact of harsh processing conditions on protein structure and function support the EFSA conclusion that risk assessors should consider the impact of food processing on the levels of the introduced protein, otherwise they may overestimate potential dietary exposure "…food products are often processed into ingredients and/or incorporated in formulated processed food products, where the new protein and/or the novel secondary gene product attrition will occur. This may result in significant reduction in the theoretical maximum daily intake (TMDI) of the novel gene product, resulting in over-estimated exposure levels and even larger margins of safety for man" (EFSA 2008).

16.7 Human Dietary Exposure Assessment for Cry Proteins in *Bt* Crops

There is a history of safe consumption of Cry proteins. As mentioned earlier, *Bt* microbial pesticides have also been added to drinking water in outdoor storage facilities in various locations around the world to control insect (mosquito) vectors of disease (WHO/ICPS 1999). In locations where water has to be stored in containers for use in drinking *Bt* microbials have been added to the drinking water to control mosquito larvae etc (WHO/IPCS 1999; Bravo et al. 2007). Because of their safety profile and long history of safe use in agriculture, *Bt* microbials have been exempted from the requirement of a setting a tolerance in countries which they have been registered (OECD 2007).

They have also been applied topically to vegetables in organic agriculture to control insect pests (WHO/IPCS 1999). Residual levels of up to 10^4 CFUs (colony forming units, i.e., viable *Bt* microbes)/gram of plant tissue have been found on fresh vegetables marketed in Europe following application of commercial *Bt* microbial formulations (Frederiksen et al. 2006). In a recent review on the safety of *Bt*, it stated: "In many regions of the world where fresh vegetable crops are marketed within a few days of harvest, these have been recently sprayed with *Bt*. This is especially true of vegetables grown using organic methods. It is quite common for vegetables treated with *Bt*, such as broccoli, tomatoes, cucumbers, cauliflower and lettuce to be eaten raw with only minimal washing. In these cases, humans are directly consuming thousands of *Bt* spores and insecticidal crystals" (Federici and Siegel 2008).

A "back-of-the-envelope" estimate of potential human dietary exposure to Cry proteins from application of commercial *Bt* microbial pesticide formulations was undertaken. It was assumed that the *Bt* microbial formulation was applied shortly before harvest and that the broccoli was consumed raw (e.g. salads) and was not cooked. If the *Bt* microbial formulation was applied weeks prior to harvest, or the broccoli was cooked, the potential dietary intake of functionally active Cry proteins would be much lower. According to WHO/GEMs, the highest acute dietary intake of broccoli in the countries that responded to the WHO survey was in the United States. The WHO/GEMs US adult acute dietary intake (97th percentile) of broccoli is 5.8 gm/kg body weight. In a recent review, it was reported that chronic intake of broccoli in the US was 8.2 g/capita/day (Latte et al. 2011). For purposes of this assessment, it was assumed that the broccoli was consumed raw and not cooked (eg. in salads). The commercial application rate for the *Bt* microbial formulation to control insect pests was based on the label directions from the supplier which could range up to 32 oz/acre. The *Bt* microbial formulation contained ~10% Cry protein (w/w) and was applied once. For purposes of the exposure calculation, it was assumed that 10% of the applied *Bt* microbial formulation was deposited on broccoli heads. Based on these assumptions, the acute adult intake of Cry protein from consumption of uncooked broccoli (heads) was estimated to range between 50–1 µg/kg body weight for acute and chronic consumption respectively. This estimated exposure is considerably higher than the previous estimate (~0.008 µg/kg body weight) of dietary intake of functionally active Cry protein from food derived from *Bt* maize (page 9). For other kinds of Bt maize, the levels of Cry proteins in grain were higher than 0.3 ppm ranging from approximately 15–115 ppm (Hammond and Cockburn 2008). Using the same assumptions (Hammond and Jez 2011), the potential human dietary intake of functionally active Cry protein for these Bt maize varieties could range up to approximately 2 ug/kg body weight/day. It seems likely that dietary exposure to functionally active Cry proteins from application of Bt microbial formulations to vegetables (shortly before harvest) could be similar to or even higher than dietary exposure from consumption of foods derived from Bt crops.

Regulatory agencies have confirmed the history of safe consumption of Cry protein residues on crops and in drinking water "The use patterns for *B. thuringiensis* may result in dietary exposure with possible residues of the bacterial spores on raw agricultural commodities. However, in the absence of any toxicological concerns, risk from the consumption of treated commodities is not expected for both the general population and infants and children" (EPA 1998) and "*Bt* has not been reported to cause adverse effects on human health when present in drinking-water or food" (WHO/IPCS 1999).

Although Cry proteins have been considered to be safe as mentioned above, even from organic agricultural use, questions still arise challenging their safety when incorporated in biotechnology-derived crops. For example, a recent paper gained attention by reporting the detection of Cry1Ab protein in the serum of non-pregnant women, pregnant women, and the cord blood of their fetuses at mean levels of detection of 0.19, 0.13 and 0.04 ng/ml respectively (Aris and Leblanc 2011). The authors used a commercially available ELISA immunoassay kit which has been validated for use in detecting Cry1Ab protein in grain/seed samples but not human serum. Considering the vastly different composition of these matrices, it cannot be overlooked that the authors did not report that they had validated the assay for use in human serum. Furthermore, the majority of the serum "detects" in the Aris and Leblanc paper were at, or below, the limit of detection (LOD) for this commercial kit for use in grain and seed. This raises serious questions about the accuracy of their data especially in light of previous reports on the topic including, (1) a commercial immunoassay kit that was not validated to detect Cry1Ab protein in porcine blood produced invalid results (Chowdhury et al. 2003a), and (2) a validated immunoassay to quantify Cry1Ab protein in plasma (LOD 1 ng/ml) was unable to detect Cry1Ab protein in any of the plasma samples collected from cows fed MON 810 maize at 70% w/w (dry matter) in the diet for 1 or 2 months (Paul et al. 2008). The amount of MON 810 (YIELDGARD Corn Borer®) maize consumed by dairy cows greatly exceeds potential human intake. In consideration of the much lower

Table 16.2 Examples of subchronic toxicity studies with *Bt* crops

Bt crop	% in diet[a]	References
Bt tomato	10	Noteborn et al. (1995)
Bt/HT[b] maize (ECB[c]/RR[d])	11/33	EFSA (2005a)
Bt/HT maize (CRW[e]/RR)	11/33	EFSA (2005b)
Bt/HT maize (ECB/CRW/RR)	11/33	EFSA (2005c)
Bt maize(ECB/CRW)	11/33	EFSA (2005d)
Bt maize (ECB)	11/33	Hammond et al. (2006a)
Bt maize (CRW)	11/33	Hammond et al. (2006b)
BT maize (ECB)	11/13	MacKenzie et al. (2007)
Bt cotton	10	Dryzga et al. (2007)
Bt rice	60	Schroder et al. (2007)
Bt/HT maize (CRW/Gluf[f])	35	Malley et al. (2007)
Bt/HT maize (CRW/RR)	11/33	Healy et al. (2008)
Bt maize (CRW)	50/70	He et al. (2008)
Bt/HT maize (ECB/CRW)	34	Appenzeller et al. (2009)

[a] percent (w/w) maize, rice or cottonseed meal added to the diet
[b] HT—herbicide tolerant
[c] ECB—European maize borer
[d] RR—Roundup Ready® (tolerant to ROUNDUP herbicide)
[e] CRW—maize rootworm
[f] Gluf—glufosinate (tolerant to glufosinate herbicide)
® registered trademark of Monsanto Technology, LLC

potential human dietary intake of MON 810 maize compared to farm animals fed high levels in the diet, and in consideration of the digestibility of Cry1Ab protein when exposed to pepsin, the reported levels of intact Cry1Ab detected in human serum must be questioned.

16.8 Toxicology Feeding Studies in Rodents Fed *Bt* Crops

In the United States and Canada, there has been no requirement to routinely feed *Bt* crops to animals to confirm their safety for human and farm animal consumption. The composition and agronomic performance of *Bt* crops has been shown in many field trials to be similar to conventional non-biotech comparators. As a consequence, following review of submitted dossiers summarizing all relevant data by registrants, the USFDA and EPA have, to date, not considered additional animal toxicology studies as necessary to confirm safety. However, 90 day rodent subchronic feeding studies were often required by the EU to confirm the safety of the first generation of *Bt* crops that were registered for import or production in Europe. Some of these studies are shown in Table 16.2.

In 2008, EFSA published a review of its safety assessment of biotech crops over the last several years, and concluded that, where the safety of the introduced protein had been confirmed and no other unintended changes had been detected, the conduct of animal feeding studies added little to the overall safety assessment. They concluded that the majority of studies they had reviewed confirmed the food safety of biotech crops. Ninety-day rat toxicology studies with Bt biotech crops are listed in Table 16.2. EFSA also acknowledged that some published toxicology studies reported adverse effects when biotech crops were fed to animals, but EFSA concluded that because of deficiencies in these studies, the results were not interpretable (EFSA 2008).

No evidence of any treatment related adverse effects were observed in 90 day rat toxicology studies carried out with *Bt* crops whether the crops contained one or more Cry proteins. The maize grain that was added to rat diets was simply ground into meal and not further processed.

In a subchronic gastrointestinal impairment model in rats, chemically-induced gastrointestinal impairment (treated with famotidine to reduce gastric acid secretion and indomethacin to cause damage to the intestinal epithelium) was induced in rats fed Cry1Ab protein in the diet (10 ppm) for 2 weeks (Onose et al. 2008). Controls were treated in the same manner except they were not fed Cry1Ab protein. It was expected that Cry1Ab protein would not have been digested due to reduction of gastric acid production allowing intact Cry1Ab protein to enter systemic circulation via the intestinal tract damaged by indomethacin pretreatment. As expected, there was no evidence of meaningful toxicological effects (changes in clinical blood parameters and histologic appearance of organs) reported in Cry1Ab dosed animals.

Additionally, there have also been a number of feeding studies in laboratory and farm animals such as poultry, swine and ruminants fed both either *Bt* microbial pesticides and/or *Bt* crops (Betz et al. 2000; Flachowsky et al. 2005a, b, 2007; Federici and Siegel 2008; WHO/IPCS 1999; McClintock et al. 1995; OECD 2007; Siegel 2001; Brake et al. 2003; EFSA 2008; Scheideler et al. 2008; McNaughton et al. 2007; Taylor et al. 2005, 2007). No evidence of adverse effects were reported in these studies as the animals responded similarly in growth and feed consumption to controls fed non-biotech derived crops.

16.9 Food and Feed Safety Benefits of *Bt* Crops

16.9.1 Reduced Insecticide Application

The adoption of *Bt* crops in various world areas has contributed to a significant reduction in chemical insecticide applications. In Burkino Faso in West Africa, the planting of *Bt* cotton has enabled farmers to reduce the number of chemical insecticide sprays during a growing season from 6–2 applications (James 2010). India, which is now the world's biggest producer of *Bt* cotton with an estimated 23.2 million acres planted in 2010, reported pesticide use has been cut at least in half. In the most comprehensive survey conducted to date (2002–2008), Indian farmers reported *Bt* cotton use prevented at least 2.4 million cases of pesticide poisoning saving $ 14 million (US dollar equivalent) in annual health costs (Kouser and Qaim 2011).

16.9.2 Reduced Mycotoxin Contamination of Grain

In addition to protecting crops from insect feeding damage, grain from *Bt* maize was found to have lower levels of the mycotoxin fumonisin based on field trials in various countries (Ostry et al. 2010; Folcher et al. 2010). Various mycotoxin producing fungi such as fusarium species exist in the environment wherever maize is grown. They can enter the maize plant through insect damaged tissue allowing the fungi to enter plant tissue (e.g., stalks, ears), producing fumonisins which are common mycotoxin contaminants of maize wherever it is grown (Miller 2001). Since Cry proteins can reduce feeding damage by controlling insect pests, they can lower the potential for fungal colonization and therefore mycotoxin contamination. Dietary exposure to fumonisin can cause a variety of adverse health effects in farm animals and possibly humans (Li et al. 2001; CAST 2003; Wang et al. 2003; Marasas et al. 2004; Wu et al. 2004), and lower levels of contamination associated with *Bt* maize should be considered a beneficial aspect of this technology.

16.10 Conclusions

There is a 50-year history of safe use and consumption of foods sprayed with commercial *Bt* microbial pesticide products and 14 year history of safe consumption of food and feed derived from *Bt* crops. Many toxicology studies conducted with *Bt* microbial pesticides and *Bt* crops have confirmed their safety for consumption. There is a history of safe consumption of Cry proteins from use of *Bt* microbial pesticides on vegetable food crops, and dietary exposures may be comparable or higher than that from consumption of foods derived from *Bt* crops. In either case, dietary intake of intact Cry proteins is very low. Use of *Bt* crops has secondary benefits by significantly reducing insecticide use, lowering insecticide exposure to applicators and improving the food security of maize grain by reducing contamination by fumonisin mycotoxins.

References

Adel-Patient K, Guimaraes VD, Paris A, Drumare MF, Ah-Leung S, Lamourette P, Nevers MC, Canlet C, Molina J, Bernard H, Créminon C, Wal JM (2010) Immunological and metabolomic impacts of administration of Cry1Ab protein and MON 810 maize in mouse. PLoS One 6(1):e16346. doi:10.1371/journal.pone.0016346

Appenzeller LM, Malley LA, MacKenzie SA, Everds N, Hoban D, Delaney B (2009) Subchronic feeding study with genetically modified stacked trait lepidopteran and coleopteran resistant (DAS Ø15Ø7 1 x DAS-59122-7) maize grain in Sprague-Dawley rats. Food Chem Toxicol 47:1512–1520
Aris A, Leblanc S (2011) Maternal and fetal exposure to pesticides associated to genetically modified foods in Eastern Townships of Quebec, Canada. Reprod Toxicol 31:528–533
Berk Z (1992) Technology of production of edible flours and protein products from soybeans. FAO Agricultural Services Bulletin No. 97
Betz FS, Hammond BG, Fuchs RL (2000) Safety and advantages of *Bacillus thuringiensis*-protected plants to control insect pests. Regul Toxicol Pharmacol 32:156–173
Bogani P, Minunni M, Spiriti M, Zavaglia M, Tombelli S (2008) Transgenes monitoring in an industrial soybean processing chain by DNA-based conventional approaches and biosensors. Food Chem 113:658–664
Bondzio A, Stumpff F, Schön J, Martens H, Einspanier R (2008) Impact of *Bacillus thuringiensis* toxin Cry1Ab on rumen epithelial cells (REC)—A new in vitro model for safety assessment of recombinant food compounds. Food Chem Toxicol 46:1976–1984
Branden C, Tooze J (2001) Introduction to protein structure. Garland Publishing, New York
Brake J, Faust MA, Stein J (2003) Evaluation of transgenic event Bt 11 hybrid maize in broiler chickens. Poult Sci 82:551–559
Bravo A, Gill SS, Soberón M (2007) Mode of action of *Bacillus thuringiensis* Cry and Cyt toxins and their potential for insect control. Toxicon 49(4):423–435
Cao S, He X, Xu W, Ran W, Liang L, Luo Y, Yuan Y, Zhang N, Zhou X, Huang K (2010) Safety assessment of Cry1C protein from genetically modified rice according to the national standards of PR China for a new food source. Regul Toxicol Pharmacol 58(3):474–481
CAST (2003) Mycotoxins. Risks in plant, animal, and human systems. Council for Agricultural Science and Technology. Task Force Report No. 139
Chowdhury EH, Kuribara H, Hino A, Sultana P, Mikami O, Shimada N, Guruge NS, Saito M, Nakajima Y (2003a) Detection of maize intrinsic and recombinant DNA fragments and Cry1Ab protein in the gastrointestinal contents of pigs fed genetically modified maize Bt11. J Anim Sci 81:2546–2551
Chowdhury EH, Shimada N, Murata H, Mikami O, Sultana P, Miyazaki S, Yoshioka M, Yamanaka N, Hirai N, Nakajima Y (2003b) Detection of Cry1Ab protein in the gastrointestinal contents but not in the visceral organs of genetically modified Bt11-fed calves. Vet Hum Toxicol 45:72–75
Codex (2009) WHO/FAO. Foods derived from modern biotechnology, 2nd edn. Rome
Codex (2010) Committee on Sampling and Detection Methods (CCMAS). Proposed draft guidelines on performance criteria and validation of methods for detection, identification, and quantification of specific DNA sequences and specific proteins in food (At step 5/8 of the procedure). ALINORM 10/33/23. Appendix III, pp 47–68
Creighton TE (1993) Proteins: structures and molecular properties. W.H. Freeman and Company, New York
Delaney B, Astwood JD, Cunny H, Eichen Coon R, Herouet-Guicheney C, MacIntosh S, Meyer LS, Privalle L, Gao Y, Mattsson J, Levine M (2008) Evaluation of protein safety in the context of agricultural biotechnology. Food Chem Toxicol 46:S71–S97
de Luis R, Lavilla M, Sánchez L, Calvo MD, Pérez M (2009) Immunochemical detection of Cry1A(b) protein in model processed foods made with transgenic maize. Eur Food Res Technol 229:15–19
Dryzga MD, Yano BL, Andrus AK, Mattsson JL (2007) Evaluation of the safety and nutritional equivalence of a genetically modified cottonseed meal in a 90 day dietary toxicity study in rats. Food Chem Toxicol 45:1994–2004
Duensing WJ, Roskens AB, Alexander RJ (2003) Chapter 11. Maize dry milling: processes, products, and applications. In: White P, Johnson L (eds) Maize chemistry and technology, 2nd edn. AAOCS, St Paul

EFSA (2005a) Opinion of the Scientific Panel on Genetically Modified Organisms on an application (Reference EFSA GMO UK 2004 01) for the placing on the market of glyphosate tolerant and insect-resistant genetically modified maize NK603 × MON810, for food and feed uses, and import and processing under Regulation (EC) No 1829/2003 from Monsanto (Question No EFSA Q-2004-086). EFSA J 309:1–22

EFSA (2005b) Opinion of the Scientific Panel on Genetically Modified Organisms on an application (Reference EFSA GMO UK 2004 06) for the placing on the market of insect-protected glyphosate tolerant genetically modified maize MON863 × NK603, for food and feed uses, and import and processing under Regulation (EC) No 1829/2003 from Monsanto (Question No EFSA Q 2004 154). EFSA J 255:1–21

EFSA (2005c) Opinion of the Scientific Panel on Genetically Modified Organisms on an application (Reference EFSA GMO BE 2004 07) for the placing on the market of insect protected glyphosate tolerant genetically modified maize MON863 × MON810 × NK603, for food and feed uses, and import and processing under Regulation (EC) No 1829/2003 from Monsanto (Question No EFSA Q 2004 159). EFSA J 256:1–25

EFSA (2005d) Opinion of the Scientific Panel on Genetically Modified Organisms on a request from the Commission related to the Notification (Reference C/DE/02/9) for the placing on the market of insect-protected genetically modified maize MON 863 x MON 810, for import and processing, under Part C of Directive 2001/18/EC from Monsanto. EFSA J 251:1–22

EFSA (2007) Opinion of the Scientific Panel on Genetically Modified Organisms on an application (Reference EFSA-GMO-NL-2005-12) for the placing on the market of insect-resistant genetically modified maize 59122, for food and feed uses, import and processing under Regulation (EC) No 1829/2003, from Pioneer Hi-Bred International, Inc. and Mycogen Seeds, c/o Dow Agrosciences LLC. EFSA J 470:1–25

EFSA (2008) Safety and nutritional assessment of GM plants and derived food and feed: the role of animal feeding trials. Food Chem Toxicol 46:S2–S70

EFSA (2011) Guidance for risk assessment of food and feed from genetically modified plants. EFSA J 9(5):2150

EPA (1998) EPA Registration Eligibility Decision (RED) *Bacillus thuringiensis* EPA 738-R-98-004

EPA (2001) *Bacillus thuringiensis* Cry1F protein and the genetic material necessary for its production in maize; exemption from the requirement for a tolerance. Fed Regist 66(109):30321

EPA (2004) *Bacillus thuringiensis* VIP3A insect control protein and the genetic material necessary for its production; notice of a filing to a pesticide petition to amend the exemption from the requirement for a tolerance for a certain pesticide chemical in the food. Fed Regist 69(178):55605

EPA (2008a) Exemption from the requirement of a tolerance for the Bacillus Thuringiensis Cry1A.105 protein in maize. 40 CFR 174.502

EPA (2008b) Exemption from the requirement of a tolerance for the Bacillus Thuringiensis Cry-2Ab2 protein in maize. 40 CFR 174.503

EPA (2010) Biopesticide Registration Action Document. *Bacillus thuringiensis* Cry1Ac Protein and the Genetic Material (Vector PV-GMIR9) Necessary for Its Production in MON 87701 (OECD Unique Identifier: MON 877Ø1-2) Soybean (PC Code 006532). U.S. Environmental Protection Agency Office of Pesticide Programs Biopesticides and Pollution Prevention Division. http://www.epa.gov/oppbppd1/biopesticides/pips/bt-cry1ac-protien.pdf. Accessed 8 Oct 2010

Federici B, Siegel J (2008) Safety assessment of *Bacillus thuringiensis* and Bt crops used in insect control. In: Hammond BG (ed) Food safety of proteins in agricultural biotechnology. CRC Press, Boca Raton

Finamore A, Roselli M, Britti S, Monastra G, Ambra R, Turrini A, Mengheri E (2008) Intestinal and peripheral immune response to MON810 maize ingestion in weaning and old mice. J Agric Food Chem 56:11533–11539

Fisher R, Rosner L (1959) Toxicology of the microbial insecticide, Thuricide. J Agric Food Chem 7:686–688

Flachowsky G, Chesson A, Aulrich K (2005a) Animal nutrition with feeds from genetically modified plants. Arch Anim Nutr 59:1–40

Flachowsky G, Halle I, Aulrich K (2005b) Long term feeding of Bt-maize—a ten generation study with quails. Arch Anim Nutr 59:449–451
Flachowsky G, Aulrich K, Bohme H, Halle I (2007) Studies on feeds from genetically modified plants (GMP)—Contributions to nutritional and safety assessment. Anim Feed Sci Technol 133:2–30
Folcher L, Delos M, Marengue E, Jarry M, Weissenberger A, Eychenne N, Regnault-Roger C (2010) Lower mycotoxin levels in Bt maize. Agron Sustain Dev 30:711–719
Frederiksen K, Rosenquist H, Jørgensen K, Wilcks A (2006) Occurrence of natural *Bacillus thuringiensis* contaminants and residues of *Bacillus thuringiensis*-based insecticides on fresh fruits and vegetables. Appl Environ Microbiol 72(5):3435–3440
Goodman R, Vieths S, Sampson HA, Hill D, Ebisawa M, Taylor SL, Ree R van (2008) Allergenicity assessment of genetically modified crops—what makes sense? Nat Biotechnol 26:73–81
Griffiths JS, Haslam SM, Yang T, Garczynski SF, Mulloy B, Morris H, Cremer PS, Dell A, Adang MJ, Aroian RV (2005) Glycolipids as receptors for *Bacillus thuringiensis* crystal toxin. Science 307:922–925
Grothaus GD, Bandla M, Currier T, Giroux R, Jenkins GR, Lipp M, Shan G, Stave JW, Pantella V (2006) Immunoassay as an analytical tool in agricultural biotechnology. AOAC Int 89(4):913–928
Guimaraes V, Drumare MF, Lereclus D, Gohar M, Lamourette P, Nevers MC, Vaisanen-Tunkelrott ML, Bernard H, Guillon B, Creminon C, Wal JM, Adel-Patient K (2010) In vitro digestion of Cry1Ab proteins and analysis of the impact on their immunoreactivity. J Agric Food Chem 58:3222–3231
Hammond B, Cockburn A (2008) The safety assessment of proteins introduced into crops developed through agricultural biotechnology: a consolidated approach to meet current and future needs. In: Hammond BG (ed) Food safety of proteins in agricultural biotechnology. CRC Press, New York
Hammond BG, Jez JM (2011) Impact of food processing on the safety assessment for proteins introduced into biotechnology-derived soybean and maize crops. Food Chem Toxicol 49(4):711–721
Hammond BG, Dudek R, Lemen JK, Nemeth MA (2006a) Results of a 90-day safety assurance study with rats fed grain from maize borer-protected maize. Food Chem Toxicol 44(7):1092–1099
Hammond B, Lemen J, Dudek R, Ward D, Jiang C, Nemeth M, Burns J (2006b) Results of a 90-day safety assurance study with rats fed grain from maize rootworm-protected maize. Food Chem Toxicol 44(2):147–160
He XY, Huang KL, Li X, Qin W, Delaney B, Luo YB (2008) Comparison of grain from maize rootworm resistant transgenic DAS 59122-7 maize with non transgenic maize grain in a 90 day feeding study in Sprague Dawley rats. Food Chem Toxicol 46:1994–2002
Healy C, Hammond B, Kirkpatrick J (2008) Results of a 90 day safety assurance study with rats fed grain from maize rootworm protected, glyphosate tolerant MON 88017 maize. Food Chem Toxicol 46:2517–2524
Herman RA, Schafer BW, Korjagin VA, Ernest AD (2003) Rapid digestion of Cry34Ab1 and Cry35Ab1 in simulated gastric fluid. J Agric Food Chem 51:6823–6827
Herman RA, Storer NP, Gao Y (2006) Acid-induced unfolding kinetics in simulated gastric digestion of proteins. Regul Toxicol Pharmacol 46:93–99
Hofmann C, Luthy P, Hutter R, Piska V (1988a) Binding of the delta endotoxin from *Bacillus thuringiensis* to brush-border membrane vesicles of the Cabbage Butterfly (Pieris brassicae). Eur J Biochem 173:85–91
Hofmann C, Vanderbruggen H, Hofte H, Van Rie J, Jansens S, Van Mellaert H (1988b) Specificity of B. thuringiensis delta-endotoxins is correlated with the presence of high affinity binding sites in the brush-border membrane of target insect midgastrointestinal tracts. Proc Natl Acad Sci USA 85:7844–7848
Huang DF, Zhang J, Song FP, Lang ZH (2007) Microbial control and biotechnology research on *Bacillus thuringiensis* in China. J Invertebr Pathol 95:175–180

James C (2010) Global Status of Commercialized Biotech GM Crops: 2010 International Service for the Acquisition of Agri-biotech Applications (ISAAA). ISAAA Briefs, brief 42

Juberg DR, Herman RA, Thomas J, Delaney B (2009) Acute and Repeated Dose (28 Day) Mouse Oral Toxicology Studies with Cry34Ab1 and Cry35Ab1 *Bt* Proteins Used in Coleopteran Resistant DAS-59122-1 Maize. Regul Toxicol Pharmacol 54:154–163

Kilara A, Sharkasi T (1986) Effects of temperature on food proteins and its implications on functional properties. Crit Rev Food Sci Nutr 23(4):323–395

Kroghsbo S, Madsen C, Poulsen M, Schroder M, Kvist PH, Taylor M, Gatehouse A, Shu Q, Knudsen L (2008) Immunotoxicological studies of genetically modified rice expressing PHA-E lectin or *Bt* toxin in Wistar rats. Toxicology 245:24–34

Kouser S, Qaim M (2011) Impact of Bt cotton on pesticide poisoning in smallholder agriculture: a panel data analysis. Ecol Econ 70(11):2105–2113

Lambert B, Buysse L, Decock C, Jansens S, Piens C, Saey B, Seurinck J, Van Audenhove K, Van Rie J, Van Vliet A, Peferoen M (1996) A *Bacillus thuringiensis* insecticidal crystal protein with a high activity against members of the family Noctuidae. Appl Environ Microbiol 62:80–86

Latte KP, Appel KE, Lampen A (2011) Health benefits and possible risks of broccoli. Food Chem Toxicol 49:3287–3309

Li FQ, Yoshizawa T, Kawamura O, Luo XY, Li YW (2001) Aflatoxins and fumonisins in maize from the high-incidence area for human hepatocellular carcinoma in Guangxi, China. J Agric Food Chem 49:4122–4126

MacKenzie SA, Lamb I, Schmidt J, Deege L, Morrisey MJ, Harper M, Layton RJ, Prochaska LM, Sanders C, Locke M, Mattsson JL, Fuentes A, Delaney B (2007) Thirteen week feeding study with transgenic maize grain containing event DASO15O7 1 in Sprague–Dawley rats. Food Chem Toxicol 45:551–562

Marasas WFO, Riley RL, Hendricks KA, Stevens VL, Sadler TW, Waes JG van, Missmer SA, Cabrera J, Torres O, Gelderblom WCA, Allegood J, Martínez C, Maddox J, Miller JD, Starr L, Sullards MC, Roman AV, Voss KA, Wang E, Merrill AH Jr (2004) Fumonisins disrupt sphingolipid metabolism, folate transport, and neural tube development in embryo culture and in vivo: a potential risk factor for human neural tube defects among populations consuming fumonisin-contaminated maize. J Nutr 134:711–716

Malley LA, Everds NE, Reynolds J, Mann PC, Lamb I, Rood T, Schmidt J, Layton RL, Prochaska LM, Hinds M, Locke M, Chui CF, Claussen F, Mattsson JL, Delaney B (2007) Subchronic feeding study of DAS 59122 7 maize grain in Sprague Dawley rats. Food Chem Toxicol 45:1277–1292

Margarit E, Reggiardo M, Vallejos R, Permingeat H (2006) Detection of *Bt* transgenic maize in foodstuffs. Food Res Int 39:250–255

McClintock J, Schaffer C, Sjoblad R (1995) A comparative review of the mammalian toxicity of *Bacillus thuringiensis*-based pesticides. Pestic Sci 45:95–105

McNaughton JL, Roberts M, Rice D, Smith B, Hinds M, Schmidt J, Locke M, Bryant A, Rood T, Layton R, Lamb I, Delaney B (2007) Feeding performance in broiler chickens fed diets containing DAS-59122-7 maize grain compared to diets containing non-transgenic maize grain. Anim Feed Sci Technol 132:227–239

Meade S, Reid E, Gerrard J (2005) The impact of processing on the nutritional quality of food proteins. J AOAC Int 88(3):904–922

Mendelsohn M, Kough J, Vaituzis Z, Matthews K (2003) Are *Bt* crops safe? Nat Biotechnol 21:1003–1009

Miller JD (2001) Factors That affect the occurrence of fumonisin. Environ Health Perspect 109(Suppl 2):321–324

Moreno-Fierros L, Garcia N, Gastrointestinal tractierrez R, Lopez-Revilla R, Vazquez-Padron RI (2000) Intranasal, rectal and intraperitoneal immunization with protoxin Cry1Ac from *Bacillus thuringiensis* induces compartmentalized serum, intestinal, vaginal and pulmonary immune responses in Balb/c mice. Microbes Infect 2:885–890

Noteborn HPJ, Bienenmann-Ploum ME, Berg JHJ van den, Alink GM, Zolla L, Reynaerts A, Pensa M, Kuiper HA (1995) Safety Assessment of *Bacillus thuringiensis* insecticidal protein

CRYIA(b) expressed in transgenic tomato. In: Engel K-H, Takeoka GR, Teranishi R (eds) Genetically modified foods: safety issues. American Chemical Society, Washington,DC
OECD (2007) Consensus document on safety information on transgenic plants expressing *Bacillus thuringiensis*—derived insect control proteins. In Joint Meeting of the Chemicals Committee and the Working Party on Chemicals, Pesticides and Biotechnology, Paris, France
Ofori-Anti AO, Ariyarathna H, Chen L, Lee HL, Pramod SN, Goodman RE (2008) Establishing objective detection limits for pepsin digestion assay used in the assessment of genetically modified foods. Regul Toxicol Pharmacol 52:94–103
Okunuki H, Teshima R, Shigeta T, Sakushima J, Akiyama H, Goda Y, Toyoda M, Sawada J (2001) Increased digestibility of two products in genetically modified food (CP-4 EPSPS and Cry-1Ab) after preheating. J Food Hyg Soc Jpn 43(2):68–73
Onose J, Imai T, Hasumura M, Ueda M, Ozeki Y, Hirose M (2008) Evaluation of subchronic toxicity of dietary administered Cry1Ab protein from *Bacillus thuringiensis var. Kurustaki* HD-1 in F344 male rats with chemically induced gastrointestinal impairment. Food Chem Toxicol 46:2184–2189
Ostry V, Ovesna J, Skarkova J, Pouchova V, Ruprich J (2010) A review on comparative data concerning Fusarium mycotoxins in *Bt* maize and non-Bt isogenic maize. Mycotox Res 26:141–145
Paul V, Steinke K, Meyer HH (2008) Development and validation of a sensitive enzyme immunoassay for surveillance of Cry1Ab toxin in bovine blood plasma of cows fed *Bt*-maize (MON 810). Anal Chim Acta 607:106–113
Rooney LW, Serna-Saldivar SO (2003) Chapter 13. Food use of whole maize and dry-milled fractions. In: While PJ, Johnson LA (eds) Maize chemistry and technology, 2nd edn. American Association of Cereal Chemists, Inc., St Paul
Sacchi VF, Parenti P, Hanozet GM, Giordana B, Luthy P, Wolfersberger MG (1986) *Bacillus thuringiensis* toxin inhibits K+-gradient-dependent amino acid transport across the brush border membrane of Pieris brassicae midgastrointestinal tract cells. FEBS Lett 204:213–218
Sanchis V (2010) From microbial sprays to insect-resistant transgenic plants: a history of the biopesticide *Bacillus thuringiensis.* A review. Agron Sustain Dev 1–15. doi:10.1051/agro/2010027
Scheideler SE, Rice D, Smith B, Dana G, Sauber T (2008) Evaluation of nutritional equivalency of maize grain from DAS-Ø15Ø7-1 (Herculex I) in the diets of laying hens. J Appl Poult Res 17:383–389
Schrøder M, Poulsen M, Wilcks A, Kroghsbo S, Miller A, Frenzel T, Danier J, Rychlik M, Emami K, Gatehouse A, Shu Q, Engel KH, Altosaar I, Knudsen I (2007) A 90-day safety study of genetically modified rice expressing Cry1Ab protein (*Bacillus thuringiensis* toxin) in Wistar rats. Food Chem Toxicol 45(3):339–349
Schultz RM, Liebman MN (2002) Chapter 3. Proteins I: composition and structure. In: Devlin TM (ed) Textbook of biochemistry with clinical correlations, 5th edn. Wiley-Liss, NY
Shimada N, Miyamoto K, Kanda K, Murata H (2006) *Bacillus thuringiensis* insecticidal Cry1Ab toxin does not affect the membrane integrity of the mammalian intestinal epithelial cells: an in vitro study. Vitro Cell Dev Biol Anim 42:45–49
Siegel JP (2001) The mammalian safety of *Bacillus thuringiensis*-based insecticides. J Invertebr Pathol 77:13
Siegel JP, Shadduck JA (1989) Safety of microbial insecticides to vertebrates, Chapter 8. In: Laird M, Lacey L, Davidson E (eds) Safety of microbial insecticides. CRC Press, Boca Raton
Soberón M, Pardo L, Muñóz-Garay C, Sánchez J, Gómez I, Porta H, Bravo A (2010) Chapter 11. Pore formation by Cry toxins. In: Anderluh G, Lakey J (eds) Proteins: membrane binding and pore formation. Landes Bioscience and Springer Science+Business Media, NY
Taylor ML, Hartnell G, Nemeth M, Karunanandaa K, George B (2005) Comparison of broiler performance when fed diets containing maize grain with insect-protected (maize rootworm and European maize borer) and herbicide-tolerant (glyphosate) traits, control maize, or commercial reference maize—revisited. Poult Sci 84:1893–1899

Taylor M, Hartnell G, Nemeth M, Lucas D, Davis S (2007) Comparison of broiler performance when fed diets containing grain from second-generation insect-protected and glyphosate-tolerant, conventional control or commercial reference. Maize. Poult Sci 86:1972–1979

Terry C, Harris N, Parkes HC (2002) Detection of genetically modified crops and their derivatives: critical steps in sample preparation and extraction. J AOAC Int 85:768–774

Thomas K, Aalbers M, Bannon GA, Bartels M, Dearman RJ, Esdaile DJ, Fu TJ, Glatt CM, Hadfield N, Hatzos C, Hefle SL, Heylings JR, Goodman RE, Henry B, Herouet C, Holsapple M, Ladics GS, Landry TD, MacIntosh SC, Rice EA, Privalle LS, Steiner HY, Teshima R, Ree R van, Woolhiser M, Zawodnyk J (2004) A multi-laboratory evaluation of a common in vitro pepsin digestion assay protocol used in assessing the safety of novel proteins. Regul Toxicol Pharmacol 39:87–98

Thomas K, Herouet-Guicheney C, Ladics G, Bannon G, Cockburn A, Crevel R, Fitzpatrick J, Mills C, Privalle L, Veiths S (2007) Evaluating the effect of food processing on the potential human allergenicity of novel proteins: international workshop report. Food Chem Toxicol 45:1116–1122

Thomas K, MacIntosh S, Bannon G, Herouet-Guicheney C, Holsapple M, Ladics G, McClain S, Vieths S, Woolhiser M, Privalle L (2009) Scientific advancement of novel protein allergenicity evaluation: an overview of the HESI Protein Allergenicity Technical Committee (2000–2008). Food Chem Toxicol 47:1041–1050

Van Rie J, Jansens S, Hofte H, Degheele D, Van Mellaert H (1989) Specificity of *Bacillus thuringiensis* delta-endotoxins, importance of specific receptors on the brush border membrane of the mid-gastrointestinal tract of target insects. Eur J Biochem 186:239–247

Van Rie J, Jansens S, Hofte H, Degheele D, Van Mellaert H (1990) Receptors on the brush border membrane of the insect midgastrointestinal tract as determininants of the specificity of *Bacillus thuringiensis* delta-endotoxins. Appl Environ Microbiol 56:1378–1385

Vazquez-Padron RI, Gonzales-Cabrera J, Garcia-Toyar C, Neri-Bazan L, Lopez-Revilla R, Hernandez M, Moreno-Fierro L, De La Riva GA (2000) Cry1Ac protoxin from *Bacillus thuringiensis* sp. kurstaki HD73 binds to surface proteins in the mouse small intestine. Biochem Biophys Res Commun 271:54–58

Wang J, Wang S, Su J, Huang T, Hu X, Yu J, Wei Z, Liang Y, Liu Y, Luo H, Sun G (2003) Food contamination of fumonisin B1 in high-risk area of esophageal and liver cancer. Toxicol Sci 72:188

WHO/GEMs Food Programme. http://www.who.int/foodsafety/chem/gems/en/. Accessed 20 Aug 2011

WHO/IPCS (International Programme on Chemical Safety) (1999) Environmental health criteria 217: microbial pest control agent *Bacillus thuringiensis*. Geneva, Switzerland

Wolfersberger MG, Hofmann C, Luthy P (1986) In: Falmagne P, Alouf JEF, Fehrenbach J, Jeljaszewics J, Thelestam M (eds) Bacterial protein toxins. Fischer, NY

Wu F, Miller JD, Casman EA (2004) The economic impact of *Bt* maize resulting from mycotoxin reduction. J Toxicol 23:393–419

Xu W, Cao S, Hea X, Luoa Y, Guoa X, Yuan Y, Huanga K (2010) Safety Assessment of Cry1Ab/Ac fusion protein. Food Chem Toxicol 47(7):1716–1721

Ziwen Y (2010) Hubei Bt Research and Development. Commercialization of *Bacillus thuringiensis* Insecticides in China. http://www.authorstream.com. Accessed 24 Sept 2010

Part IV
Other *Bacillus* Species in Biotechnology

Chapter 17
The Most Important *Bacillus* Species in Biotechnology

Noura Raddadi, Elena Crotti, Eleonora Rolli, Ramona Marasco, Fabio Fava and Daniele Daffonchio

Abstract Aerobic spore forming *Bacillus* and related bacteria have been known for their important impact on human activities. By including bacteria characterized by a high diversity with regard to the G+C content and genetic and metabolic capabilities, *Bacillus* and relatives encompass both pathogenic and beneficial bacteria. The second group is represented by several species that have been implicated in various industrial applications. These include industrial production of enzymes with great interest in detergent and food sectors; the production of primary metabolites such as vitamins and ribonucleosides; of secondary metabolites including bacteriocins and biosurfactants and of plant growth promoting formulations. Moreover, recent studies have shown that the aerobic spore formers can produce fine chemicals with interesting biotechnological applications like for example carotenoid pigments and a variety of biopolymers including poly-γ-glutamic and lactic acids. These findings open perspectives for new biotechnological applications of *Bacillus* and related species. In this chapter, we discuss the evolution and ecology of spore formers within the genus *Bacillus* and report a literature review of the various biotechnological applications of the most important species including some now affiliated to related genera.

Keywords *Bacillus* biotechnology · *Bacillus* enzymes · Secondary metabolites · Biopolymers · Plant growth promoting inoculants

D. Daffonchio (✉) · E. Crotti · E. Rolli · R. Marasco
Dipartimento di Scienze e Tecnologie Alimentare e Microbiologiche (DISTAM),
Università Degli Studi di Milano, 20133 Milano, via Celoria 2, Italy
e-mail: daniele.daffonchio@unimi.it

N. Raddadi · F. Fava
Dipartimento di Ingegneria Civile, Ambientale e dei Materiali (DICAM)-
Unità di Ricerca di Biotecnologie Ambientali e Bioraffinerie,
Università di Bologna, 40131 Bologna, via Terracini 28, Italy

E. Sansinenea (ed.), *Bacillus thuringiensis Biotechnology*,
DOI 10.1007/978-94-007-3021-2_17,

17.1 Introduction

The genus *Bacillus* encompasses bacteria that are Gram-positive, low G+C, rod-shaped endospore-formers and are aerobes or facultative anaerobes. *Bacillus* species are ubiquitous and widely diffused in the environment including soil, alkaline environments, paintings, hydrothermal vents, insect gut and seawater (Yazdani et al. 2009; Ettoumi et al. 2009; Aizawa et al. 2010; Alcaraz et al. 2010 and references there in; Fan et al. 2011). The presence of the bacteria in these very diverse environments reflects their broad and versatile metabolic capabilities, which makes of them very important microorganisms with a high potential for the industry. Indeed, with the exception of some species in the *B. cereus* group that is known to have a bivalent face with regard to its impact on human activity comprising both very useful (for instance the worldwide used biocontrol agent *B. thuringiensis*) and even fatal pathogenic (*B. anthracis* and cereulide-producing *B. cereus*) species (Daffonchio et al. 2000; Raddadi et al. 2009, 2010), several species in the *Bacillus* genus are economically very important being employed for the production of several molecules and other products for food, pharmaceutical, environmental and agricultural industries (Schallmey et al. 2004; Chen et al. 2007; Vary et al. 2007; Coutte et al. 2010; Yamashiro et al. 2011; Pérez-García et al. 2011; Sheremet et al. 2011; Huang et al. 2011; Eppinger et al. 2011; Zhang et al. 2011a).

In this chapter we will give a survey on the most important biotechnological processes that have as effectors bacterial species in the genus *Bacillus*. Before highlighting the main biotechnological applications in which these species have been implicated, we will briefly introduce the taxonomy, the ecology, the evolution and the natural variation that characterizes *Bacillus* genus.

17.1.1 Taxonomy, Ecology, Evolution and Natural Variation of the Bacillus Genus

The genus *Bacillus* (Kingdom Bacteria; Phylum Firmicutes; Class Bacilli; Order Bacillales; Family Bacillaceae) is characterized by a very high diversity with %G+C content ranging from 35% (*B. cereus* group) to 46% (*B. licheniformis*), genome sizes between 3.35 to more than 5.5 Mb (Alcaraz et al. 2008, 2010) and species that differ radically in lifestyles and metabolic properties (Maughan and Van der Auwera 2011). Actually, more than 108 *Bacillus* genomes have been sequenced most of which are of strains of clinical interest. Phylogenetic analysis based on 20 core genome sequences evidenced four major groups: (i) *B. clausii-B. halodurans*, (ii) *B. subtilis-B. licheniformis-B. pumilus*, (iii) *B. anthracis-B. thuringiensis-B. cereus*, and (iv) a novel group, *Bacillus* sp. NRRLB-14911-*B. coahuilensis*-m3-13 which were isolated from shallow waters exposed to high radiation and oligotrophic conditions. The analysis indicated that less than one third of the *B. subtilis* genes are conserved across other Bacilli and that most variation was shown to occur in genes that are needed to respond to environmental signals, suggesting that Bacilli are genetically specialized to allow for the occupation of diverse habitats and niches. On the other hand, the novel group (iv) has the largest proportion of signal transduction and translation and ribosome structure gene content as well as DNA replication, recombination and repair genes; while genes codifying for transcription factors; energy production and conversion were less represented which is congruent with the necessity of these bacteria to adapt to continuous fluctuations and high irradiation characterizing the environment from which they were isolated.

The adaptation to different environmental habitats has shaped the genomes of the different Bacilli. For example, it has been shown that members of the soil inhabiting *B. subtilis* group have genomes harboring many genetic determinants for the metabolism of carbohydrates, on the opposite to bacteria of the *B. cereus* group which are rich in genes for protein metabolism. This suggested that the latter evolved by adapting to nutrient rich environments like animal guts or animal tissues and fluids rather than the soil plant-environment (Alcaraz et al. 2010; Jensen et al. 2003). According to this and other observations, it has been proposed that the primary environmental niche for bacteria of the *B. cereus* group is the invertebrate intestinal tract (Margulis et al. 1998; Jensen et al. 2003; Swiecicka and Mahillon 2006). In particular Jensen et al. (2003) proposed that the *B. cereus* group species are normally associated to the intestine of invertebrates with a symbiotic life style. Occasionally they can colonize other animal hosts provisionally changing their usual life style from a symbiotic to a pathogenic one. This can makes *B. thuringiensis* virulent for certain insects, arthropods and nematodes, and *B. anthracis* and *B. cereus* for mammals including humans.

A specialization on different non-usual animal hosts differently shaped the evolution of the species in the *B. cereus* group. A clear example is that of *B. anthracis* for which a specialization as a lethal pathogen for mammal initiated the divergence from the other members of the group that is nowadays observed. A recent evolutionary pathway have been proposed for *B. anthracis*, that has been hypothesized to be derived rather recently from *B. cereus* i.e. between 13,000–26,000 years ago, through acquisition of plasmids resulting in the actual pXO plasmid pattern responsible of lethal virulence for mammals (Keim et al. 1997). Indeed, the *B. anthracis* genome is rather similar to that of certain *B. cereus* strains and the anthrax virulence factors are the only phenotypic traits discriminating from *B cereus*. *B. anthracis* still retains the typical virulence factors of *B. cereus* but it has been counter selected for the downregulation of those factors in favor of the expression of the pXO genes involved in the anthrax virulence (Agaisse et al. 1999; Mignot et al. 2001).

There is a high natural intraspecific variation within the *Bacillus* genus (Daffonchio et al. 2000; Daffonchio et al. 2003; Cherif et al. 2003). By using different techniques many works have shown a marked variability in different species. For instance in the case of *B. cereus*, *B. thuringiensis* and *B. anthracis* the variability observed among strains in a species is larger than the average variability among the species. In the different clustering procedures of the fingerprinting profiles many strains of a given species result intermixed with those of the other species suggesting that the three species indeed are only one. For instance, Daffonchio et al. (2000) showed that in a collection of strains of the six species of the *B. cereus* group the homoduplex-heteroduplex fingerprinting of the 16S–23S ribosomal spacers (ITS-HHP, internal transcribed spacers-homoduplex heteroduplex polymorphisms) identified many ITS-HHP types several of which corresponded to specific phenotypes such as *B. anthracis* or serotypes of *B. thuringiensis*. However, there was not a clear separation of the three species since some strains of a species were clustered with those of another. For example *B. anthracis* appeared as a lineage of *B. cereus*. In general the cluster analysis of the fingerprinting showed two main groups: one including *B. thuringiensis* and *B. cereus* with *B. anthracis* within this second and a second group including the other species of the *B. cereus* group, *B. mycoides*, *B. weihenstephanensis*, and *B. pseudomycoides*. According to the recent evolutionary origin of *B. anthracis*, this species showed to be the most monomorphic.

In general, it can be concluded that the diversity and the genetic variability of the species within *Bacillus* and the related genera is rather high making the natural reservoir of these bacteria an enormous genetic resource for the development of novel strains or enzymes for new applications in environmental, agricultural and industrial biotechnology.

17.2 Most Important Biotechnological Applications of *Bacillus* Species

17.2.1 Bacillus Enzymes

Bacteria in the genus *Bacillus* are characterized by a high capacity of their secretion systems and are producers of a variety of interesting hydrolytic extracellular enzymes. These traits are of great interest from a commercial perspective since produced proteins are easier to purify from the culture medium rather than from cytoplasm and they should present a reduced risk of co-purification of endotoxins as in the case of Gram-negative bacteria. Several *Bacillus* species are widely used for the industrial production of enzymes for the detergent, textile, food, feed and beverage industries. For example, *Bacillus* alkaline serine proteases represent the main detergent enzyme on the market. These enzymes are produced by the species *B. clausii*, *B. amyloliquefaciens* and *B. halodurans* (Schallmey et al. 2004; Saeki et al. 2007). It is however, very important to note that, the production of high number of extracellular proteases that have a high proteolytic activity towards foreign proteins (Pohl and Harwood 2010) constitute a limitation for the use of *Bacillus* spp. as hosts for the production of heterologous proteins with the exception of *B. megaterium* which does not possess excreted alkaline proteases (Vary et al. 2007).

Among the enzymes of interest in the food industry, amylases (α- and β-amylases), pullulanases, β-glucanase are extensively employed in the brewing and bakery industries as well as in the production of glucose and fructose syrups (Schallmey et al. 2004 and references therein).

B. megaterium and *B. stearothermophilus* are also able to produce extracellular α-galactosidase, an enzyme with broad applications in beet sugar, pulp and paper industries, soya food and animal feed processing as well as in medicine like for example the treatment of Fabray's disease, and xenotransplantation. This enzyme is also used for the treatment of soymilk in order to remove flatulence causing factors (Gote et al. 2004; Patil et al. 2010).

Bacillus spp. are also important producers of enzymes able to hydrolyse cellulosic material such as cellulases and xylanases (Aizawa et al. 2010). Xylanases have extensive applications in different industries especially in paper and pulp industry, in the bioconversion of agricultural wastes, in fruit softening and juice clarifying, in the texture improvement of bakery products and the production of xylose, in the textile industry and the improvement of nutritional value of poultry feed. A high production of xylanase by a newly isolated *B. pumilus* under solid state fermentation using wheat bran as a substrate has been reported recently (Nagar et al. 2011).

Other examples of *Bacillus* enzymes with industrial importance include chitinases produced by *B. thuringiensis* and that have a very important role in the synergism with other insecticidal metabolites, esterases which are a class of lipases very interesting in detergent industry, levansucrases used for the production of the biopolymer levan (Schallmey et al. 2004 and references therein; Rairakhwada et al. 2010).

17.2.2 Primary and Secondary Metabolites

By definition, primary metabolites are those produced during the exponential growth phase of the microorganism while secondary metabolites are those produced during the stationary phase. While the first are necessary for the microorganism growth and survival, the second are facultative and usually play subsidiary roles like for defense from abiotic and biotic stresses such as for example pigments, antibiotics and bacteriocins.

Bacillus secondary metabolites have been well detailed in other chapters of this book (Chap. 18 and 19). Here we just mention that they include, among others, surfactants and bacteriocins. Microbial surfactants are surface active molecules that exhibit several biological activities including surface activity as well as anti-cellular and anti-enzymatic activities. Among them, lipopeptide biosurfactants (LPBSs), molecules having in their structure a hydrophobic fatty acid portion linked to a hydrophilic peptide chain and synthesized in a ribosome-independent manner, have received increasing interest over the years. LPBSs, among which the well studied surfactin produced by *B. subtilis*, have potential applications in agricultural, chemical, food, and pharmaceutical industries as well as in bioremediation of contaminated sites (Roongsawang et al. 2010; Liu et al. 2010; Marvasi et al. 2010; Coutte et al. 2010; Saimmai et al. 2011). *Bacillus* spp. are also known to produce another kind of secondary metabolites: bacteriocins. These are ribosomally synthesized antimicrobial peptides, that, with the exception of coagulin and a class II pediocin-like bacteriocin are members of class I bacteriocins or lantibiotics. These antimicrobial peptides have great potential for biotechnological applications including food, agricultural, and pharmaceutical industries to prevent or control spoilage and pathogenic microorganisms (Cherif et al. 2001; Lee and Kim 2011; Abriouel et al. 2011).

Bacillus spp. have a metabolism exhibiting a high flux of the pentose phosphate pathway which makes them efficient producers of purine nucleosides and vitamins. *B. subtilis*, *B. amyloliquefaciens*, *B. licheniformis*, *B. megateriun* and *B. pumilis* have been widely exploited as producers of such primary metabolites (Yu et al. 2011; Sheremet et al. 2011; Li et al. 2011; Zhang et al. 2011a; Eppinger et al. 2011). Among *Bacillus* vitamins cobalamin, riboflavin, folic acid, and biotin are the most exploited (Shi et al. 2009; Kariluoto et al. 2010; Duan et al. 2010; Biedendieck et al. 2010). *Bacillus megaterium* has been also employed together with *Ketogulonicigenium vulgare* in a two-step fermentation process for the production of 2-keto-L-gulonic acid (2-KLG), the precursor of L-ascorbic acid commonly known as vitamin C (Zhang et al. 2011b). The inosine and guanosine nucleosides are precursors for synthesis of the flavor enhancers inosine monophosphate (IMP) and guanosine monophosphate (GMP). Industrial production of purine ribonucleosides is industrially performed by *B. subtilis* and *B. amyloliquefaciens* (Asahara et al. 2010; Li et al. 2011; Sheremet et al. 2011).

17.2.3 Biopolymers

To date, there is a growing interest in the use of bio-based polymers, i.e. polymers that are totally produced by microorganisms, those obtained by chemical synthesis starting from a monomer of microbial origin or polymers obtained by enzymatic polymerization of chemically synthesized monomers. This is because of the growing concern about the non-renewable origin of most of the polymers used as well as the impact the petroleum-based polymers could have on the environment. Biopolymers have broad areas of application ranging from food packaging to cosmetics and medicines as drug carriers. Microorganisms have been reported to produce a large variety of polymers and in specific polyglutammic acid (PGA), polylactic acid (PLA), polyhydroxyalcanoates (PHA) and exopolysaccharides (EPS) are among the biopolymers produced by *Bacillus* spp.

Poly-γ-glutamic acid (PGA), poly-ε-lysine (ε-PL) and cyanophycin (a polymer of α-aspartic and arginine residues) are three poly amino acids occurring in nature. ε-PL is produced by bacteria in the genus *Streptomyces* (Shih et al. 2011), while cyanophycin are produced by *Cyanobacteria* (Solaiman et al. 2011). PGA is an anionic water-soluble biopolymer that has broad applications in different sectors including among others the food, pharmaceutical and cosmetics areas as well as in wastewater treatment where it is used as flocculant or heavy metal chelator (For review please see Bajaj and Singhal (2011)). PGA is produced predominantly by bacteria belonging to *Bacillus* sp. and in particular by *B. licheniformis*, *B. amyloliquifaciens* and *B. subtilis* (Cao et al. 2011; Yamashiro et al. 2011).

Besides its use in food, chemical, cosmetics and pharmaceutical industries, the fermentation product lactic acid is receiving a high attention in relation to its use as precursor for the production of bio-based plastics (polylactic acid (PLA), a biodegradable and environmental friendly plastic) as an alternative to petroleum-based plastics. Actually, the optically pure lactic acid is mainly produced through fermentation of sugars by homofermentative thermophilic lactic acid bacteria (Abdel-Rahman et al. 2011). Some thermo-tolerant *Bacillus*, like *Bacillus coagulans* and *Bacillus licheniformis* has emerged as new lactic acid producers due to non-sterilized fermentation. Recent studies reported on the isolation, from environmental samples, of acid- and thermo-tolerant (i.e. capable of growing at pH 5.5 and temperatures of 50–55°C) *B. coagulans* strains able to grow and ferment lignocellulosic biomass to L(+)-lactic acid (Patel et al. 2006; Ou et al. 2011). Another strain of *B. licheniformis* was able to grow and produce lactic acid on mineral salt medium (Wang et al. 2011). The use of such *Bacillus* strains constitutes a good opportunity to reduce the cost of lactic acid for bio-based plastics production and replace food products as carbohydrate feedstocks with carbohydrates from non-food sources like lignocellulose. Another advantage in using such isolates is related to the fermentation conditions that could be performed at high temperature which help minimizing contamination by undesirable mesophilic microorganisms.

With regard to exopolysaccharides, levan production by *Bacillus* spp. has been reported. In specific, the commercial natto starter, *Bacillus subtilis* (natto) Takahashi, has recently been shown to produce up to 60 g/L of levan in sucrose medium in batch fermentation (Shih et al. 2011). Levan is synthesized from sucrose by transfructosylation reaction that is catalyzed by the enzyme levansucrase while glucose is released into the medium. This microbial biopolymer has attracted great attention owing to its broad industrial applications which include mainly the food (stabilizer, emulsifier and thickener agent) and pharmaceutical (hypocholesterolemic, antitumor and encapsulating agent) areas. Recently, Shih et al. (2011) developed a sequential fermentation process in which the by-product glucose was used as substrate for the production of poly-l-lysine by a *Streptomyces albulus* strain.

Other examples of exopolysaccharides produced by *Bacillus* spp. include a fucose-containing exopolysaccharide from a cellulase-producing *B. megaterium* strain RB-05 grown on jute as substrate (Chowdhury et al. 2011) and polyhydroxyalcanoates (PHA) (Valappil et al. 2007; Shrivastav et al. 2010).

Bacillus spp. also produces a wide range of depolymerizing enzymes such as the peptidoglycan hydrolases (autolysins) that hydrolyze specific bonds in the protective and shape-maintaining cell wall peptidoglycan (Raddadi et al. 2004, 2005). On the basis of their cleavage specificity, autolysins are classified as: N-acetylmuramidase, N-acetylglucosaminidase, N-acetylmuramyl-L-alanine amidase and endopeptidase. It has been shown that several *Bacillus* species including *B. thuringiensis* and the other species of the *B. cereus* group are capable of producing a wide range of peptidoglycan hydrolases that use for the completion of the cellular division and the spore formation and germination (Raddadi et al. 2004, 2005). Peptidoglycan hydrolases have interesting perspectives in several fields of biotechnology such as microbial biomass digestion for energy production purposes and in bio-control of bacteria mediated diseases.

17.2.4 Functional Foods

Functional foods/feeds can be defined as foods that have, beyond basic nutritional functions, beneficial effects on the host health when used as dietary supplements. They comprise probiotics (live microbes), prebiotics (nutrients that enhance the growth of the probiotic microbiota in the gut) and nutraceuticals (a food or part of it that helps in the prevention and/or treatment of disease(s) and/or disorder(s)) (Cencic and Chingwaru 2010). *Bacillus* species have been used for more than 50 years as probiotics and owing to their spore formation they have gained much interest. Indeed, on the opposite to the commonly consumed probiotics, i.e. *Lactobacillus* species which do not form spores, *Bacillus* spp. have as main advantage the commodity for the storage and the resistance to low gastric pH. Spore probiotics have been used in humans as dietary supplements, in animals as growth promoters and antimicrobial agents and in aquaculture for enhancing the growth and disease-resistance of fish. These include *B. clausii*, *B. subtilis*, *B. licheniformis*, *B. cereus* and *B. coagulans*. The probiotic effect of *Bacillus* spores is due to their capacity to germinate in the gastro-intestinal tract and produce bacteriocins or other antibiotics with antagonistic activity against pathogenic bacteria, to produce enzymes like for example the serine protease Nattokinase that is produced by *B. subtilis* var. Natto at high levels and that has been shown to reduce blood coagulation by fibrinolysis and to elicit the immune system of the host (Ripamonti et al. 2009; Williams et al. 2009; Nayak 2010; Cutting 2011; Endres et al. 2011).

Recently, bacteria in the genus *Bacillus* have been reported to produce carotenoid pigments (le Duc et al. 2006; Yoshida et al. 2009; Khaneja et al. 2010; Perez-Fons et al. 2011) that are high-value fine chemicals with attractive biotechnological properties. These pigments, also produced by fungi and higher plants, are considered the main and most abundant pigment group. They exist in their hydrocarbon forms as carotenes (α-carotene, β-carotene and lycopene), or in oxygenated derivative forms as xanthophylls (such as lutein, canthaxanthin, and astaxanthin) and appear yellow, orange, or red. Beyond their use as food colorants, and owing to their biological properties, i.e. anti-cancer, anti-inflammatory, and anti-oxidative activities, carotenoids are used commercially as food and feed supplements and, more recently, as nutraceuticals for cosmetic and pharmaceutical purposes.

Generally, the commercial carotenoid production is mostly based on extraction from plant tissues like marigold plants (Tagetes flowers) (Fernández-Sevilla et al. 2010) or chemical synthesis. Both the approach have however several drawbacks. For example limitations related to the natural colorant production systems include the dependence on the seasonal supply of raw materials, the variations of the composition of the extracted pigments in relation to the cultivar and climatic conditions and the difficulty of obtaining pure compounds and in high yields. The disadvantages of the approach of chemical synthesis are related to the production of stereo isomers not found in the natural product and to contamination with reaction intermediates/products. Moreover, there are emerging safety concerns with increasing application of synthetic coloring agents in food. These considerations together with the development of the industry and hence the increasing demand for sustainable production of pigments make the microbial production of these compounds of great potential in terms of both the efficiency of production and the diversity of structures. Currently, microbial resources of carotenoids include the unicellular algae *Dunaliella salina*, *Spirulina* and *Haematococcus*, the filamentous fungus *Blakeslea trispora* and non-carotenogenic metabolically engineered cells of *Escherichia coli*, *B. subtilis; Pichia pastoris*, *Saccharomyces cerevisiae*, *Candida utilis*, and *Zymomonas mobilis* (Das et al. 2007; Yoshida et al. 2009). Recent studies have demonstrated that pigmented (yellow, orange and pink) bacilli are ubiquitous and are related with known species such as *B. marisflavi*, *B. indicus*, *B. firmus*, *B. altitudinis* and *B. safensis* and contain new carotenoid biosynthetic pathways that may have industrial importance. Moreover, the authors report on the presence of a water-soluble pigment that may also be a carotenoid. This finding is important in the industrial sector mainly for the production of foods colorants; helping to overcome some drawbacks related to the use of carotenoids as natural colorants and which include their poor solubility in aqueous solutions and instability due to their hydrophobic nature (le Duc et al. 2006; Yoshida et al. 2009; Khaneja et al. 2010; Perez-Fons et al. 2011).

17.2.5 Plant Growth Promoting Inoculants

Plant growth promoting rhizobacteria (PGPR) exert their beneficial effect on plant growth through several mechanisms. They can promote the growth of the plant either directly or indirectly. Direct plant promotion is performed by providing nutrients such as phosphate solubilization, iron chelation by siderophores; or by producing phytohormones like auxins (indole-3-acetic acid, IAA). Indirect promotion of the plant growth includes a biocontrol activity which protects the host plant against pests including phytopathogenic bacteria, fungi and insects (Raddadi et al. 2008, 2009).

Biopesticides are included, as a crop protection tool, in the Integrated Pest Management (IPM) practices that nowadays are gaining more and more acceptance over the world (Chandler et al. 2011). Among biological control agents, the most useful ones are represented by some aerobic spore-forming bacteria, due to their insecticidal and anti-microbial properties, coped with growth promotion and defense stimulation capabilities. Members of the genus *Bacillus* have been used to formulate *Bacillus*-based products which constitute the most important class of microbial products used commercially worldwide in the crop protection field (Pérez-García et al. 2011). The most widely used microbial biopesticide, representing nearly the 70% of the global market of biopesticides, is the entomopathogenic bacterium *B. thuringiensis* that has become, to date, the leading biological insecticide employed commercially to control agricultural pests (Chandler et al. 2011). During the sporulation *B. thuringiensis* produces shaped protein crystals that are selectively toxic to several species of different invertebrate phyla. Generally, these crystals comprise one or more proteins (Cry or Cyt toxins), also known as δ-endotoxins. Once ingested by the invertebrate target, the Cry proteins are solubilized in the host gut, the activated through cleavage by host digestive proteolytic enzymes, and then they interact with the constituents of the surface of the host intestinal epithelium cells, in order to be inserted into the cell membrane forming a pore. On the other hand, Cyt proteins are able to lyse a wide range of cells *in vitro* and, even if they have a different structure than Cry ones, they are still pore-forming proteins, like Cry proteins (see Chap. 2 of this book for more information; Bravo et al. 2007). Noteworthy it is that, beyond Cry and Cyt crystals, *B. thuringiensis* produces other insecticidal compounds, like phospholipases, thermolabile toxins, β-exotoxins, proteases, chitinases, vegetative insecticide proteins (Vip) and the antifungal antibiotic zwitteriomycin A (Raddadi et al. 2009). All of these components may contribute to enhance *B. thuringiensis* toxicity and to delay the appearance of *B. thuringiensis*-resistant insects.

The commercialized products foresee different kinds of formulations, like liquid concentrates, wettable powders, and ready-to use dusts and granulates (Soberón et al. 2009). Even if a large number of Cry proteins have been studied, the commercialized ones, in sprays or in transgenic crops engineered with the Cry toxin of *B. thuringiensis*, account only for a dozen (Cry1Aa, Cry1Ab, Cry1Ac, Cry1C, Cry1D, Cry1E, Cry1F, Cry2Aa, Cry2Ab, Cry3A, Cry3B and Cry34/Cry35; (Bravo and Soberón 2008). Products like Biobit, Condor, Cutlass, Dipel, Full-Bac, Javelin and M-Peril are based on *B. thuringiensis* var. *kurstaki*, active against leaf-feeding caterpillars and defoliators; products like Certan, Agree and Xentari are composed of *B. thuringiensis* var. *aizawaii*, active against other lepidopterans, like the Indian meal moth larvae. Products like M-Trak, Foil and Novodor act against coleopteran insects and are based on *B. thuringiensis* var. *san diego* and *B. thuringiensis* var. *tenebrionis*, while Vectobac, Teknar, Bactimos, Skeetal and Mosquito Attack are based on *B. thuringiensis* var. *israelensis* (Bti) to control mosquitoes and black flies (Soberón et al. 2009). Other *Bacillus* based plant growth promoting spore formulations include those based on *B. amyloliquefaciens* FZB24 and *B. amyloliquefaciens* FZB42.

On the other hand Bti has been successfully employed in the eradication program of the river blindness launched in 1983 in 11 countries of Western Africa. Beyond the genera *Anopheles* and *Aedes*, vectors of malaria and viruses, respectively, Bti is active against the black fly *Simulium damnosum*, vector of the nematodes that cause onchocerciasis (Becker 2000). In 2007 it has been estimated that Bti applications protected more than 80% of this region, while less than 20% was protected by the chemical larvicide, temephos. No cases of Bti-resistant black flies have been reported at that time (Guillet et al. 1990).

Another entomopathogenic *Bacillus* is represented by *Bacillus spahericus* that, along the years, has been successfully used to control mosquito-borne diseases, like dengue and malaria *B. spahericus* is commonly present in soil and water and produces two types of toxins, Btx and Mtx, which are active against mosquito larvae (El-Bendary 2006). Btx is a crystal protein, produced during the sporulation, and comprises Bin A (41.9 KDa) and Bin B (51.4 KDa) components. For this reason it has been also defined as a binary toxin (Baumann et al. 1991; Charles et al. 1996; Humphreys and Berry 1998). On the other hand, Mtx toxins are associated to *B. spahericus* cell membrane, being produced during the vegetative growth of

B. spahericus and include three types, Mtx1, Mtx2 and Mtx3. *B. spahericus* strains with a high toxicity produce Btx toxin and may synthesize one or more Mtx toxins (Liu et al. 1996). Like *B. thuringiensis*, *B. spahericus* is considered as completely safe for not-target organisms, including humans and animals, and for the environment.

Beyond insecticidal and nematocidal activity, members of the genus *Bacillus* own fungicidal activity. This has been exploited in the formulation of *Bacillus*-based products active against fungi. On the market several *Bacillus*-based biofungicidal commercial products are available, based on *B. amyloliquefaciens*, *B. licheniformis*, *B. pumilus*, and *B. subtilis* (Pérez-García et al. 2011). They are employed to control fungal diseases, like root diseases (such as tomato damping-off, avocado root rot, and wheat take-all), foliar diseases (cucurbit and strawberry powdery mildews) and postharvest diseases. In the next years it is expected a great increase in this field of application for biofungicidal Bacilli. Few examples are Avogreen and Biobest based on *B. subtilis* or Ballad Plus and Sonata based on *B. pumilus* (Pérez-García et al. 2011).

The attention of researchers is also focused on the antimicrobial properties of bacilli. For instance, recent studies proposed *B. subtilis*, *B. pumilus*, *B. licheniformis*, *B. cereus*, *B. megaterium* and *Brevibacillus laterosporus* as interesting candidates for the control of the Gram positive bacterium *Paenibacillus larvae*, the etiological agent of American Foulbrood disease (AFB), a serious disease that affects honeybees, causing economical damages in apiculture (Alippi and Reynaldi 2006). These strains have been shown to inhibit on plates the growth of *P. larvae*. In apiculture great efforts are made on the employments of bacilli and bacilli-related species to counteract honeybee pathogens and parasites.

Recently *Br. laterosporus*, previously known as *Bacillus laterosporus* and reclassified as *Br. laterosporus* in 1996 (Shida et al. 1996), has been shown to exert an antagonistic activity toward phytopathogenic fungi, such as *Fusarium oxysporum* f. sp. *cicero*, *Fusarium semitectum*, *Magnaporthe grisea* and *Rhizoctonia oryzae*, and the Gram positive bacterium *Staphylococcus aureus*. *Br. laterosporus* has been used in greenhouse trials to control rice blast disease, caused by *Magnaporthe grisea*, resulting able to reduce disease incidence by 30–67% (Saikia et al. 2011).

In the last few decades a great attention has been directed toward *Bacillus* biopesticides, and their application in the agro-ecosystem. This has allowed the development of *Bacillus*-based products and transgenic plants. However the studies and applications need a further deepening; many characteristics owned by *Bacillus* and their potential applications are yet poorly understood. As instance, *B. thuringiensis*-engineered crops are best suited for high volume crops, like maize, cotton and rice. *B. thuringiensis* foliar sprays remain as part of IPM only for small market sectors. Moreover, *B. thuringiensis* based products foresee high production cost and in developing countries this limits their use with a preferential use of cheaper synthetic insecticides, despite the risk to human health and the environment (Crickmore 2006). Further studies are needed to broaden *B. thuringiensis* insecticidal spectrum, to improve performances, or to find new *Bacillus* species and/or strains with biopesticide properties.

PGPR can also enhance the plant growth through the elicitation of induced systemic resistance (ISR). Several strains of the species *B. amyloliquefaciens*, *B. subtilis*, *B. pasteurii*, *B. cereus*, *B. pumilus*, *B. mycoides* and *B. sphaericus* may determine significant reductions in the incidence or severity of various diseases on a diversity of hosts. The mechanism of elicitation of ISR by these strains has been demonstrated in greenhouse and/or field trials on different plants resulting in their protection against fungal and bacterial pathogens, systemic viruses and root-knot nematodes (Chen et al. 2007; Choudhary and Johri 2009; Pérez-García et al. 2011).

17.3 Conclusions and Perspectives

Bacteria in the genus *Bacillus* could be classified into two major groups in relation to their type of impact on human activities, i.e. the *B. cereus* group where most of the human pathogens/opportunistic pathogens are located and the *B. subtilis* group which encompasses soil bacteria. Most of the biotechnologically important *Bacillus* species are located within this second group and has been implicated in several industrial productions including primary metabolites, enzymes, antimicrobial metabolites, functional foods, biopolymers and plant growth promoting formulations.

In addition to the above mentioned biotechnological applications, *Bacillus* spp. strains have several other potential applications such as environmental biotechnology. In this context, Bacteria from the genus *Bacillus* able to degrade and completely remove the natural estrogen 17β-estradiol, a major endocrine disruptor, from wastewater have been isolated from activated sludge collected from a wastewater treatment plant and characterized. In particular, one isolate was shown to maintain high degradation capacity even at initial concentrations of the steroid up to 50 mg/l which highlight its potential in biological remediation applications (Jiang et al. 2010). From oil-contaminated tropical marine sediments, *Bacillus naphthovorans*, a strain that was able to grow with naphthalene as sole carbon source has been isolated and characterized (Zhuang et al. 2002). Bacteria belonging to the genus *Bacillus* were isolated from the rock of an oil reservoir located in a deep-water production basin in Brazil. The strains were shown to be able to degrade different oil types, to grow in the presence of carbazole and n-hexadecane polyalphaolefin (PAO) as the only carbon sources and to produce key enzymes involved in aromatic hydrocarbons biodegradation process (da Cunha et al. 2006).

References

Abdel-Rahman MA, Tashiro Y, Zendo T, Hanada K, Shibata K, Sonomoto K (2011) Efficient homofermentative L-(+)-lactic acid production from xylose by a novel lactic acid bacterium, *Enterococcus mundtii* QU 25. Appl Environ Microbiol 77:1892–1895

Abriouel H, Franz CM, Ben Omar N, Gálvez A (2011) Diversity and applications of *Bacillus* bacteriocins. FEMS Microbiol Rev 35:201–232

Agaisse H, Gominet M, Okstad OA, Kolstø AB, Lereclus D (1999) PlcR is a pleiotropic regulator of extracellular virulence factor gene expression in *Bacillus thuringiensis*. Mol Microbiol 32:1043–1053

Aizawa T, Urai M, Iwabuchi N, Nakajima M, Sunairi M (2010) *Bacillus trypoxylicola* sp. nov. xylanase-producing alkaliphilic bacteria isolated from the guts of Japanese horned beetle larvae (*Trypoxylus dichotomus septentrionalis*). Int J Syst Evol Microbiol 60:61–66

Alcaraz LD, Olmedo G, Bonilla G, Cerritos R, Hernández G, Cruz A, Ramírez E, Putonti C, Jiménez B, Martínez E, López V, Arvizu JL, Ayala F, Razo F, Caballero J, Siefert J, Eguiarte L, Vielle JP, Martínez O, Souza V, Herrera-Estrella A, Herrera-Estrella L (2008) The genome of *Bacillus coahuilensis* reveals adaptations essential for survival in the relic of an ancient marine environment. Proc Natl Acad Sci USA 105:5803–5808

Alcaraz LD, Moreno-Hagelsieb G, Eguiarte LE, Souza V, Herrera-Estrella L, Olmedo G (2010) Understanding the evolutionary relationships and major traits of Bacillus through comparative genomics. BMC Genomics 11:332

Alippi AM, Reynaldi FJ (2006) Inhibition of the growth of *Paenibacillus larvae*, the causal agent of American foulbrood of honeybees, by selected strains of aerobic spore-forming bacteria isolated from apiarian sources. J Invertebr Pathol 91:141–146

Asahara T, Mori Y, Zakataeva NP, Livshits VA, Yoshida K, Matsuno K (2010) Accumulation of gene-targeted *Bacillus subtilis* mutations that enhance fermentative inosine production. Appl Microbiol Biotechnol 87:2195–2207

Bajaj I, Singhal R (2011) Poly (glutamic acid)-an emerging biopolymer of commercial interest. Bioresour Technol 102:5551–5561

Baumann P, Clark M, Baumann L, Broadwell AH (1991) *Bacillus sphaericus* as a mosquito pathogen: properties of the organism and its toxins. Microbiol Rev 55:425–436

Becker N (2000) Bacterial control of vector-mosquitoes and black flies. In: Charles JF, Delécluse A, Nielsen-LeRoux C (eds) Entomopathogenic bacteria: from laboratory to field application. Kluwer Academic, Dordrecht, p 383

Biedendieck R, Malten M, Barg H, Bunk B, Martens JH, Deery E, Leech H, Warren MJ, Jahn D (2010) Metabolic engineering of cobalamin (vitamin B12) production in *Bacillus megaterium*. Microb Biotechnol 3:24–37. doi:10.1111/j.1751-7915.2009.00125.x

Bravo A, Soberón M (2008) How to cope with insect resistance to Bt toxins? Trends Biotechnol 26:573–579

Bravo A, Gill SS, Soberón M (2007) Mode of action of *Bacillus thuringiensis* Cry and Cyt toxins and their potential for insect control. Toxicon 49:423–435

Cao M, Geng W, Liu L, Song C, Xie H, Guo W, Jin Y, Wang S (2011) Glutamic acid independent production of poly-γ-glutamic acid by *Bacillus amyloliquefaciens* LL3 and cloning of pgsBCA genes. Bioresour Technol 102:4251–4257

Cencic A, Chingwaru W (2010) The role of functional foods, nutraceuticals, and food supplements in intestinal health. Nutrients 2:611–625

Chandler D, Bailey AS, Tatchell GM, Davidson G, Greaves J, Grant W (2011) The development, regulation and use of biopesticides for integrated pest management. Philos Trans R Soc B 366:1987–1998

Charles JF, Nielsen-LeRoux C, Delecluse A (1996) *Bacillus sphaericus* toxins: molecular biology and mode of action. Annu Rev Entomol 41:451–472

Chen XH, Koumoutsi A, Scholz R, Eisenreich A, Schneider K, Heinemeyer I, Morgenstern B, Voss B, Hess WR, Reva O, Junge H, Voigt B, Jungblut PR, Vater J, Süssmuth R, Liesegang H,

Strittmatter A, Gottschalk G, Borriss R (2007) Comparative analysis of the complete genome sequence of the plant growth-promoting bacterium *Bacillus amyloliquefaciens* FZB42. Nat Biotechnol 25:1007–1014

Cherif A, Ouzari H, Daffonchio D, Cherif H, Ben Slama K, Hassen A, Jaoua S, Boudabous A (2001) Thuricin 7: a novel bacteriocin produced by Bacillus thuringiensis BMG1.7, a new strain isolated from soil. Lett Appl Microbiol 32:243–247

Cherif A, Borin S, Rizzi A, Ouzari H, Boudabous A, Daffonchio D (2003) *Bacillus anthracis* diverges from related clades of the *Bacillus cereus* group in 16S–23S rDNA internal transcribed spacers containing tRNA genes. Appl Environ Microbiol 69:33–40

Choudhary DK, Johri BN (2009) Interactions of *Bacillus* spp. and plants with special reference to induced systemic resistance (ISR). Microbiol Res 164:493–513

Chowdhury SR, Basak RK, Sen R, Adhikari B (2011) Production of extracellular polysaccharide by *Bacillus megaterium* RB-05 using jute as substrate. Bioresour Technol 102:6629–6632

Coutte F, Lecouturier D, Yahia SA, Leclère V, Béchet M, Jacques P, Dhulster P (2010) Production of surfactin and fengycin by *Bacillus subtilis* in a bubbleless membrane bioreactor. Appl Microbiol Biotechnol 87:499–507

Crickmore N (2006) Beyond the spore—past and future developments of *Bacillus thuringiensis* as a biopesticide. J Appl Microbiol 101:616–619

Cutting SM (2011) *Bacillus* probiotics. Food Microbiol 28:214–220

da Cunha CD, Rosado AS, Sebastián GV, Seldin L, Weid I von der (2006) Oil biodegradation by *Bacillus* strains isolated from the rock of an oil reservoir located in a deep-water production basin in Brazil. Appl Microbiol Biotechnol 73:949–959

Daffonchio D, Cherif A, Borin S (2000) Homoduplex and heteroduplex polymorphisms of the amplified ribosomal 16S–23S internal transcribed spacers describe genetic relationships in the "*Bacillus cereus* group". Appl Environ Microbiol 66:5460–5468

Daffonchio D, Cherif A, Brusetti L, Rizzi A, Mora D, Boudabous A, Borin S (2003) Nature of polymorphisms in 16S–23S rRNA gene intergenic transcribed spacer fingerprinting of bacillus and related genera. Appl Environ Microbiol 69:5128–5137

Das A, Yoon SH, Lee SH, Kim JY, Oh DK, Kim SW (2007) An update on microbial carotenoid production: application of recent metabolic engineering tools. Appl Microbiol Biotechnol 77:505–512

Duan Y, Chen T, Chen X, Zhao X (2010) Overexpression of glucose-6-phosphate dehydrogenase enhances riboflavin production in *Bacillus subtilis*. Appl Microbiol Biotechnol 85:1907–1914

El-Bendary MA (2006) *Bacillus thuringiensis* and *Bacillus sphaericus* biopesticides production. J Basic Microbiol 46:158–170

Endres JR, Qureshi I, Farber T, Hauswirth J, Hirka G, Pasics I, Schauss AG (2011) One-year chronic oral toxicity with combined reproduction toxicity study of a novel probiotic, *Bacillus coagulans*, as a food ingredient. Food Chem Toxicol 49:1174–1182

Eppinger M, Bunk B, Johns MA, Edirisinghe JN, Kutumbaka KK, Koenig SS, Huot Creasy H, Rosovitz MJ, Riley DR, Daugherty S, Martin M, Elbourne LD, Paulsen I, Biedendieck R, Braun C, Grayburn S, Dhingra S, Lukyanchuk V, Ball B, Ul-Qamar R, Seibel J, Bremer E, Jahn D, Ravel J, Vary PS (2011) Genome sequences of the biotechnologically important *B. megaterium* strains QM B1551 and DSM319. J Bacteriol 193:4199–4213

Ettoumi B, Raddadi N, Borin S, Daffonchio D, Boudabous A, Cherif A (2009) Diversity and phylogeny of culturable spore-forming Bacilli isolated from marine sediments. J Basic Microbiol 49:S13–S23

Fan L, Bo S, Chen H, Ye W, Kleinschmidt K, Baumann HI, Imhoff JF, Kleine M, Cai D (2011) Genome sequence of *Bacillus subtilis subsp. spizizenii* gtP20b, isolated from the Indian ocean. J Bacteriol 193:1276–1277

Fernández-Sevilla JM, Acién Fernández FG, Molina Grima E (2010) Biotechnological production of lutein and its applications. Appl Microbiol Biotechnol 86:27–40

Gote M, Umlakar H, Khan I, Khire J (2004) Thermostable α-galactosidase from Bacillus stearothermophilus (NCIM 5146) and its application in the removal of flatulence causing factors from soymilk. Process Biochem 39:1723–1729

Guillet P, Kurstack DC, Philippon B, Meyer R (1990) In: De Barjac H, Sutherland DJ (eds) Bacterial control of mosquitoes and blackflies. Rutgers University Press, Piscataway, NJ, pp 187–190
Huang B, Qin P, Xu Z, Zhu R, Meng Y (2011) Effects of CaCl2 on viscosity of culture broth, and on activities of enzymes around the 2-oxoglutarate branch, in *Bacillus subtilis* CGMCC 2108 producing poly-(γ-glutamic acid). Bioresour Technol 102:3595–3598
Humphreys MJ, Berry C (1998) Variants of the *Bacillus sphaericus* binary toxins: Implications for differential toxicity of strains. J Invertebr Pathol 71:184–185
Jensen GB, Hansen BM, Heilenberg J, Mahillon J (2003) The hidden lifestyles of *Bacillus cereus* and relatives. Environ Microbiol 5:631–640
Jiang L, Yang J, Chen J (2010) Isolation and characteristics of 17beta-estradiol-degrading *Bacillus* spp. strains from activated sludge. Biodegradation 21:729–736
Kariluoto S, Edelmann M, Herranen M, Lampi AM, Shmelev A, Salovaara H, Korhola M, Piironen V (2010) Production of folate by bacteria isolated from oat bran. Int J Food Microbiol 143:41–47
Keim P, Kalif A, Schupp J, Hill K, Travis SE, Richmond K, Adair DM, Hugh-Jones M, Jackson PJ (1997) Molecular evolution and diversity in *Bacillus anthracis* as detected by amplified fragment length polymorphism markers. J Bacteriol 179:818–824
Khaneja R, Perez-Fons L, Fakhry S, Baccigalupi L, Steiger S, To E, Sandmann G, Dong TC, Ricca E, Fraser PD, Cutting SM (2010) Carotenoids found in *Bacillus*. J Appl Microbiol 108:1889–1902
le Duc H, Fraser PD, Tam NK, Cutting SM (2006) Carotenoids present in halotolerant *Bacillus* spore formers. FEMS Microbiol Lett 255:215–224
Lee H, Kim HY (2011) Lantibiotics, class I bacteriocins from the genus Bacillus. J Microbiol Biotechnol 21:229–235
Li H, Zhang G, Deng A, Chen N, Wen T (2011) De novo engineering and metabolic flux analysis of inosine biosynthesis in *Bacillus subtilis*. Biotechnol Lett 33:1575–1580
Liu JW, Porter AG, Wee BY, Thanabalu T (1996) New gene from nine *Bacillus sphaericus* strains encoding highly conserved 35.8 kilodalton mosquicidal toxin. Appl Environ Microbiol 62:2174–2176
Liu X, Ren B, Chen M, Wang H, Kokare CR, Zhou X, Wang J, Dai H, Song F, Liu M, Wang J, Wang S, Zhang L (2010) Production and characterization of a group of bioemulsifiers from the marine *Bacillus velezensis* strain H3. Appl Microbiol Biotechnol 87:1881–1893
Margulis L, Jorgensen JZ, Dolan S, Kolchinski R, Rainey FA, Lo SC (1998) The *Arthromitus* stage of *Bacillus cereus*: intestinal symbionts of animals. Proc Natl Acad Sci USA 95:1236–1241
Marvasi M, Visscher PT, Casillas Martinez L (2010) Exopolymeric substances (EPS) from *Bacillus subtilis*: polymers and genes encoding their synthesis. FEMS Microbiol Lett 313:1–9
Maughan H, Van der Auwera G (2011) *Bacillus* taxonomy in the genomic era finds phenotypes to be essential though often misleading. Infect Genet Evol 11:789–797
Mignot T, Mock M, Robichon D, Landier A, Lereclus D, Fouet A (2001) The incompatibility between the PlcR- and AtxA-controlled regulons may have selected a nonsense mutation in *Bacillus anthracis*. Mol Microbiol 42:1189–1198
Nagar S, Mittal A, Kumar D, Kumar L, Kuhad RC, Gupta VK (2011) Hyper production of alkali stable xylanase in lesser duration by *Bacillus pumilus* SV-85S using wheat bran under solid state fermentation. N Biotechnol 28:581–587
Nayak SK (2010) Probiotics and immunity: a fish perspective. Fish Shellfish Immunol 29:2–14
Ou MS, Ingram LO, Shanmugam KT (2011) L: (+)-Lactic acid production from non-food carbohydrates by thermotolerant *Bacillus coagulans*. J Ind Microbiol Biotechnol 38:599–605
Patel MA, Ou MS, Harbrucker R, Aldrich HC, Buszko ML, Ingram LO, Shanmugam KT (2006) Isolation and characterization of acid-tolerant, thermophilic bacteria for effective fermentation of biomass-derived sugars to lactic acid. Appl Environ Microbiol 72:3228–3235
Patil AG, K PK, Mulimani VH, Veeranagouda Y, Lee K (2010) alpha-Galactosidase from *Bacillus megaterium* VHM1 and its application in removal of flatulence-causing factors from soymilk. J Microbiol Biotechnol 20:1546–1554

Perez-Fons L, Steiger S, Khaneja R, Bramley PM, Cutting SM, Sandmann G, Fraser PD (2011) Identification and the developmental formation of carotenoid pigments in the yellow/orange *Bacillus* spore-formers. Biochim Biophys Acta 1811:177–185
Pérez-García A, Romero D, Vicente A de (2011) Plant protection and growth stimulation by microorganisms: biotechnological applications of Bacilli in agriculture. Curr Opin Biotechnol 22:187–193
Pohl S, Harwood CR (2010) Heterologous protein secretion by *Bacillus* species from the cradle to the grave. Adv Appl Microbiol 73:1–25
Raddadi N, Cherif A, Mora D, Ouzari H, Boudabous A, Molinari F, Daffonchio D (2004) The autolytic phenotype of *Bacillus thuringiensis*. J Appl Microbiol 97:158–168
Raddadi N, Cherif A, Mora D, Brusetti L, Borin S, Boudabous A, Molinari F, Daffonchio D (2005) The autolytic phenotype of the *Bacillus cereus* group. J Appl Microbiol 99:1070–1081
Raddadi N, Cherif A, Boudabous A, Daffonchio D (2008) Screening of plant growth promoting traits of *Bacillus thuringiensis*. Ann Microbiol 58:47–52
Raddadi N, Belaouis A, Tamagnini I, Hansen BM, Hendriksen NB, Boudabous A, Cherif A, Daffonchio D (2009) Characterization of polyvalent and safe *Bacillus thuringiensis* strains with potential use for biocontrol. J Basic Microbiol 49:293–303
Raddadi N, Rizzi A, Brusetti L, Borin S, Tamagnini I, Daffonchio D (2010) Bacillus. In: Liu D (ed) Molecular detection of foodborne pathogens. CRC Press, Boca Raton, pp 129–144
Rairakhwada D, Seo JW, Seo MY, Kwon O, Rhee SK, Kim CH (2010) Gene cloning, characterization, and heterologous expression of levansucrase from *Bacillus amyloliquefaciens*. J Ind Microbiol Biotechnol 37:195–204
Ripamonti B, Agazzi A, Baldi A, Balzaretti C, Bersani C, Pirani S, Rebucci R, Savoini G, Stella S, Stenico A, Domeneghini C (2009) Administration of *Bacillus coagulans* in calves: recovery from faecal samples and evaluation of functional aspects of spores. Vet Res Commun 33:991–1001
Roongsawang N, Washio K, Morikawa M (2010) Diversity of nonribosomal peptide synthetases involved in the biosynthesis of lipopeptide biosurfactants. Int J Mol Sci 12:141–172
Saeki K, Ozaki K, Kobayashi T, Ito S (2007) Detergent alkaline proteases: enzymatic properties, genes, and crystal structures. J Biosci Bioeng 103:501–508
Saikia R, Gogoi DK, Mazumder S, Yadav A, Sarma RK, Bora TC, Gogoi BK (2011) *Brevibacillus laterosporus* strain BPM3, a potential biocontrol agent isolated from a natural hot water spring of Assam, India. Microbiol Res 166:216–225
Saimmai A, Sobhon V, Maneerat S (2011) Molasses as a whole medium for biosurfactants production by *Bacillus* strains and their application. Appl Biochem Biotechnol 165:315–335
Schallmey M, Singh A, Ward OP (2004) Developments in the use of *Bacillus* species for industrial production. Can J Microbiol 50:1–17
Sheremet AS, Gronskiy SV, Akhmadyshin RA, Novikova AE, Livshits VA, Shakulov RS, Zakataeva NP (2011) Enhancement of extracellular purine nucleoside accumulation by *Bacillus* strains through genetic modifications of genes involved in nucleoside export. J Ind Microbiol Biotechnol 38:65–70
Shi S, Chen T, Zhang Z, Chen X, Zhao X (2009) Transcriptome analysis guided metabolic engineering of *Bacillus subtilis* for riboflavin production. Metab Eng 11:243–252
Shida O, Takagi H, Kadowaki K, Komagata K (1996) Proposal for two new genera, Brevibacillus gen. nov. and Aneurinibacillus gen. nov. Int J Syst Bacteriol 46:939–946
Shih IL, Wang TC, Chou SZ, Lee GD (2011) Sequential production of two biopolymers-levan and poly-ε-lysine by microbial fermentation. Bioresour Technol 102:3966–3969
Shrivastav A, Mishra SK, Shethia B, Pancha I, Jain D, Mishra S (2010) Isolation of promising bacterial strains from soil and marine environment for polyhydroxyalkanoates (PHAs) production utilizing *Jatropha* biodiesel byproduct. Int J Biol Macromol 47:283–287
Soberón M, Gill SS, Bravo A (2009) Signaling versus punching hole: how do *Bacillus thuringiensis* toxins kill insect midgut cells? Cell Mol Life Sci 66:1337–1349
Solaiman DK, Garcia RA, Ashby RD, Piazza GJ, Steinbüchel A (2011) Rendered-protein hydrolysates for microbial synthesis of cyanophycin biopolymer. N Biotechnol 28:552–558

Swiecicka I, Mahillon J (2006) Diversity of commensal *Bacillus cereus sensu lato* isolated from the common sow bug (*Porcellio scaber*, Isopoda). FEMS Microbiol Ecol 56:132–140
Valappil SP, Boccaccini AR, Bucke C, Roy I (2007) Polyhydroxyalkanoates in Gram-positive bacteria: insights from the genera *Bacillus* and *Streptomyces*. Antonie Van Leeuwenhoek 91:1–17
Vary PS, Biedendieck R, Fuerch T, Meinhardt F, Rohde M, Deckwer WD, Jahn D (2007) *Bacillus megaterium*-from simple soil bacterium to industrial protein production host. Appl Microbiol Biotechnol 76:957–967
Wang Q, Zhao X, Chamu J, Shanmugam KT (2011) Isolation, characterization and evolution of a new thermophilic *Bacillus licheniformis* for lactic acid production in mineral salts medium. Bioresour Technol 102:8152–8158
Williams LD, Burdock GA, Jiménez G, Castillo M (2009) Literature review on the safety of Toyocerin, a non-toxigenic and non-pathogenic *Bacillus cereus* var. toyoi preparation. Regul Toxicol Pharmacol 55:236–246
Yamashiro D, Yoshioka M, Ashiuchi M (2011) *Bacillus subtilis* pgsE (Formerly ywtC) stimulates poly-γ-glutamate production in the presence of zinc. Biotechnol Bioeng 108:226–230
Yazdani M, Naderi-Manesh H, Khajeh K, Soudi MR, Asghari SM, Sharifzadeh M (2009) Isolation and characterization of a novel gamma-radiation-resistant bacterium from hot spring in Iran. J Basic Microbiol 49:119–127
Yoshida K, Ueda S, Maeda I (2009) Carotenoid production in *Bacillus subtilis* achieved by metabolic engineering. Biotechnol Lett 31:1789–1793
Yu WB, Gao SH, Yin CY, Zhou Y, Ye BC (2011) Comparative transcriptome analysis of *Bacillus subtilis* responding to dissolved oxygen in adenosine fermentation. PLoS One 6:e20092. doi:10.1371/journal.pone.0020092
Zhang G, Deng A, Xu Q, Liang Y, Chen N, Wen T (2011a) Complete genome sequence of *Bacillus amyloliquefaciens* TA208, a strain for industrial production of guanosine and ribavirin. J Bacteriol 193:3142–3143
Zhang J, Zhou J, Liu J, Chen K, Liu L, Chen J (2011b) Development of chemically defined media supporting high cell density growth of *Ketogulonicigenium vulgare* and *Bacillus megaterium*. Bioresour Technol 102:4807–4814
Zhuang WQ, Zhuang WQ, Maszenan AM, Tay ST (2002) *Bacillus naphthovorans* sp. nov. from oil-contaminated tropical marine sediments and its role in naphthalene biodegradation. Appl Microbiol Biotechnol 58:547–553

Chapter 18
Secondary Metabolites of *Bacillus*: Potentials in Biotechnology

Ines Chaabouni, Amel Guesmi and Ameur Cherif

Abstract The huge diversity characterizing the *Bacillus* species at the taxonomic level, is also noticeable for their metabolic features. These bacteria are able to produce a wide range of secondary metabolites with very different natures and structures and displaying broad spectra of activities. These metabolites; including antibiotics, pigments, toxins, growth promoters (animals and plants), effectors of ecological competition, pheromones, enzyme inhibitors and others bioactive compounds; are originally designed to enable the bacterium to survive in its natural environment (Stein 2005). In general, these metabolites serve as; (i) competitive weapons used against other bacteria, fungi, amoebae, plants, insects and large animals; (ii) metal transporting agents; (iii) symbiosis effectors between microbes and plants, nematodes, insects and higher animals; (iv) sexual hormones; and (v) as differentiation factors (Demain and Fang 2000). This wide variability of the structure and activity of the secondary compounds expands the potential industrial importance of the genus *Bacillus* and its related genera (Sansinenea and Ortiz 2011). Besides, *Bacillus* species form spores that can be easily formulated and have high viability compared with vegetative cells. Finally they are commonly diffused in the environment including soil (Sansinenea and Ortiz 2011).

Keywords Secondary metabolites · Bacteriocins · Protein crystals · Siderophores · Polyketides

A. Cherif (✉)
Laboratoire de Biotechnologie et de Valorisation des Bio-Géo Ressources, Institut Supérieur de Biotechnologie, Université de la Manouba, Biotechpole de Sidi Thabet, Sidi Thabet, 2020 Ariana, Tunisia
e-mail: cherif.ameur@gmail.com

I. Chaabouni · A. Guesmi · A. Cherif
Laboratoire Microorganismes et Biomolécules Actives, Faculté des Sciences de Tunis, Université de Tunis, El Manar, 2092 Tunis, Tunisia

I. Chaabouni
Centre National des Sciences et Technologies Nucléaires, Biotechpole Sidi Thabet, Sidi Thabet, 2020 Ariana, Tunisia

E. Sansinenea (ed.), *Bacillus thuringiensis Biotechnology*,
DOI 10.1007/978-94-007-3021-2_18,

18.1 Introduction

The huge diversity characterizing the *Bacillus* species at the taxonomic level, is also noticeable for their metabolic features. These bacteria are able to produce a wide range of secondary metabolites with very different natures and structures and displaying broad spectra of activities. These metabolites; including antibiotics, pigments, toxins, growth promoters (animals and plants), effectors of ecological competition, pheromones, enzyme inhibitors and others bioactive compounds; are originally designed to enable the bacterium to survive in its natural environment (Stein 2005). In general, these metabolites serve as; (i) competitive weapons used against other bacteria, fungi, amoebae, plants, insects and large animals; (ii) metal transporting agents; (iii) symbiosis effectors between microbes and plants, nematodes, insects and higher animals; (iv) sexual hormones; and (v) as differentiation factors (Demain and Fang 2000). This wide variability of the structure and activity of the secondary compounds expands the potential industrial importance of the genus *Bacillus* and its related genera (Sansinenea and Ortiz 2011). Besides, *Bacillus* species form spores that can be easily formulated and have high viability compared with vegetative cells. Finally they are commonly diffused in the environment including soil (Sansinenea and Ortiz 2011).

In this chapter, we will summarize the most important secondary metabolites produced by members of genus *Bacillus*. Bacteriocins of *B. subtilis* and *B. thuringiensis* and other metabolites (as zwittermicin A, siderophore, surfactin …) are described and the potential application of antimicrobial peptides in food, agriculture and pharmaceutical industries are discussed. This biotechnological potential will be highlighted and the safety evaluation of the metabolites and the producer species will be discussed.

18.2 The Most Important Secondary Metabolites of *Bacillus* Species

18.2.1 Bacteriocins from Bacillus Species

Bacteriocins are bacterial ribosomally synthesized antimicrobial peptides lethal to bacteria other than the producing strain, usually against bacteria closely related to the producer (De Vuyst and Vandamme 1994; Riley and Wertz 2002). Compared to those produced by lactic acid bacteria, bacteriocins from the genus *Bacillus* have been relatively less recognized despite their broad antimicrobial spectra and high activity. Nonetheless, bacteriocins from *Bacillus* spp. may have great potential for application in food, agriculture, and pharmaceutical industries to prevent or control spoilage and pathogenic microorganisms (Lee and Kim 2010).

Several species of the *Bacillus* genus are bacteriocin-producers, such as *B. subtilis* which produces subtilin (Jansen and Hirschmann 1944) and subtilosin (Zheng and Slavik 1999), *B. coagulans*, *B. megaterium* and *B. thermoleovorans* producing respectively coagulin (Hyronimus et al. 1998), megacin (Von Tersch and Carlton 1983) and thermoleovorin (Novotny and Perry 1992). Also, several bacteriocins associated with bacteria from the *Bacillus cereus* group have been studied and totally or partially characterized. These include bacteriocins produced by *B. thuringiensis* such as thuricin from strain HD2 (Favret and Yousten 1989), tochicin from strain HD868 (Paik et al. 1997), thuricin 7 from strain BMG1.7 (Cherif et al. 2001), entomocin 9 from strain subspecies *entomocidus* HD9 (Cherif et al. 2003) and entomocin 110 from strain HD110 (Cherif et al. 2008). Many bacteriocins have as well been partially characterized from *B. cereus* species such as cerein 7 from strain BC7 (Oscariz et al. 1999), cerein 8A from strain 8A (Bizani et al. 2005), and the bacteriocin-like inhibitory substance (BLIS) from the *B. cereus* type strain ATCC 14579T (Risøen et al. 2004). Subtilin, the well studied and investigated bacteriocin from *B. subtilis*, will be detailed later in the chapter.

18.2.1.1 Classification

Bacteriocins are usually divided into four classes based on their biochemical and genetic properties (Klaenhammer 1993; Nes et al. 1996). Both class I and II bacteriocins are small (3–10 kDa), cationic, amphiphilic, membrane-active peptides (Lee and Kim 2010). Class I bacteriocins (or lantibiotics) characterized by modification of pre-final peptide with specific enzymes leading to formation of uncommon amino acid residues such as dehydroalanine and dehydrobutyrine, and formation of intramolecular lantionine and β-methyl-lanthionine bridges. These modifications are essential for the antimicrobial activity of the final peptide (Sablon et al. 2000; Quadri 2003).

Bacteriocines of Class II are unmodified peptides and nonlantibiotic heat stable. They can be subdivided into three groups: (i) class IIa, Listeria-active peptides with the consensus sequence -Y-G-N-G-V-X-C- near the N-terminus; (ii) class IIb, two peptide bacteriocins in which both components are required for antimicrobial activity; and (iii) class IIc, thiol-activated peptides requiring reduced cysteine residues for activity (Lee and Kim 2010). Class III bacteriocins includes thermolabile and high molecular mass proteins (>30 kDa), while Class IV bacteriocins are complex peptides associated with other chemical moieties essential for activity.

18.2.1.2 Mode of Action

The antibiotic activity of bacteriocins from Gram-positive bacteria is based on the interaction with the bacterial membranes, mainly composed of negatively charged cardiolipin, phosphatidylglycerol, or phosphatidylserine. Therefore, bacteriocins which most of them are small cationic can be electrostatically attracted to bacterial cell membranes and form pores in the target cells, disrupting membrane potentials and causing cell death (Oscariz et al. 2000). Although bacteriocin antimicrobial activity relies on pore formation, the spectrum of activity depends on the peptide; this observation implies that specific receptor molecules on the surface of target cells may generate differences in antimicrobial activity (Lee and Kim 2010).

18.2.1.3 Regulation of Bacteriocin Production

Production of bacteriocins is an inducible trait in several Gram-positive bacteria. The gene cluster for bacteriocin synthesis is composed of structural genes and accessory genes necessary for bacteriocin transport, regulation and processing to produce the mature form (Nes et al. 1996). Bacteriocins are initially synthesized as premature peptides with a leader or signal sequence at the N-terminal. After being modified and transported out of producer cells, bacteriocins are enzymatically cleaved to remove the leader sequence and to yield the mature peptide. Transcription of accessory proteins is co-regulated with the production of bacteriocins by signal transduction system in a cell density-dependent manner or quorum sensing (Hyungjae and Kim 2011).

Quorum sensing, or cell-to-cell communication, depends on the production, diffusion, and recognition of small signal molecules or inducers (Miller and Bassler 2001; Fuqua et al. 2001). It is associated with sporulation and cell differentiation, biofilm formation, virulence response, production of antibiotics, antimicrobial peptides and toxins, development of genetic competence, conjugative plasmid transfer and other physiological events. Two different types of inducers are involved in communication between bacterial cells: lactone analogs in Gram-negative and small peptides in Gram-positive bacteria (Kleerebezem and Quadri 2001). Most of bacteriocins produced by gram-positive bacteria have been known to be induced by themselves or other peptides (Lee and Kim 2010). Using these induction molecules a serial of enzymatic reactions were involved. When the level of an inducer reaches a threshold in the environment, the inducer from a cell can induce other cells, and subsequently the induced cells produce the bacteriocin at a high level in a short period of time (Hyungjae and Kim 2011).

18.2.1.4 Potential Applications

18.2.1.4.1 Food Applications

Several bacteriocins produced by different groups of bacteria offer potential applications in food and agriculture industries which are suffering from considerable economic loss caused by pathogenic micro-organisms such as *Listeria monocytogenes* and *Staphylococcus* spp (Lee and Kim 2010). The use of bacteriocins in the food industry can help to reduce the addition of chemical preservatives as well as the intensity of heat treatments (Gálvez et al. 2007). Bacteriocins produced by lactic acid bacteria (LAB) have been the subject of intensive investigation in recent years due to their potential use as natural preservatives (Sansinenea and Ortiz 2011). Currently, only nisinA is approved by the US Food and Drug Administration for application as a natural preservative in food. In contrast, *Bacillus* bacteriocins are increasingly becoming more important due to their sometimes broader spectra of inhibition (as compared with most LAB bacteriocins), which may include Gram-negative bacteria, yeasts or fungi, in addition to Gram-positive species, some of which are known to be pathogenic to humans and/or animals (Abriouel et al. 2011). Indeed, the antimicrobial activity of *Bacillus* bacteriocins against a variety of pathogens is as high as that of nisin and other bacteriocins from LAB. This factor in combination with their broad antimicrobial spectra makes bacteriocins from *Bacillus* spp. promising for application in the food, and agriculture industries. *Bacillus* bacteriocins may be advantageous for the development of more efficient biologically safe antimicrobial compounds to avoid the extensive use of antibiotics by agricultural and pharmaceutical industries which lead to the emergence of antibiotic-resistant pathogens (Thomson and Bonomo 2005).

Recently, the European Food Safety Authority (EFSA) introduced the concept of qualified presumption of safety (QPS) for acceptability of bacteria in foods (EFSA 2007a, 2008). *Bacillus cereus* strains have already been approved by EFSA for animal feed (EFSA 2004, 2005, 2007b). The qualification concerning QPS for *Bacillus* species is modified to 'absence of food poisoning toxins, absence of surfactant activities and absence of enterotoxic activities' (EFSA 2008). Nevertheless, bacteriocin-producing strains or their bacteriocin preparations could still be used in food preservation provided they meet the criteria established by EFSA.

Bacteriocins from *Bacillus* have a potential preservative application in different food substrates like in dairy products such as milk and cheeses (Sharma et al. 2009a, b). Two representative examples are bacillocin 490 and cerein 8A; the later has been tested in dairy products (milk and soft cheese) to control the development of *L. monocytogenes* (Bizani and Brandelli 2002; Bizani et al. 2005). In addition, *Bacillus* strains play a central role in the manufacture of alkaline-fermented foods and beverages (Wang and Hesseltine 1982; Odunfa and Oyeyiola 1985; Yokotsuka 1985; Wang and Fung 1996) and as starter culture such as in the case of *B. subtilis* used for the production of the traditional West African condiment dawadawa by fermenting soybeans (Terlabie et al. 2006). Application of bacteriocin-producing strains in these food substrates may offer new opportunities in food biopreservation.

18.2.1.4.2 Applications in Environment and Agriculture

Many of the bacteriocins produced by *Bacillus* species could inhibit plant-pathogenic bacteria and could be applied in the biological control of plant diseases. Plant growth-promoting rhizobacteria (PGPR) are of great interest for application in agriculture.

Species of *Bacillus* are known to act as promoting agents for plant growth and promoting disease resistance in plants, as exemplified by the polypeptide produced by *B. thuringiensis* strain NEB17 (Smith et al. 2008) which was isolated from soybean root nodules (Bai et al. 2002). Bacteriocins displaying antifungal activities could be applied in the biocontrol of plant decay and postharvest control of fruits and vegetables. An overview of the biotechnological potential and application of *Bacillus*-based products in agriculture was recently published by Perez-Garcia et al. (2011).

In the petroleum industry domain, *Bacillus* strains with antimicrobial activity against sulphate-reducing bacteria (SRB) have been isolated from oil reservoirs. This is the case of strain H2O-1 of *B. firmus*, able to withstand the environmental conditions during oil drilling and to produce a small peptide bacteriocin stable to heat and alkaline pH. Hence, this type of bacteriocin offers potential use as a biocide in the petroleum industry for controlling the problems associated with the formation of biofilms and bio-corrosion (Korenblum et al. 2008).

18.2.1.4.3 Applications in Human and Animal Health

Bacteriocin-producing *Bacillus* strains may find application as probiotics for human use, due to their inhibitory activity against intestinal pathogens such as *Clostridium perfringens*, *C. difficile* and others. For example the probiotic strain *B. clausii* O/C produces inhibitory substances towards *Staphylococcus aureus*, *Enterococcus faecium* and *C. difficile* (Urdaci et al. 2004), and thuricin CD specifically targets *C. difficile* (Hill et al. 2009).

Moreover, Bacteriocin producing *Bacilli* could be also used as probiotics in livestock to improve animal health and inhibit pathogenic bacteria. A strain of *B. licheniformis* producing lichenin was reported to exhibit an antibacterial effect against *S. bovis* and *Eubacterium ruminantium*, and to possess remarkable hydrolytic activities against various polysaccharides (Pattnaik et al. 2001). Therefore, it offers potential applications to improve rumen fermentation due to its role as a digestive aid and due to the production of antimicrobial properties (Pattnaik et al. 2001). *Bacillus cereus* ssp. *toyoi* is also used as probiotic supplements in animal feed and for human dietary supplements as well as in registered medicines (Lodemann et al. 2008).

The heat stability and ability of the probiotic to survive the gastric barrier makes them attractive as food additives, and this use is now being taken forward (Cutting 2010). The emergence of multidrug-resistant pathogens and the restriction on the use of antibiotics as growth promoters in animal feed have drawn attention to the search for possible alternatives. In this regard much interest has been focused on bacteriocins due to their great potential applications in medicine. Some studies have evaluated the therapeutic efficacy of bacteriocins in animal disease treatment (Xie et al. 2009).

Bacteriocins from *Bacillus* may also have potential applications as natural contraceptives. One example is subtilosin A, which shows spermicidal activity against spermatozoa from humans and a variety of farm animals (Sutyak et al. 2008a, b).

18.2.2 Main Secondary Metabolites of *Bacillus subtilis*

18.2.2.1 Subtilin

Subtilin is a bacteriocin of class I produced by *Bacillus subtilis* ATCC 6633. Its chemical structure was first unraveled by Gross et al. (1973) and its structural gene was isolated in 1988 (Banerjee and Hansen 1988). Subtilin is composed of 32 amino acids, eight of which are modified and its structure is very similar to that of nisin, produced by lactic acid bacteria and which has been applied as a food preservative in dairy product including cheese (Lee and Kim 2010). Subtilin shows antimicrobial activity in the nanomolar range against a broad spectrum of Gram-positive bacteria. Its antibacterial activity results from permeabilisation of the cytoplasmic membrane of sensitive bacteria (Schuller et al. 1989; Parisot et al. 2008). Subtilin production is controlled both by culture density in a quorum-sensing mechanism in which subtilin plays a pheromone type role and in response to the growth phase (Stein et al. 2002). The gene cluster for subtilin biosynthesis includes two genes required for the regulatory system: *spaR*, encoding a response regulator homologous protein; and *spaK*, a sensor kinase. It has been suggested that subtilin is sensed by histidine protein kinase (*spaK*) and the signal is transferred to the response regulator (spar) by phosphorylation (Stein et al. 2003).

18.2.2.2 Biosurfactant: Surfactin

Surfactin is a cyclic lipopeptide biosurfactant and is produced from *Bacillus subtilis* as a result of non-ribosomal biosynthesis catalyzed by a large multienzyme peptide synthetase complex called the surfactin synthetase (Das et al. 2008). It possesses antimicrobial, antiviral, antitumor, hemolytic, blood anticoagulant, and fibrinolytic activities (Al-Ajlani et al. 2007). These properties of therapeutic and commercial importance make it a truly versatile biomolecule. However, the commercial potential of this surface active molecule could not be fully realized, particularly as a therapeutic agent, mainly because of its hemolytic property (Sen 2010). Surfactin which is one of strongest biosurfactants (Nicolas 2003) has numerous environmental and biotechnological applications (Solaiman 2005). It is used as an enhanced oil recovery through a reduction of the interfacial tension between the oil and water interfaces, or by mediating changes in the wettability index of the system (Schaller et al. 2004). Surfactin has shown particular utility in bioremediation by using it as biosurfactant into environments where bioremediation/biodegradation rates of organic pollutants is to be enhanced (Franzetti et al. 2010). The applications of this biosurfactant in the environmental industries are promising due to its biodegradability, low toxicity and effectiveness in enhancing biodegradation and solubilisation of low solubility compounds (Mulligan 2005). Surfactin is also considered as a biocontrol agent against phythopathogens and insects (Bais et al. 2004; Assie et al. 2002). However, despite the immense potential for commercial, therapeutic and environmental applications of surfactin, its use is currently limited by the high cost of production and by limited understanding of its interactions with cells and with the abiotic environment (Banat et al. 2010).

18.2.3 Main Secondary Metabolites of Bacillus thuringiensis

Bacillus thuringiensis is widely used in agriculture as bioinsecticide for the control of many insect pests. It produces a characteristic proteinaceous crystalline toxin (δ-endotoxin) with a specific activity against certain insect species, and other extracellular compounds such as β-exotoxins, chitinase, antibiotic, anti-fungal (Stabb et al. 1994), anti-cancer cell activities (Yamashita et al. 2000), bacteriocins (Cherif et al. 2003), autolysins (Raddadi et al. 2004, 2005), lactonases (Dong et al. 2002) and siderophores (Bode 2009). In this section, agents responsible for the entomopathogenic and the antifungal activities we will be reviewed.

18.2.3.1 Protein Crystals: The δ-Endotoxins and the β-Exotoxins

Bacillus thuringiensis produces during the sporulation phase, large amount of entomopathogenic proteins that form a parasporal crystal. When ingested by susceptible insect larvae, these insecticidal proteins, known as Cry protoxins or δ-endotoxin of 130–140 kDa, are solubilised and proteolytically digested to yield the active toxic form of about 60 kDa. The activated toxins bind to specific receptors in the epithelial insect midgut leading to pores formation and a loss of normal membrane function (Luthy and Wolfersberger 2000; Schwart and Laprade 2000). As a result of the membrane permeability disruption, epithelial cells lyses and feeding activity is paralyzed. Finally, insect larvae die of starvation, septicaemia or a combination of both (Porcar and Juárez-Pérez 2003). Cry proteins are active against lepidopteran, dipteran and coleopteran insect larvae and have been shown to target nematodes as well, including the intestinal parasite *Nippostrongylus brasiliensis* (Schnepf et al. 1998; Wei et al. 2003; Kotze et al. 2005). This ability to control crop pests has made *B. thuringiensis* one of the most used microorganisms in agriculture as formulated bioinsecticides or in transgenic plants expressing the δ-endotoxin *cry* genes (Roh et al. 2007).

In addition to the δ-endotoxin many isolates of *B. thuringiensis* produce other protein crystals secreted into the culture medium such as β-exotoxin, a non proteinaceous toxin which does not have specific receptor and thus, may have detrimental effects on non-target organisms. Even thought to make part of its potential pathogenic traits, β-exotoxins enhance the insecticidal activity of the bacterium. β-exotoxins inhibits the synthesis of RNA by competing with ATP for binding sites, thereby affecting insect moulting and pupation, and in some cases having teratogenic effects. This toxin is particularly active against dipteran, coleopteran, and lepidopteran species, but it is also active against beneficial species such as the honeybee *Apis mellifera* (Espinasse et al. 2002, 2004).

18.2.3.2 Zwittermicin A

Zwittermicin A is a linear aminopolyol antibiotic, which represents a new class of antibiotics. This compound was firstly isolated in *B. cereus* for its ability to suppress a fungal plant disease (Silo-Suh et al. 1998). Moreover, it has an unusual chemical structure that includes a D-amino acid and ethanolamine and glycolyl moieties, as well as unusual terminal amide that generated from the modification of the non-proteinogenic amino acid ureidoalanine (Sansinenea and Ortiz 2011). Studies on its synergistic effect with the entomopathogenic activity showed that Zwittermicin A enhances the insecticidal activity of the *B. thuringiensis* endotoxin (Zhou et al. 2008). Zwittermicin A has proved difficult to isolate in substantial quantities due to its highly-polar, charged nature at physiological pH and sensitivity to alkaline conditions. This difficulty in obtaining the compound from natural sources underscores an outstanding need for practical syntheses of zwittermicin A (Rogers and Molinski 2009).

18.3 Other Metabolites and Activities of *Bacillus* Species

18.3.1 Siderophore

Plant growth and reproduction can be severely affected by various biotic and abiotic stresses. Among the abiotic stresses, iron (Fe) deficiency constitutes a major factor leading to a reduction in crop yield, especially in calcareous soils in which the solubility of Fe is extremely low (Kobayashi et al. 2005).

One of the most widely utilized mechanisms of microbial iron acquisition is the production and excretion of siderophores, low molecular weight iron chelators that bind ferric iron with extremely high affinity and shuttle it into the cells (Zawadzka et al. 2009). Thus the presence of siderophore-producing micro-organisms in the rhizosphere contributes to plant health by complexing iron and making it less available to phytopathogenes that are generally not able to produce comparable Fe-transport systems (Chen et al. 2009). It has been reported that under sterile soil system, plants show iron-deficiency symptoms and have fairly low iron level in roots which suggests the role of soil microbial activity in iron acquisition and plant growth (Raddadi et al. 2007). Siderophore production was initially reported for Gram-negative bacteria and possibly implicated in their virulence potential. In *Bacilli*, several siderophores were detected, mainly schizokinen from *B. megaterium* and other from *B. subtilis*, *B. licheniformis* and the *B. cereus* group: Bacillibactin and petrobactin are produced by *B. cereus* (Park et al. 2005), *B. anthracis*

(Cendrowski et al. 2004) and *B. thuringiensis* (Wilson et al. 2006). This character in the case of *B. thuringiensis* could be relevant for biocontrol of phytopathogenic fungi because of competition effects for iron, but also for providing the plant with iron (Raddadi et al. 2007).

18.3.2 Polyketides

Polyketides are the other dominant family of secondary metabolites having relevant bioactivities. They are biosynthesized through decarboxylative condensation of malonyl-CoA derived extender units in a similar process to fatty acid synthesis (a Claisen condensation). Their biosynthetic pathway follows generally the same logic as in non ribosomally synthesized peptides and requires at least three domains. Essential domains of the modules harboured in bacterial type I polyketides synthase are acyl transferase (AT), ketosynthase and an acyl carrier protein (ACP) which needs to be activated by Ppant-transferse (Chen et al. 2009). Polyketide represented by a wide range of active family molecules are currently used in human and animal health and in agriculture as antimicrobial (erythromycin, rifamycin), antifungal, immunosuppressant and antitumor agents. Polyketides are actually difficult to synthesise chemically which make their heterologous expression a promising alternative to the use of natural producers. Heterologous production systems can facilitate analysis of the catalytic properties of polyketide producing enzymes. Examples of heterologous hosts for polyketide protein production include bacteria such as *Escherichia coli*, *Bacillus subtilis*, and *Streptomyces coelicolor* (Rude and Khosla 2004).

Several polyketides were found to possess interesting activities. Bacillaene, produced by *B. subtilis* and *B. amyloliquifaciens*, inhibits prokaryotic protein synthesis by means of unknown mechanism, and exhibits high bacteriostatic activity against a wide spectrum of bacteria. Difficilin, another polyketides, was shown to be promising in its suppressive action against plant pathogen (Hamdache et al. 2011).

18.3.3 AHL-Lactonases

In Gram-negative bacteria, the Quorum-Sensing (complex communication system) rely on the interaction of N-acylhomoserine lactones (AHLs), molecules that share identical homoserine lactone rings but vary in length and the substitution of the acyl side chain. AHL signals are involved in the regulation of a range of important biological functions, including luminescence, antibiotic production, plasmid transfer, motility, virulence, biofilm formation and functional coordination among microbial communities (Whitehead et al. 2001; Zhang 2003; Federle and Bassler 2003). The quorum-sensing pathways could be subverted by expressing "quorum quenching" enzymes that can hydrolyze AHL-signaling molecules. Dong et al. (2002) studies have shown that *B. thuringiensis* strains, producing AHL-lactonase suppress the Quorum-Sensing dependent virulence of the plant bacterial pathogen *E. carotovora* through signal interference. These findings, illustrate the promising potential to explore the microbial antagonistic mechanisms such as signal interference, for the control and prevention of infectious diseases. Interfering in signaling pathways of pathogenic bacteria and suppressing their virulence potential, represent a serious alternative to the use of antibiotics and prevent from resistance appearance.

18.3.4 Plant Growth Promoting Potential

Rhizosphere, phylloplane and free-living soil bacteria with plant growth promoting (PGP) activities have received considerable attention in this last decade. A range of mechanisms has been found to be associated with this activity, some of which facilitate indirectly the growth of plant by producing secondary metabolites that antagonises bacterial and fungal pathogens (Ahmad et al. 2008) or by biosynthesis of different plant growth hormones (auxin, gibberellins, cytokinin). It has been reported that *B. thuringiensis* harbours and expresses several plant growth promoting determinants that could enhance the plant growth. *B. thuringiensis* isolates are capable to hydrolyze mineral phosphates, release indole −3-acetic acid and produce the ACC deaminase, suggesting that they can interfere with plant hormone balance in a complex way (Raddadi et al. 2008). Also representatives of the *B. subtilis*, *B. amyloliquefaciens* and *B. cereus* has been reported to synthesize auxin, gibberellins, cytokinin, abscisic acid and antibiotic compounds with antifungal and antibacterial activities like fengycin, surfactin, iturin A and bacillibactin (Arguelles-Arias et al. 2009; Hamdache et al. 2011). In addition to these two main mechanisms, PGP activity could be ensured through nitrogen fixation in a non-symbiotic manner, such as in the case of *Paenibacillus polymyxa*, *P. macerans* and *P. azotofixans*.

Volatile compounds enhancing PGP activity were also reported from *Bacilli*. Several species produces volatile isoprene, sporulenes (A, B, and C) and pumilin. Sporulenes was shown to be implicated in sporulation by protecting *Bacilli* spores against oxidative stress. Pumilin, produced by *B. pumilus*, is active against Gram-positive bacteria and also show high cytotoxic activity in bioassays against human cancer cell. The isocoumarins represent an interesting group of phenolic compounds produced by *Bacilli*. They exhibited antibacterial activity, suppress inflammation and ulcer activity, and some of them have a potent inhibitor of protein synthesis in mammalian cells (Hamdache et al. 2011). Other volatile compounds, i.e. 2,3-butanediol and acetoin, released by *B. pumilus*, *B. pasteurii*, *B. subtilis* and *B. amyloliquifaciens*, were also shown to significantly enhance plant growth (Ryu et al. 2003).

18.4 Secondary Metabolites from Marine *Bacilli*

Most of the *Bacillus* species screened for production of secondary metabolites are isolated from terrestrial habitats. During the last two decades, the increasing needs for new bioactive metabolites stimulated to look for unconventional new sources of bioactive molecules. Recent studies demonstrated the huge biodiversity of *Bacillus* species in marine environment (Ettoumi et al. 2009; Sass et al. 2008). Marine sediments and superficial waters have yielded microorganisms closely related to terrestrial ones with some highly unexpected metabolites (Kelecom 2002). *Halobacillus salinus*, a marine Gram-positive bacterium, secretes secondary metabolites capable of quenching quorum sensing-controlled behaviours in several Gram negative strains (Teasdale et al. 2009). Cereulide and homocereulide, obtained from *B. cereus* isolated from the surface of a snail in Japan, are cyclodespeptides of both L-and D-amino acids with potent cytotoxic activity against P388 leukaemic cell lines. *Bacillus* sp. Isolated from deep waters in the western Pacific were able to produce an aminoglucose with high antibacterial activity (Pietra 1997). In addition, antimicrobial activity was detected in *Bacillus* isolated from marine sponge, well known to produce a number of biologically active secondary metabolites. Several studies have demonstrated that microbial isolates associated with sponges produced the same compounds isolated from sponges (Wang 2006). Other secondary metabolites with novel structures and biotechnological activities with attractive industrial features will be identified from marine *Bacilli*.

18.5 Safety Evaluation and Risk Assessment

With all the different activities listed above, *Bacillus* species are considered as an inexhaustible reservoir of active biomolecules with potential biotechnological applications in diverse domains. However, the "safe" status of the producing species or their metabolites is not systematically verified. Indeed, several *Bacillus* species are pathogens to human and animals or are able to express virulence traits, in addition to the production of resistant endospores (Granum 2007; Stenfors Arnesen et al. 2008). These considerations stress the safety challenge of the use of *Bacillus*-based products in the food, agriculture and pharmaceutical industries and point out the need of a careful safety evaluation prior to any application (Hong et al. 2005) and taking into account the intraspecies divergent virulence characteristics (Abriouel et al. 2011). We focus in this section on *B. thuringiensis* species, one of the most successfully commercialised bioinsecticide worldwide.

Bacillus thuringiensis is used as biopesticide since more than 50 yrs owing to its entomopathogenic activity. Moreover, *B. thuringiensis* represent a potential biocontrol agent against different fungal and bacterial species. However, this species is closely related to *B. anthracis*, the etiological agent of anthrax (Turnbuel et al. 1992) and *B. cereus*, the opportunistic human pathogen (Lund and Granum 1997).

B. cereus causes food poisoning and other infections (Beecher et al. 2000). Two principal types of food poisoning caused by *B. cereus*, emetic and diarrhoeal, have been described. The emetic type which is characterized by nausea and vomiting is effected by cereulide while diarrhoeal types which are characterized by abdominal pain and diarrhea are attributed to enterotoxins (Hansen and Hendriksen 2001). In addition to the phylogenetic relatedness, it has to be mentioned that *B. thuringiensis* has been isolated from outbreaks of gastrointestinal diseases and several genetic determinants of virulence, typical of *B. cereus*, have been found in most *B. thuringiensis* strains even those used in commercial formulations (Hansen and Hendriksen 2001; Yang et al. 2003; Hoffmaster et al. 2004). Hence, it is important for the bio-insecticide industry to consider that although *B. thuringiensis* insecticide has been used for many years, introduction of *B. thuringiensis* spores into the human food chain through the application of this bacterial species to crops, followed by spore regermination may cause a risk of food-borne poisoning cases. To reduce such a risk, it is urgent to focus efforts on the isolation of safe, non enterotoxigenic *B. thuringiensis* strains even thought that it is difficult to find such isolates in nature (Raddadi et al. 2007). The screening for polyvalent biocontrol and safe *B. thuringiensis* strain with capacity to protect plants against phytopathogenic insects, fungi and bacteria and without enterotoxigenicity, cytotoxicity and psychrotrophic characteristics could be very promising in field application (Raddadi et al. 2009).

18.6 Conclusion and Perspectives

Bacillus species represent a rich source of secondary metabolites that exhibit strong antifungal and antibacterial activities and enable the bacterium to survive in its natural environment (Stein 2005). The wide variability of these secondary metabolites with broad biological activities has great potential of applications in the agriculture, food and pharmaceutical industries to prevent or control spoilage and pathogenic microorganisms. Besides, the sporulation capacity of *Bacillus* strains, confer an ecological advantage and ability to survive and grow in different ecosystems and make easier products formulation.

Apart of the recognised entomopathogenic activity mediated mainly by the δ-endotoxins, and the antimicrobial activity due to the production of bacteriocins and antibiotics, *Bacillus* species were shown to possess other interesting capabilities. This battery of secondary metabolites, with broad or narrow and specific spectra of activities, is in general complementary to the overall anti-biological potential of *Bacilli*. The exploitation of such a potential, is yet limited to some fields principally in agriculture, which point out the need to urgently undertake feasibility and applicability researches on existing molecules rather then the detection of new metabolites. The application limitation is also caused by the "safe" status not yet recognised for *Bacilli* unlike lactic acid bacteria, which are generally regarded as safe or "GRAS" (O'Sullivan et al. 2002). This general perception of risky microorganisms is persistent despite a safe use history in some traditional food and industry (Pedersen et al. 2002).

In this perspective, the development of *Bacillus*-based probiotics for animal feeding and formulations for plant growth, constitute a promising applied research field. Selected *Bacilli* should have polyvalent activity with several expressed secondary metabolites, and a "safe" profile. An example of this approach was developed for honeybees where a safe and polyvalent strain of B. thuringiensis was selected. This bio-control strain enhances the honeybee larval growth, inhibits the pathogen bacterium *Paenibacillus* larvae that cause American foulbrood disease, and is able to counteract the invasion of the fungal and lepidopteran pathogens of bees (Cherif et al. 2007).

References

Abriouel H, Franz CM, Ben Omar N, Gálvez A (2011) Diversity and applications of *Bacillus* bacteriocins. FEMS Microbiol Rev 35:201–232

Ahmad F, Ahmad I, Khan MS (2008) Screening of free-living rhizospheric bacteria for their multiple plant growth promoting activities. Microbiol Res 163:173–181

Al-Ajlani MM, Sheikh MA, Ahmad Z, Hasnain S (2007) Production of surfactin from *Bacillus subtilis* MZ-7 grown on pharmamedia commercial medium. Microb Cell Fact 6:17

Arguelles-Arias A, Ongena M, Halimi B, Lara Y, Brans A, Joris B, Fickers P (2009) *Bacillus amyloliquefaciens* GA1 as a source of potent antibiotics and other secondary metabolites for biocontrol of plant pathogens. Microb Cell Fact 8:63

Assie LK, Deleu M, Arnaud L, Paquot M, Thonart P, Gaspar CH, Haubruge E (2002) Insecticide activity of surfactins and iturins from a biopesticide *Bacillus subtilis* Cohn (S499 strain). Meded Rijksuniv Gent Fak Landbouwkd Toegep Biol Wet 67:647–655

Bai Y, D'Aoust F, Smith DL, Driscoll BT (2002) Isolation of plant growth-promoting *Bacillus* strains from soybean root nodules. Can J Microbiol 48:230–238

Bais HP, Fall R, Vivanco JM (2004) Biocontrol of *Bacillus subtilis* against infection of Arabidopsis roots by *Pseudomonas syringae* is facilitated by biofilm formation and surfactin production. Plant Physiol 134:307–319

Banat IM, Franzetti A, Gandolfi I, Bestetti G, Martinotti MG, Fracchia L, Smyth TJ, Marchant R (2010) Microbial biosurfactants production, applications and future potential. Appl Microbiol Biotechnol 87:427–444

Banerjee S, Hansen JN (1988) Structure and expression of a gene encoding the precursor of subtilin, a small protein antibiotic. J Biol Chem 263:9508–9514

Beecher DJ, Olsen TW, Somers EB, Wong ACL (2000) Evidence for contribution of tripartite hemolysin BL, phosphatidylcholine- preferring phospholipase C, and collagenase to virulence of *Bacillus cereus* endophthalmitis. Infect Immun 68:5269–5276

Bizani D, Brandelli A (2002) Characterization of a bacteriocin produced by a newly isolated *Bacillus* sp. strain 8A. J Appl Microbiol 93:512–519

Bizani D, Motta AS, Morrissy JAC, Terra RM, Souto AA, Brandelli A (2005) Antibacterial activity of cerein 8A, a bacteriocin-like peptide produced by *Bacillus cereus*. Int Microbiol 8:125–131

Bode HB (2009) Entomopathogenic bacteria as a source of secondary metabolites. Curr Opin Chem Biol 13:224–230

Cendrowski S, MacArthur W, Hanna P (2004) *Bacillus anthracis* requires siderophore biosynthesis for growth in macrophages and mouse virulence. Mol Microbiol 51:407–417

Chen XH, Koumoutsia A, Scholza R, Schneiderb K, Vaterb J, Süssmuthb R, Piel J, Borrissa R (2009) Genome analysis of *Bacillus amyloliquefaciens* FZB42 reveals its potential for biocontrol of plant pathogens. J Biotechnol 140:27–37

Cherif A, Ouzari H, Daffonchio D, Cherif H, Ben Slama K, Hassen A, Jaoua S, Boudabous A (2001) Thuricin 7: a novel bacteriocin produced by *Bacillus thuringiensis* BMG1.7, a new strain isolated from soil. Lett Appl Microbiol 32:243–247

Cherif A, Chehimi S, Limem F, Hansen BM, Hendriksen NB, Daffonchio D, Boudabous A (2003) Detection and characterization of the novel bacteriocin entomocin 9, and safety evaluation of its producer, *Bacillus thuringiensis* subsp. *Entomocidus* HD9. J Appl Microbiol 95:990–1000

Cherif A, Hamdi C, Essanaa J, Rezgui W, Raddadi N, Rizzi A, Daffonchio S, Barbouche N, Boudabous A (2007) Versatility of *Bacillus thuringiensis* in biocontrol: perspectives and limitations of current applications. Entomol Res 37(S1):A34–A35

Cherif A, Rezgui W, Raddadi N, Daffonchio D, Boudabous A (2008) Characterization and partial purification of entomocin 110, a newly identified bacteriocin from *Bacillus thuringiensis* subsp. *Entomocidus* HD110. Microbiol Res 163:684–692

Cutting SM (2010) *Bacillus* probiotics. Food Microbiol 28:214–220

Das P, Mukherjee S, Sen R (2008) Genetic regulations of the biosynthesis of microbial surfactants: an overview. Biotechnol Genet Eng Rev 25:165–185

Demain AL (1999) Pharmaceutically active secondary metabolites of microorganisms. Appl Microbiol Biotechnol 52:455–463

Demain AL, Fang A (2000) The natural functions of secondary metabolites. Adv Biochem Eng Biotechnol 69:1–39

De Vuyst L, Vandamme EJ (1994) Nisin, a lantibiotic produced by *Lactococcus lactis* subsp *lactis*: properties, biosynthesis, fermentation and application. In: De Vuyst L, Vandamme EJ (eds) Bacteriocins of lactic acid bacteria. Microbiology, genetics and applications. Chapman & Hall, London, pp 151–221

Dong YH, Gusti AR, Zhang Q, Xu JL, Zhang LH (2002) Identification of quorum-quencing N-acyl homoserine lactonases from *Bacillus* species. Appl Environ Microbiol 68:1754–1759

EFSA (2004) Opinion of the scientific panel on additives and products or substances used in animal feed (FEEDAP) on the efficacy of product Toyocerin for pigs in fattening. EFSA J 62:1–5

EFSA (2005) Scientific opinion of the panel on additives and products or substances used in animal feed (FEEDAP) on the modification of terms of authorisation of the microorganism preparation of *Bacillus cereus* var. *toyoi* (NCIMB 40112/CNCMI-1012) (Toyocerins) authorised as a feed additive in accordance with Directive 70/524/EEC. EFSA J 288:1–7

EFSA (2007a) Introduction of a qualified presumption of safety (QPS) approach for assessment of selected microorganisms referred to EFSA. EFSA J 587:1–16

EFSA (2007b) Scientific opinion of the panel on additives and products or substances used in animal feed (FEEDAP) on the safety and efficacy of Toyocerins (*Bacillus cereus* var. *toyoi*) as feed additive for turkeys. EFSA J 549:1–11

EFSA (2008) The maintenance of the list of QPS microorganisms intentionally added to foods or feeds. Scientific opinion of the panel on biological hazards. EFSA J 923:1–48

Entian KD, Vos WM (1996) Genetics of subtilin and nisin biosyntheses of lantibiotics. Antonie Van Leeuwenhoek 69:109–117

Espinasse S, Gohar M, Lereclus D, Sanchis V (2002) An ABC transporter from *Bacillus thuringiensis* is essential for beta-exotoxin I production. J Bacteriol 184:5848–5854

Espinasse S, Gohar M, Lereclus D, Sanchis V (2004) An extracytoplasmic-function sigma factor is involved in a pathway controlling beta-exotoxin I production in *Bacillus thuringiensis* subsp. *thuringiensis* strain 407-1. J Bacteriol 186:3108–3116

Ettoumi B, Raddadi N, Borin S, Daffonchio D, Boudabous A, Cherif A (2009) Diversity and phylogeny of culturable spore-forming *Bacilli* isolated from marine sediments. J Basic Microbiol 49:1–11

Favret ME, Yousten AA (1989) Thuricin: the bacteriocin produced by *Bacillus thuringiensis*. J Invertebr Pathol 53:206–216

Federle MJ, Bassler BL (2003) Interspecies communication in bacteria. J Clin Invest 112:1291–1299

Franzetti A, Tamburini E, Banat IM (2010) Applications of biological surface active compounds in remediation technologies. Adv Exp Med Biol 672:121–134

Fuqua C, Parsek MR, Greenberg EP (2001) Regulation of gene expression by cell-to-cell communication: acyl-homoserine lactone quorum sensing. Annu Rev Genet 35:439–468

Gálvez A, Abriouel H, López RL, Ben Omar N (2007) Bacteriocin-based strategies for food biopreservation. Int J Food Microbiol 120:51–70

Granum PE (2007) *Bacillus cereus*. In: Doyle MP, Beuchat LR (eds) Food microbiology, fundamentals and frontiers, 3rd edn. ASM Press, Washington, pp 445–455

Gross E, Kiltz HH, Nebelin E (1973) Subtilin, VI: structure of subtilin. Hoppe Seylers Z Physiol Chem 354:810–822

Hamdache A, Lamarti A, Aleu J, Collado IG (2011) Non-peptide metabolites from the genus *Bacillus*. J Nat Prod 74:893–899

Hansen BM, Handriksen NB (2001) Detection of enterotoxic *Bacillus cereus* and *Bacillus thuringiensis* strains by PCR analysis. Appl Environ Microbiol 67:185–189

Hill C, Rea M, Ross P (2009) Thuricin CD, an antimicrobial for specifically targeting *Clostridium difficile*. Patent: WO 2009068656-A1 13 04-JUN-2009; TEAGASC, The Agriculture and Food Development Authority (IE), University College Cork-National University of Ireland, Cork (IE)

Hoffmaster AR, Ravel J, Rasko DA, Chapman GD, Chute MD, Marston CK, De BK, Sacchi CT, Fitzgerald C, Mayer LW, Maiden MCJ, Priest FG, Barker M, Jiang L, Cer RZ, Rilstone J, Peterson SN, Weyant RS, Galloway DR, Rea TD, Popovic T, Fraser CM (2004) Identification of anthrax toxin genes in a *Bacillus cereus* associated with an illness resembling inhalation anthrax. Proc Natl Acad Sci USA 101:8449–8454

Hong HA, Duc le H, Cutting SM (2005) The use of bacterial spore formers as probiotics. FEMS Microbiol Rev 29:813–835

Hyronimus B, Le Merrec C, Urdaci MC (1998) Coagulin, a bacteriocin-like inhibitory substance produced by *Bacillus coagulans* I4. J Appl Microbiol 85:42–50

Hyungjae L, Kim HY (2011) Lantibiotics, class I bacteriocins from the genus *Bacillus*. J Microbiol Biotechnol 21:229–235

Jansen EF, Hirschmann DJ (1944) Subtilin, an antibacterial substance of *Bacillus subtilis*.: culturing conditions and properties. Arch Biochem 4:297–309

Karlovsky P (2008) Secondary metabolites in soil ecology. Soil Biol 14:1–19

Kelecom A (2002) Secondary metabolites from marine microorganisms. An Acad Bras Cienc 74:151–170

Klaenhammer TR (1993) Genetics of bacteriocins produced by lactic acid bacteria. FEMS Microbiol Rev 12:39–856

Kleerebezem M, Quadri LE (2001) Peptide pheromone-dependent regulation of antimicrobial peptide production in Gram-positive bacteria: a case of multicellular behavior. Peptides 22:1579–1596

Kobayashi T, Suzuki M, Inoue H, Itai RN, Takahashi M, Nakanishi H, Mori S, Nishizawa NK (2005) Expression of iron-acquisition-related genes in iron-deficient rice is co-ordinately induced by partially conserved iron-deficiency-responsive elements. J Exp Bot 56:1305–1316

Korenblum E, Sebastián GV, Paiva MM, Coutinho CM, Magalhães FC, Peyton BM, Seldin L (2008) Action of antimicrobial substances produced by different oil reservoir *Bacillus* strains against biofilm formation. Appl Microbiol Biotechnol 79:97–103

Kotze AC, O'Grady J, Gough JM, Pearson R, Bagnall NH, Kemp DH, Akhurst RJ (2005) Toxicity of *Bacillus thuringiensis* to parasitic and free-living life-stages of nematode parasites of livestock. Int J Parasitol 35:1013–1022

Lee H, Kim HY (2010) Lantibiotics, class I bacteriocins from the genus *Bacillus*. J Microbiol Biotechnol 21:229–235

Lodemann U, Lorenz BM, Weyrauch KD, Martens H (2008) Effects of *Bacillus cereus* var. *toyoi* as probiotic feed supplement on intestinal transport and barrier function in piglets. Arch Anim Nutr 62:87–106
Lund T, Granum PE (1997) Comparison of biological effect of the two different enterotoxin complexes isolated from three different strains of *Bacillus cereus*. Microbiology 143:3329–3336
Luthy P, Wolfersberger MG (2000) Pathogenisis of *Bacillus thuringiensis* toxins. In: Charles JF, Delécluse A, Nielsen-LeRoux C (eds) Entomopathogenic bacteria: from laboratory to field application. Kluwer Academic, Dordrecht, pp 167–180
Miller MB, Bassler BL (2001) Quorum sensing in bacteria. Annu Rev Microbiol 55:165–199
Mulligan CN (2005) Environmental applications for biosurfactants. Environ Pollut 133:183–198
Nes IF, Diep DB, Havarstein LS, Brurberg MB, Eijsink V, Holo H (1996) Biosynthesis of bactericins in lactic acid bacteria. Antonie Van Leeuwenhoek 70:113–128
Nes IF, Yoon SS, Diep DB (2007) Ribosomally synthesiszed antimicrobial peptides (bacteriocins) in lactic acid bacteria: a review. Food Sci Biotechnol 16:675–690
Nicolas JP (2003) Molecular dynamics simulation of surfactin molecules at the water hexane interface. Biophys J 85:1377–1391
Novotny JF, Perry JJ (1992) Characterization of bacteriocins from two strains of *Bacillus thermoleovorans*, a thermophilic hydrocarbon-utilizing species. Appl Environ Microbiol 58:2393–2396
O'Sullivan L, Ross RP, Hill C (2002) Potential of bacteriocin producing lactic acid bacteria for improvements in food safety and quality. Biochimie 84:593–604
Odunfa SA, Oyeyiola GF (1985) Microbiological study of the fermentation of ugba, a Nigerian indigenous fermented food flavour. J Plant Foods 6:155–163
Oscariz JC, Lasa I, Pisabarro AG (1999) Detection and characterization of cerein 7, a new bacteriocin produced by *Bacillus cereus* with a broad spectrum of activity. FEMS Microbiol Lett 178:337–341
Oscariz JC, Pisabarro AG (2000) Characterization and mechanism of action of cerein 7, a bacteriocin produced by *Bacillus cereus* Bc7. J Appl Microbiol 89:361–369
Paik HD, Bae SS, Pan JG (1997) Identification and partial characterization of tochicin, a bacteriocin produced by *Bacillus thuringiensis* subsp. *tochigiensis*. J Ind Microbiol Biotechnol 19:294–298
Parisot J, Carey S, Breukink E, Chan WC, Narbad A, Bonev B (2008) Molecular mechanism of target recognition by subtilin, a Class I lanthionine antibiotic. Antimicrob Agents Chemother 52:612–618
Park RY, Choi MH, Sun HY, Shin SH (2005) Production of catechol-siderophore and utilization of transferrin-bound iron in *Bacillus cereus*. Biol Pharm Bull 28:1132–1135
Pattnaik P, Kaushik JK, Grover S, Batish VK (2001) Purification and characterization of a bacteriocin-like compound (Lichenin) produced anaerobically by *Bacillus licheniformis* isolated from water buffalo. J Appl Microbiol 91:636–645
Pedersen PB, Bjrnvad ME, Rasmussen MD, Petersen JN (2002) Cytotoxic potential of industrial strains of *Bacillus* sp. Regul Toxicol Pharm 36:155–161
Perez-Garcia A, Romero D, Vicente A de (2011) Plant protection and growth stimulation by microorganisms: biotechnological applications of *Bacilli* in agriculture. Curr Opin Biotechnol 22:187–193
Pietra F (1997) Secondary metabolites from marine microorganisms: bacteria, protozoa, algae and fungi. Achievements and prospects. Nat Prod Rep 14:453–464
Porcar M, Juárez-Pérez V (2003) PCR-based identification of *Bacillus thuringiensis* pesticidal crystal genes. FEMS Microbiol Rev 26:419–432
Quadri LEN (2003) Regulation of class II bacteriocin production by cell-cell signaling. J Microbiol 41:175–182
Raddadi N, Cherif A, Mora D, Ouzari H, Boudabous A, Molinari F, Daffonchio D (2004) The autolytic phenotype of *Bacillus thuringiensis*. J Appl Microbiol 97:158–168
Raddadi N, Cherif A, Mora D, Brusetti L, Borin S, Boudabous A, Daffonchio D (2005) The autolytic phenotype of the *Bacillus cereus* group. J Appl Microbiol 99:1070–1081

Raddadi N, Cherif A, Ouzari H, Marzorati M, Lorenzo Brusetti L, Boudabous A, Daffonchio D (2007) *Bacillus thuringiensis* beyond insect biocontrol: plant growth promotion and biosafety of polyvalent strains. Ann Microbiol 57:481–494

Raddadi N, Cherif A, Boudabous A, Daffonchio D (2008) Screening of plant growth promoting traits of *Bacillus thuringiensis*. Ann Microbiol 58:47–52

Raddadi N, Belaouis A, Tamagnini I, Hansen BM, Hendriksen NB, Boudabous A, Cherif A, Daffonchio D (2009) Characterization of polyvalent and safe *Bacillus thuringiensis* strains with potential use for biocontrol. J Basic Microbiol 48:1–11

Riley MA, Wertz J (2002) Bacteriocins: evolution, ecology and application. Annu Rev Microbiol 56:117–137

Risøen PA, Rønning P, Hegna IK, Kolstø AB (2004) Characterization of a broad range antimicrobial substance from *Bacillus cereus*. J Appl Microbiol 96:648–655

Rogers EW, Molinski TF (2009) Asymmetric synthesis of diastereomeric diaminoheptanetetraols. A proposal for the configuration of (+)-zwittermicin a. Org Lett 9:437–440

Roh JY, Choi JY, Li MS, Jin BR, Je HE (2007) *Bacillus thuringiensis* as a specific, safe, and effective tool for insect pest control. J Microbiol Biotechnol 17:547–559

Rude MA, Khosla C (2004) Engineered biosynthesis of polyketides in heterologous hosts. Chem Eng Sci 59:4693–4701

Ryu CM, Farag MA, Hi CH, Reddy MS, Wei HX, Paré PW, Kloepper JW (2003) Bacterial volatiles promote growth in Arabidopsis. Proc Natl Acad Sci USA 100:4927–4932

Sablon E, Contreras B, Vandamme E (2000) Antimicrobial peptides of lactic acid bacteria: mode of action, genetics and biosynthesis. Adv Biochem Eng Biotechnol 68:21–60

Sansinenea E, Ortiz A (2011) Secondary metabolites of soil *Bacillus* spp. Biotechnol Lett 33:1523–1538

Sass AM, Mckew BA, Sass H, Fichtel J, Timmis KN, Mcgenity TJ (2008) Diversity of *Bacillus* like organisms isolated from deep-sea hypersaline anoxic sediments. Saline Syst 4(8):1–11

Schaller KD, Fox SL, Bruhn DF, Noah KS, Bala GA (2004) Characterization of surfactin from *Bacillus subtilis* for application as an agent for enhanced oil recovery. Appl Biochem Biotechnol 113–116:827–836

Schnepf E, Crickmore N, Van Rie J, Lereclus D, Baum J, Feitelson J, Zeigler DR, Dean DH (1998) *Bacillus thuringiensis* and its pesticidal crystal proteins. Microbiol Mol Biol Rev 62:775–806

Schuller F, Benz R, Sahl HG (1989) The peptide antibiotic subtilin acts by formation of voltage-dependent multi-state pores in bacterial and artificial membranes. Eur J Biochem 182:181–186

Schwart JL, Laprade R (2000) Membrane permeabilisation by *Bacillus thuringiensis* toxins: protein formation and pore insertion. In: Charles JF, Delécluse A, Nielsen-LeRoux C (eds) Entomopathogenic bacteria: from laboratory to field application. Kluwer Academic Publishers, Dordrecht, pp 199–218

Sen R (2010) Surfactin: biosynthesis, genetics and potential applications. Adv Exp Med Biol 672:316–323

Sharma N, Attri A, Gautam N (2009a) Purification and characterization of bacteriocin like substance produced from *Bacillus lentus* with perspective of a new biopreservative for food preservation. Pak J Sci Ind Res 52:191–199

Sharma N, Kapoor G, Gautam N, Neopaney B (2009b) Characterization of partially purified bacteriocin of *Bacillus* sp. MTCC 43 isolated from rhizosphere of radish (*Raphanus sativus*) and its application as a potential food biopreservative. J Sci Ind Res 68:881–886

Silo-Suh LA, Stabb EV, Raffel SJ, Handelsman J (1998) Target range of zwittermicin A, an aminopolyol antibiotic from *Bacillus cereus*. Curr Microbiol 37:6–11

Smith D, Lee KD, Gray E, Souleimanov A, Zhou X (2008) Use of bacteriocins for promoting plant growth and disease resistance. US Patent Application number: 20080248953

Solaiman D (2005) Applications of microbial biosurfactants. Inform 16:408–410

Stabb EV, Jacobson LM, Handelsman J (1994) Zwittermycin A-producing strains of *Bacillus cereus* from diverse soils. Appl Environ Microbiol 60:4404–4412

Stein T (2005) *Bacillus subtilis* antibiotics: structures, syntheses and specific functions. Mol Microbiol 56:845–857

Stein T, Borchert S, Kiesau P, Heinzmann S, Kloss S, Klein C, Helfrich M, Entian KD (2002) Dual control of subtilin biosynthesis and immunity in *Bacillus subtilis*. Mol Microbiol 44:403–416

Stein T, Heinzmann S, Kiesau P, Himmel B, Entian KD (2003) The spa-box for transcriptional activation of subtilin biosynthesis and immunity in *Bacillus subtilis*. Mol Microbiol 47:1627–1636

Stenfors Arnesen LP, Fagerlund A, Granum PE (2008) From soil to gut: *Bacillus cereus* and its food poisoning toxins. FEMS Microbiol Rev 32:579–606

Sutyak KE, Anderson RA, Dover SE, Feathergill KA, Aroutcheva AA, Faro S, Chikindas ML (2008a) Spermicidal activity of the safe natural antimicrobial peptide subtilosin. Infect Dis Obstet Gynecol 2008:540–758

Sutyak KE, Wirawan RE, Aroutcheva AA, Chikindas ML (2008b) Isolation of the *Bacillus subtilis* antimicrobial peptide subtilosin from the dairy product-derived *Bacillus amyloliquefaciens*. J Appl Microbiol 104:1067–1074

Teasdale ME, Liu J, Wallace J, Akhlaghi F, Rowley DC (2009) Secondary metabolites produced by the marine bacterium *Halobacillus salinus* that inhibit quorum sensing-controlled phenotypes in Gram-negative bacteria. Appl Environ Microbiol 75:567–572

Terlabie NN, Sakyi-Dawson E, Amoa-Awua WK (2006) The comparative ability of four isolates of *Bacillus subtilis* to ferment soybeans into dawadawa. Int J Food Microbiol 106:145–152

Thomson JM, Bonomo RA (2005) The threat of antibiotic resistance in Gram-negative pathogenic bacteria: Betaβ-lactams in peril! Curr Opin Microbiol 8:518–524

Turnbuell PC, Hutson RA, Ward MJ, Jones MN, Quinn CP, Finnie NJ, Duggleby SJ, Kramer JM, Melling J (1992) *Bacillus anthracis* but not always anthrax. J Appl Bacteriol 72:21–28

Urdaci MC, Bressolier P, Pinchuk I (2004) *Bacillus clausii* probiotic strains: antimicrobial and immunomodulatory activities. J Clin Gastroenterol 38:S86–S90

Von Tersch MA, Carlton BC (1983) Bacteriocin from *Bacillus megaterium* ATCC 19213: comparative studies with megacin A-216. J Bacteriol 155:872–877

Wang G (2006) Diversity and biotechnological potential of the sponge-associated microbial consortia. J Ind Microbiol Biotechnol 33:545–551

Wang HL, Hesseltine CW (1982) Oriental fermented foods. In: Reed R (ed) Prescott and Dunn's industrial microbiology. AVI Publishing Company, Hartford, pp 492–538

Wang J, Fung DY (1996) Alkaline-fermented foods: a review with emphasis on pidan fermentation. Crit Rev Microbiol 22:101–138

Wei JZ, Hale K, Carta L, Platzer E, Wong C, Fang SC, Aroian RV (2003) *Bacillus thuringiensis* crystal proteins that target nematodes. Proc Natl Acad Sci USA 100:2760–2765

Whitehead NA, Barnard AM, Slater H, Simpson NJ, Salmond GP (2001) Quorum-sensing in Gram-negative bacteria. FEMS Microbiol Rev 25:365–404

Wilson MK, Abergel RJ, Raymond KN, Arceneaux JEL, Byers BR (2006) Siderophores of *Bacillus anthracis, Bacillus cereus* and *Bacillus thuringiensis*. Biochem Biophys Res Commun 348:320–325

Wulff EG, Mguni CM, Mansfeld-Giese K, Fels J, Luberck M, Hockenhull J (2002) Biochemical and molecular characterization of *bacillus amyloliquefaciens*, *B. subtilis* and *B. pumilus* isolates with distinct antagonistic potential against *Xanthomonas campestris* pv. *campestris*. Plant Pathol 51:574–584

Xie J, Zhang R, Shang C, Guo Y (2009) Isolation and characterization of a bacteriocin produced by an isolated *Bacillus subtilis* LFB112 that exhibits antimicrobial activity against domestic animal pathogens. Afr J Biotechnol 8:5611–5619

Yamashita S, Akao T, Mizuki E, Saitoh H, Higuchi K, Park YS, Kim HS, Ohba M (2000) Characterization of the anticancer- cell parasporal proteins of a *Bacillus thuringiensis* isolate. Can J Microbiol 46:913–919

Yang CY, Pang JC, Kao SS, Tsen HY (2003) Enterotoxigenicity and cytotoxicity of *Bacillus thuringiensis* strains and development of a process for Cry1Ac production. J Agric Food Chem 51:100–105

Yokotsuka T (1985) Fermented protein foods in the orient, with emphasis on shoyu and mMiso in Japan. In: Wood BJB (ed) Microbiology of fermented foods, vol 1. Elsevier, London, pp 263–293

Zawadzka AM, Abergel RJ, Nichiporuk R, Andersen UN, Raymond KN (2009) Siderophore-mediated iron acquisition systems in *Bacillus cereus*: identification of receptors for anthrax virulence-associated petrobactin. Biochemistry 48:3645–3657
Zhang LH (2003) Quorum quenching and proactive host defense. Trends Plant Sci 8:238–244
Zheng G, Slavik MF (1999) Isolation, partial purification and characterization of a bacteriocin produced by a newly isolated *Bacillus subtilis* strain. Lett Appl Microbiol 28:363–367
Zhou Y, Choi YL, Sun M, Yu Z (2008) Novel roles of *Bacillus thuringiensis* to control plant diseases. Appl Microbiol Biotechnol 80:563–572

Chapter 19
Future Challenges and Prospects of *Bacillus thuringiensis*

J. E. Barboza-Corona, N. M. de la Fuente-Salcido and M. F. León-Galván

Abstract *Bacillus thuringiensis* is the most important entomopathogenic microorganism, owing to its insecticidal crystal proteins (Cry) that are nontoxic to the human beings and represent the hallmark of this bacterium. Most of the studies performed with this microorganism are focused to understand the role of Cry proteins in toxicity, mainly because these components constituted the toxic molecules of its commercial products. *B. thuringiensis* produces several metabolites with potential applied uses, in particular, chitinolytic enzymes and bacteriocins are two types of proteins produced by different subspecies of this microorganism that could expand the perspective of application of this extraordinary bacterium. In this chapter, we review the different kinds of chitinases that are synthesized by *B. thuringiensis*, their roles in nature, and their applications in environment, agriculture and food industry. Additionally, we analyze bacteriocins of *B. thuringiensis* reported to date, how to enhance their production, and the methods for screening the bacteriocin activity. Finally, the future challenges and prospects of the antimicrobial peptides as biopreservatives, antibiotics, and nodulation factors are showed.

Keywords Chitinases · Metabolites · Food industry · Bacteriocins · Potential applications

19.1 Introduction

Many species of the genera *Bacillus* are of practical importance in agriculture, food, environment and pharmaceutical industries. For example, *B. thuringiensis* and *B. sphaericus* are two insecticidal bacteria that have been used widely for insect pest

J. E. Barboza-Corona (✉) · M. F. León-Galván
División Ciencias de la Vida, Departamento de Alimentos, Universidad de Guanajuato Campus Irapuato-Salamanca, 36500 Irapuato, Guanajuato, Mexico
e-mail: josebar@ugto.mx

N. M. de la Fuente-Salcido
División Ciencias de la Vida, Departamento de Alimentos, Universidad de Guanajuato Campus Irapuato-Salamanca, 36500 Irapuato, Guanajuato, Mexico

Escuela de Ciencias Biológicas, Universidad Autónoma de Coahuila, 27104 Torreón, Coahuila, Mexico

E. Sansinenea (ed.), *Bacillus thuringiensis Biotechnology*,
DOI 10.1007/978-94-007-3021-2_19,

and vector control owing to their capacity to produce crystalline inclusions with insecticidal properties (Park et al. 2006). *B. cereus* is a food-borne pathogenic bacterium responsible for gastrointestinal diseases (i.e. emetic and diarrheal syndromes) and eye infection, *B. antracis* causes lethal diseases (anthrax) (Barboza-Corona et al. 2007; Ibrahim et al. 2010). As *Bacillus* species have the capacity to secret proteins of industrial relevance, such as *B. subtilis* and *B. licheniformis*, have been used for expressing heterologous proteins generally recognized as safe (GRAS) for human use, for example α-amylase, pullulanase, and α-acetolactate, decarboxylase, α-amylase, respectively (Olempska-Beer et al. 2006), or proteins with potential applied use, for example, ice structuring proteins (antifreeze proteins) that can be used as additive in cryogenic preservation of food products as they produce a noncolligative depression of the freezing point (Yeh et al. 2007).

In particular, *B. thuringiensis* is the most important entomopathogenic microorganism, owing to its insecticidal crystal proteins (Cry) that are nontoxic to the human beings and represent the hallmark of this bacterium (Siegel 2001; Park et al. 2006). As shown in other chapters of this book, Cry proteins have been exhaustively studied by different research teams worldwide, but most of those works have been focused to screen strains that produce Cry proteins with new or higher activity, to optimize or generate new culture media, to increment their production or expand the spectrum of activity by engineering Cry, to develop transgenic plants, to study the mechanism of how insects develop resistance to the Cry proteins and to search strategies to reduce or eliminate it (Tamez-Guerra et al. 2008; Park and Federici 2009). Alternatively, the potential of *B. thuringiensis* to secrete proteins to the culture medium, with the capacity to inhibit food-borne pathogenic bacteria or phytopathogenic microorganisms (i.e. bacteria, fungi) and also to release hydrolytic enzymes that can generate derivates with antimicrobial activity that could be used in food preservation or medicine, has augmented the interest of this extraordinary bacterium to expand its use not only as bioinsecticide but also as a machinery for the production of enzymes or other proteins with potential applied use (De la Fuente-Salcido and Barboza-Corona 2006).

In this chapter, the future challenges and prospects of *B. thuringiensis* not only as bioinsecticide but also as a bacterium able to generate molecules (i.e. chitinases, antimicrobial peptides) with use in food industry are showed.

19.2 Chitinases of *B. thuringiensis* in Environment and Food Industry

19.2.1 Chitin and Chitinases

Chitin is an insoluble linear β-(1–4)-linked polymer of *N*-acetylglucosamine (GlcNAc) that is associated with proteins to form a variety of structures including the exoskeleton of many invertebrates (i.e. crustaceans, insects and spiders) and the cell wall of fungi and algae. Bacteria, plants, fungi, invertebrates and vertebrates are known to produce chitinases, which are categorized as endo- and exochitinases. The endochitinases hydrolyze chitin randomly on internal points of the chitin polymer, whereas the exochitinases begin their progressive action from the nonreducing end of chitin releasing chitobiose or GlcNAc units, which are subsequently hydrolyzed by respectively chitobiosidases and *N*-acetylglucosaminidases. The combination of both types of enzymes, results in a synergistic effect that increases the production of GlcNAc, which is involved in bioenergetics and synthesis of diverse cellular structures. Additionally, chitinolytic enzymes play different roles in nature, for example, maintaining a steady state in the ocean by transforming chitin to soluble biological material ready to be used, protecting both plants and humans against fungal infections and also begin involved in cuticle turnover in arthropods (Chernin et al. 1995; Morales de la Vega et al. 2006; Bhattacharya et al. 2007; Barboza-Corona et al. 2008).

19.2.2 Chitinases of *B. thuringiensis*

In particular, *B. thuringiensis* produces small quantities of chitinases, sometimes difficult to detect with traditional methods using chitin as carbon and nitrogen sources (i.e. halo formation). To date, at least 38 chitinase genes from different subspecies of *B. thuringiensis* such as *kenyea, pakistani, colmeneri, canadiensis, entomocidus, kurstaki, israelensis* and *konkukian* have been reported in the GenBank database. From those, 55% were cloned in China, 18% in Mexico and the rest in other countries including USA, Egypt, Turkey, Thailand, India, Pakistan, Tunisia and Korea (http://www.ncbi.nlm.nih.gov). Most of *B. thuringiensis* chitinases showed identities to each other from 93.8–99.6%. Based on amino acid sequence alignments, chitinases of *B. thuringiensis* were recently classified in four groups (Fig. 19.1). In Group 1 (Fig. 19.1aA) were included ChiA-HD73, ChiA74 and other 9 chitinases from *B. thuringiensis*. Chitinases of groups 2 (Fig. 19.1aB) and 3 (Fig. 19.1aC), contain an additional sequence in the N-terminus of, respectively, 12 (i.e. Chi74, ChiABtisra, ChiABtkurst, and ChiABtsot) or 25 amino acids (i.e. ChiAHa), not present in those of the Group 1 class. Chitinase synthesized by *B. thuringiensis* subsp. *pakistani* (ChiABtpakist) was placed in Group 4 (Fig. 19.1aD). Interestingly, the main difference between the first group and the fourth group (i.e. ChiABtpakist) was found in a fragment of 118 amino acids (Val-331 to Glu-448) presents in the first but lacking in the latter. Also, at the C-terminus of ChiABt from *B. thuringiensis* subsp. *pakistani* (Group 4) was found a sequence of 93 residues (Tyr-543 to Cys-635) that was not present in Group 1 chitinases. ChiABtc, a chitinase produced by *B. thuringiensis* subsp. *colmeneri* (Gen Bank accession number EF103273) was not included in a category because of its low identity (~13%) with other chitinases of *B. thuringiensis* (Barboza-Corona et al. 2008).

Chitinases of *B. thuringiensis* have a modular structure composed by a catalytic domain and a putative chitin-binding domain (ChBD), typical structure characteristics of enzymes that degrade biopolymers such as chitin and cellulose. The catalytic domain is identical on the majority of chitinases of *B. thuringiensis*, but showed a

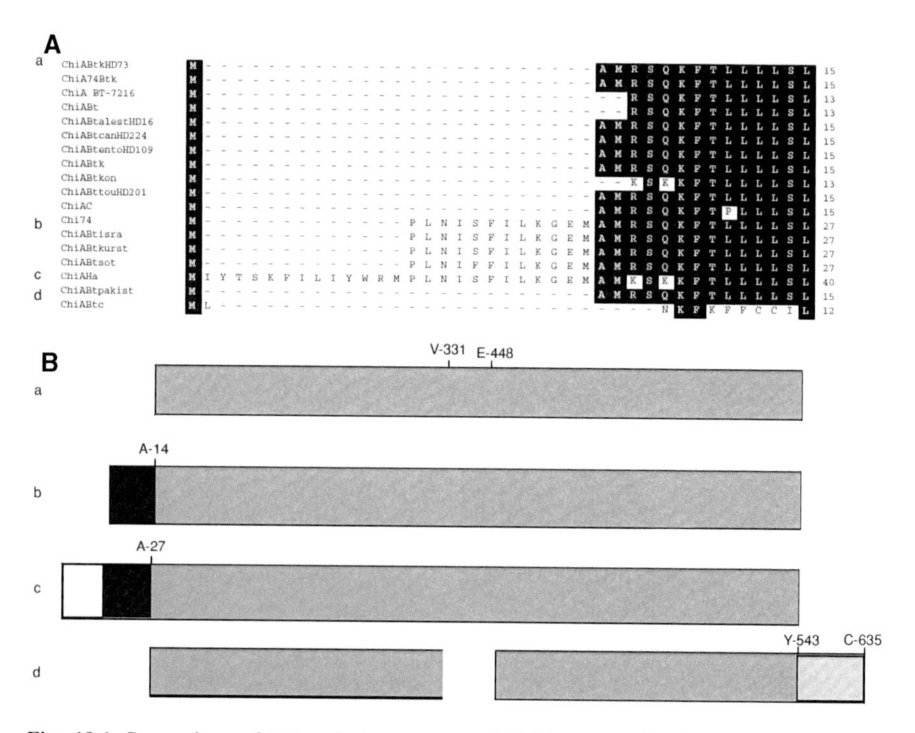

Fig. 19.1 Comparison of N-terminal sequences of chitinases produced by *B. thuringiensis*. **A** According to the amino acid residues located in the N-terminus, chitinases from *B. thuringiensis* were grouped in four groups: **a** Group 1 contains MAMRSQKFTLLLL. **b** Group 2 includes those chitinases that begin with MPLNISFILKGEM. **c** In Group 3, ChiAHa, a chitinase from *B. thuringiensis* Al Hakam was included; the sequence of this enzyme begins with MIYTSKFILIYWRM. **d** ChiABtpakist was the only chitinase in Group 4, and it has the same N-terminus that chitinase from type 1, however has two marked differences showed in Figure B. Chitinase from *B. thuringiensis* subsp. *colmeneri* (ChiABtc) was not included in any Group as it is too divergent from the other chitinases. **B** Schematic representation of the different chitinases groups from *B. thuringiensis*. (From Barboza-Corona et al. 2008)

lower degree of homology with the catalytic domain of other bacterial chitinases such as *B. circulans* Chi41, *B. subtilis* Chi, *Serratia marcescens* Chi and *Enterobacter agglomerans* ChiA. Chitinases of *B. thuringiensis* contain a catalytic domain active-site motif (Asp-Gly-Val-Asp-Leu-Asp-Trp-Glu) characteristic belonging to family 18 of glycosyl hydrolases. The chitin binding domain (ChBD) of *B. thuringiensis* chitinases shows identities of ~99% to each other and contains aromatic residues (e.g. Trp-591, Tyr-595 and Trp-626 in ChiA-HD73) that are highly conserved not only in the chitinases from *B. thuringiensis* but also in other bacterial chitinases such as those produced by *S. marcescens* and *B. circulans* where are directly implicated on binding of chitin. Additionally, endochitinases such as ChiA74, following its catalytic domain, have two fibronectin-like domains (FLDs) that contain conserved aromatic amino acids that may be implicated in substrate attachment (Watanabe et al. 1993; Thamthiankul et al. 2001; Barboza-Corona et al. 2003, 2008).

19.2.3 Chitinases of *B. thuringiensis* and the Environment

Chitinases of *B. thuringiensis* play important roles in nature and can be used for different environment purposes, for example: (i) in the degradation of shrimp wastes, (ii) improving the insecticidal activity of Cry or (iii) in the control of phytopathogenic fungi. The last two options, implicate alternatives to reduce the use of chemicals.

Shrimps represent traditional dish in seafood restaurants. However, the non-edible wastes such as the "heads" and exoskeleton represent 50–60% of the volume. These wastes are composed of a complex of chitin-protein and minerals. *B. thuringiensis* secretes a battery of chitinases and proteases, that can act synergistically to hydrolyzes shrimp wastes that can be used in pharmacy, medicine, food, water treatment, as ingredient in culture media, in pigment extraction, in the formulation of animal food, for reducing nematode population in soils and to produce single-cell protein (Rojas-Avelizapa et al. 1999).

In addition, it has been demonstrated that chitinases of *B. thuringiensis* destroy the peritrophic membrane, a film-like structure constituted by a complex of chitin-protein that protects and separates food from midgut tissues in the insects, increasing the insecticidal activity of Cry proteins. For biotechnological purposes, an increment in the chitinase production is desirable. In this regard, hyperexpression of homologues chitinases in *B. thuringiensis* has been performed using a genetic system currently used to increase the Cry production (Park et al. 1998). For example *B. thuringiensis* HD-73 was transformed with the endochitinase gene *chiA74* under the control of a strong promoter (*pcytA)* and a 5′ mRNA stabilizing (STAB-SD) sequence (HD-73-pEBchiA74). Expression levels were compared with those observed from the wild type strain (HD-73) and the recombinant HD-73 strain expressing *chiA74* under the control of its native promoter (HD-73-pEHchiA74) (Fig. 19.2). The chitinolytic activity of HD-73-pEBchiA74 was markedly elevated, being ~58- and 362-fold higher than HD-73-pEHchiA74 and parental strain HD-73 respectively, representing the highest levels of chitinase expression in recombinant *B. thuringiensis* reported to date. Although a significant increment in chitinase production was observed, a reduction in crystal size and sporulation was detected, also low or no increase in the insecticidal activity was observed (Casique-Arroyo et al. 2007; Barboza-Corona et al. 2009b). Further studies will be necessary to optimize the chitinase and Cry production in order to detect a synergistic effect. Recently, co-crystallization of Cry-chitinases has been performed in acrystalliferous strains of *B. thuringiensis*, although instability of chimeric crystals and low synergistic effect were observed (Driss et al. 2011). Additionally, it has been shown that chitinases of *B. thuringiensis* have inhibitory activity against phytopathogenic fungi such as *Fusarium* sp. and *Sclerotium rolfsii* (Morales de la Vega et al. 2006).

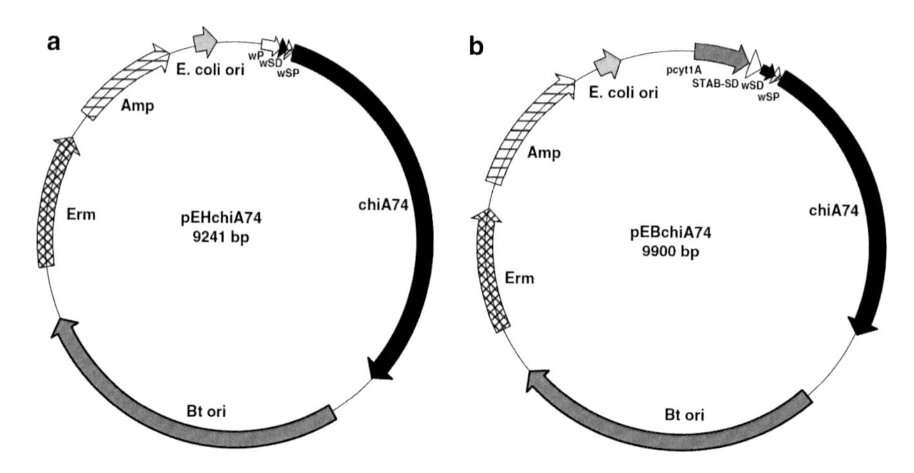

Fig. 19.2 Map of recombinant plasmids pEHchiA74, pEBchiA74 containing the *chiA74* gene under the control of wild promoter and *pcyt1A*-STAB sequence respectively. To construct the pEHchiA74 the *chiA74* gene including its native promoter was ligated into the *Sal*I and *Pst*I sites of the pHT3101 plasmid. To generate the pEBchiA74, the *chiA74* open reading frame was inserted into the *Sal*I and *Pst*I sites of the pPF-CH vector. Bt ori, *B. thuringiensis* origin of replication; Erm, erythromycin resistance gene; Amp, ampicillin resistance gene; *E. coli* ori, *E. coli* origin of replication; wP, wild promoter; pcyt1A, *cytA* promoters; STAB-SD, mRNA stabilizing sequence and ribosome binding site; wSD, wild Shine Dalgarno sequence; wSP, wild signal peptide sequence. (From Barboza-Corona et al. 2009b)

19.2.4 Chitinases of B. thuringiensis in the Generation of Chitin Derivates Oligosaccharides Toxic to Food-borne Pathogenic Bacteria

Oligosaccharides (OGS) are organic molecules containing from 2–25 monosaccharide subunits. OGS have important and diverse roles in molecular processes such as intercellular communication, immune recognition, and microbial pathogenesis, and as nodulation factors and architectural components in the structural biology of cells. Additionally, OGS could be used to maintain food quality by functioning as a preservative and could be employed as bioactive additives with medical purposes that include enhancing mineral absorption, suppressing the proliferation and the migration of tumor-induced cells and hepatocellular carcinoma, and as precursors in the synthesis of short chain fatty acids that could play an important role in the prevention of colon cancer (Suzuki et al. 1986; Barreteau et al. 2006; Weijers et al. 2008).

Fortunately, and for biotechnological purposes, wild chitinases genes of *B. thuringiensis* with their *cis* regulatory elements have been expressed and their corresponding proteins secreted by *Escherichia coli*. This phenomenon has several advantages over intracellular production e.g. protein can increase its biological activity, it can have correct folding and processing, higher stability and solubility and allow downstream process (Barboza-Corona et al. 2003, 2008; Yamabhai et al. 2008). Although most of studies about chitinases of *B. thuringiensis* have been focused on using these enzymes to improve their insecticidal or fungicidal activity,

Table 19.1 Inhibitory activity of chitin-derived oligosaccharides produced by ChiA74 of *Bacillus thuringiensis* against bacteria of clinical importance

Pathogenic bacteria	Inhibitory activity*
Bacillus cereus	16.5
Listeria innocua	14.9
Escherichia coli	16.4
Staphylococcus xylosus	37.6
Salmonella species	14.9
Staphylococcus aureus	14.9
Pseudomonas aeruginosa	34.5
Shigella flexneri	16.4
Proteus vulgaris	31.4

*One chitin oligosaccharide inhibition unit (IU) was defined as equal to 1 mm^2 of the zone of inhibition of growth of the bacterium tested. (Modified from Ortiz-Rodriguez et al. 2010)

the capacity to be secreted in *E. coli* enable them to be consider as alterative in the generation of chitin-derived OGS.

In particular ChiA74 is an endochitinase produced by a Mexican strain of *Bacillus thuringiensis* subsp. *kenyea* (Barboza-Corona et al. 2003). When the intact *chiA74* gene with its *cis* elements was cloned into high and moderately-high copy number *E. coli* expression vectors, the highest level of chitinase production was found at ~20 h. Functionally secreted ChiA74 was produced and the endochitinase cleaved substrate colloidal chitin to produce OGS with 3, 5 and 6 degrees of polymerization. The enzyme was active for an extended period of incubation (24 h), but its activity showed a decrement of 73% and 87%, respectively, after 24 h of incubation at 37°C and 55°C. OGS showed inhibitory activity against bacteria of clinical importance such as *Bacillus cereus* and *Staphylococcus aureus* (Table 19.1), which suggests a high potential to be used as food preservatives (Barboza-Corona et al. 2008; Ortiz-Rodríguez et al. 2010). Unfortunately, enzymes produced in a non GRAS bacterium make those biomolecules not suitable for using in products consumed by human beings. Recently, ChiA74 and other bacterial chitinases were expressed in a strain of *E. coli* recognized as safe and demonstrated the functionality of the enzymes to generate chitin-OGS with antimicrobial activity and its potentiality to be used as food preservatives (Data not published).

19.3 Bacteriocins of *Bacillus thuringiensis*

19.3.1 Generalities of Bacteriocins Synthesized by Bacillus thuringiensis

Bacteriocins are natural peptides that are synthesized and secreted by bacteria. These peptides, which often contain unusual amino acids made by modifying residues

Table 19.2 Bacteriocins synthesized by *Bacillus* spp. (Modified from Abriouel et al. 2011)

Class and subclass	Bacteriocin	Bacillus spp
Class I		
Post-translationally modified peptides		
Subclass I.1	Subtilin	*B. subtilis*
Single-peptide, elongated lantibiotics	Ericin S, Ericin A	*B. subtilis* A1/3
Subclass I.2	Sublancin 168	*B. subtilis* strain 168
Other single-peptide lantibiotics	Mersacidin	*Bacillus* sp. strain HIL Y-85
	Paenibacillin	*Paenibacillus polymyxa* OSY-DF
Subclass I.3	Haloduracin	*B. halodurans* C-125
Two peptide lantiobiotics	Lichenicidin	*B. licheniformis* DSM 13
Subclass I.4		
Other post-translationally modified peptides	Subtilosin A	*B. subtilis* strain 168
Class II		
Nonmodified peptides		
Subclass II.1	Coagulin, SRCAM 37, SRCAM 602, SCAM 1580	*B. coagulans* I_4
Pediocin-like peptides		
Subclass II.2	Thuricin H	*B. thuringiensis* SF361
Thuricin-like peptides	Thuricin S	*B. thuringiensis* subsp *entomocidus* strain HD198
	Thuricin 17	*B. thuringiensis* strain BMG1.7
	Bacthuricin F4	*B. thuringiensis* subsp. *kurstaki* strain BUPM4
	Cerein MRX1	*B. cereus* MRX1
Subclass II.3	Cerein 7A	*B. cereus* Bc7
Other linear peptides	Cerein 7B	*B. licheniformis*
	Lichenin	*B. thuringiensis* strain B439
	Thuricin 439	
Class III	Megacin A-216	*B. megaterium* 216
Large proteins	Megacin A- 19213	*B. megaterium* ATCC 19213

prescribed by the genetic code, inhibit the growth of closely or not closely related species thereby eliminating or significantly reducing competition for available nutrients (Barboza-Corona et al. 2007). Although bacteriocins of lactic acid bacteria (LAB) are the most studied worldwide, recently investigators are looking for new natural antimicrobial peptides that satisfy the consumer demands for minimally processed foods and diminish the resistance developed by conventional antibiotics. In this regard, the interest for studying the antimicrobial peptides produced by *Bacillus* spp with novel properties has been increased during the last years.

Recently it was proposed an innovative classification of the bacteriocins produced by *Bacillus* species based on their similarities with bacteriocins of lactic acid bacteria (LAB) (Abriouel et al. 2011) (Table 19.2). In this classification, bacteriocins of *B. thuringiensis* (bacthuricin F4, thuricin H and thuricins S and 17) were grouped in the thuricin-like peptides, subclass II.2, together with the cerein MRX1, as they share the same conserved region DWTXWSXL in the N-terminus. Other bacteriocins of *B. thuringiensis* whose amino acid sequences have not been determined were not included. Currently, 17 bacteriocins of *B. thuringiensis* have been reported (Table 19.3), however only the complete amino acid sequence of the Thuricin CD, a two-component antimicrobial with unusual structure, is known (Rea et al. 2010; Murphy et al. 2011). Recently it was carried out a bioinformatic analysis that could allow the bacteriocin gene ubication in the *B. thuringiensis* genome by looking for *in silico* the most highly conserved components of gene clusters that might be associated to bacteriocins (Murphy et al. 2011). Once bacteriocin sequence is known, prediction of the physicochemical properties and comparisons with other bacteriocins can be carried out using a platform for bacteriocin characterization (i.e. BACTIBASE) (Hammami et al. 2010).

19.3.2 *Methods for Screening Bacteriocins of Bacillus thuringiensis*

At present, the methods for determining bacteriocin activity are based on enzymatic and non-enzymatic procedures. Enzymatic methods rely on immediate measurement of intracellular enzymes released after cellular lysis upon exposure to bacteriocins. For example, elevated levels of lactate dehydrogenase (LDH) have been observed following lysis of susceptible strains by lactococcins (Morgan et al. 1995).

In the non-enzymatic method, conventional techniques are the well-diffusion or disc diffusion methods, where bacteriocins are added in wells dug into solid culture media, or on sterilized paper discs placed on the surface of agar pre-inoculated with an indicator strain (Tagg and McGiven 1971; Bhunia et al. 1988). However, those protocols are time-consuming (i.e. 2–3 days) and can provide errors in quantitative analyses that result from inconsistencies in the geometric progression of halos as bacteriocin concentration increases (Delgado et al. 2005). Non-enzymatic methods where bacteriocins are tested against indicator bacteria growing in liquid medium also present limitations, for example, even though these methods eliminate problems associated with agar diffusion methods, cell sedimentation, interference by sample color, and sigmoidal curve relationships between the bacteriocin concentration and inhibitory responses can adversely influence the accuracy of results (Kumazaki and Ishii 1982; Mary-Harting et al. 1972; Delgado et al. 2005).

Recently it was developed a fluorogenic method for determining the inhibitory activity of bacteriocins produced by *B. thuringienis*. In this protocol the membrane damage induced by bacteriocins (pore formation) causes the entrance of berberine into the cells. Berberine fluoresces immediately when it binds to different biomolecules (e.g. DNA and glycosaminoglycans); and bacteriocin activity is measured as relative fluorescence (Fig. 19.3). In the fluorogenic method, 1 h or less is required to determine the presence or absence of bacteriocins in a sample. This protocol can be used for screening rapidly not only novel wild bacteriocins but also transformants expressing bacteriocins genes (De la Fuente-Salcido et al. 2007).

Table 19.3 Bacteriocins of *Bacillus thuringiensis*

Bacteriocin	Molecular mass (kDa)	Producer strain	Sensitive strains	References
Bacthuricin F103	~11	*B. thuringiensis* BUPM 103	*Listeria monocytogenes Bacillus cereus*, *Agrobacterium tumefaciens*	Kamoun et al. (2011)
Thuricin CD	Trn-α 2.76 Trn-β 2.86	*B. thuringiensis* DPC 6431	*Clostridium difficile*	Rea et al. (2010)
Thuricin H	3.14	*B. thuringiensis* SF361	*Bacillus cereus* F4552	Lee et al. (2009)
Entomocin 110	4.8	*B. thuringiensis* serovar *thuringiensis* HD 2	Different species of *Bacillus*, *Listeria monocytogenes*, *Paenibacillus larvae*	Cherif et al. (2008)
Thuricin S	2.76 2.86	*B. thuringiensis* subsp *entomocidus* HD 198	*Bacillus thuringiensis* subsp. *darmastadiensis* 10T	Chehimi et al. (2007)
		Grupo A		
Morricin 269		*B. thuringiensis* subsp. *morrisoni* LBIT 269	Different species of *B. thuringiensis*, *B. cereus*,	Barboza-Corona et al. (2007)
Kurstacin 287		*B. thuringiensis* subsp. *kurstaki* LBIT 287	*Listeria innocua*, *Vibrio cholerae*, *Staphylococcus aureus*, *S. xylosus*, *Shigella flexneri*, *Salmonella sp*,	
		Grupo B		
Kenyacin 404	~10	*B. thuringiensis* subsp. *kenyae* LBIT 404	*Streptococcus pyogenes*, *E. coli*, *Klebsiella pneumoniae*.	De la Fuente-Salcido et al. (2008)
Entomocin 420		*B. thuringiensis* subsp. *Entomocidus* LBIT 420	*Proteus vulgaris*, *Enterobacter cloacae*, *Enterococcus faecium*,	
Tolworthcin 524		*B. thuringiensis* subsp *tolworthi* LBIT 524	*Pseudomonas aeruginosa*, *Rizophus sp.*, *Fusarium oxysporum*, *Mucor rouxii IM80*, *Trichoderma sp. SH1*, *Trichoderma sp. SD3*	
Thuricin 17	3.172	*B. thuringiensis* NBEB 17*	Different species of *B. thuringiensis*, *B. cereus*, *E. coli* MM294	Gray et al. (2006)
Bacthuricin F4	3.16	*B. thuringiensis* subsp *kurstaki* BUPM4	Different species of *Bacillus*	Kamoun et al. (2005)
Entomocin 9	12.4	*B. thuringiensis* ssp *entomocidus* HD9	*B. thuringiensis*, *L. monocytogenes*, *P. aeruginosa*	Cherif et al. (2003)
Thuricin 439	2.92 and 2.80	*B. thuringiensis* B 439	*B. thuringiensis*, *B. cereus*, *L. innocua 4202*	Ahern et al. (2003)

Table 19.3 (continued)

Bacteriocin	Molecular mass (kDa)	Producer strain	Sensitive strains	References
Thuricin 7	11.6	*B. thuringiensis* BMG 1.7	*B. thuringiensis, B. cereus, B. micoides, Str. pyogenes, L. monocytogenes, B. weihenstephanensis*	Cherif et al. (2001)
Tochicin	10.5	*B. thuringiensis* subsp *tochigiensis* HD868	Different species of *B. thuringiensis B. cereus, Leuconostoc mesenteroides*	Paik et al. (1997)
Thuricin	>950	*B. thuringiensis* serovar *thuringiensis* HD2	Different species of *Bacillus*	Favret and Younsten (1989)

19.3.3 Enhanced Synthesis of Bacteriocins from *Bacillus thuringiensis*

Proteins overproduction is a crucial concern for industry, because higher yields per unit of culture lower the production costs (Jan et al. 2001). In this regard, different strategies have been developed to improve the production of proteins with biotechnological potential uses (e.g. insecticidal proteins, antimicrobial peptides, chitinases). As bacteriocins of *B. thuringiensis* are produced in low amount, it is important to develop strategies in order to increase the biosynthesis of these promising peptides. Optimizing the bacteriocin yield either through enhanced fermentation or genetic engineering techniques is required for large scale production and applied use of these antimicrobial peptides, particularly those produced by *Lactobacillus* and *Bacillus* species. Bacteriocin synthesis of *B. thuringiensis* can be augmented without using genetic engineering. It was recently reported that the synthesis of five bacteriocins (Morricin 269, Kurstacin 287, Kenyacin 404, Entomocin 420, and Tolworthcin 524) from Mexican strains of *B. thuringiensis* is independent of the presence of a target inducing bacterium. However when the strains were co-cultured in presence of proteins synthesized and secreted or released by *B. cereus*, strains of *B. thuringiensis* increases the synthesis of the bacteriocins by quorum-sensing (De la Fuente-Salcido et al. 2008). Additionally, an increment in the synthesis of the antimicrobial peptides has been obtained by modification of the culture physical conditions, as pH, temperature, aeration, in induced or non-induced bacteria (Kamoun et al. 2009; Martínez-Cardeñas et al. 2012), demonstrating that pH is probably the most important physical factor that affected not only growth of *B. thuringiensis* strains, but also bacteriocin production (Martínez-Cardeñas et al. 2012).

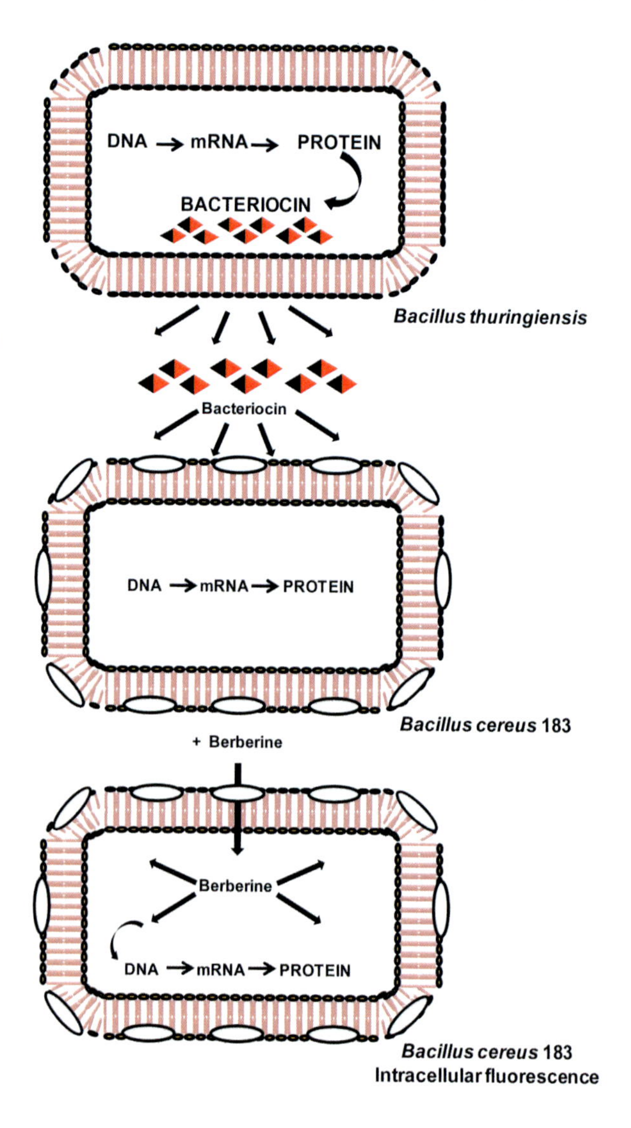

Fig. 19.3 Detection of bacteriocin activity with a fluorogenic method using berberine as substrate. Bacteriocins induce membrane damage, causing the entrance of berberine into the cells. Berberine fluoresces immediately when it binds to different biomolecules (e.g. DNA), then bacteriocin activity is measured as relative fluorescence

19.3.4 Potential Applications of *Bacillus thuringiensis* Bacteriocins

Bacillus thuringiensis is an excellent example of metabolic "machinery" that can produce biomolecules with a wide range of applications (Table 19.4).

Biopreservation is a modern technology implemented in food processing and microbiological food-safety for extending the shelf life by the use of natural antimicrobial compounds. Nisin, a bacteriocin synthesized by *Lactococcus lactis* is currently the only bacteriocin with use in food industry and has been employed

Table 19.4 Potential use of bacteriocins produced by *Bacillus thuringiensis*

Application area	Bacteriocin	Potential use
Food industry Biopreservative	Bacthuricine F103, Thuricin S, Thuricin H, Entomocin 9, Thuricin 7, Tochicin, Thuricin	Biopreservative against *Listeria monocytogenes, B. cereus*
Food industry Biopreservative Human health Control clinical pathogens	Thuricin 7	Biopreservative in dairy products and raw milk against *B. weihenstephanensis* Human diseases as strep throat and skin infection (*Streptococcus pyogenes*)
Human health Control foodborne pathogens	Thuricin CD	Specific treatment for *Clostridium difficile* (diarrhea outbreaks)
Human health Control clinical pathogens	Entomocin 9	Antimicrobial agent against *Pseudomonas aeruginosa* nosocomial agent of diseases ranging from pneumonia, septicemia and urinary tract infections
Human health Control foodborne and clinical pathogens	Morricin 269 Kurstacin 287 Kenyacion 404 Entomocin 420 Tolworthcin 524	Antimicrobial against *P. aeruginosa, S. pyogenes* (outbreaks of throat and scarlet fever), *Shigella flexneri, Salmonella sp., E. coli, Vibrio cholerae* (diarrheal and emetic syndromes)
Animal health Clinical Biotherapeutic agent		Control of subclinical mastitis in dairy herds (*S. aureus*)
Environmental application Biotherapeutic		Control pathogens on crops (*Salmonella* spp in vegetables)
Environmental application Biocontrol agent	Entomocin 110	Agent protector against *Paenibacillus larvae* causal agent of american foulbrood of honeybees (pollinators insect)
Environmental application Plant protection	Morricin 269 Kurstacin 287 Kenyacion 404 Entomocin 420 Tolworthcin 524	Antifungal action against phytopathogenic fungi
Environmental application Plant protection	Bacthuricin F103	Antimicrobial against phytopathogenic bacteria the major causal agents of disease crops
Environmental application Plant growth promoting	Thuricin 17	Plant growth promoting the increment in root nodulation (soybean)
Environmental application Biocontrol agent	Thuricin H, Thuricin S, Bacthuricin F4, Thuricin 7, Tochicin, Thuricin	*Bacillus* species

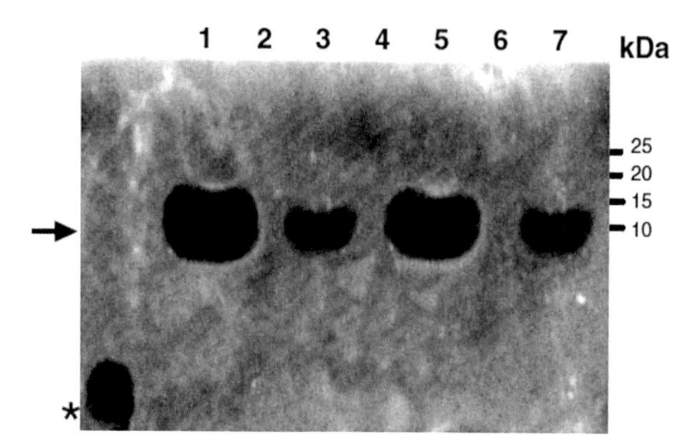

Fig. 19.4 Direct detection of the antibacterial activity of *B. thuringiensis* bacteriocins against *Bacillus cereus* 183. Assays were performed using bacteriocins fractionated in SDS-PAGE and overlaid *B. cereus* 183 culture. Lanes *1* and *5*, Tolworthcin 524; lanes *3* and *7*, Morricin 269. Samples were supplemented with (lanes *1* and *3*) or without (lanes *5* and *7*) β-mercaptoethanol. As negative controls, sample buffer with or without β-mercaptoethanol, respectively, lanes *2* and *4* and lane *6*, were also loaded in the gel. The arrow indicates the relative position (10 kDa) of the growth inhibition zones observed after overnight incubation at 28°C. The inhibition zone in the lower part (*) was obtained with an aliquot of the same Tolworthcin 524 sample dot-blotted after electrophoresis. Similar results were obtained for Kurstacin 287, Kenyacin 404 and Entomocin 420. BenchMark prestained protein ladder (Invitrogen) was used to estimate the molecular masses of bacteriocins. (From Barboza-Corona et al. 2007)

for the last two decades as a food preservative. It has been shown that bacteriocins of *B. thuringiensis* have a wide or narrow spectrum of activity against pathogenic bacteria (Cherif et al. 2008; De la Fuente-Salcido et al. 2008; Rea et al. 2010). Both types of activity are important and could have different uses. For example, bacteriocins with wide activity could be used not only as antibiotics but also as biopreservatives in the control of food-borne pathogenic bacteria (Barboza-Corona et al. 2007; De la Fuente-Salcido et al. 2008). Entomocin 110 and Thuricin 7 could be used respectively to control *Listeria monocytogenes* and preventing spoilage of raw milk and dairy products caused by *B. weihenstephanensis* (Cherif et al. 2001, 2008). Morricin 269, Kurstacin 287, Kenyacin 404, Entomocin 420, Tolworthcin 524 bacteriocins of ~10 kDa (Fig. 19.4), and also Thuricin 17 could be used as sanitizers in food processing industries or as preservatives or bioactive component in food package to inhibit the growth of bacteria that secret toxins and produce diarrhea such as *B. cereus* (Gray et al. 2006; Barboza-Corona et al. 2007; Galvez et al. 2007). Thuricin CD, a bacteriocin with a narrow activity, could be employed as an antibiotic in the treatment of diarrhea associate with *Clostridium difficile*, avoiding the use of broad spectrum antimicrobials that can led to the emerge of multi-drug resistant pathogens (Murphy et al. 2011).

Additionally, bacteriocins represent attractive control agents to prevent mastitis in animals. On this regard, five bacteriocins from Mexican strains of *B. thuringiensis* were evaluated for their antimicrobial effect against *S. aureus* isolated from infected cows. This assessment demonstrated that all bacteria isolated were susceptible to the five bacteriocins, mainly to Morricin 269 and Kurstacin 287 followed by Kenyacin 404, Entomocin 420 and Tolworthcin 524. It is important to remark that *S. aureus* strains tested were mainly resistant to traditional antibiotics used in the mastitis treatment (Barboza-Corona et al. 2009a).

Alternatively, pathogenic microorganisms can be transferred in fresh vegetables and fruits, representing not only a risk to human health but also an important cause of millionaire losses for farmers and traders. Recently *Salmonella* spp was obtained from fresh lettuce commercialized in markets and super markets in Mexico. These strains were resistant to antibiotic commonly used in the salmonellosis treatment. Five bacteriocins produced by Mexican strains of *B. thuringiensis* were tested against those isolates, demonstrating that *Salmonella* strains were susceptible to the novel bacteriocins. This result suggests the potential use of bacteriocins synthesized by Mexican strain of *B. thuringiensis* as biotherapeutic agents for reducing or destroying populations of *Salmonella* spp in vegetables crops or in the sanitization of post-harvest products (Castañeda-Ramírez et al. 2011).

In addition, bacteriocins can play an important role as nodulation factors. For example, when Thuricin 17, a bacteriocin isolate from *B. thuringiensis* NEB17 was applied to leaves or roots of soybean and corn, the growth of both plants was stimulated (Lee et al. 2009). Bacteriocins can also be used to control pest that affect honeybees, for example, Entomocin 110 has growth inhibitory activity to *Paenibacillus* larvae, the causal agent of American foul-brood of honeybees (*A. mellifera* L.) (Cherif et al. 2008). Finally, bacteriocins of *B. thuringiensis* have growth inhibitory activity against the phytopathogenic bacteria, for example Bacthuricin F103 synthesized by *B. thuringiensis* strain BUPM103 showed inhibitory effect to *Agrobacterium tumefaciens* (Kamoun et al. 2011).

References

Abriouel H, Franz C, Ben Omar N, Gálvez A (2011) Diversity and applications of *Bacillus* bacteriocins. FEMS Microbiol Rev 35:201–232

Ahern M, Verschueren S, Van Sinderen D (2003) Isolation and characterisation of a novel bacteriocin produced by *Bacillus thuringiensis* strain B439. FEMS Microbiol Lett 220:127–131

Barboza-Corona JE, Nieto-Mazzocco E, Velázquez-Robledo R, Salcedo-Hernández R, Bautista M, Jiménez B, Ibarra JE (2003) Cloning, sequencing and expression of the chitinase gene *chiA74* from *Bacillus thuringiensis*. Appl Environ Microbiol 69:1023–1029

Barboza-Corona JE, Vázquez-Acosta H, Bideshi D, Salcedo-Hernández R (2007) Bacteriocin-like inhibitor substances production by Mexican strains of *Bacillus thuringiensis*. Arch Microbiol 187:117–126

Barboza-Corona JE, Reyes-Rios DM, Salcedo-Hernández R, Bideshi D (2008) Molecular and biochemical characterization of an endochitinase (ChiA-HD73) from *Bacillus thuringiensis* subsp. *kurstaki* HD-73. Mol Biotechnol 39:29–37

Barboza-Corona JE, la Fuente-Salcido N de, Alva-Murillo N, Ochoa-Zarzosa A, Lopez-Meza JE (2009a) Activity of bacteriocins synthesized by *Bacillus thuringiensis* against *Staphylococcus aureus* isolates associated to bovine mastitis. Vet Microbiol 138(1):179–183

Barboza-Corona JE, Ortiz-Rodríguez T, de la Fuente-Salcido N, Ibarra J, Bideshi DK, Salcedo-Hernández R (2009b) Hyperproduction of chitinase influences crystal toxin synthesis and sporulation of *Bacillus thuringiensis*. Antonie Van Leeuwenhoek 96:31–42

Barreteau H, Delattre C, Michaud P (2006) Production of oligosaccharides as promising new food additive generation. Food Technol Biotechnol 44:323–333

Bhattacharya D, Nagpure A, Gupta RK (2007) Bacterial chitinases: properties and potential. Crit Rev Biotechnol 27:21–28

Bhunia AK, Johnson MC, Ray B (1988) Purification, characterization and antimicrobial spectrum of a bacteriocin produced by *Pediococcus acidolactici*. J Appl Bacteriol 65:261–268

Casique-Arroyo G, Bideshi D, Salcedo-Hernández R, Barboza-Corona JE (2007) Development of a recombinant strain of *Bacillus thuringiensis* subsp. *kurstaki* HD-73 that produces the endochitinase ChiA74. Antonie Van Leeuwenhoek 92:1–9

Castañeda-Ramírez C, Cortes-Rodríguez V, de la Fuente-Salcido N, Bideshi DK, Barboza-Corona JE (2011) Isolation of *Salmonella* spp. from lettuce and evaluation of its susceptibility to novel bacteriocins synthesized by *Bacillus thuringiensis* and antibiotics. J Food Protect 74:274–278

Chehimi S, Delalande F, Sablé S, Hajlaoui MR, Van Dorsselaer A, Limam F, Pons AM (2007) Purification and partial amino acid sequence of thuricin S, a new anti-Listeria bacteriocin from *Bacillus thuringiensis*. Can J Microbiol 53:284–290

Cherif A, Ouzari H, Daffonchio D, Cherif H, Ben Slama K, Hansen A, Jaous S, Boudabous A (2001) Thuricin 7: a novel bacteriocin produced by *Bacillus thuringiensis* BMG1.7, a new strain isolated from soil. Lett Appl Microbiol 32:243–247

Cherif A, Chehimi S, Limem F, Hansen BM, Hendriksen NB, Daffonchio D, Boudabbous A (2003) Detection and characterization of the novel bacteriocin entomocin 9, and safety evaluation of its producer, *Bacillus thuringiensis* subsp. *Entomocidus* HD9. J Appl Microbiol 95:990–1000

Cherif A, Rezgui W, Raddadi N, Daffonchio D, Boudabous A (2008) Characterization and partial purification of entomocin 110, a newly identified bacteriocin from *Bacillus thuringiensis* subsp. *Entomocidus* HD110. Microbiol Res 163:684–692

Chernin L, Ismailov Z, Haran S, Chet I (1995) Chitinolytic *Enterobacter agglomerans* antagonistic to fungal plant pathogens. Appl Environ Microbiol 61:1720–1726

De la Fuente Salcido NM, Barboza-Corona JE (2006) Duality of *Bacillus thuringiensis*: bioinsecticide and biopreservative. Cienc Acierta 6:27–28. (In Spanish)

De la Fuente-Salcido N, Salcedo-Hernández R, Alanis-Guzmán MG, Bideshi DK, Barboza-Corona JE (2007) A new rapid fluorogenic method for measuring bacteriocin activity. J Microbiol Methods 70:196–199

De la Fuente-Salcido N, Alanís-Guzmán MG, Bideshi DK, Salcedo-Hernández R, Bautista-Justo M, Barboza-Corona JE (2008) Enhanced synthesis and antimicrobial activities of bacteriocins produced by Mexican strains of *Bacillus thuringiensis*. Arch Microbiol 190:633–640

Delgado A, Brito D, Fevereiro P, Tenreiro R, Peres C (2005) Bioactivity quantification of crude bacteriocin solutions. J Microbiol Methods 62:121–124

Driss F, Rouis S, Azzouz H, Tounsi S, Zouari N, Jaoua S (2011) Integration of a recombinant chitinase into *Bacillus thuringiensis* parasporal insecticidal cristal. Curr Microbiol 62:281–288

Favret ME, Yousten AA (1989) Thuricin: the bacteriocin produced by *Bacillus thuringiensis*. J Invertebr Pathol 53:206–216

Gálvez A, Abriouel H, Lucas López R, Ben Omar N (2007) Bacteriocin-based strategies for food biopreservation. Int J Food Microbiol 120:51–70

Gray EJ, Lee KD, Souleimanov AM, Di Falco MR, Zhou X, Ly A, Charles TC, Driscoll BT, Smith DL (2006) A novel bacteriocin, thuricin 17, produced by plant growth promoting rhizobacteria strain *Bacillus thuringiensis* NEB17: isolation and classification. J Appl Microbiol 100:545–554

Hammami R, Zouhir A, Le Lay C, Ben Hamida J, Fliss I (2010) BACTIBASE second release: a database and tool platform for bacteriocin characterization. BMC Microbiol 10:22. d

Ibrahim MA, Griko N, Junker M, Bulla LA (2010) *Bacillus thuringiensis*. A genomic and proteomics perspective. Bioeng Bugs 1:31–50
Jan J, Valle F, Bolivar F, Merino E (2001) Construction of protein overproducer strains in *Bacillus subtilis* by an integrative approach. Appl Microbiol Biotechnol 55:69–75
Kamoun F, Mejdoub H, Aouissaoui H, Reinbolt J, Hammami A, Jaoua, S (2005) Purification, amino acid sequence and characterization of Bacthuricin F4, a new bacteriocin produced by *Bacillus thuringiensis*. J Appl Microbiol 98:881–888. d
Kamoun F, Zouari N, Saadaoui I, Jaoua S (2009) Improvement of *Bacillus thuringiensis* bacteriocin production through culture conditions optimization. Prep Biochem Biotechnol 39:400–412
Kamoun F, Ben Fguira I, Ben Hassen NB, Mejdoub H, Lereclus D, Jaoua S (2011) Purification and Characterization of a New *Bacillus thuringiensis* Bacteriocin active against *Listeria monocytogenes, Bacillus cereus* and *Agrobacterium tumefaciens*. Appl Biochem Biotechnol 165:300–314
Kumazaki T, Ishii S (1982) A simple photometric method for determination of the activity of pyrocin R 1. J Biochem 91:817–823
Lee KD, Gray EJ, Mabood F, Jung WJ, Charles T, Clark SRD, Ly A, Souleimanov A, Zhou X, Smith DL (2009) The class IId bacteriocin thuricin-17 increases plant growth. Planta 229:747–755
Martínez-Cardeñas JA, De la Fuente-Salcido NM, Salcedo-Hernández R, Bideshi DK, Barboza-Corona JE (2012) Effects of physical culture parameters on bacteriocin production by Mexican strains of *Bacillus thuringiensis* after cellular induction. J Ind Microbiol Biotechnol 39:183–189
Mayr-Harting A, Herges AJ, Berkeley CW (1972) Methods for studying bacteriocins. In: Norris JR, Ribbons DW (eds) Methods in microbiology. Academic Press, New York
Morales de la Vega L, Barboza-Corona JE, Aguilar-Uscanga MG, Ramírez-Lepe M (2006) Purification and characterization of an exochitinase from *Bacillus thuringiensis* ssp. *aizawai* and its action against phytopathogenic fungi. Can J Microbiol 52:651–657
Morgan S, Ross RP, Hill C (1995) Bacteriolytic activity caused by the presence of a novel lactococcal plasmid encoding lactococcins A, B, and M. Appl Environ Microbiol 61:2995–3001
Murphy K, O'Sullivan O, Rea MC, Cotter PD, Ross RP, Hill C (2011) Genome mining for radical SAM protein determinants reveals multiple sactibiotic-like gene clusters. PLoS One 6:e20852
Olempska-Beer ZS, Merker RI, Ditto MD, DiNovi MJ (2006) Food-processing enzymes from recombinant microorganism- a review. Regul Toxicol Pharmacol 45:144–158
Ortiz-Rodríguez T, de la Fuente-Salcido N, Bideshi DK, Salcedo-Hernández R, Barboza-Corona JE (2010) Generation of chitin-derived oligosaccharides toxic to pathogenic bacteria using ChiA74, an endochitinase native to *Bacillus thuringiensis*. Lett Appl Microbiol 51:184–190
Paik HD, Bae SS, Park SH, Pan JG (1997) Identication and partial characterization of tochicin, a bacteriocin produced by *Bacillus thuringiensis* subsp. *tochigiensis*. J Ind Microbiol Biotechnol 19:294–298
Park HW, Federici BA (2009) Chapter 12. Genetic engineering of bacteria to improve efficacy using the insecticidal proteins of *Bacillus* species. In: Stock SP, Vandenberg J, Glazer I, Boemare N (eds) Insect pathogens-molecular approaches and techniques. CABI Publishing, Cambridge, pp 275–305
Park HW, Ge B, Bauer LS, Federici BA (1998) Optimization of Cry3A yields in *Bacillus thuringiensis* by use of sporulation-dependent promoters in combination with the STAB-SD mRNA sequence. Appl Environ Microbiol 64:3932–3938
Park HW, Federici BA, Sakano Y (2006) Inclusion proteins from other insecticidal bacteria. In: Shively JM (ed) Microbiology monographs, Volume 1, "Inclusions in Prokaryotes". Springer, Heidelberg, pp 321–330
Rea MC, Sit CS, Clayton E, O'Connor PM, Whittal RM, Zheng J, Veredas JC, Ross P, Hill C (2010) Thuricin CD, a posttranslationally modified bacteriocin with a narrow spectrum of activity against *Clostridium difficile*. Proc Natl Acad Sci USA 107:9352–9357

Rojas Avelizapa LI, Cruz-Camarillo R, Guerrero MI, Rodríguez-Vázquez R, Ibarra JE (1999) Selection and characterization of proteo-chitinolytic strain of *Bacillus thuringiensis* able to grow in shrimp waste media. World J Microbiol Biotechnol 15:299–308

Siegel JP (2001) The mammalian safety of *Bacillus thuringiensis*-based insecticides. J Invertebr Pathol 77:13–21

Suzuki K, Mikami T, Okawa Y, Tokoro A, Suzuki S, Suzuki M (1986) Antitumor effect of hexa-*N*-acetylchitohexaose and chitohexaose. Carbohydr Res 151:403–408

Tagg JR, McGiven AR (1971) Assay system for bacteriocins. Appl Microbiol 21:943

Tamez-Guerra P, Valadez-Lira JA, Alcocer-Gonzalez JM, Oppert B, Gomez-Flores R, Tamez-Guerra R, Rodriguez-Padilla C (2008) Detection of gene encoding antimicrobial peptides in Mexican strains of *Trichoplusia ni* (Hübner) exposed to *Bacillus thuringiensis*. J Invertebr Pathol 98:218–227

Thamthiankul S, Suan-Ngay S, Tantimavanich S, Panbangred W (2001) Chitinase from *Bacillus thuringiensis* subsp. *pakistani*. Appl Microbiol Biotechnol 56:395–401

Watanabe T, Kobori K, Miyashita K, Fujii T, Sakai M, Uchida M, Tamaka H (1993) Identification of glutamic acid 201 and aspartic 200 in chitinase A1 of *Bacillus circulans* WL-12 as essential residues for chitinase activity. J Biol Chem 268:18567–18572

Weijers CA, Franssen MC, Visser GM (2008) Glycosyltransferase-catalyzed synthesis of bioactive oligosaccharides. Biotechnol Adv 26:436–456

Yamabhai M, Emrat S, Sukasem S, Pesatcha P, Jaruseranee N, Buranabanyat M (2008) Secretion of recombinant *Bacillus* hydrolytic enzymes using *Escherichia coli* expression systems. J Biotechnol 133:50–57

Yeh CM, Wang JP, Su FS (2007) Extracellular production of a novel ice structuring protein by *Bacillus subtilis*-a case of recombinant food peptide additive production. Food Biotechnol 21:119–128

Index

E. Sansinenea (ed.), *Bacillus thuringiensis Biotechnology*,
DOI 10.1007/978-94-007-3021-2, © Springer Science+Business Media B.V. 2012

Q

R